POLYMERIC BUILDING MATERIALS

POLYMERIC BUILDING MATERIALS

DOREL FELDMAN BEng., PhD(Eng.), DSc

Centre for Building Studies, Concordia University,
Montreal, Quebec, Canada

ELSEVIER APPLIED SCIENCE
LONDON and NEW YORK

ELSEVIER SCIENCE PUBLISHERS LTD
Crown House, Linton Road, Barking, Essex IG11 8JU, England

Sole distributor in the USA and Canada
ELSEVIER SCIENCE PUBLISHING CO., INC.
655 Avenue of the Americas, New York, NY 10010, USA

WITH 165 TABLES AND 210 ILLUSTRATIONS

© 1989 ELSEVIER SCIENCE PUBLISHERS LTD

British Library Cataloguing in Publication Data

Feldman, Dorel
    Polymeric building materials
    1. Building materials: Plastics. use
    I. Title
    691'.92

Library of Congress Cataloging-in-Publication Data

Feldman, Dorel.
    Polymeric building materials.

    Bibliography: p.
    Includes index.
    1. Polymers and polymerization. 2. Building
materials. I. Title.
    TA455.P58F39 1989   624.1'892   88-21354

    ISBN 1-85166-269-3

Typeset and Printed in Northern Ireland by The Universities Press (Belfast) Ltd.

To Canada, Country of Freedom

For the 10th Anniversary of the Centre for
Building Studies of Concordia University

Preface

Besides such well known inorganic engineering materials as ceramics, glasses and cements, organic polymers are finding an increasingly important role in different industrial sectors and as such are considered materials of the future. Following the evolution of organic synthetic polymers, the most promising market for these new engineering materials is seen to be in the building and construction industries.

Three decades ago there existed only a few well developed manufacturing methods and related process equipments to convert organic polymers from the state of latex, suspension or powder, into a final marketable product (e.g. fibres, plastics, rubbers, adhesives, etc.). Today there are many relatively inexpensive technologies, which are both rapid and reliable, for extruding, spinning, casting, moulding, stamping, and forming polymers into interesting, useful and readily marketable goods.

Of course, the existence of numerous applications in a variety of areas and the availability of standardized and automated methods of production, stimulate the synthesis and development of new members to the organic family.

Currently, more than 100×10^6 t/year of synthetic polymers are produced in the world in a wide variety of products for different applications. Their consumption as building materials is close to 3.6×10^6 t.

Europe continues to be a few years ahead of North America in terms of using polymers in building and construction. The reason for this is both a difference between the nature of the construction industry in these areas and price differences for raw materials (e.g. in Europe wood is a scarcity and the cost of steel, steel alloys, aluminium, and other metals is generally higher than in North America). As a result, polymer building product technologies are generally more advanced in Europe, and design concepts tend to be more innovative.

Technical literature classifies the polymers used in building as:

—auxiliaries to other materials,
—non-structural,
—semi-structural, and
—structural.

Auxiliaries include: adhesives and binders, for such applications as laminated timber and other wood derivatives; sealants; gaskets; decorative and protective coatings.

Non-structural uses (i.e. unstressed or relatively unstressed applications) constitute by far the greatest volume. They include such applications as: floor and wall coverings; decorative laminates; insulation; vapour barriers; plumbing fixtures; hardware; glazing; and lighting.

For most semi-structural and structural applications, in which both strength and stiffness are of major importance, the mechanical performance characteristics of unmodified polymers are frequently inadequate and consequently more complex materials must be used. Polymer composites have been developed to meet strength and stiffness requirements by reinforcing the base polymer with different kinds of fillers, fibres and other similar materials.

The book, organized in eight chapters, emphasizes the complex role of polymers in the building industry. It is an outgrowth of polymer courses taught by the author at the Centre for Building Studies and the Department of Chemistry of Concordia University, Montreal, as well as some papers presented by the author at conferences dealing with polymer applications in industry.

The book is designed as an introduction to polymer applications in the building industry for students of building, civil and chemical engineering, and of materials science.

Chapter 1 is an introduction to polymer chemistry and technology. It provides information on different classes of high polymers and the manner in which they are synthesized. Some physical and mechanical characteristics of polymers are underlined as well as the procedures used for their processing. Chapter 2 covers composites, treating them as multiphase systems which have useful properties and applications. The technologies used in their production are also given. Chapter 3 is devoted to a certain group of relatively new composites, referred to as polymer–concrete composite systems. The various concrete composite systems discussed are: polymer impregnated concrete (PIC), polymer cement concrete (PCC), polymer concrete (PC), and fibre reinforced polymer concretes. These concrete composite systems show, for the most part, improved strength and durability in comparison to plain concrete. Polymer foams and their functions as thermal and acoustical insulating materials are developed in Chapter 4. Here, the main properties and applications of different groups of thermoplastic and thermosetting foams are discussed. Polymers used as adhesives and sealants are presented in Chapter 5. A brief discussion is given on the theories of adhesion and the performance and application of the most important groups of adhesives and sealants are also discussed. Chapter 6 deals with the role of macromolecular compounds in solar energy conservation in buildings. In Chapter 7 a brief review is given of new types of polymer-based roofing and flooring materials. The changes in properties due to ageing and different forms of degradation (e.g. thermal degradation, photodegradation, chemical degradation, bio-degradation, radiodegradation, etc.) and weathering, constitute the subject of Chapter 8. Results from research projects completed by the author in specific areas of interest are presented in Chapters 5 and 6.

The author thanks Ms L. M. Beznaczuk, PhD student, and Mrs Gloria Miller, secretary, for their technical help. The author is particularly grateful to Mr M. Lacasse, PhD student, for his assistance with the preparation of the manuscript. Thanks are also due to Elsevier Science Publishers for their cooperation and to the many publishers who allowed the use of data and illustrations from their publications, and who are listed separately in the acknowledgements. Special thanks go to my family.

D. FELDMAN
Montreal

Contents

Preface vii
Acknowledgements xiii
List of Abbreviations xv

Chapter 1 **Polymers** 1
    Introduction 1
    Molecular Weight (MW) 5
    Polydispersity 8
    Synthesis of Polymers 9
    Classification 19
    Physical Structure 21
    Morphological Changes in Polymers 31
    Mechanical Properties 36
    Thermal Properties 41
    Weathering and other Properties 44
    Permeability 46
    Toxicity 49
    Flammability 49
    Polymer Processing 51
    Plastic Technology 52
    References 71

Chapter 2 **Composites** 74
    Polymer Composites 77
    Types of Reinforcing Agents 113
    The Coupling Agent 119
    Fibre Reinforced Composites (FRC) 124
    Glass Fibre Manufacture 131
    Carbon Fibres 138
    Hybrid Systems 143
    Composite Technology 160
    Composites in Construction 187
    Sandwich Panels 194
    References 200

Chapter 3 **Polymer–Concrete Composites** 207
    Polymer Impregnated Concrete 208
    Polymer Cement Concrete 233

Polymer Concrete 250
Fibre Reinforced Concrete 262
References 278

Chapter 4 **Polymer Foams** 283
Introduction 283
Foaming (Blowing) Agents 287
Cellular Polymer Manufacture 290
Special Foams 296
Thermoplastic and Thermosetting Foams 304
Flammability Aspects of Foams 343
References 352

Chapter 5 **Adhesives and Sealants** 356
Introduction 356
Adhesives 358
Sealants 391
References 433

Chapter 6 **Polymers in Solar Energy Conservation** . . . 438
Covers for Solar Collectors 438
Encapsulation 457
Pipes 469
Coatings, Adhesives, Sealants 470
Polymers for Solar Ponds 473
References 475

Chapter 7 **Polymer Applications in Roofing and Flooring** . . 478
Roofing 478
Flooring 490
References 493

Chapter 8 **Polymer Degradation** 496
Mechanical Corrosion 498
Polymer Service Life in Polluted Atmospheres 500
Thermal Degradation 503
Photodegradation 520
Polymer radiodegradation 539
Chemical degradation 546
Biodegradation 549
Mechanical degradation 552
References 554

Index 557

Acknowledgements

The publisher joins with the author in thanking those individuals, publishers, societies and companies for permission to use photographs, diagrams, tables, etc., which have either appeared elsewhere or are modifications of published materials:

Academic Press
Akademie Verlag DDR
American Chemical Society
ACS Rubber Division (Akron)
ASTM Special Technical Publications, ASM International Technical Journals
Brooks/Cole Engineering Division
Butterworth Scientific Ltd
Cahner Books
Chapman and Hall
Chemical Publishing Co.
Communication Channels Inc.
M. Dekker Inc.
E. I. Dupont de Nemours and Co.
Edward Arnold (Publishers) Ltd
Ellis Horwood Ltd
Elsevier Applied Science Publishers Ltd
Evode Roofing Ltd
Federation of Society of Coating Technology
Findlay Publications Ltd
General Motors Technical Center
Gordon & Breach Science Publishers Ltd
C. Hanser Verlag
Hayden Book Co. Inc.
Hemisphere Publ. Corp.
Hüthig & Wepf Verlag, Basel
Industry Media Inc.
Institut für Spanende Technologie

R. Krieger
Litton Educational Publishing
Longman Scientific & Technical
McGraw-Hill Book Co.
Metallurgical Society
NACE, National Association of Corrosion Engineering
NASA—Langley Research Center
North Holland Publishing Co.
Pergamon Press
Plenum Press
Prentice Hall Inc.
SIRA Ltd, Scientific Instrument Research Association
Society of Chemical Industry
Society of Plastics Engineers
Springer Verlag GmbH & Co. KG
Surrey University Press, Blackie Publ. Group
Technical Research Center Finland
Technomic Publishing
Textile Research Institute
The Metallurgical Society of AIME
The Plastics and Rubber Institute
G. Thieme Verlag
Van Nostrand Reinhold Co.
Zechner & Huethig Verlag GmbH

W. D. Bascom, J. Beaudoin, M. J. Berry, A. Blaga, S. K.
Brauman, P. J. Briggs, J. A. Brydson, W. F. Carole, J. P. Critchly,
E. F. Cuddihy, C. F. Cullis, R. S. Davidson, A. Davis, H. W.
Dursch, R. Feldman, R. P. Fynn, H. P. Garg, M. J. Hitch, W.
Horn, A. J. Kinloch, J. M. Klosowski, J. L. Koenig, R. H.
Leitheiser, K. J. Lewis, R. E. Lodrigan, R. M. Luck, G. L. Nelson,
A. J. Majumdar, P. Maslow, F. R. Mayo, P. D. Metz, J. V.
Milewski, K. L. Mittal, R. J. Morgan, Y. Ohama, R. B. Pettit, A.
V. Pocius, S. L. Pohlman, H. R. Ray, B. Rogowski, C. A. Rude, P.
Schissel, J. C. Seferis, R. B. Seymour, M. A. Shulman, T. Sugama,
S. S. P. Sung, W. G. Wilhelm, K. B. Wishman, R. M. Woodley, W.
W. Wright.

                                                        D. F.

List of Abbreviations

ABS	Acrylonitrile–butadiene–styrene.
AIBN	2,2′,-azo-bis(isobutyronitrile).
AMVN	2,2′,-azo-bis(2,4-dimethyl valeronitrile).
AN	Acrylonitrile.
ASTM	American Society for Testing and Materials.
AT	Acrylic terpolymer.
B	Butadiene.
BA	Butyl acrylate.
BMC	Bulk moulding compound.
BP	Benzoyl peroxide.
CAB	Cellulose acetate butyrate.
CPCS	Consumer Product Safety Commission.
CPE	Chlorinated polyethylene.
CPVC	Chlorinated poly(vinyl chloride).
CSP	Chopped strand mat.
DC	Degree of crystallinity.
DCPD	Dicyclopentadiene.
DNA	Deoxyribonucleic acid.
DOP	Dioctylphthalate.
DP	Degree of polymerization.
DSC	Differential scanning calorimetry
DTA	Differential thermal analysis.
EDXA	Energy dispersive X-ray analysis.
EMA	Ethylene–methacrylate copolymer.
ENB	5-Ethylidene-2-norbornene.
EP	Ethylene–propylene copolymer.
EPDM	Ethylene–propylene diene monomer.
EPS	Expanded polystyrene.
EVA	Ethylene–vinyl acetate copolymer.

FEP   Fluorinated ethylene–propylene copolymer.
FRC   Fibre reinforced composite.
FRP   Fibre reinforced plastic.

GF   Glass fibre.
GFP   Glass fibre reinforced plastic.
GRP   Glass reinforced plastic.
GRS   Butadiene–styrene rubber.

HDPE   High density polyethylene.

LDPE   Low density polyethylene.
LOI   Limiting oxygen index.
LP   Lauroyl peroxide.

MA   Methyl acrylate.
MDPE   Medium density polyethylene.
MEKP   Methyl ethyl ketone peroxide.
MMA   Methylmethacrylate.
MVT   Moisture vapour transmission.

NBR   Acrylonitrile–butadiene rubber.
NBS   National Bureau of Standards
NIBD   Nickel dibutyl dithiocarbamate.

PA   Polyamide.
PAB-T   Poly(p-aminobenzhydrazide terephthalamide).
PAN   Polyacrylonitrile.
PC   Polycarbonate.
PCC   Polymer cement concrete.
PCM   Phase change material.
PCO   Polycarbonate.
PCTFE   Poly(chloro-trifluoroethylene).
PE   Polyethylene.
PET   Poly(ethylene terephthalate).
PF   Phenol-formaldehyde.
PIB   Polyisobutylene.
PIC   Polymer impregnated concrete.
PIR   Polyisocyanurate.
PMMA   Poly(methyl methacrylate).

PMVK	Poly(methyl vinyl ketone).
P-nBA	Poly(n-butylacrylate).
PP	Polypropylene.
PPB	Poly(p-benzamide).
PPCC	Polymer Portland cement concrete.
PPO	Poly(phenylene oxide).
PPT	Poly(p-phenylene terephthalamide).
PS	Polystyrene.
PSO	Polysiloxane.
PTFE	Poly(tetrafluoroethylene).
PU	Polyurethane.
PVA	Poly(vinyl alcohol).
PVAc	Poly(vinyl acetate).
PVC	Poly(vinyl chloride).
PVDC	Poly(vinylidene chloride).
PVDF	Poly(vinylidene fluoride).
PVF	Poly(vinyl fluoride).
RRIM	Reinforced reaction injection moulding.
RTV	Room temperature vulcanization.
S	Styrene.
SA	Stearic acid.
SAN	Styrene–acrylonitrile copolymer.
SBR	Styrene–butadiene rubber
SBS	Styrene–butadiene–styrene block copolymer.
SEM	Scanning electron microscopy.
SIS	Styrene–isoprene–styrene block copolymer.
SMC	Sheet moulding compound.
T_β	Temperature of the highest secondary transition.
T_g	Glass transition temperature.
T_m	Melting temperature.
TDI	Toluene 2,4-diisocyanate.
TGA	Thermogravimetric analysis
TMC	Thick moulding compound.
TMPTMA	Trimethylol propane trimethacrylate.
UF	Urea-formaldehyde.
UHF	Ultra high frequency.

UP   Unsaturated polyester.
UPVC   Unplasticized poly(vinyl chloride).
UTS   Ultimate tensile strength.
UV   Ultra-violet light.

VAc   Vinyl acetate.
VAc-VC   Vinyl acetate–vinyl chloride copolymer.
VC   Vinyl chloride.

Chapter 1

Polymers

INTRODUCTION

Polymers have become an increasingly important group of the general field of engineering materials. Their range of properties and applications is at least as broad as that of other major classes of materials, and ease of fabrication frequently makes it possible to produce finished items very economically. Some important industries such as those for fibres, rubbers, plastics, adhesives, sealants and caulking compounds are based on polymers.

These materials vary widely from thin low viscous liquids and soft elastic rubbers to hard, strong solids, but they have many similar fundamental structures, chemical, physical and mechanical properties.

Polymer science is divided into biological and non-biological materials. Biopolymers (biological polymers) form the very foundation of life and provide much of the food on which man exists.

Non-biological polymers are primarily the synthetic materials used for fibres, plastics, elastomers, adhesives, sealants, coatings and the inorganic polymers, such as polysiloxanes, polyboranes, sulphur, etc. Now these materials are truly indispensable to mankind, being essential for clothing, shelter, transportation and communication, as well as to the conveniences of modern living.

Most of these materials consist of the following elements: C, H, N and O, and are therefore classified as organic polymers. However, in some cases other elements such as Si, S, B, P, F and Cl are present in certain proportions and influence the ultimate properties of the products. Nevertheless, this group of compounds is referred to as organic polymers or organic macromolecules.

1

Together with metals and ceramics, polymers represent the essential engineering materials in the construction of buildings, vehicles, engines, household articles of all kinds, etc. The rapid growth of these new engineering materials during the last three decades is due to the following main factors:

(1) The *availability of basic raw materials* for their production, that means, coal, oil, wood, agriculture and forestry wastes.
(2) The *ensemble of technical properties* specific for polymers such as light weight, chemical stability, elasticity, etc.
(3) *Easy processing* and the knowledge of efficient processing methods such as extrusion, thermal forming, injection moulding, calendering, casting, etc.

A polymer is a large molecule—a macromolecule—build up by the repetition of small, simple chemical units called monomers; some of the simplest building units such as: ethylene, propylene, isobutylene, butadiene, adhesives, sealants, coatings, are by-products of the manufacture of the gasoline and luboils, and are available in large quantities. Other monomers are simple derivatives of ethylene, benzene, formaldehyde, phenol and other basic organic chemicals.

The similarity between fibres, plastics and rubbers can most readily be grasped from a consideration of the structures of the macromolecules. Such structures may be extremely complicated but some generalizations are possible that aid greatly in understanding the close alliance between polymeric materials.

Such concepts are very useful in understanding the behaviour and explaining the properties of high polymers. The main geometrical shapes can be divided into the following groups:

—linear and branched polymers,
—bidimensional networks,
—three-dimensional networks.

Some polymers may have star, umbrella or other shapes. Linear polymers are long chains with a high degree of asymmetry.

$$\ldots -A-A-A-A-A-A- \ldots$$

(A being the mer, the unit of the chain)

A branched polymer is a backbone chain with side branches, the number and length of which may vary widely

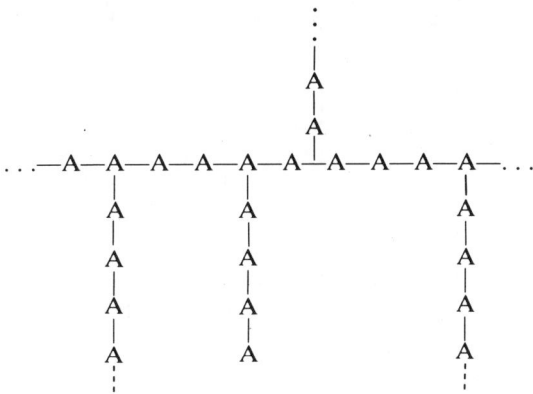

The model for bidimensional polymers is the graphite lattice (Fig. 1.1).

During phenolic resin synthesis, in stage B these kinds of networks are obtained.

High cross-linked or three-dimensional (or space) polymers consist of long chains connected into a three-dimensional network by chemical cross-links.

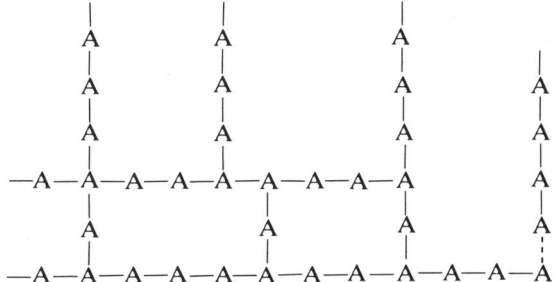

The model for three-dimensional polymer networks is the diamond structure (Fig. 1.2).

In diamond there are a lot of strong bonds between carbon atoms; these bonds are oriented in all spatial directions. Their presence in high numbers explains the high density and also other mechanical properties.

Polymer characteristics are strongly related to their geometrical shape.

A linear polymer is normally *thermoplastic* and relatively soluble, although certain varieties show low solubility because of the extreme

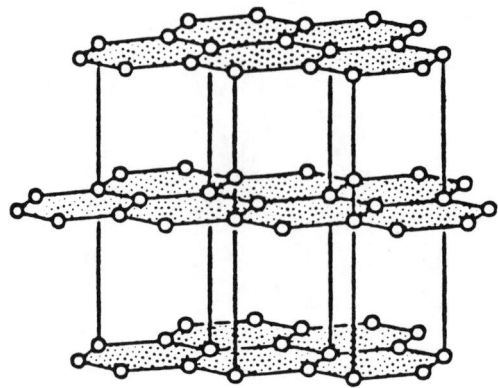

Fig. 1.1. Graphite lattice [1.1]. Carbon atoms in basal planes; van der Waals forces shown by vertical lines between basal planes. (Reprinted with permission from *Makromolekulare Chemie*, 3rd edn., by K. H. Meyer & H. Mark, 1953. Copyright Akademische Verlagsgesellschaft Geest & Portig K. G.).

length of chains or the chemical groups (different from that of the solvent) attached to the chains. Fibres are usually formed from this type of linear polymer.

In the case of branched polymers the amount of lateral chains may be very small and have little effect on the properties of the resulting polymer, or it may be quite extensive and be a principal factor influencing properties such as solubility, softening point, workability, etc.

The structure of bidimensional polymers is usually built up by cross-linking linear polymers by means of a chemical reaction. The

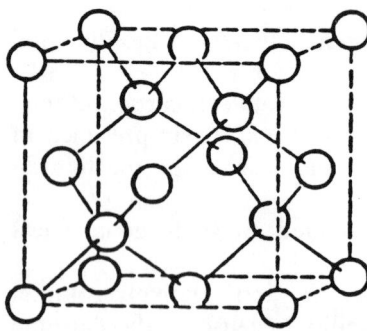

Fig. 1.2. Diamond lattice. Covalent bonding [1.1]. (Reprinted with permission from *Makromolekulare Chemie*, 3rd edn., by K. H. Meyer & H. Mark, 1953. Copyright Akademische Verlagsgesellschaft Geest & Portig K. G.)

Table 1.1.

Main differences between low and high molecular mass substances

Properties	Low molecular substances	High polymers
Molecular weight (MW)	Low, <10 000	>10 000
States of matter	Vapour, liquid, solid	Liquid, solid
Solubility	Most soluble	Some are not soluble
Fusibility	Generally fusible	Some are not fusible
Homogeneity of molecules	Homogeneous (identical)	Polydisperse (heterogeneous)
Fibre forming ability	Impossible	Possible for linear ones

effect of even a small number of cross-linkages on properties may be enormous. An increase in the degree of cross-linking increases the mechanical properties, the softening point, and decreases the solubility.

Three-dimensional polymers consist of long chains connected up into a three-dimensional network. A high frequency of cross-linking is characteristic of *thermoset* (or thermosetting) polymers. This type of structure is usually termed a *space structure*. The insolubility, in-fusibility, strength and low extensibility of thermoset polymers are attributed to the large number of cross-linkages (high degree of cross-linking) which make it almost impossible to separate segments of the structure.

We may obtain three-dimensional structures in stage C of the synthesis of phenol-formaldehyde, urea-formaldehyde or other poly-mers. Fibres do not contain this structure.

Some important differences between low molecular substances and high polymers are presented in Table 1.1.

MOLECULAR WEIGHT (MW)

The important factor in the application of polymers is their molecular mass. The terms macromolecule, giant molecule, high polymer, polymer, indicate that the molecules of this kind of materials are large and hence consist of a great number of units (mers).

One of the most important characteristics of these long chain molecules is their degree of polymerization (DP); it represents the number of units in a given macromolecule.

$$DP = \frac{\text{MW of the polymer}}{\text{MW of the structural unit (mer)}}$$

In the case of $-[-CH_2-CH_2-]_n-$, n represents DP, that means the number of units in the macromolecular chain.

As the MW of the polymer is related to the degree of polymerization, it can be defined as:

$$MW = \text{monomer mass} \times DP$$

During the course of the synthesis, the polymer chains grow to different lengths, giving a product consisting of a mixture of macromolecules of a wide range of molecular weights.

To characterize a polymer, we have to use certain average values of molecular weight [1.1–1.3]. These averages can be obtained by several different methods; they are:

—number average molecular weight, $\bar{M}_n$
—weight average molecular weight, $\bar{M}_w$
—Z, average molecular weight, $\bar{M}_z$
—viscosity average molecular weight, $\bar{M}_v$

$\bar{M}_n$ is the weight of a polymer sample divided directly by the number of molecules which exist in this sample

$$\bar{M}_n = \frac{\sum\limits_{i=1}^{n} n_i M_i}{\sum\limits_{i=1}^{n} n_i} \tag{1.1}$$

where n_i is the number of macromolecules in the fraction i having the mean molecular mass M_i.

$\bar{M}_n$ value may be obtained from such measurements as osmotic pressure, boiling point elevation, and freezing point depression. In all these methods the number of molecules for each fraction is counted in a known mass of the polymer and through Avogadro's number the number average molecular weight is estimated. Table 1.2 presents some $\bar{M}_n$ values.

Mathematically, the weight average molecular weight $\bar{M}_w$ is repre-

Table 1.2.
Typical number average molecular weights of some commercial polymers [1.3]

Polymer	$\bar{M}_n$
LDPE	20 000
HDPE (Standard Oil process)	15 000
Nylon	20 000
Polyvinyl chloride	40 000
Polypropylene (Zeigler process)	40 000
Polyethylene terephthalate	20 000

Reprinted with permission from *Organic Polymer Chemistry* by K. J. Saunders, 1973. Copyright Chapman and Hall, London.

sented by the following relation:

$$\bar{M}_w = \frac{\sum\limits_{i=1}^{n} n_i M_i}{\sum\limits_{i=1}^{n} n_i M_i} \tag{1.2}$$

The weight average molecular mass is usually determined by light scattering, which depends on the size and of the mass of the molecule.

$$\bar{M}_z = \frac{\sum\limits_{i=1}^{n} n_i M_i^3}{\sum\limits_{i=1}^{n} n_i M_i^2} \tag{1.3}$$

The viscosity average molecular mass ($\bar{M}_v$) is defined by Mark–Houwink equation:

$$[\eta] = k\bar{M}_v^{\alpha} \tag{1.4}$$

where $[\eta]$ is the intrinsic viscosity, k and α are constants which depend on the nature of the binary system polymer and solvent (Table 1.2).

$$[\eta] = \lim_{c \to 0} \frac{\eta_{sp}}{c} \tag{1.5}$$

$$\eta_{sp} = \eta_{rel} - 1 = \frac{\eta}{\eta_o} - 1 = \frac{\eta - \eta_o}{\eta_o} \tag{1.6}$$

where:

η_{sp} = specific viscosity
η_{rel} = relative viscosity
η = the viscosity of the solution
η_o = the viscosity of the solvent
c = the concentration of the polymer in the
    solution, g/100 ml.

Often the ratio $\bar{M}_w/\bar{M}_n$ is used to determine the spread of the molecular weight distribution of the polymer, that means the polydispersity.

For a narrow polydispersity the ratio $\bar{M}_w/\bar{M}_n$ is close to one, but for broad molecular mass distribution, it may be as high as 3–10.

POLYDISPERSITY

In addition to the degree of polymerization of a high polymer, the distribution of the various chain lengths has an important effect on the properties of the material.

Generally all natural and synthetic polymers are heterogeneous in respect to the lengths of their chains. Some proteins are less polydisperse and recently a new group transfer polymerization process has provided more homogeneous polymers.

Some analytical methods such as fractional precipitation or gel permeation chromatography (GPC) are employed often for the determination of molecular weight distribution of various polymers.

Frequently we represent the polymer polydispersity by differential and integral distribution curves.

Typical differential distribution curves are represented in Figs. 1.3 and 1.4.

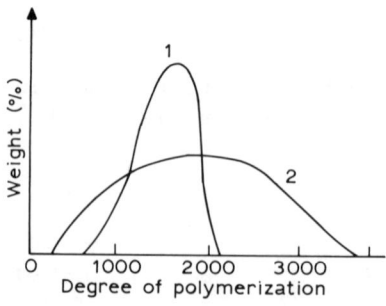

Fig. 1.3. Differential distribution curves.

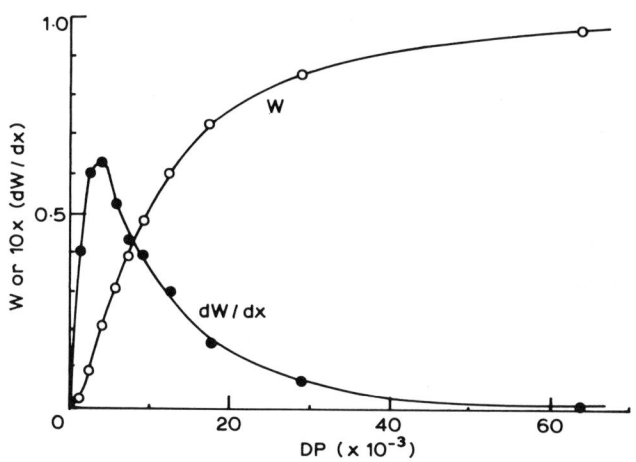

Fig. 1.4. Differential and cumulative diagrams [1.4]. (Reprinted with permission from *Principles of Polymer Systems*, 2nd edn., by F. Rodriguez, 1982. Copyright Hemisphere Publ. Co.)

In Fig. 1.3 curve 1 reflects a lower polydispersity (lower $\bar{M}_w/\bar{M}_n$) than curve 2 (higher $\bar{M}_w/\bar{M}_n$),

As was mentioned, information on MW distribution is commonly presented as a graph in which the relative number of molecules or the weight of molecules within a narrow range of molecular weights is plotted against the MW, with the curve describing all species present in the sample. Curves representing both types are shown in Fig. 1.5; these have been calculated from an ideal equation representing the distribution expected in some types of polymerization processes.

SYNTHESIS OF POLYMERS

Most synthetic polymers are produced from low molecular mass compounds (called monomers), by:

—addition polymerization;
—condensation polymerization (polycondensation);
—ring opening polymerization;

as well as by chemical transformation of other synthetic polymers.

Some polymers (artificial polymers) can be done by the chemical

Polymeric Building Materials

Table 1.3.
Parameters for the Mark–Houwink equation [1.4]

Polymer	Solvent	Temperature (°C)	$K' \times 10^5$	α
Cellulose triacetate	Acetone	25	8·97	0·90
SBR rubber	Benzene	25	54	—
Natural rubber	Benzene	30	18·5	0·66
	n-propyl ketone	14·5	119	0·50
Polyacrylamide	Water	30	68	0·66
Polyacrylonitrile	Dimethyl formamide	25	23·3	0·75
Poly(dimethylsiloxane)	Toluene	20	20·0	0·66
Polyethylene	Decalin	135	62	0·70
Polyisobutylene	Benzene	24	107	0·50
	Benzene	40	43	0·60
	Cyclohexane	30	27·6	0·69
Poly(methyl methacrylate)	Toluene	25	7·1	0·73
Polystyrene				
atactic	Toluene	30	11·0	0·725
isotactic	Toluene	30	10·6	0·725
Poly(vinyl acetate)	Benzene	30	22	0·65
	Ethyl n-butyl ketone	29	92·9	0·50
Poly(vinyl chloride)	Tetrahydrofuran	20	3·63	0·92

modification of some natural ones such as cellulose, proteins, natural rubber, etc.

A proper classification and definition should be based on features typical of the true mechanism of the reaction. The main aspect in all such processes is the propagation of a polymer chain by which the polymeric product is formed. As a consequence, classification of polymer-forming processes should be based on the specific mechanism of the macromolecular chain growth.

Addition Polymerization

During addition polymerization, reactive sites exist only at the end (or ends) of the growing polymer chain; the chain growth takes place by the reaction between the active site of the macromolecular chain and the monomer molecules.

For the activation of such processes some small amounts (0·1%) of

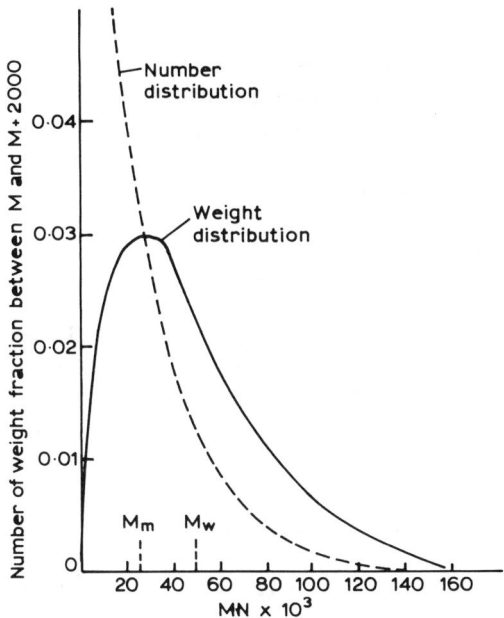

Fig. 1.5. Most probable MW distribution, calculated for a polymer having MW = 50 000 and $M_w/M_n = 2$ [1.2]. (Reprinted with permission from *Characterization of Polymers*. Encyclopedia Reprints by M. B. Bikales (Ed.), 1971. Copyright John Wiley.)

initiators (in case of a radical mechanism) or catalysts (for ionic mechanisms) are used.

Schematically the addition polymerization which takes place by a radical mechanism, known as radical polymerization, may be represented as follows [1.5]:

$$I \xrightarrow{\text{decomposition}} 2R^{\bullet} \left.\right\}$$
$$R^{\bullet} + M \rightarrow M^{\bullet}$$ Initiation

$$M^{\bullet} + M \rightarrow M - M^{\bullet} \text{ (or } M_2^{\bullet})$$
$$M_2^{\bullet} + M \rightarrow M_3^{\bullet}$$ Propagation
----------------------------- (or chain growth)
$$M_{n-1}^{\bullet} + M \rightarrow M_n^{\bullet}$$
$$M_n^{\bullet} + M_m^{\bullet} \rightarrow M_{m+n} \text{ or } P_{m+n}$$ Termination by coupling of macroradicals

where:

$$I = \text{initiator molecule}$$
$$R^{\bullet} = \text{radical obtained from I (free radical)}$$
$$M = \text{monomer molecule}$$
$$M^{\bullet} = \text{monomer radical (free radical)}$$
$$M_2^{\bullet} = \text{dimer radical}$$
$$M_3^{\bullet} = \text{trimer radical}$$
$$M_{n-1}^{\bullet}, M_m^{\bullet}, M_n^{\bullet} = \text{macroradicals}$$
$$P_m, M_m - M_n = \text{inactive polymer macromolecules.}$$

Each act of chain growth involves the disappearance of one monomer molecule. The reactive site at the end of the polymeric chain is regenerated after each elementary act of polymerization, and the number of reactive sites before and after the growth act remains constant. Thus, in the kinetics sense of the term, addition polymerization is a typical chain process (*chain polymerization*). Monomers such as the following are made of unsaturated molecules.

$$\begin{array}{ccc}
CH_2{=}CH_2 & CH_2{=}CH & CH_2{=}CH \\
\text{(ethylene)} & | & | \\
& CH_3 & Cl \\
& \text{(propylene)} & \text{(vinyl chloride)}
\end{array}$$

The reactive species, which may be a free radical, cation or anion, adds to a monomer molecule by opening the Π_1-bond to form a new radical, cation or anion centre as the case may be. The process is repeated as many more monomer molecules are successively added to continuously propagate the reactive centre.

$$R^{\bullet} + CH_2{=}CH \longrightarrow R{-}CH_2{-}\overset{H}{\underset{X}{C}}^{\bullet} \overset{CH_2{=}CH}{\underset{X}{\longrightarrow}} R{-}CH_2{-}CH{-}CH_2{-}\overset{H}{\underset{X}{C}}^{\bullet}$$

$$+ m - 1 \overset{CH_2{=}CH}{\underset{X}{}} \longrightarrow R{-}CH_2{-}CH{-}[CH_2{-}CH]_{m-1}{-}CH_2{-}CH^{\bullet}$$

Polymer growth is terminated at some point by destruction of the

Table 1.4.

Types of chain polymerization undergone by various unsaturated monomers
[1.6]

Monomers	Type of initiation		
	Radical	Cationic	Anionic
Ethylene	+	−	+
Styrene	+	−	+
Halogenated olefins (vinyl chloride)	+	−	+
Acrylates	+	−	+
Methacrylates	+	−	+
Acrylonitrile	+	−	+
Vinyl ethers	+	−	+

Reprinted with permission from *Principles of Polymerization,* 2nd edn, by G. Odian, 1981. Copyright John Wiley.

reactive centre by an appropriate reaction depending on the type of reactive centre and the particular reaction conditions.

Chain polymerization proceeds by a distinctly different mechanism from polycondensation, known also as step polymerization (Table 1.4). The most significant difference is that high molecular weight polymer is formed immediately in chain polymerization. A radical, anionic or cationic reactive centre, once produced, adds many monomer molecules in the chain reaction and grows rapidly to a large size. The monomer concentration decreases throughout the course of the process as the number of macromolecules increases. At any instant the reaction mixture contains monomer, high polymer and the growing radicals. The MW of the polymer is relatively unchanged during the polymerization, although the overall percentage conversion of monomer to polymer increases with reaction time [1.6].

The situation is quite different for step polymerization. Whereas only monomer and the propagation species can react with each other in chain polymerization, any two molecular species present can react in step polymerization. Monomer disappears much faster in step polymerization as one proceeds slowly to produce dimer, trimer, tetramer, and so on. The MW increases throughout the course of the reaction and high MW polymer is not obtained until the end of polymerization. Long reaction times are necessary for both high percentage conversion and high molecular mass (Fig. 1.6).

Catalysts are used for the ionic chain polymerization which produces

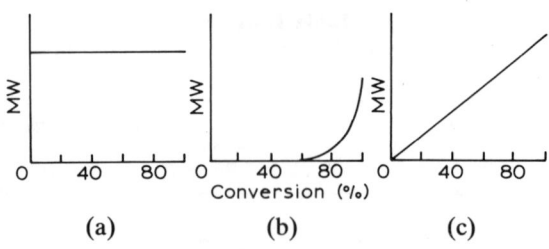

Fig. 1.6. Variation of MW with conversion. (a) Chain polymerization, (b) step polymerization, (c) non-terminating ionic chain polymerization and protein synthesis [1.6]. (Reprinted with permission from *Principles of Polymerization*, 2nd edn., by G. Odian, 1981. Copyright John Wiley.)

during the process macroions instead of macroradicals.

$$M + AK \rightarrow AM^- \qquad \text{anionic activation}$$
$$K^+$$

$$M + AK \rightarrow KM^+ \qquad \text{cationic activation}$$
$$A^-$$

$$AM^- + M \rightarrow AM_2^- \rightarrow \dots \quad AM_n^-$$
$$K^+ \qquad K^+ + (n-2)M \quad K^+$$
$$\text{(macroanion)}$$

$$KM^+ + M \rightarrow AM_2^+ \rightarrow \dots \qquad M_n^+$$
$$A^- \qquad A^- + (n-2)M \quad A^-$$
$$\text{(macrocation)}$$

$$AK = \text{catalyst}$$

Example of cationic polymerization (of isobutylene);

$$BF_3 \cdot H_2O + CH_2\!\!=\!\!\underset{\underset{CH_3}{|}}{\overset{\overset{CH_3}{|}}{C}} \longrightarrow CH_3\!\!-\!\!\underset{\underset{CH_3}{|}}{\overset{\overset{CH_3}{|}}{C^+}}[BF_3 \cdot OH]^- \qquad \text{Initiation}$$

$$CH_3\!\!-\!\!\underset{\underset{CH_3}{|}}{\overset{\overset{CH_3}{|}}{C^+}}[BF_3 \cdot OH]^- + CH_2\!\!=\!\!\underset{\underset{CH_3}{|}}{\overset{\overset{CH_3}{|}}{C}} \longrightarrow CH_3\!\!-\!\!\underset{\underset{CH_3}{|}}{\overset{\overset{CH_3}{|}}{C}}\!\!-\!\!CH_2\!\!-\!\!\underset{\underset{CH_3}{|}}{\overset{\overset{CH_3}{|}}{C^+}}[BF_3 \cdot OH]^-$$

Propagation

$$\cdots CH_2\text{—}\underset{\underset{CH_3}{|}}{\overset{\overset{CH_3}{|}}{C}}{}^+[BF_3\cdot OH]^- \longrightarrow \cdots\text{—}CH_2\text{—}\underset{\underset{CH_2}{\|}}{\overset{\overset{CH_3}{|}}{C}} + H^+[BF_3\cdot OH]^- \quad \text{Termination}$$

Example of anionic polymerization (of styrene):

$$BuLi + CH_2\text{=}\underset{\underset{C_6H_5}{|}}{CH} \longrightarrow BuCH_2\text{—}\underset{\underset{C_6H_5}{|}}{CH}{}^-Li^+ \qquad Bu = C_4H_9 \quad \text{Initiation}$$

$$BuCH_2\text{—}\underset{\underset{C_6H_5}{|}}{CH}{}^-Li^+ + CH_2\text{=}\underset{\underset{C_6H_5}{|}}{CH} \longrightarrow BuCH_2\text{—}\underset{\underset{C_6H_5}{|}}{CH}\text{—}CH_2\text{—}\underset{\underset{C_6H_5}{|}}{CH}{}^-Li^+ \quad \text{Propagation}$$

The final step is decomposition of the BuLi by the addition of water or alcohol:

$$BuCH_2\text{—}\underset{\underset{C_6H_5}{|}}{CH}\text{----}CH_2\text{—}\underset{\underset{C_6H_5}{|}}{CH}{}^-Li^+ \longrightarrow CH_3\text{—}\underset{\underset{C_6H_5}{|}}{CH}\text{----}CH\text{=}\underset{\underset{C_6H_5}{|}}{CH} + BuLi$$

Termination

Some examples of polymers which can be obtained by addition polymerization are those of ethylene, vinyl chloride, styrene, etc.

$$CH_2\text{=}CH_2 \longrightarrow [CH_2\text{—}CH_2]_n$$
(ethylene)   poly-ethylene

$$CH_2\text{=}\underset{\underset{Cl}{|}}{CH} \longrightarrow [\text{—}CH_2\text{—}\underset{\underset{Cl}{|}}{CH}\text{—}]_n$$
vinyl chloride   poly(vinyl chloride)

$$CH_2\text{=}\underset{\underset{C_6H_5}{|}}{CH} \longrightarrow [\text{—}CH_2\text{—}\underset{\underset{C_6H_5}{|}}{CH}\text{—}]_n$$
Styrene   poly(styrene)

A particular kind of addition polymerization is *copolymerization*. It is the simultaneous polymerization of two or more chemically different monomers that react to form a polymer containing both monomers bonded in one chain. For instance, Saran is a copolymer made with vinyl chloride and vinylidene chloride:

$$n\text{CH}_2\!\!=\!\!\underset{\underset{\text{Cl}}{|}}{\text{CH}} + n\text{CH}_2\!\!=\!\!\text{CCl}_2 \longrightarrow \cdots -\left[-\text{CH}_2\!-\!\underset{\underset{\text{Cl}}{|}}{\text{CH}}\!-\!\text{CH}_2\!-\!\underset{\underset{\text{Cl}}{|}}{\overset{\overset{\text{Cl}}{|}}{\text{C}}}\!-\right]_n\!\!-\cdots$$

(vinyl chloride) + (vinylidene chloride)

GRS rubber is the product of copolymerization of butadiene and styrene:

$$n\text{CH}_2\!\!=\!\!\text{CH}\!-\!\text{CH}\!\!=\!\!\text{CH}_2 + n\text{CH}_2\!\!=\!\!\underset{\underset{\text{C}_6\text{H}_5}{|}}{\text{CH}} \rightarrow$$

(butadiene)   (styrene)

$$\cdots -[-\text{CH}_2\!-\!\text{CH}\!\!=\!\!\text{CH}\!-\!\text{CH}_2\!-\!\text{CH}_2\!-\!\underset{\underset{\text{C}_6\text{H}_5}{|}}{\text{CH}}-]_n\!-\cdots$$

ABS is a terpolymer which results from the copolymerization (terpolymerization) of acrylonitrile (A), butadiene (B) and sytrene (S).

Depending on the reactivities of monomers, let's say X and Y and the conditions of the process, a few kinds of arrangement are possible which lead to [1.8]:

(a) ...—X Y XX YYY X Y XXXX—... Random copolymer
(b) ...—XXXX YYYYY XXXXX—... Block copolymer
(c) ...—X Y X Y X Y X Y X Y—... Alternating copolymer.

A graft copolymer is essentially a branched-chain structure having side chains composed of one type of monomer attached to the backbone chain made of another monomer.

$$
\begin{array}{c}
\vdots \\
| \\
\text{Y} \\
| \\
\text{Y} \\
|
\end{array}
$$

(d) $\cdots -\text{X}-\text{X}-\text{X}-\text{X}-\text{X}-\text{X}-\text{X}-\text{X}-\text{X}-\text{X}-\cdots$

$$
\begin{array}{c}
| \\
\text{Y} \\
| \\
\text{Y} \\
| \\
\text{Y} \\
| \\
\vdots
\end{array}
$$

In all these cases the properties are different and strongly related to the microstructure of the copolymer.

Condensation Polymerization (Polycondensation, Stepwise Polymerization)

It is important to realize that in chain polymerization, propagation to final molecular mass is very rapid. In the styrene polymerization, for example, we might start a bulk polymerization and stop it after only 1% of the monomer is converted to polymer. Analysis would show that the mixture consisted of 99% unreacted monomer and 1% of a high polymer. Conversion of the rest of the monomer to polymer would not affect the already formed product, which, with the exception of some 'living' polymer systems, is inactive (dead). We cannot ordinarily make a low molecular mass polymer by addition polymerization and then increase its molecular weight by more of the same reaction. This is a major point of difference between chain and stepwise polymerization.

A second distinction is that usually polymers formed at early stages of conversion in polycondensation are not dead but can react as easily as monomers. The principle of equal reactivity regardless of molecular mass is fundamental to stepwise polymerization also.

A third characteristic of this process is the formation in addition to the high polymer of a low molecular weight compound such as water, hydrochloric acid, alcohol, etc.

The common attribute of each polycondensation process is the reaction of two bifunctional or polyfunctional monomers with one another. Usually two different monomers are employed, each with functional groups capable of reacting with the other. Thus dibasic acids react with polyalcohols to give polyesters. Polycondensation polymers may be either thermoplastic or thermosetting. Normally, thermoplastics are produced if the monomers used are exclusively difunctional. Polyfunctionality is required for the formation of thermosetting polymers. If approximately equal molar quantities of tri- and difunctional monomers react, a tight three-dimensional network is produced that is characteristic of the thermosets. The following reactions between dicarboxylic acid and dihydroxy alcohols, resulting in polyesters, illustrate this process.

$$a—A—a + b—B—b \rightleftharpoons a—A—B—b + ab$$

a and b being the functional groups. A and B are the main parts of the monomer molecules, a—A—a and b—B—b being the monomers.

The resultant molecule a—A—B—b reacts again in the same way with a hydroxy alcohol molecule or dicarboxylic acid; the process repeats itself until linear chains of indefinite length are formed:

$$a—A—B—b + a—A—a \rightleftarrows a—A—B—A—a + ab$$

ab is the low molecular mass compound which results in the synthesis of the high polymers through the polycondensation process.

$$a—A—B—A—a + b—B—b \rightleftarrows a—A—B—A—B—b + ab$$

Chain propagation occurs step by step, the polycondensation process being based on a stepwise intermolecular mechanism.

As a result of this process we obtain a macromolecular compound which possesses at its end functional groups which belong to one or both initial monomers.

$$a—A—B—A—B—...—B—A—B—A—B—b$$

The polycondensation of dicarboxylic acid and a dihydroxy alcohol may be more properly written in the following way:

HOOC—R—COOH + HO—R' – OH

$$\rightleftarrows HOOC—R—CO—O—R'—OH + H_2O$$

The resultant molecule reacts again in the same way with the dihydroxy alcohol molecule; the process repeats itself until linear chains of indefinite length are formed.

HO—R'—OH + HOOC—RCO—O—R'—OH

$$\rightleftarrows HO—R'OOC—R—COOR'—OH + H_2O$$

R and R' stand as A and B in the former example for organic groups such as CH_2, $(CH_2)_n$ and others. The resulting polymer is a linear one that, depending on the degree of polymerization, may range from a viscous liquid to a rigid solid; usually it is represented as.

HO—R'—O—[OC—R—CO—O—R'O—]$_n$—OC – R – COOH

$$n = DP$$

Because of its relatively symmetrical structure and the presence of numerous polar groups, such a polymer is a good fibre-forming material.

Other reactions of polycondensation which result in the formation of long linear chains are used in the manufacture of nylon (polyamide), polycarbonate, etc. Phenol-formaldehyde, urea-formaldehyde, melamine-formaldehyde, which in the last stage (C) are three-dimensional polymers, are also obtained by polycondensation processes.

CLASSIFICATION

There are many schemes of classification possible. One of the most used classes all polymers as *synthetic* (man-made) or *natural*. Synthetic polymers could be further classified taking into account different criteria such as:

—monomer type;
—preparative techniques;
—polymer structure;
—physical properties;
—processing techniques;
—end uses, etc.

A classification according to the thermal behaviour might first divide polymers into thermoplastics and thermosets and carry on from there.

A classification by end uses might attempt to associate polymer groups with specific industries such as: rubber industry, fibre industry, film industry, etc. These classifications are useful in a practical sense, but offer little or no insight into the scientific principles.

A more comprehensive classification divides polymers into the following groups:

—organic polymers;
—element-organic or semi-organic polymers;
—mineral polymers.

The first group includes compounds containing, apart from carbon atoms, hydrogen, oxygen, nitrogen, sulphur and halogen atoms, even if the O, N or S is in the backbone chain. Examples are: poly(vinyl alcohol), plexiglas, polyurethane, etc.

$$\cdots-CH_2-CH-CH_2-CH-CH_2-CH-\cdots$$
$$\quad\quad\quad | \quad\quad\quad | \quad\quad\quad |$$
$$\quad\quad\quad OH \quad\quad OH \quad\quad OH$$

poly(vinyl alcohol) (PVA)

$$\quad\quad CH_3 \quad\quad CH_3 \quad\quad CH_3$$
$$\quad\quad | \quad\quad\quad | \quad\quad\quad |$$
$$\cdots-CH_2-C-CH_2-C-CH_2-C-\cdots$$
$$\quad\quad | \quad\quad\quad | \quad\quad\quad |$$
$$\quad\quad COOCH_3 \; COOCH_3 \; COOCH_3$$

plexiglas [poly-(methyl methacrylate)] (PMMA)

$$\ldots-OC-NH-(CH_2)_m-NH-CO-O-(CH_2)_n-O-\ldots$$

polyurethane (PU)

The group of element-organic polymers includes:

—compounds whose chains contain carbon atoms and heteroatoms (except for C, N, S and O atoms);
—compounds with mineral chains if they contain side groups with carbon atoms connected directly to the chain such as poly(-dimethyl siloxane):

$$\quad\quad CH_3 \quad\quad CH_3 \quad\quad CH_3$$
$$\quad\quad | \quad\quad\quad | \quad\quad\quad |$$
$$\cdots-Si-O-Si-O-Si-O-\cdots$$
$$\quad\quad | \quad\quad\quad | \quad\quad\quad |$$
$$\quad\quad CH_3 \quad\quad CH_3 \quad\quad CH_3$$

—compounds whose main chains consist of carbon atoms and whose side groups contain hetero-atoms (except for N, S, O and halogen atoms) connected directly to the carbon atoms of the chain.

Mineral polymers do not contain carbon atoms; they have been studied very little so far, and for this reason it is difficult to divide them into classes at present.

All the elements of group IV can form linear chains analogous to those of polyethylene:

$$\quad H \;\; H \;\; H \;\; H \;\; H \quad\quad\quad\quad H \;\; H \;\; H \;\; H$$
$$\quad | \;\; | \;\; | \;\; | \;\; | \quad\quad\quad\quad | \;\; | \;\; | \;\; |$$
$$\cdots-Si-Si-Si-Si-Si-\cdots \quad\quad \cdots-Ge-Ge-Ge-Ge-$$
$$\quad | \;\; | \;\; | \;\; | \;\; | \quad\quad\quad\quad | \;\; | \;\; | \;\; |$$
$$\quad H \;\; H \;\; H \;\; H \;\; H \quad\quad\quad\quad H \;\; H \;\; H \;\; H$$

Polysilanes   Polygermanes

In most silicates the atoms in the chain are linked by covalent bonds, and the chains are connected to each other by ionic bonds:

$$\cdots-\overset{\overset{\textstyle O^-}{|}}{\underset{\underset{\textstyle O^-}{|}}{Si}}-O-\overset{\overset{\textstyle O^-}{|}}{\underset{\underset{\textstyle O^-}{|}}{Si}}-O-\overset{\overset{\textstyle O^-}{|}}{\underset{\underset{\textstyle O^-}{|}}{Si}}-O-\cdots$$

Such a chain is known as a pyroxene chain.

Polymeric silicates may be of lamellar or three-dimensional crystalline structure (e.g. quartz). Talc and some varieties of natural asbestos, such as chrysotile asbestos, are lamellar silicates.

Silicate glasses, whose main constituent is SiO_2, have a polymeric structure. The presence of metallic atoms (K, Na, Ca, etc.) in glass disturbs its crystalline structure and that is why silicate glasses are amorphous under normal conditions.

Other noteworthy polymeric compounds of this type are *cement* (the generic name given to numerous mineral binders), which consists mainly of various silicates, and *concrete* [1.1].

Some natural aluminosilicates of lamellar structure are clays such as kaolinites, montmorillonites and zeolites. Asbestos fibres are based on chains of Si and O and lateral groups with MgO.

Among the group of natural polymers with industrial applications can be mentioned: natural rubber, polysaccharides such as cellulose, starch, agar, etc., and proteins such as casein, keratins, collagen, gelatin.

PHYSICAL STRUCTURE

So far some characteristics of the microstructure of polymers such as geometry, molecular weight, polydispersity, synthesis, have been discussed. Their physical structure (macrostructure) reflects the type of aggregation of the polymeric chains.

If a macromolecule has a sufficiently regular structure it may be capable of some degree of crystallization. Crystallization is limited to certain linear or slightly branched polymers with a high regularity of the chain. Polyethylene and polyamides are crystalline polymers, their macromolecules having a high regularity and symmetry.

$$\cdots-CH_2-CH_2-CH_2-CH_2-\cdots \text{ polyethylene (PE)}$$

$$\cdots-OC(CH_2)_m-CO-NH-(CH_2)_n-NH-\cdots \text{ polyamide (PA)}$$

Crystalline regions in a polymer have large effects on their physico-mechanical properties such as strength, density, stiffness, melting temperature, clarity, etc. Polymers like the fibre-forming ones, have a large number of crystalline domains. The most direct evidence of this fact is provided by X-ray and electron diffraction studies. The X-ray patterns of crystalline polymers show both sharp features associated with regions of three-dimensional order, and more diffuse features characteristic of molecularly disordered substances (amorphous). Both order (crystalline) and short distance order regions (amorphous) coexist in most crystalline polymers. Additional evidence to this effect comes from other polymer properties such as densities, which are intermediate between those calculated for completely crystalline and amorphous species.

A complication in polymer systems compared to those of low molecular mass substances arises in that polymer X-ray photographs, such as the powder diagram in Fig. 1.7, show the presence of both liquid-like and crystalline-like diffraction patterns. The sharp

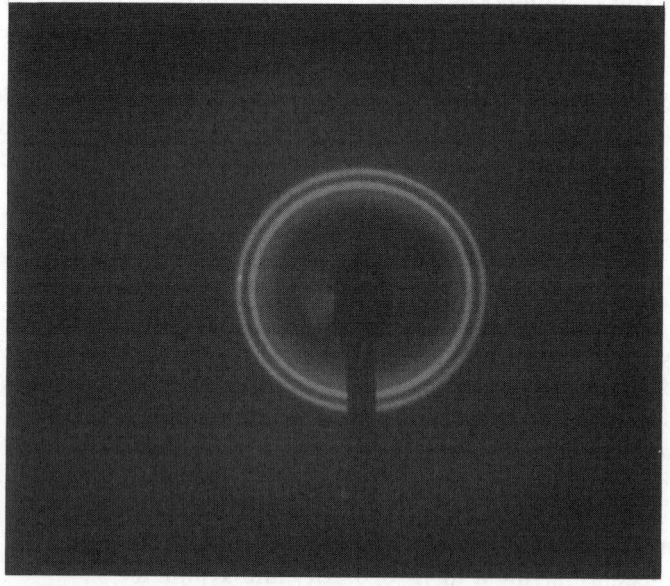

Fig. 1.7. X-ray powder diffraction pattern showing sharp crystal rings and diffuse amorphous bands. Compression moulded linear PE [1.9]. (Courtesy P. H. Geil.)

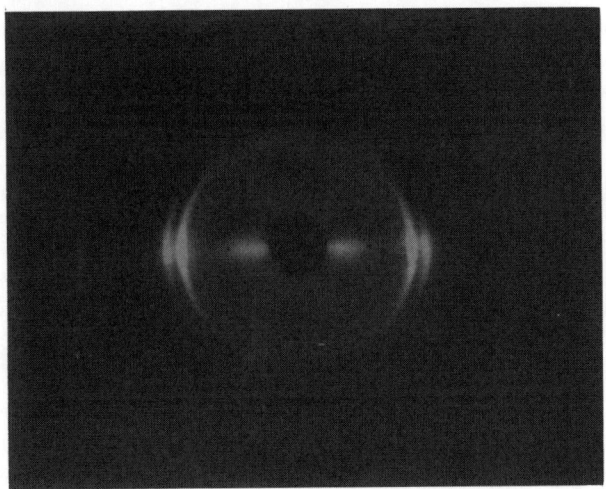

Fig. 1.8. X-ray fibre diagram showing the arcs and spots resulting from partial orientation [1.9]. (Reprinted with permission from *Experiments in Polymer Science* by F. A. Collins, J. Bares & F. W. Billmeyer Jr, 1973. Copyright John Wiley.)

diffraction rings are characteristic of regions of three-dimensional order at least several hundred angstroms in size, while the more diffuse features are associated with the presence of disordered regions similar to liquids of polymer melts.

The presence of both types of pattern from the same sample was used historically to argue that polymers were biphasic, consisting of distinct crystalline domains surrounded by an amorphous phase. Polymers crystallize with a wide variety of unit cells.

When a polymer sample is increasingly oriented, the macro-molecules become parallel, the rings of the powder diagram first break up into arcs, which then narrow to the relatively sharp spots normally seen in a fibre pattern. An example is shown in Fig. 1.8.

In the early days of polymer science, those familiar with crystals of low molecular mass compounds failed to realize that a macromolecule, by virtue of a regular repeating structure, runs through many consecutive unit cells. Figure 1.9 represents a unit cell of a crystalline polyethylene (HDPE). The carbon atoms of the macromolecule form a plane zigzag chain. The C–C distance in the chain equals 1·54 Å, the C–C–C valency angle is 109·5°, the unit cell dimensions are $a = 7·41$;

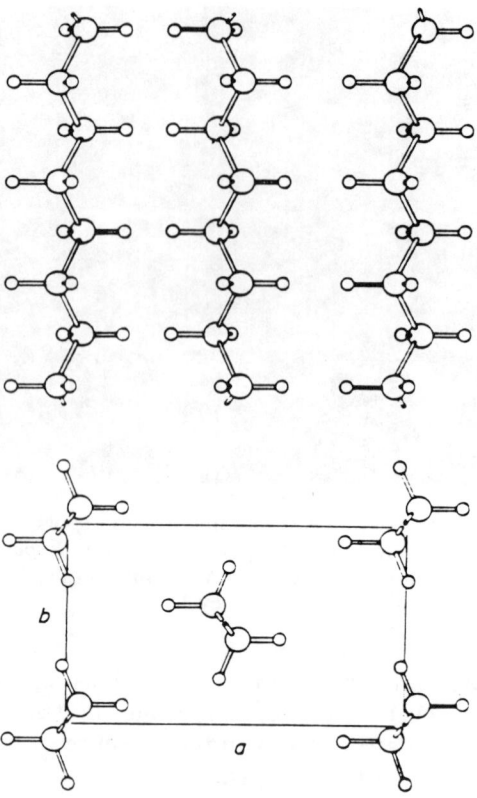

Fig. 1.9. Unit cell of PE [1.4]. (Reprinted with permission from *Principles of Polymer Systems,* 2nd edn., by F. Rodriguez, 1982. Copyright Hemisphere Publ. Co.)

$b = 4.94$; $c = 2.55$ Å. All the macromolecules in the polymer are parallel to one another and extend along the c axis of the unit cell [1.4, 1.10]. The highest degree of crystallinity is normally encountered in natural fibres, especially those in which the cellulose chain is deposited parallel to the fibre axis. The proposed structure is shown in Fig. 1.10. In this crystal-structure model, the chains run alternately in one direction and then the other. The crystalline domains in cellulose, although fairly long in the direction of the fibre axis, are not very wide in the perpendicular directions. There is no discontinuity between the crystalline regions and the surrounding amorphous regions. The long

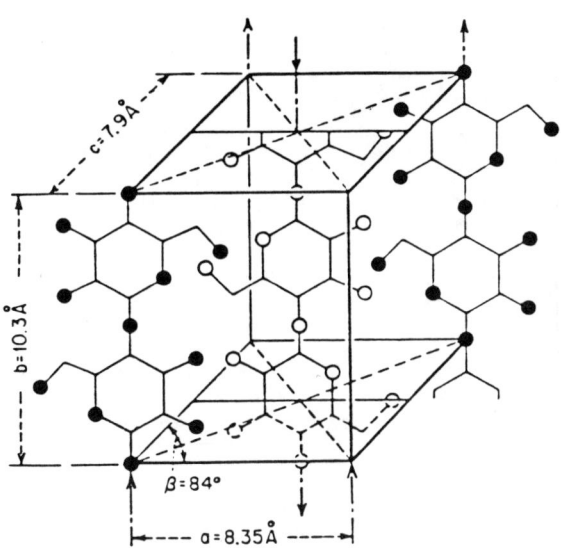

Fig. 1.10. Crystal unit of cellulose [1.1]. (Reprinted with permission from *Makromolekulare Chemie* by K. H. Meyer and H. Mark, 3rd edn., 1953. Copyright Akademische Verlagsgesellschaft Geest & Portig K. G.)

thread-like chains made of glucose units may extend from an amorphous domain, through a crystalline one (crystallite or micelle), and into another amorphous region.

X-ray diffraction experiments with spinnable polymers utilize the symmetry provided by their orientation, and provide most of the information we have about the atomic and molecular arrangements within polymer crystals. An example is the diffractometer scan of linear PE (Fig. 1.11), such a scan being a plot of diffracted intensity versus diffraction angle.

Extensive work on copolymers of all kinds fully confirms the importance of structural regularity on crystallization tendency and, consequently, on properties. Those copolymers built by a regular alternation of the two components, let's say, X and Y, which can be represented by . . . —XYXYXYXYXYXY— . . . show a distinct tendency to crystallize, whereas the corresponding copolymer with random geometric distribution of X and Y,

. . . —XYYXXXYXXYYYYXX— . . .

Polymeric Building Materials

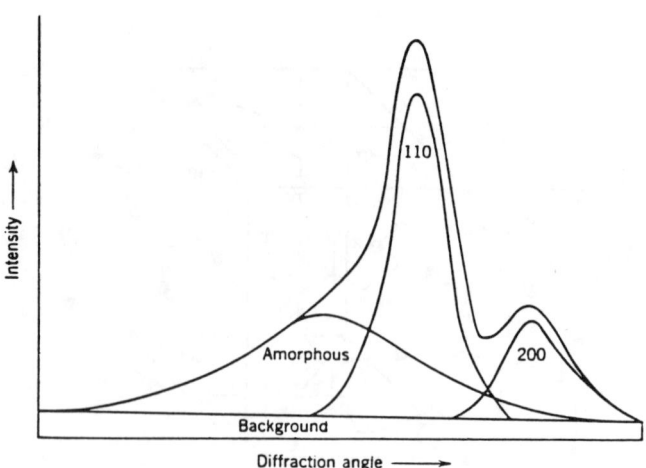

Fig. 1.11. Diffractometer scan of linear PE showing resolution into contributions from the (110) and (200) crystalline reflections, the amorphous peak and
the background [1.10]. (Reprinted with permission from *Textbook of Polymer
Science*, 3rd edn., by F. W. Billmeyer Jr, 1984. Copyright John Wiley.)

is intrinsically amorphous and represents non-rigid, soluble, and
low-softening products.

In the classical concept, crystallinity is expressed by the presence in
the polymer of crystallites which generally extend in length to several
thousands of angstroms. Only portions of the long chain molecules
make up a given crystallite, the remaining portion being either part of
the amorphous phase or part of other adjacent crystallites. Thus, one
macromolecule can be part of more than one crystallite. Such a
product which contains crystalline and amorphous regions is generally
termed a semi-crystalline polymer. Such a structure can be schematically illustrated as shown in Fig. 1.12.

This is a two-phase model in the sense that one can recognize two
types of structural arrangement: namely crystalline regions, which are
small crystallites or micelles with the molecules regularly packed in a
crystal lattice, distributed in a matrix which is the disordered (amorphous or non-crystalline) domain. Individual macromolecules pass
through a number of crystalline regions, alternating with segments in
the amorphous region. At the edge of each crystalline micelle the
chains form a fringe as they diverge into the non-crystalline material
[1.4].

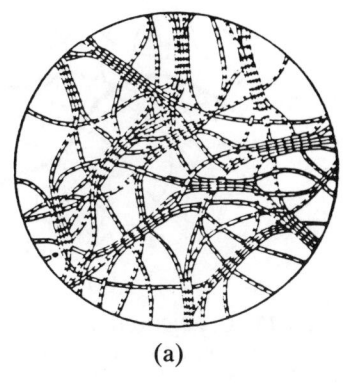

(a)

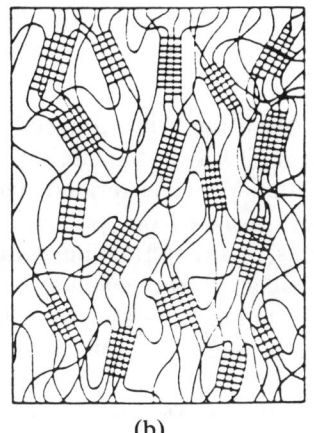

Fig. 1.12. (a) View of fringed micelle structure as originally proposed by Herrmann and Gerngross (Hermann & Gerngross, *Kautschuk*, **8** (1932) 181, (b) typical view of fringed micelle structure in an oriented polymer [1.11]. (Hearle, *J. Polym. Sci. C*, **20** (1967) 215.)

(b)

The two-phase model explains many features of polymer behaviour. The crystallite provides cohesion, stability and strength, and maintains the orientation in oriented structures. The amorphous region lowers the density, provides freedom for deformation, chemical reactivity, accessibility for water and additives, and allows a path for diffusion. Furthermore there is a plausible mode of formation. If one imagines crystallization to start in many places, then the same molecules may be trapped in several growing crystallites. Each crystallite continues growing until the growth is blocked because the molecules are trapped in different places and the disordered tangle left between them cannot be sorted out any further. The sequence is illustrated in Fig. 1.13.

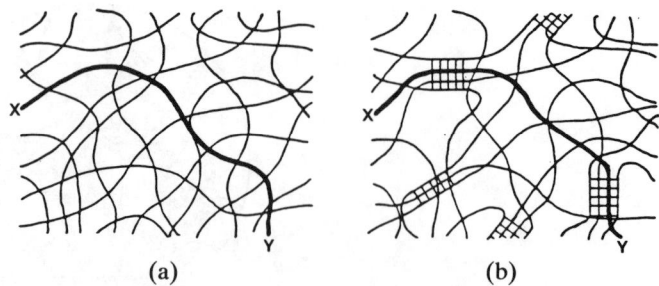

(a)   (b)

Fig. 1.13. Crystallization of amorphous polymer. (a) Starting at various places leads to a fringed micelle structure. (b) The chain XY is shown trapped in two micelles and is thus unable to crystallize any more [1.11]. (Reprinted with permission from *Polymers and their Properties,* Vol. 1, 1982, by J. W. S. Hearle. Copyright Ellis Horwood Ltd.)

After the fringed-micelle theory, the *folded-chain lamella* theory arose; it was in the late 1950s when polymer single crystals in the form of thin platelets termed 'lamella', measuring about $10\,000\,\text{Å} \times 100\,\text{Å}$ were grown from polymer solutions. Contrary to previous expectations, X-ray diffraction patterns showed the polymer chain axes to be parallel to the smaller dimension of the platelet. Since polymer macromolecules are much longer than $100\,\text{Å}$, the polymer molecules are presumed to fold back and forth on themselves in an accordion-like manner in the process of crystallization.

Different polymers have different properties and are synthesized and used differently because of varying degrees of crystallinity. The extent of crystallinity developed in a polymer sample is a consequence of both thermodynamic and kinetic factors. The degree of crystallinity of a polymer depends on whether its structure is conducive to packing into the crystalline state and on the magnitude of the intermolecular forces. Packing is facilitated for polymers that have:

—microstructural regularity,
—compactness,
—streamlining,
—some degree of flexibility,
—short structural units,
—polar groups which will provide the secondary forces.

The stronger the secondary forces, the greater will be the driving

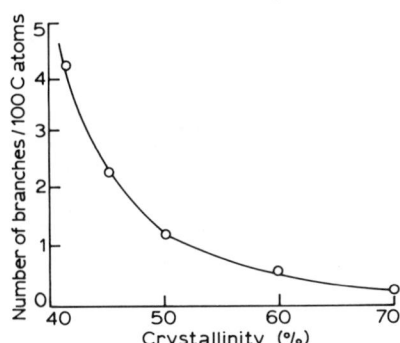

Fig. 1.14. Crystallinity of PE versus chain branching [1.12]. (Reprinted with permission from *Structure and Properties of Oriented Polymers* by I. M. Ward, 1975. Copyright Applied Science Publishers.)

force for the ordering and crystallization of the macromolecular chains.

Some polymers have a high degree of crystallinity because their microstructure is conducive to a high degree of packing, while others are crystalline because of strong polar groups able to realize strong secondary forces.

Unlike polyethylene which has a single and regular symmetric chain, polystyrene, PVC and poly(methyl methacrylate) usually show very poor crystallization tendencies. Large size substituents and branches lead to difficulties in packing. Cross-linking completely prevents crystallization. High flexibility as in natural and synthetic rubbers leads to an inability of the macromolecules to pack in normal thermal conditions; they are completely amorphous polymers [1.12].

In the case of PE which has a very simple microstructure, the degree of crystallinity is largely controlled by the number and distribution of branches along the main chain. This is shown in Fig. 1.14.

The two-phase system undergoes continual rearrangement under the influence of different manufacturing processes, tests and final end use applications. The particular state of the material at any one of these stages will depend on its previous environmental history.

Before considering the polymer properties it is important to have a visual conception of the organizational options available to a crystalline polymer. Isotactic polypropylene (PP) is used herein as a model system to decribe the common thread that exists between structure, properties and applications. Thus, initially, a qualitative fictional description of this polymer's organizational character is necessary if a realistic visual working model of the structural elements and their behaviour is to be developed [1.13].

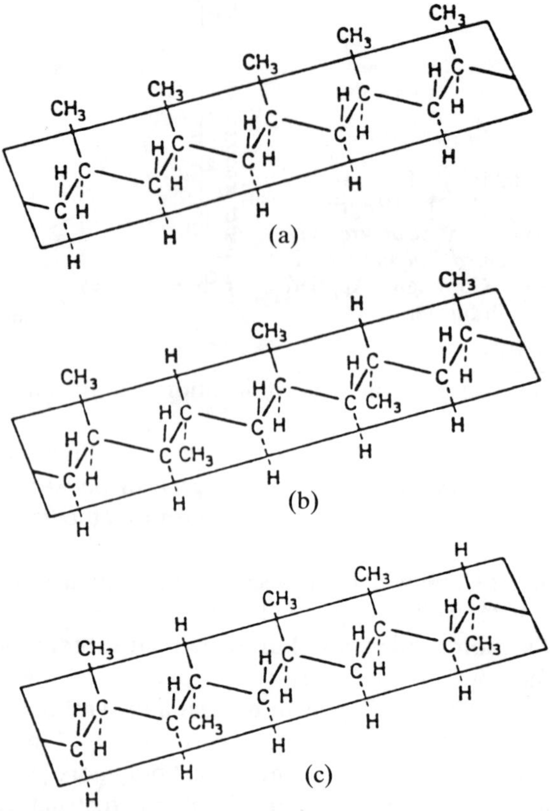

Fig. 1.15. Schematic representation of the spatial disposition of methyl groups in: (a) isotactic PP, (b) syndiotactic PP and (c) atactic PP chain segments [1.13]. (Reprinted with permission from *Structured Polymer Properties* by R. J. Samuels, 1974. Copyright John Wiley.)

The term *isotactic* identifies the particular spatial position of the PP methyl groups along the PP carbon chain backbone (Fig. 1.15). In the isotactic case the methyl groups all fall on the same side of the main carbon backbone (a). This is in contrast to the syndiotactic case where each alternating methyl group is on the opposite side of the carbon backbone (b). Finally, random placement of the methyl groups along the main chain leads to atactic PP (c). As a consequence of different conformations the properties are different; isotactic and syndiotactic PP are able to form crystalline polymers and especially the isotactic one is recommended for the production of fibres; in contrast, atactic PP is amorphous, has a higher solubility and is not suitable for fibres.

MORPHOLOGICAL CHANGES IN POLYMERS

Most synthetic polymers show a characteristic sequence of changes as they are heated. At low temperatures all linear polymers are in a glassy state. As the temperature is raised, a certain point is reached at which the polymer changes from the glassy state to another physical state, the rubbery, gum-like, and finally liquid state with no clear demarcation between the different phases. On the other hand, crystalline polymers remain flexible and thermoplastic above T_g until the temperature reaches the crystalline melting temperature (T_m). At this temperature crystalline macromolecular compounds melt to a viscous liquid. The crystalline melting phenomenon occurs when sections of adjacent chains are packed together in a regular array, and the melting point represents the temperature at which these 'micro-crystallites' are thermally disrupted. The different characteristics of amorphous and crystalline polymers at various temperatures are illustrated in Fig. 1.16. Extensive cross-linking may distort this picture and mask the transitions.

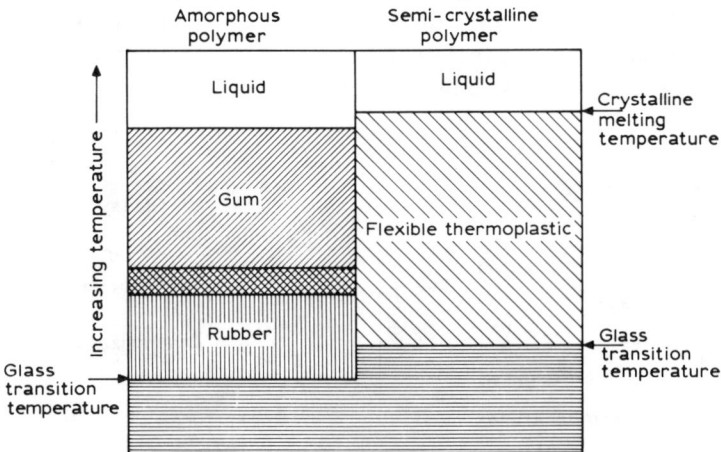

Fig. 1.16. Different transition behaviours [1.14]. (Reprinted with permission from *Contemporary Polymer Chemistry* by H. R. Allcock & F. W. Lampe, 1981. Copyright Prentice Hall Inc.)

Most polymers are neither classical solids nor liquids. They are viscoelastic materials. The viscoelastic state has the characteristics of both the solid and liquid states [1.14].

Consider a piece of lightly cross-linked natural rubber. At rest on a bench, it has all characteristics of a solid; that means fixed volume, definite shape, no evidence of liquid flow. But if we stretch the material or apply pressure to one part of it, it will change shape like a liquid; after we release the tension or pressure, it will revert to its original shape. These are some of the characteristics of the viscoelastic state, and these unusual properties account for the valuable properties of macromolecular compounds. Having these properties, high polymers can be used for many applications where conventional solids or liquids would be unsuitable.

The liquid phase is the state of substances at temperatures above their melting points and of all solid amorphous substances (e.g. ordinary silicate glass). Since silicate glass has no crystal lattice, all solid amorphous bodies are said to be glassy or glasses. That means in the solid physical state we may discuss glassy and crystalline bodies.

The Glassy State

A convenient method of discriminating between the processes of glass formation and of crystallization is dilatometry. There is no abrupt volume change at T_g whilst crystallization is accompanied by a volume change [1.12]. The former thus constitutes a second-order transition, whilst the latter gives a first-order transition on a volume–temperature plot. This distinction is shown in Fig. 1.17.

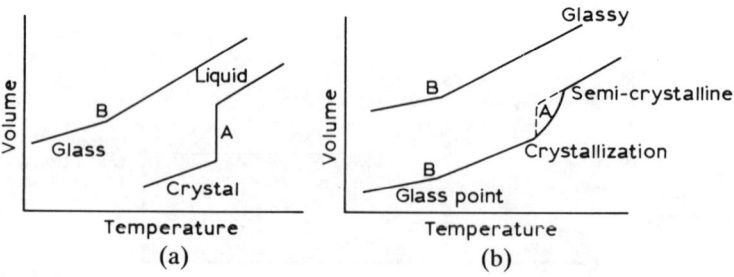

Fig. 1.17. First and second order transitions. (a) Behaviour of low molecular weight compounds. (b) Behaviour of glassy and semicrystalline polymers [1.12]. (Reprinted with permission from *Structure and Properties of Oriented Polymers* by I. M. Ward, 1975. Copyright Applied Science Publishers.)

Two typical glassy polymers are poly(methyl methacrylate) (PMMA) and polystyrene (PS). Although they have some flexibility as thin samples, they are not elastomeric. At room temperature both are well below their T_g ($\simeq 100°C$). Because most glassy high polymers are used as structural materials, rigidity and resistance to creep, high impact strength and high T_g are the most desirable properties. The fundamental problem with a glassy material is that the conversion of the impact into the breakage of bonds is one of the few mechanisms available for dissipation of that energy. By contrast, an elastomer absorbs the energy into harmless molecular motions. In the glassy state, the polymer molecules are rigidly fixed in place, and such impact-absorbing molecular mobility is not present. Hence, the material shatters.

The Rubbery State (High-elastic)

Elastomers are polymers that at normal temperature are well above their T_g. Most are amorphous. The high elastic characteristics become lost if the temperature is high enough to induce gum-like behaviour. Cross-links between the chains maintain the elastomeric character of the product at high temperatures.

The Liquid State (Viscoelastic Fluid)

If cross-links are absent, both amorphous and microcrystalline polymers melt at high temperatures. The melting process allows the chains to separate from each other and permits viscous flow to occur readily. Few high polymers are used as technological materials in the molten state. Decomposition reactions occur at or above the high temperatures required for melting. However, the molten state is used extensively for the fabrication of items based on polymers [1.14].

Glass Transition Temperature

Whether a polymer exhibits glass transition temperature (T_g) and melting temperature (T_m) or only one of them depends on its morphology. Amorphous polymers show only T_g. A completely crystalline one—if it existed—would show only a T_m. Semicrystalline macromolecular compounds exhibit both the crystalline melting and glass transition temperature. The two thermal transitions are conveniently measured by changes in properties such as specific volume and heat capacity.

Glass transition is the passage of a mobile liquid into the solid state with no change of phase (Fig. 1.18), i.e. with the retention of

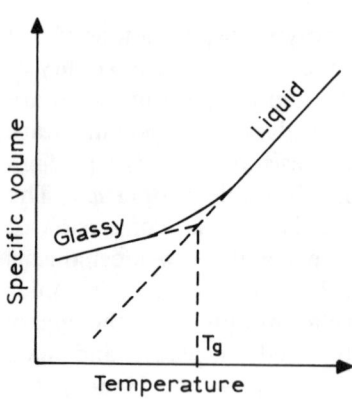

Fig. 1.18. Temperature dependence of specific volume of a macromolecular compound.

short-range order; hence, glass transition is not a phase transition. Crystallization is the transition from a state of short-range order to one of long-range order, i.e. the process of formation of a new phase. It is therefore a phase transition of the first order. T_g is characteristic of a particular polymer in much the same way that a melting temperature is characteristic of common low molecular mass products. T_g varies with the microstructure of the polymers, with the types of side groups (polar or not), the disposition of side groups, the molecular mass, etc. Table 1.5 lists the values of T_g and T_m for different polymers.

Low molecular mass substances pass from a glassy into a liquid state; amorphous and semi-crystalline high polymers pass from the glassy into a high elastic state, and on further heating, into a viscoelastic fluid state. The transition of a macromolecular compound from the high elastic to the viscoelastic fluid state is fairly distinct on deformation versus temperature graphs. The method consisting of measuring the temperature dependence of the deformation of a polymer is called the thermomechanical method.

The thermomechanical curve (Fig. 1.19) of an amorphous linear high polymer has three regions corresponding to three physical states:

—glassy,
—high elastic, and
—viscoelastic fluid state.

In the glassy state not very high stresses cause small strains; the high elastic state is characterized by large recoverable deformations. On these deformations is superimposed the flow deformation which

Table 1.5.

Glass transition temperatures (T_g) and crystalline melting temperatures (T_m) for selected polymers[a] [1.14]

Polymer	T_g (°C)	T_m (°C)
Polystyrene (isotactic)	100	240
Poly(m-methyl styrene) (isotactic)	70	215
Poly(methyl methacrylate) (atactic)	114	—
Poly(methyl methacrylate) (isotactic)	48	160
Poly(methyl methacrylate) (syndiotactic)	126	200
Poly(cis-1,4-isoprene)	−67	36
Poly(trans-1,4-isoprene)	−68	74
Poly(dimethylsiloxane)	−123	−29
Poly(dichlorophosphazene)	−63	−10
Poly[bis(trifluoroethoxy)phosphazene]	−66	242
Poly[bis(ethoxy)phosphazene]	−84	—
Poly(trifluoroethoxypentafluoropropoxy-phosphazene) rubber	−77	—
Polyacrylonitrile	85	317
Nylon 66	45	267
Poly(ethylene terephthalate)	17	285
Polyethylene	−20	141
Polyethylene	−107	95

[a] A compilation of T_g and T_m values can be found in O. G. Lewis, *Physical Constants of Linear Homopolymers* (New York: Springer-Verlag, 1968). Reprinted with permission from *Contemporary Polymer Chemistry* by H. R. Allcock and F. W. Lampe, 1981. Copyright Prentice Hall Inc.

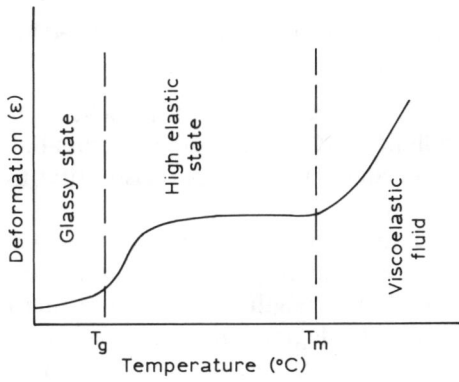

Fig. 1.19. Thermomechanical curve of an amorphous linear high polymer.

increases with rising temperature. At fairly high temperatures relative movements of entire chains become easy enough to enable what is known as true flow of the polymer to occur. This transition from the high elastic to the viscoelastic fluid state is accompanied by a sharp increase in deformation.

As in the case of T_g, the T_m is not a definite point, but the average temperature of the range within which true flow of the polymer develops.

To transform a polymer into a useful plastic or fibre, we have to process it at a temperature higher than T_m or T_g depending on its physical structure (semi-crystalline or amorphous).

MECHANICAL PROPERTIES

The mechanical properties of high polymers are of interest in all applications where polymers are used as structural materials. However, the prime consideration in determining the general utility of a polymer is its mechanical behaviour, that is, its deformation and flow characteristics under stress. Four important qualities characterize the stress–strain behaviour of a macromolecular compound:

(1) Modulus: the resistance to deformation as measured by the initial stress divided by $\Delta L/L$.
(2) Ultimate strength or tensile strength: the stress required to rupture the sample.
(3) Ultimate elongation: the extent of elongation at the point where the sample ruptures.
(4) Elastic elongation: the elasticity as measured by the extent of reversible elongation.

Polymers vary widely in their mechanical behaviour depending on the degree of crystallinity (DC), the degree of cross-linking, the values of T_g and T_m, the molecular mass and polydispersity; that means their macro- and microstructure.

High strength and low extensibility are obtained in polymers by having high degrees of crystallinity or cross-linking or a high value of T_g. High elasticity and low strength in macromolecular compounds are synonymous with a low degree of cross-linking, low degree of crystallinity and low T_g values. The temperature limits of utility of a high polymer are governed by its crystalline T_m or T_g for amorphous

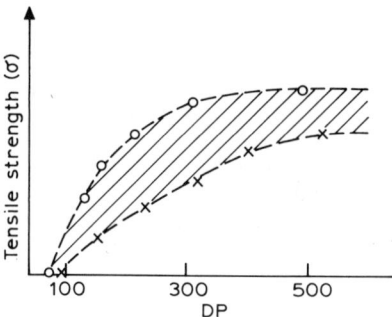

Fig. 1.20. Tensile strength (σ) versus DP; ×—weakest polymers; ○—strongest polymers [1.15]. (Reprinted with permission from *Plasticizers, Stabilizers and Fillers* by P. D. Ritchie, 1972. Copyright The Plastics Institute.)

polymers; strength is lost above T_g for an amorphous polymer and above T_m for a crystalline one [1.6].

The tensile strength of a material is determined as a rule, from its stress–strain curves. It is the limiting stress at which the specimen fractures. This definition is generally accepted and one usually speaks of ultimate strength which in the case of various polymers lies between 5000 and 10 000 N/cm². With increasing degree of polymerization the strength of a material first increases and then becomes constant (Fig. 1.20).

The stress–strain test is probably the most widely used mechanical test for engineering materials. Stress–strain performance of high polymers varies widely with sample history, changes in temperature, speed of deformation, etc. The type of stress–strain curve obtained from high polymers is frequently used to subdivide polymeric materials into five classes, for example:

—soft and weak,
—hard and brittle,
—soft and tough,
—hard and strong, and
—hard and tough [1.16].

The corresponding stress–strain curves are illustrated in Fig. 1.21.

Tensile strength and extension of break indicate, respectively, the maximum stress and maximum strain that a material can withstand. Curves illustrating the wide range of stress–strain relations of some polymers are shown in Figs. 1.22 and 1.23. Figure 1.22 presents average results from normal continuous filaments (fibres). In high tenacity fibres the extension at break is lower and with cut-staple fibres

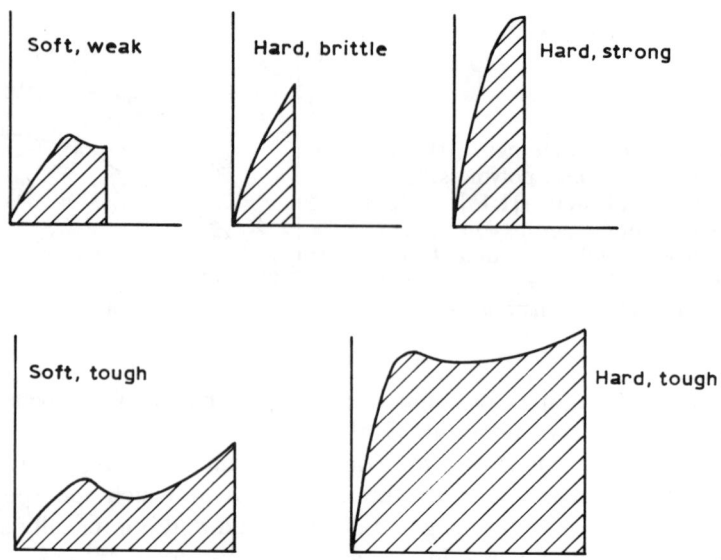

Fig. 1.21. Classes of high polymers judged by their stress–strain behaviour [1.5]. (Reprinted with permission from *Structure and Properties of Polymers* by H. V. Boenig, 1973. Copyright G. Thieme Publ.)

the strength is usually reduced while the extension at break is increased. Polymers follow curves of the same order as those of weaker fibres, but the curves terminate (rupture point) at lower values.

A distinction can be made between brittle and tough high polymers; at the usual rate of loading the former, even though having high strength, have very low extensibility [polystyrene, poly(methyl methacrylate)], whereas tough elastics have relatively high extensibility [polyethylene, plasticized poly(vinyl chloride)], and hence require much more energy to produce rupture, this energy being represented by the area under the stress–strain curve. Figures 1.22 and 1.23 present the stress–strain curves of some polymers to produce rupture, this energy being represented by the area under the stress–strain curve.

Depending on the particular combination of properties, a specific polymer will be used as an elastomer (rubber-like product), rigid or flexible plastic or as a fibre. Commonly encountered articles which typify these uses of polymers are rubber bands, plastic eyeglass lenses

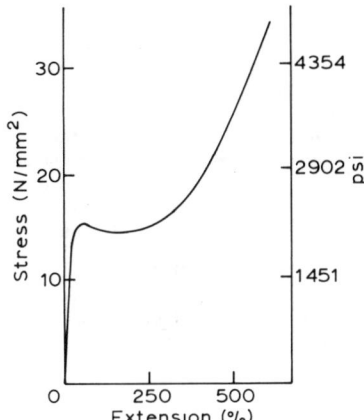

Fig. 1.22. Stress–strain curve in extension to break of partially crystalline polymers. Unoriented LDPE [1.11]. (After Boenig, 1966.)

or door frames (rigid plastic), packaging films (flexible plastic), clothing (fibre), etc. Table 1.6 shows the uses of many of the common macromolecular compounds as elastomers, plastics or fibres. Some are used in more than one category because certain mechanical properties can be altered by appropriate chemical or physical procedures, for example, by modifying the orientation (crystallinity) or adding plasticizers. Some macromolecular compounds are used as both plastics and fibres, others as both elastomers and plastics.

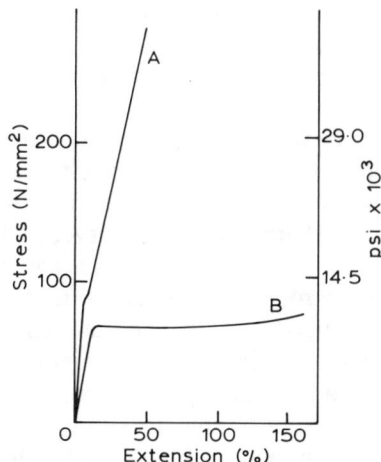

Fig. 1.23. Stress–strain curves in extension to break of partially crystalline polymers. Polycarbonate film which has been uniaxially stretched (a) parallel, (b) perpendicular to orientation [1.11]. (After Schnell, 1964.)

Table 1.6.
Uses of polymers [1.6]

Elastomers	*Plastics*	*Fibres*
Polyisoprene	Polyethylene	
Polyisobutylene	Polytetrafluoroethylene	
	Poly(methyl methacrylate)	
	Phenol-formaldehyde	
	Urea-formaldehyde	
____Polystyrene____		
____Poly(vinyl chloride)____		
____Polyurethane		
____Polysiloxane		
		____Polyamide____
		____Polyester____
		____Cellulosics____
		____Polypropylene____

Forced rubbery deformation develops in a very characteristic manner. At a certain cross-section of specimen a narrowing suddenly appears, which then grows at the expense of the gradually diminishing initial thick part of the specimen. The formation of such a 'neck' corresponds to the transition to the horizontal (plastic deformation) part of the stress–strain diagram. From this point to the end of the horizontal part of the curve the specimen is non-uniform, because it has a thick and a thin part, i.e. cross-sections of two different areas, the initial one and that corresponding to the neck, coexist in the specimen. After complete conversion of the specimen into the neck, i.e. at the end of the horizontal section, the specimen again becomes uniform and is subsequently extended as a whole. This is accompanied by a sharp rise of the curve, and very soon failure occurs.

Naturally, the development of forced rubbery deformation causes anisotropy in the specimen. If the initial glassy specimen is isotropic, the neck and, hence, the entire specimen at the last stage of the extension process is a highly oriented glassy body possessing pronounced mechanical, optical, and other kinds of anisotropy. This anisotropy is due to straightening of the macromolecules during the development of forced rubber-like elasticity and remains as long as the

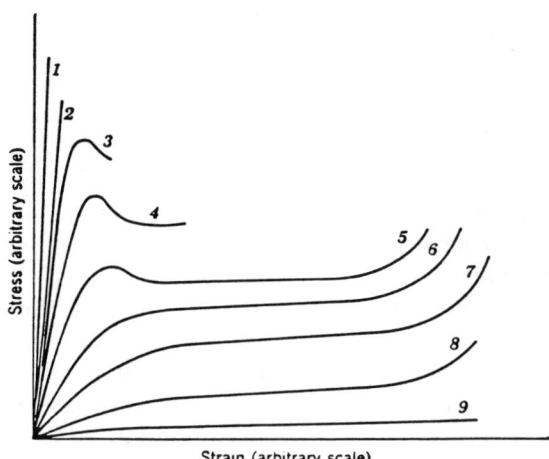

Fig. 1.24. Stress–strain diagram of solid amorphous (glassy) polymers at different temperatures (temperature increases with ascending numbers of the curves) [1.2]. (Reprinted with permission from *Characterization of Polymers*. Encyclopedia Reprints by W. M. Bikales, Ed., 1971. Copyright John Wiley.)

forced rubbery deformation attained is preserved. It disappears after the specimen has contracted to its initial size on heating [1.24].

A characteristic feature of the rise of forced rubber-like elasticity is the small maximum preceding the horizontal section of the diagram. Other explanations have been offered to account for the appearance of the neck owing to breakdown of the elements of supermolecular structure at sufficiently high stresses, making possible development of larger deformations at lower stresses.

Mechanical properties are highly influenced by temperature, as we may see from the stress–strain diagram (Fig. 1.24).

THERMAL PROPERTIES

As well as the mechanical properties, before using a macromolecular compound as a building material we have to know some of its other properties such as:

—thermal properties,
—the effect of the environment,

—the effect of high energy irradiation,
—ageing and weathering,
—permeability,
—toxicity,
—flammability.

If a polymer is heated to a sufficiently high temperature reversible and irreversible changes in its structure will occur. These modifications may either be undesirable, or they can be useful.

Undesirable modifications mean chain cracking, the formation of low molecular weight products, discoloration or any other changes in desirable properties which the macromolecular compound possessed before heat exposure. On the other hand, the reversible softening of thermoplastics permits them to be converted into useful shapes [1.5].

The thermal stability of a polymer is defined by the temperature range over which it retains useful properties.

As already mentioned, the mechanical properties of thermoplastics are temperature dependent. For polymers such as polyamides, poly-acetals and polyethylenes which have a crystalline superstructure, the crystalline melting point is an upper reference temperature, but form stability is achieved only at a much lower temperature. For PVC, polystyrene, polycarbonate and other amorphous polymers, softening can occur over a wide range of temperature.

Few thermoplastics are recommended for prolonged use above 100°C although polypropylene articles can withstand repeated steam sterilization at about 125°C.

At 150°C some polymers such as PE, PS and PVC are molten, but glass-filled polyamide has a useful mechanical life of about 1 year in air at this temperature.

Polytetrafluoroethylene (PTFE, Teflon), and it is claimed, poly-phenylene sulphide, can withstand 260°C in air almost indefinitely, although the former is very soft and flexible at this temperature [1.17].

Thermal conductivities (K factor) of macromolecular compounds have very low values and are therefore of interest to the industry with its concern for the fabrication and application of engineering materials (thermal insulators). The values of a number of polymers are listed in Table 1.7.

Thermal expansion of polymers is relatively large, they tend to expand or contract more with temperature changes than do metals. This must be considered in the design and use of polymer parts,

Table 1.7.
Thermal conductivity of polymers [1.5]

Polymer	Thermal conductivity $10^{-4} cal/s\ cm\ °C$
ABS	4·5–8·0
Poly(methylmethacrylate)	4·0–6·0 (plexiglas)
Epoxy (unfilled)	4·0–5·0
PTFE	6·0
Nylons (PA)	5·1–5·8
Polyesters (rigid, thermosetting)	4·0
LD PE	8·0
HD PE	11–12.4
PP	2·8
PS	2·4–3·3
PVC (rigid)	3·0–7·0

Reprinted with permission from *Structure and Properties of Polymers* by H. V. Boenig, 1973. Copyright G. Thieme Publishers.

particularly when employed in conjunction with other engineering materials. Specific heat values for some polymers are given in Table 1.8.

As was mentioned, polymers are good thermal insulators, and may also have outstanding electrical insulation properties. The values of the thermal conductivities of this group of polymers are much lower than those of metals. At ambient temperature unfilled polymers have conductivities normally in the range 0·15–0·13 W/m°C, and expanded polymers (plastic foams) have even lower values, for example, 0·03 W/m°C in the case of PS foam ($d = 25\ g/dm^3$), whereas the

Table 1.8.
Specific heat values for polymers [cal/g°C] [1.5]

Polymer	C_p	Polymer	C_p
Polyethylene	0·55	Poly(vinyl chloride)	0·23
Polypropylene	0·46	Poly(vinyl fluoride)	0·24
Polystyrene	0·32	Poly(vinylidene chloride)	0·32
Poly(methyl methacrylate)	0·35	Poly(vinyl acetate)	0·39
Polycarbonate	0·30	Poly(vinyl alcohol)	0·40
Acetal polymers	0·35	Polytetrafluoroethylene	0·25

Reprinted with permission from *Structure and Properties of Polymers* by H. V. Boenig, 1973. Copyright G. Thieme Publishers.

thermal conductivity of aluminium is about 240 W/m°C and that of copper is about 385 W/m°C.

WEATHERING AND OTHER PROPERTIES

Weathering and ageing of polymers depend on the following factors:

—chemical environment, which may include atmospheric oxygen, acidic fumes, acidic rain, moisture;
—heat and thermal shock;
—ultra-violet light;
—high energy radiation.

The resistance to weathering depends on the type of polymer, its composition and superstructure (the amorphous polymers are more sensitive), and on the ensemble of conditions of exposure [1.18–1.20].

Ageing may be defined as the process of deterioration of engineering materials resulting from the combined effects of atmospheric radiation, heat, oxygen, water, micro-organisms and other atmospheric factors (e.g. gases, pollutants, etc.). In a commercial polymeric product there are also a number of additives—with different functions—which may be affected by the above factors.

Since different macromolecular compounds and ingredients respond in different ways to the influence of environmental factors, weathering behaviour can be very specific.

A serious current problem for the polymer technologist is to be able to predict the weathering and ageing behaviour of a polymer over a prolonged period of time, often 20 years or more. For this reason it is desirable that some reliable accelerated weathering test should exist.

The best performers in this respect are poly(methyl methacrylate) and other acrylic polymers which are stable in the presence of ultra-violet light; outdoor signs made of cast polyacrylic sheet have survived more than 30 years' exposure without surface degradation or discoloration.

PVC compounds also prove to have good ultra-violet resistance. In cladding panels for buildings and as a plasticized PVC covering for fence wire, PVC compounds have been used continuously out of doors for more than 20 years. Some fading and yellowing occur on ageing of PVC. On continuous exposure to ultra-violet light, many polymers deteriorate. However, the useful life of many such polymers can be

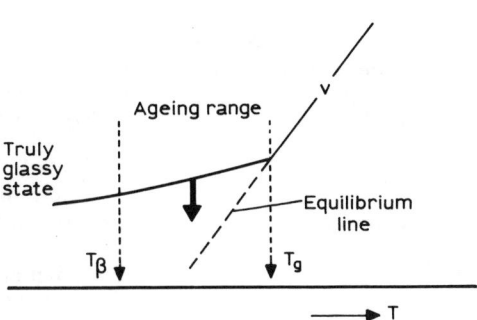

Fig. 1.25. Origin of ageing. T_g is the glass transition temperature, T_β the temperature of the highest secondary transition and v the specific volume [1.21]. (Reprinted with permission from *Polym. Engng Sci.*, **17** (1977) 3. Copyright Society of Plastics Engineers.)

improved by a factor of ten or more by incorporation of some special additives known as photostabilizers.

Some thermoplastics, such as polystyrene, poly(methyl-pentene), should not be used for outdoor applications, unless the surface is protected by painting, metalizing, or other suitable finishing.

Ageing cannot be ignored in the testing of polymers, particularly in the prediction of their long term behaviour, a property very important in construction. A study was done in 1977 on the ageing of about 35 high polymers and organic glasses. The main conclusions of the study were:

—Ageing occurs in broad temperature ranges below T_g. Generally, the ageing range runs from T_g down to the first secondary transition (T_β) presented in Fig. 1.25. Thus for PVC it runs from +70 to about −50°C, for polycarbonate (PC) from 150 to about −100°C, etc. The ageing range of many polymers includes the temperature range of practical interest [1.21].
—In the ageing range defined above, ageing is a very important phenomenon. At least the small strain mechanical properties of the material are determined mainly by time. Ageing has to be considered in the testing of polymers, particularly in the prediction of their long term behaviour.
—In the ageing range all polymers age in the same way, and even their mechanical behaviour at small strains is very similar. There is little difference in behaviour between PVC, PMMA, PC, PS, PET, etc.;

Table 1.9.

Behaviour of polymers subjected to high energy radiation [1.27]

Polymers that cross-link	*Polymers that degrade*
Polyethylene	Poly(methyl methacrylate)
Poly(acrylic acid)	Poly(tetrafluoroethylene)
Poly(methyl acrylate)	Polyisolentilene
Polyacrylamide	Poly(α-methyl styrene)
Natural rubber	Poly(methacrylic acid)
Polychloroprene	Poly(vinylidene chloride)
Polydimethylsiloxane	Polypropylene
Styrene–acrylonitrile copolymers	

Reprinted with permission from *Radiation Chemistry of Monomers, Polymers and Plastics* by J. E. Wilson, 1974. Copyright M. Dekker.

their properties are primarily determined by their amorphous superstructure.

Polymer degradation due to various factors is the subject of chapter 8.

Effect of High Energy Radiation

In analogy with thermal and light actions, high energy radiation (X-rays, γ-rays, etc.) may also lead to degradation or cross-linking depending on the structure of the polymer. This factor is especially important in relation to the building of nuclear plants, radiochemistry laboratories, etc. [1.22–1.26].

Table 1.9 lists some polymers that cross-link and some that degrade, depending on the features of their microstructure. Most polymers of monosubstituted ethylene cross-link, and most polymers of di-substituted ethylenes degrade. Exceptions are polypropylene which degrades, and PVC which according to the conditions, either degrades or cross-links. Although PTFE and PMMA have a good stability to ultra-violet light, they are both easily degraded by high energy radiation.

PERMEABILITY

Polymers are often employed as protective coatings, vapour barriers, sealants, caulking compounds, proof against gases and vapours;

for this reason, their permeability, i.e. ability to allow gases and vapours to pass through them is a very important property.

Gas permeability depends both on the nature of the polymer and the nature of the gas. Diffusion through a polymer occurs by the small molecules passing through voids and other gaps between the polymer molecules. The diffusion rate will therefore depend to a large extent on the size of the small molecules and the size of the gaps.

It is customary to compare polymers by their water permeability. Experimental data show that with respect to water permeability most macromolecular compounds are intermediate between mineral solids and common liquids, but some materials based on polymers, like paper and textiles, are more water-permeable than liquids. The coefficients of water permeability of polymers at 20°C are in the range 10^{-10}–10^{-7} cm^3 × cm/(cm^2, s . atm).

The gas permeability (P) of polymers depends greatly on their structure (Table 1.10).

The diffusion coefficient (D) of one material through another is defined by the equation:

$$F = -D \frac{\delta c}{\delta x}$$

F being the weight of the diffusing material crossing unit area of the other material per unit time, and the differential is the concentration gradient in weight per ml, per cm, at right angles to the unit area considered [1.28].

The permeability of polymers to gases, vapours and liquids is important for many applications, in which either high barrier properties are required for protective purposes, or very selective permeation is required for use of the polymer as a membrane for efficient separation of permeating mixtures. In either case, the transport of penetrant molecules through a polymer normally occurs by an activated-diffusion process. The condensed penetrant is considered to dissolve in the polymer surface layer, migrate through the bulk material by a cooperative ('activated') polymer segmental motion-penetrant molecule jump process under the influence of a concentration-gradient driving force, and evaporate from the other surface of the membrane. After a relatively short transient-state buildup, steady-state flow is attained with a constant rate of transmission of penetrant provided constant pressure or concentration difference is maintained across the film.

Table 1.10.

Permeability data for various polymers values for $P \times 10^{10}$ cm^3 s/mm/cm^2/cmHg [1.28]^a

Polymer	Permeability			90% RH $H_2O(25°C)$	Ratios to N_2 perm. = G value			Nature of polymer
	$N_2(30°C)$	$O_2(30°C)$	$CO_2(30°C)$		P_{O_2}/P_{N_2}	P_{CO_2}/P_{N_2}	P_{H_2O}/P_{N_2}	
Poly(vinylidene chloride)	0·0094	0·053	0·29	14	5·6	31	1 400	Crystalline
PCTFE	0·03	0·10	0·72	2·9	3·3	24	97	Crystalline
Poly(ethylene terephthalate)	0·05	0·22	1·53	1 300	4·4	31	26 000	Crystalline
Rubber hydrochloride (Pliofilm ND)	0·08	0·30	1·7	240	3·8	21	3 000	Crystalline
Nylon 6	0·10	0·38	1·6	7 000	3·8	16	70 000	Crystalline
PVC (unplasticized)	0·40	1·20	10	1 560	3·0	25	3 900	Slight crystalline
Cellulose acetate	2·8	7·8	68	75 000	2·8	24	2 680	Glassy
Polyethylene ($d = 0.954$–0.960)	2·7	10·8	35	130	3·9	13	48	Crystalline
Polyethylene ($d = 0.922$)	19	55	352	800	2·9	19	42	Some crystalline
Polystyrene	2·9	11	88	1 200	3·8	30	4 100	Glassy
Polypropylene	—	23	92	680	—	—	—	Crystalline
Butyl rubber	3·12	13·0	51·8	—	4·1	16·2	—	Rubbery
Methyl rubber	4·8	21·1	75	—	4·4	15·6	—	Rubbery
Polybutadiene	64·5	191	1 380	—	3·0	21·4	—	Rubbery
Natural rubber	80·8	233	1 310	—	2·9	16·2	—	Rubbery

^{a} Original sources: F. A. Paine, *J. Roy. Inst. Chem.*, **86** (1962) 263; R. Lefaux, *Practical Toxicology of Plastics*. Illife, London, 1968.

Hence permeability is the product of solubility and diffusion; it is possible to write the next equation where the solubility obeys Henry's law,

$$P = D \times S$$

where S is the solubility coefficient.

In any durability prediction of a polymer, it is important to understand how sorbed low-molecular weight molecules modify the deformation and failure processes and mechanical response. Sorbed substances generally cause failure in polymers by inducing crazing and cracking at stresses much lower than those observed in their absence. This phenomenon is complex and not completely understood on a molecular level.

TOXICITY

Unfortunately a number of monomers, chemicals and additives used in the polymer processing industry have a tendency to be dermatic, including certain halogenated aromatic materials, formaldehyde and aliphatic amines. Most toxicity problems associated with the finished product arise from the nature of the additives and seldom from the polymer.

Some scientists evaluated the performance of the indices of toxicity of thermal decomposition products of PVC, polychloroprene and polycarbonate. The most irritating and the most stressful pyrolysis products were associated with polycarbonate decomposition products. The actute lethality for polychloroprene was higher than that of the other two polymers by a factor of 4.

Recently there have been discussions on possible links between the content of vinylchloride in PVC and a form of cancer of the liver.

FLAMMABILITY

Most synthetic organic polymers burn in a manner different from that of the natural polymers such as wood, paper, cotton, etc. Some synthetics burn slower, some faster: some give off more smoke, some less, some melt and flow when subjected to heat while others char over extensively.

Polymeric Building Materials

The general magnitude of combustibility is of the same order for both synthetic and natural organic polymers. Both burn yet both can be used safely without undue risk.

It is possible to retard burning by the use of suitable additives, although these usually generate smoke and, under non-burning conditions, have a negative effect on the mechanical and other properties.

The reasons for the differences in combustion between polymers are various but in particular two factors may be noted:

(a) the higher the hydrogen to carbon ratio in the polymer the greater is the tendency to burn;

(b) some polymers while burning emit blanketing gases that suppress burning.

Table 1.11.

Collected data for limiting oxygen index for a variety of polymers (unfilled) [1.28]

Polymer	Limiting oxygen index
Polyacetal	15
Poly(methyl methacrylate)	17
Polypropylene	17
Polyethylene	17
Poly(butylene terephthalate)	18
Polystyrene	18
Polycarbonate of bis-phenol A	26
Polyimide (Ciba-Geigy P13N)	32
Polyarylate (Solvay Arylef)	34
Polyethersulphone (ICI Victrex)	34–38
Polyether ether ketone (ICI)	35
Poly(vinyl chloride)	45
Poly(vinyl fluoride)	44
Poly(phenylene sulphide)	44–53
Friedel Crafts resins	55
Poly(vinylidene chloride)	60
Poly(carborane siloxane)	62
Polytetrafluoroethylene	95

Note: % of oxygen in air = 20·9.
Polymers below line burn with increasing difficulty as the LOI increases.
Reprinted with permission from *Plastics Materials*, 4th edn. by J. A. Brydson, 1982. Copyright J. A. Brydson.

Table 1.12.

Effects of some metal oxide–halogen systems on the thermal stability and flammability of ABS copolymer [1.29]

Metal oxide(s)[a]	Increase in T_i % (K)	Increase in LOI
$Sb_2O_3(7\cdot5)$	35	15·6
$SnO_2(7\cdot5)$	35	9·2
$Fe_2O_3(7\cdot5)$	18	11·8
$Al_2O_3(7\cdot5)$	31	9·0
$Al_2O_3.3H_2O(7\cdot5)$	42	5·7
$Sb_2O_3(2\cdot5) + SnO_2(2\cdot5)$	13	17·8
$Sb_2O_3(2\cdot5) + Fe_2O_3(2\cdot5)$	14	19·5
$Sb_2O_3(2\cdot5) + Al_2O_3(2\cdot5)$	40	17·3
$Sb_2O_3(2\cdot5) + Al_2O_3.3H_2O(2\cdot5)$	33	15·5

[a] Figures in parentheses show wt% of metal oxides; polymer 70 wt%, decabromobiphenyl to 100 wt%.
Reprinted with permission from C. F. Cullis, Thermal stability and flammability of organic polymers, published in *Br. Polym. J.*, **16**, Dec. 1984, 253.

Whilst the limiting oxygen index (LOI) test is quite fundamental it does not characterize the burning behaviour of the polymer. Table 1.11 presents some LOI values for polymers.

Recently C. F. Cullis [1.29] has studied the influence of some oxides on the flammability of ABS copolymer. Some results are presented in Table 1.12.

Flame-retardant additives do not normally act by increasing the thermal stability of organic polymers. Nevertheless, information regarding the relative stabilities of macromolecular substrates and additives can throw useful light on the probable mode of action of flame-retardants on the combustion of polymers.

POLYMER PROCESSING

Polymers are used for an almost infinite number and variety of purposes, and new applications are constantly appearing. Gradually, over a period of several decades, many macromolecular compounds have become construction materials just as are ceramics, metals and wood. In producing relatively small objects they rival the metals in variety and number of applications.

Polymers are processed with a view to obtaining plastics, synthetic fibres, coatings, adhesives, rubbers, sealants, etc. No other class of engineering materials is applied in such a variety of ways or has such widely divergent end uses.

Many polymers are almost as readily machinable as metals, so that metal forming operations such as cutting, drilling, etc., are common in the production of finished plastic items. Methods of joining, utilizing adhesive bonding are almost as common as the analogous welding operation in the field of metals, and forming operations such as cutting, drilling, drawing, extrusion, rolling and moulding, etc., are common in the production of finished plastic items.

Since forming methods are not confined to rigid solids and liquid melts but can be extended to non-rigid fluids or rubbery materials, the variety of forming methods is even greater than with metals.

PLASTIC TECHNOLOGY

Since the very early stages of the development of the polymer industry it was realized that useful products could only be obtained if certain additives were incorporated into the polymer. This process is normally known as 'compounding'.

The term *additives* is used to describe those materials which are physically dispersed in a polymer mass without affecting significantly the molecular structure of the polymer. Catalysts, cross-linking agents used in thermosetting systems are included.

Additives used in plastic processing are normally classified according to their specific function, rather than on a chemical basis. Table 1.13 presents the most important group of additives used in plastic processing [1.30]. Each ingredient is used according to the needs of processing or to provide some properties for service conditions.

Complete *compatibility* (i.e. mutual miscibility at molecular level), *permanence* and *mobility* or *diffusibility* of the additive molecules within the polymer mass are essential if the action of the additive is such that any or all the molecules of the system are to interact with each other.

In general, additives should have the following features unless by virtue of their function such requirements are included:

—they should be efficient in their function;
—they should be stable under processing and under service conditions;

Table 1.13.

(i) Additives which assist processing	(a) Processing stabilizers (internal)
	(b) Lubricants (external)
	(c) Processing aids and flow promoters
	(d) Thixotropic agents
(ii) Additives which modify the bulk mechanical properties	(a) Plasticizers or flexibilizers
	(b) Reinforcing fillers
	(c) Toughening agents
(iii) Additives used to reduce formulation costs	(a) Particulate fillers
	(b) Diluents and extenders
(iv) Surface properties modifiers	(a) Antistatic agents
	(b) Slip additives
	(c) Anti-wear additives
	(d) Anti-block additives
	(e) Adhesion promoters
(v) Optical properties modifiers	(a) Pigments and dyes
	(b) Nucleating agents
(vi) Anti-ageing additives	(a) Anti-oxidants
	(b) UV stabilizers
	(c) Fungicides
(viii) Others	(a) Blowing agents
	(b) Flame retardants

—they should be cheap;
—they should not adversely affect the properties of the macromolecular compound;
—they should be non-toxic and not impart taste or odour.

One of the most important class of additives for plastic is that of *stabilizers* (thermal- and photo-stabilizers).

Degradation problems can arise when polymers are exposed to heat, light, weathering, high energy radiation, and micro-organisms. The mechanisms of breakdown of the different types of polymer and the degree of resistance to this attack are quite different. Degradation is generally noticed as a change in colour and/or a modification in physical properties such as surface cracking, loss in mechanical properties, etc. [1.31].

The entry of plastics into construction has meant that a great increase is required in the useful service life of these type of polymers under the degradative tendencies of the environment.

As was mentioned in the introduction, typical building applications are in roofing, flooring, guttering, plumbing, frames, underground

Table 1.14.
Spectral sensitivity of some polymers [1.18]

Polymer	Wavelength (nm)
Polyester	315
Polystyrene	340
Polyethylene	300
Polybutadiene (ABS)	380
PVC	310 and 370
PVA	280
Polycarbonate	280–305, 330–360
PPO	370
Polysulphone	320
Aromatic polyamide (Kevlar)	450

Reprinted with permission from *Weathering of Polymers* by A. Davis and D. Sim, 1983. Copyright Applied Science Publishers.

piping, etc. Resistance to oxidation, UV light and moisture are required for the first examples whereas underground piping must be resistant to attack by fungi and bacteria.

The UV light reaching the earth's surface generally has a minimum wavelength of about 2900 Å, although this depends on several factors, for example, the time of the day, seasons, and the geographical location.

The UV range of light reaching the earth's surface extends to approximately 4000 Å and accounts for 5% of the total energy content of the radiation falling on the earth's surface.

Table 1.14 shows the wavelength at which maximum degradation occurs in the case of different high polymers.

It has been found possible to incorporate into polymers additives which will absorb UV light and therefore act as a screen, thus preventing degradation.

Substances capable of absorbing strongly in the UV region will provide good protection for plastics used for outdoor purposes (Fig. 1.26). Carbon black can absorb over the entire range of UV and visible radiation and transforms the absorbed energy into less harmful infrared radiations.

Additives which can reflect or scatter UV and visible radiation will also offer some protection, but are less effective than absorbers.

As a result of mechanical shearing, exposure to radiation (UV, γ-rays, X-rays, etc.), attack by metal ions such as those of copper and

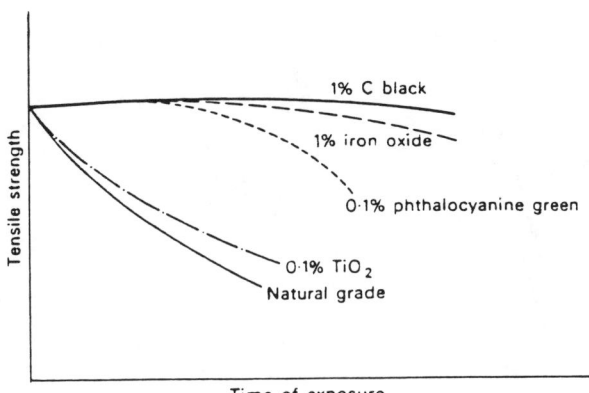

Time of exposure

Fig. 1.26. Effects of pigments on weathering resistance of polyethylene [1.30]. (Reprinted with permission from *The Role of Additives in Plastics* by L. Mascia, 1974. Copyright Edward Arnold.)

manganese, chemicals, as well as other possible mechanisms, a macromolecule breaks down into smaller particles losing the specific properties of the high polymers.

macromolecule   macroradicals

In Fig. 1.27 is presented the schematic relationship showing effect of pro-oxidants, anti-oxidants and oxidation retardant on the oxygen uptake of a polymer.

Whilst stabilizers against dehydrochlorination have found use in many chlorine-containing polymers their main application has been with PVC and vinyl chloride copolymers.

Many polymers are too stiff and brittle and have too low extensibility to be useful as plastics. They require plasticization to increase flexibility and extensibility or decrease brittleness at room temperature or below [1.32, 1.33].

Practically, plasticization consists in adding to the macromolecular compound various liquids or solids called plasticizers. Theoretically plasticization consists essentially in altering the viscosity of the system, increasing the flexibility of its macromolecules and the mobility of its supermolecular structure. Plasticizers added to a polymer affect all its physical and mechanical properties, such as: T_g, T_m, strength, elasticity, brittleness, dielectric losses, etc.

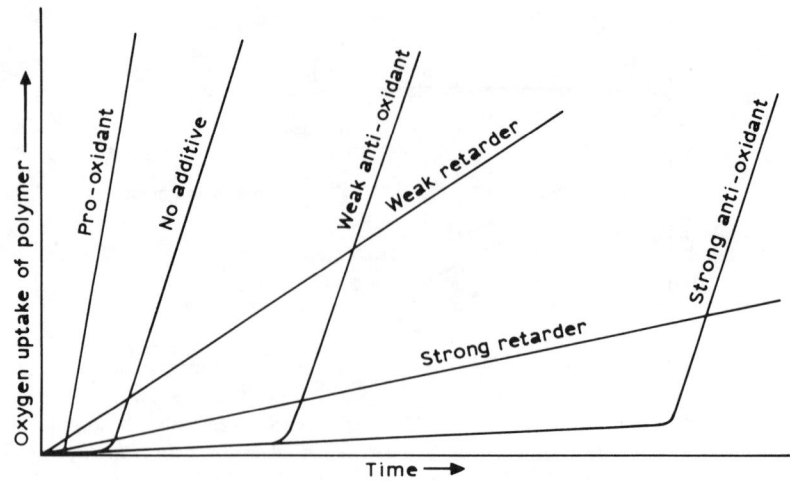

Fig. 1.27. Schematic relationship showing effect of pro-oxidants, anti-oxidants and oxidation retarders on the oxygen uptake of a polymer [1.28]. (Reprinted with permission from *Plastics Materials,* 4th edn., 1982, by J. A. Brydson. Copyright J. A. Brydson.)

Figure 1.28 shows the effect of increasing plasticizer concentration on the temperature dependence of modulus. As shown, increasing plasticizer content causes the transition from the high-modulus (glassy) plateau region to the low modulus (rubbery) plateau region to occur at progressively lower temperatures [1.34].

Representative structures for some kinds of plasticizers for PVC are illustrated in Table 1.15.

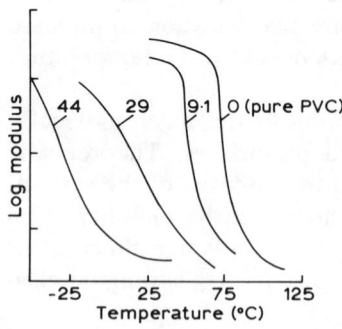

Fig. 1.28. Effect of increasing the concentration of DOP on the modulus–temperature dependence of plasticized PVC [1.34]. (Reprinted with permission from *Plastics Engineering,* **39** (1983) 37. Copyright Society of Plastics Engineers.)

Table 1.15.
Chemical structures of some plasticizers for PVC

Plasticizer	*Chemical structure*
Dialkyl phthalate	(benzene ring)–C(=O)–OR, –C(=O)–OR
Aliphatic diester	$RO-C(=O)-(CH_2)_n-C(=O)-OR$
Trialkyl phosphate	$RO-P(=O)(OR)-OR$
Trialkyl trimellitate	$RO-C(=O)-$(benzene ring)$-C(=O)-OR, -C(=O)-OR$
Polycaprolactone	$-[-C(=O)-(CH_2)_5-O]_n-$

Fillers have always played an important role in plastic processing, where they are usually classified as reinforcing, semi-reinforcing, and diluent, although there is no sharp division between these categories. Fillers are incorporated in plastics to improve general properties, to introduce particular characteristics, or to reduce the cost of the compound, although its physical properties may also be modified.

The different routes by means of which a polymer and additives can be converted to finished products are illustrated in Fig. 1.29.

It will be understood that not every polymer–additive combination can be put through every possible sequence of processing treatments.

The various processes may be broadly classified as: mixing, compounding and shaping.

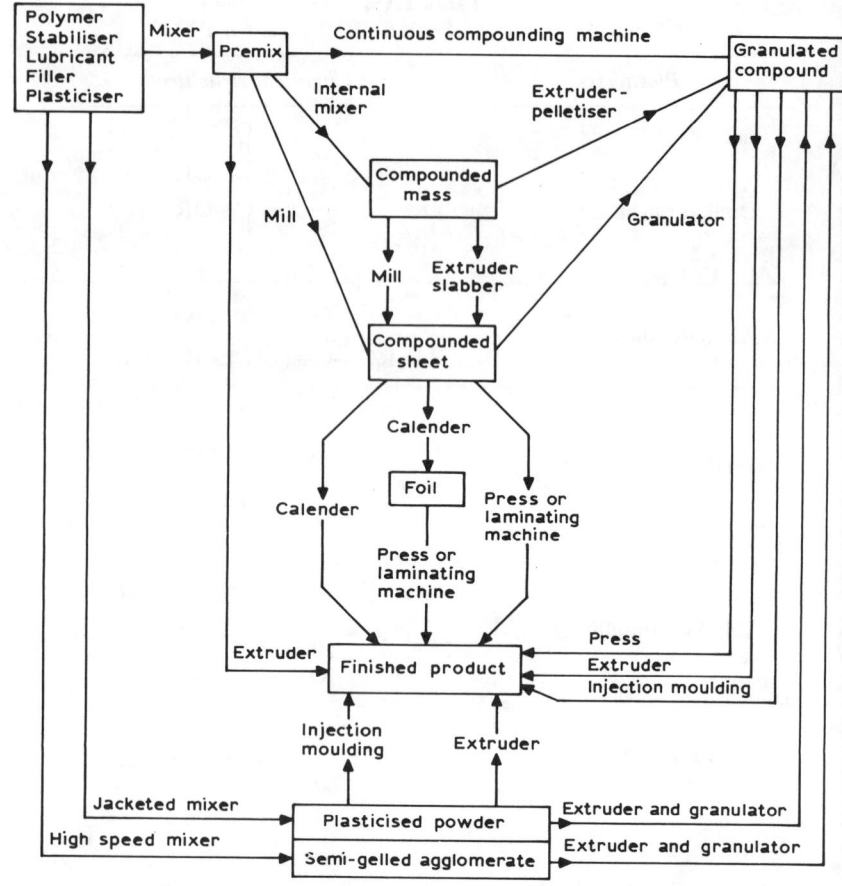

Fig. 1.29. Routes for conversion of PVC resins to finished products [1.35]. (Reprinted with permission from *Vinyl and Allied Polymers,* 2nd edn., 1972, by G. Mathews. Copyright The Plastics and Rubber Institute.)

Plastic fabrication technologies are largely determined by the rheological properties of the polymer in question. A primary consideration is whether the material is thermoplastic, i.e. retains the ability to flow at elevated temperature for relatively long times, or thermosetting, i.e. subject to cross-linking reactions at the temperatures necessary to induce flow, so that the ability to flow is rather quickly lost in favour of form stability. The softening point, stability, and of course, the size and shape of the end product are other factors of importance in selecting processing technologies. Most of them involve heating the macromolecular compound until it becomes soft, shaping the softened material and then cooling it in the required form.

In most forming operations the high polymer is subjected to intermittent shearing and stressing, usually at molten or nearly molten states. This results in the deformation and orientation of the polymer chains providing a product with a specific molecular configuration and morphology. This is particularly strongly marked during processes which produce films and fibres such as drawing, blowing, calendering, rolling, spinning, etc.

Most forming operations, especially extrusion, casting and moulding, involve melting or softening the polymer by heating it to such a temperature at which it will flow usually through a narrow nozzle under pressure to fill the cavity of the mould. At such conditions, most macromolecular compounds are highly viscous materials having viscosities between 10^3 and 10^8 Pa s, depending on the MW, temperature, pressure, the presence of fillers and other ingredients which may considerably affect the rheological characteristics of the results. Usually, polymer melts behave as pseudoplastic materials whose viscosity decreases rapidly with increasing shearing rate, Furthermore, they exhibit viscoelastic behaviour that becomes more and more pronounced on gradual cooling of the polymer during moulding or extrusion.

During the flow the whole polymer chains cannot slide over one another completely, but movement occurs by sequential motion. The movement of the entire chain length is restricted because of numerous entanglements between the chains.

One of the key advantages of the plastics industry is the ability to provide accurate components with excellent surface at low cost and high speed. Part of this success is due to the lower temperatures at which polymers are liquid compared with metals or ceramics.

The process to be used is largely dependent on whether we have a thermoplastic or thermosetting material.

In Table 1.16 are listed representative groups of plastics, and we can see that extrusion and injection moulding are predominant for thermoplastic polymers, and compression and transfer moulding are most important for the thermosets.

Extrusion

A wide variety of shapes can be made by extrusion including rods, channels and other structural shapes, tubing and hose, sheeting, films, etc.

Polymeric Building Materials

Table 1.16.
Processing methods for various polymers

Polymer	Compression		Injection moulding		Extrusion (°C)	Transfer moulding	
	Pressure (MN/m² (psi))	Temp. (°C)	(MN/m² (psi))	(°C)		Pressure (MN/m² (psi))	Temp. (°C)
			Thermoplastic				
PE			35–150 (5 076–21 758)	135–143	80–94		
PP			70–150 (5 076–21 758)	204–288	193–221		
PS			70–165 (10 152–23 933)	162–243	90–107		
PVC	3·5–14 (507–2 030)	140–175	50–100 (7 252–14 505)	160–175	162–204		
ABS			42–210 (6 092–30 461)	218–260			
Polyamides			70–140 (10 152–20 307)	270–340	270–300		
Acrylics			70–140 (10 152–20 307)	160–260	175–230		
			Thermosetting				
Phenolics	10–14 (1 450–2 030)	140–195	22–56 (3 191–8 123)	160–170		14–70 (2 030–10 152)	135–170
Urea–melamine	14–35 (2 030–5 076)	150–170				40–140 (4 802–20 307)	150–165
Polyesters						7–35 (1 015–5 076)	120–180
Epoxies						0·7–14 (101–2 030)	140–180

In the extrusion process, polymer is propelled continuously along a screw through regions of high temperature and pressure where it is melted and compressed, and finally forced through a die shaped to give the final object (Fig. 1.30).

The screw, which is the main element of the extruder, is divided into several sections, each one with a specific purpose. The *first section* picks up the finely divided (pellets) polymer from a hopper and propels it into the extruder cylinder. In the *compression section,* the loosely packed feed is compacted, melted and formed into a continuous stream of molten plastic. Although much heat is generated by friction, some external heat must be applied. The *metering section* builds up sufficient pressure in the polymer melt to force the plastic through the rest of the cylinder and out of the die [1.31].

Modern trends in extruder usage include the twin screw or multiple-screw (suitable for processing polymers in powder state) in which two screws turn side by side in opposite directions, providing more working of the melt; and the vented extruder having one opening or vent at some point along the screw which can be opened or led to vacuum to extract volatiles from the polymer melt.

With respect to plastic items for the construction industry, extrusion is used to produce:

—profiles, such as gaskets and sealing strips (from plasticized PVC and ethylene–vinyl acetate copolymer), hollow-section fencing, window frames, architraves, skirting boards, and curtain rails from rigid PVC;
—flat sheet, such as ABS and toughened polystyrene sheet for subsequent thermo-forming into hulls and dairy-product containers respectively;
—pipe, such as PE and rigid PVC pressure pipes for cold water services, and polypropylene pressure pipe for use in chemical plant;
—rigid PVC or ABS tubes for plumbing;
—corrugated sheet in translucent PVC for roof lights;
—thermoplastic polyester films for decorative, electrical and drawing office use;
—PE, PVC or other polymer films for greenhouses;
—wire covering, cables, etc.

It is possible to co-extrude two compatible materials to form a composite; window frames can be made from hollow rigid PVC sections and co-extruded with flexible PVC weather-sealing strip.

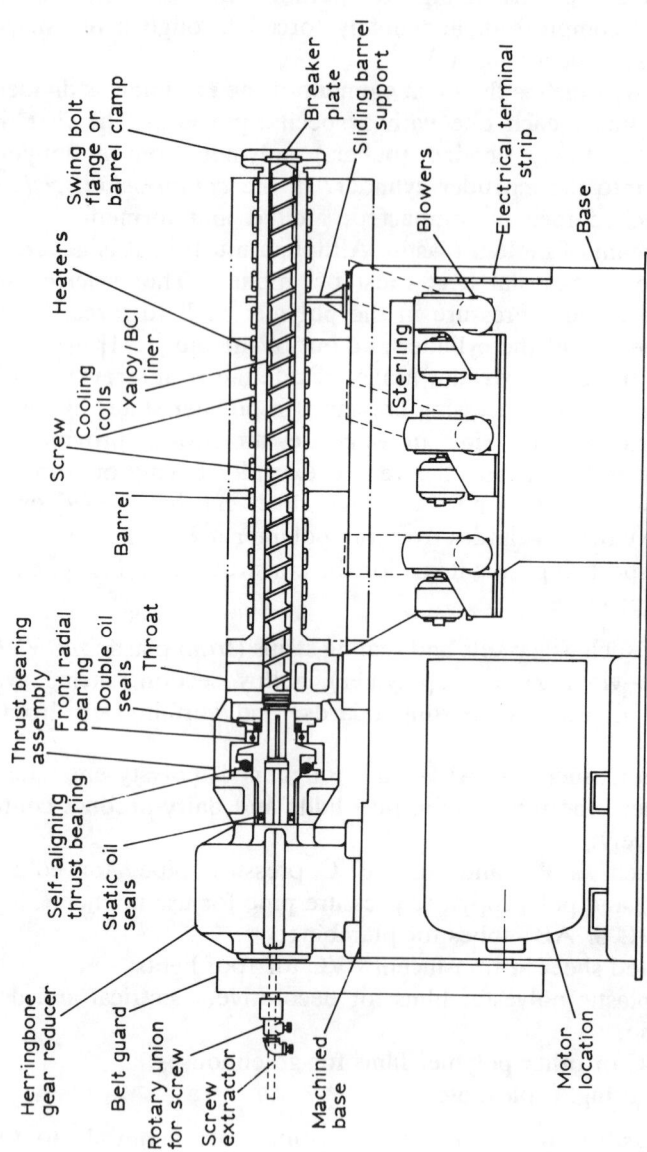

Fig. 1.30. Schematic; components of single screw extruder [1.36]. (Courtesy Sterling Extruder Corporation.)

Extrusion of pipe, rod and other shapes of constant cross-section is similar to the metal process. However, it is necessary to cool the part to near the T_g to gain dimensional stability. This is done by running the part into cooling water or by an air blast.

Injection Moulding

The injection moulding process consists essentially in transforming granular thermoplastic material into a fluid mass by the application of heat and forcing it under high pressure, usually averaging 70–250 MN/m^2, into a mould, where it becomes hard.

The operations of all standard commercial injection moulding presses are similar. The plastic material is placed in a hopper from which it is fed in a predetermined quality to a heated chamber located in front of a hydraulically actuated piston (Fig. 1.31).

After the mould is closed the piston forces the plastic through an electrically heated cylinder around a torpedo, and then through a nozzle, where it is pressed through the sprue opening in the front half of the die. The plastic then flows through runners in the die, through a gate, and into the cavities. The pressure holding the die closed and that used to inject the plastic are applied until the moulded article is ejected either automatically or manually.

The outstanding advantage of injection moulding is speed. Cycle times of 10–30 s are common. Examples of large injection moulded objects include polyethylene mechanical-handling pallets, 1·5 m long street lamp housings produced from the glass reinforced polyamide, complete chains produced from ABS and glass reinforced polyamide, etc.

Calendering

Calendering is a process used for the continuous manufacture of sheet or film from PVC or rubbers; it is little used with PE because of the difficulty in obtaining a smooth sheet. Usually in this operation a thick sheet is passed between pairs of highly polished heated rolls under high pressure. A calender consists in general of three, four or more rolls rotating as shown in Fig. 1.32.

The plastic fed mass is squeezed out into a film that passes around one or more additional rolls and emerges as a continuous film or sheet. Four roll, inverted L and Z types are the most usual.

Fabric or paper may be fed through the opening between the last

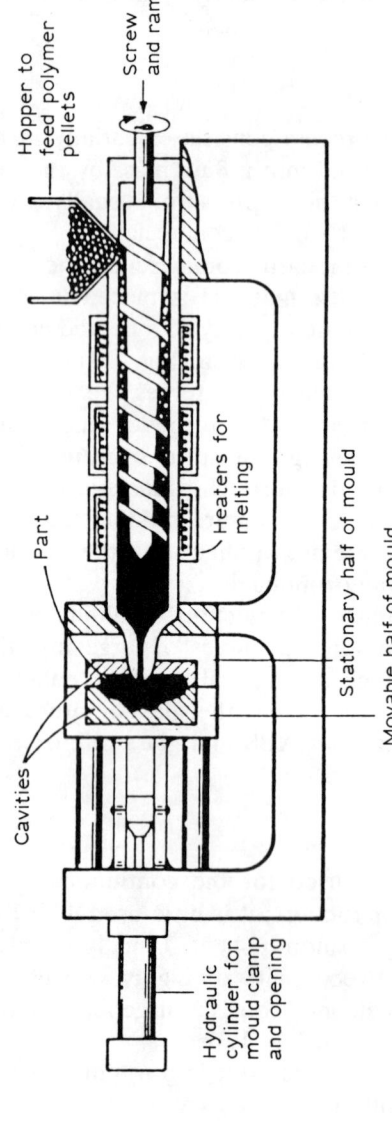

Fig. 1.31. Injection moulding [1.37]. (Reprinted with permission from *Engineering Materials; Properties and Selection*, 2nd edn., 1983, by K. Budinski. Copyright Prentice-Hall Inc.)

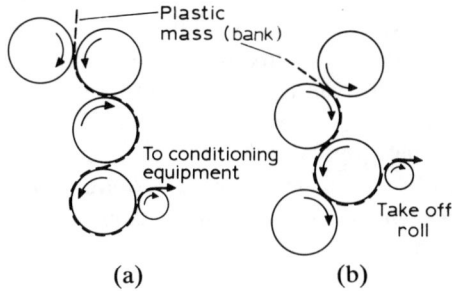

Fig. 1.32. Four roll calender; inverted (a) L and (b) Z type [1.31].

two rolls, and the film pressed into the surface of the web. When this procedure is applied, the calender becomes a coating machine.

Careful control of roll and speed ratios, roll temperatures, size of openings and compounding of the feed stock is necessary to provide smooth operation of a calender.

By calendering a mixture of granular polymer chips of varying colour, it is possible to produce unusual decorative effects, such as marblization in the product; this technique is widely employed in the manufacture of flooring compositions.

The first stage of a calendering process is the compounding of the polymer mix fed to the calender rolls. Where polyethylene is being calendered, the mix may be a simple hot melt, but in the case of PVC the mix is more complex. One method is to pre-mix the polymer in a ribbon blender with additives such as:

—stabilizers,
—plasticizers,
—fillers, etc.,

and then pass the blend to an internal mixer where the mass is gelled for about 5–10 min at about 160°C (for PVC); the gelated lumps so formed are made into a rough sheet on a two roll mill and sheet fed to the calender. A major problem in calendering is the control of gauge (thickness). Transverse variations due to roll (bowl) bending may be reduced or partially compensated for by the use of bowls of greater diameter/width ratio, by cambering the bowls, by bowl cross alignment or by deliberate bowl bending. Longitudinal variation may be largely administered by preloading the rolls to prevent the journals floating in their bearings. For technical reasons it is more convenient to preload on Z-type calenders than on L- and inverted L-type calenders. Roll temperatures are in the range 140–200°C. Large-scale leathercloth manufacture is today commonly carried out using calendering techniques.

The sheets obtained by calendering are used as they are or are sent to the next operation such as pressing, cooling, stretching, doubling, etc. Figure 1.33 shows a schematic diagram of a typical PVC calendering system.

Compression Moulding

In this procedure the plastic is placed between stationary and movable parts of a mould (Fig. 1.34). The mould is closed, and heat and pressure are applied so that the material becomes a homogeneous mass. Conditions such as pressure and temperature are dependent upon the rheological properties of the polymer.

If the processed polymer is a thermoplastic, the mould is cooled, the pressure released and the moulded article removed. If the material is a thermosetting polymer, the mould doesn't need to be cooled at the end of the cycle as the polymer will have set and can no longer flow or distort.

Spinnable Polymers

One of the most important forms of polymeric materials is the fibre, which can be described as a flexible, macroscopically homogeneous body, with a high length-to-thickness ratio and a small cross-section.

The principal naturally occurring fibres used in the textile industry are cotton, linen, jute, silk, asbestos, wool, etc. With the exception of asbestos, natural fibres are based on the natural organic polymers such as proteins and cellulose.

Man-made fibres are manufactured from naturally occurring polymers, synthetic polymers and mineral substances. Glass fibre is the only mineral man-made fibre in common use today, although other inorganic and metallic fibres are being developed, principally for fibre-reinforced composite materials.

Polyamide, polyester, acrylic, polyolefin and polyurethane homopolymers and copolymers are the most extensively used in fibre manufacture.

The organization of the polymer chains in three-dimensional space determines to a large extent the chemical, physical and particularly the

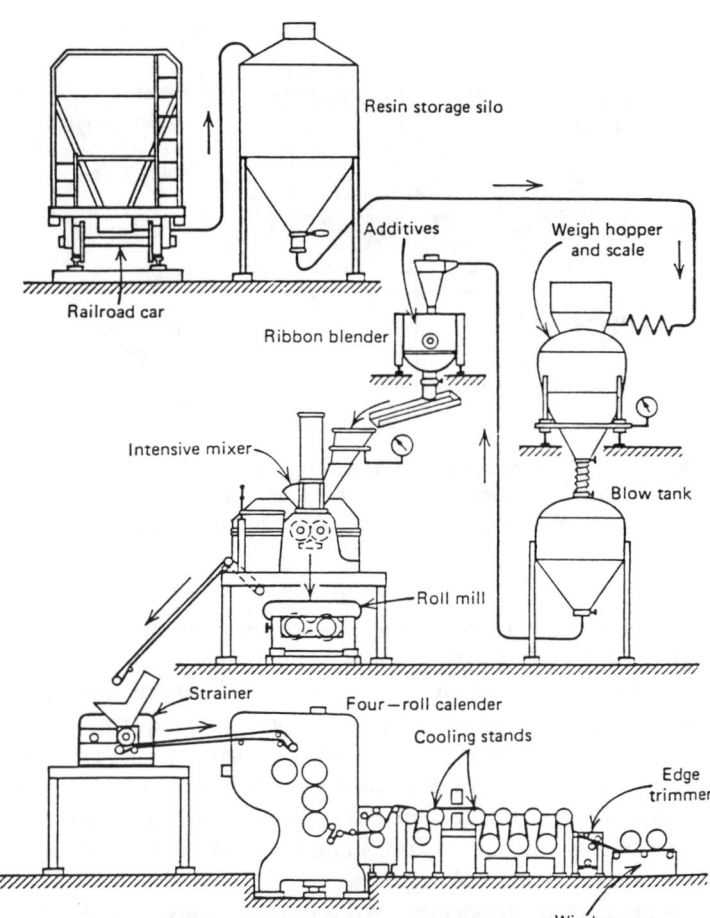

Fig. 1.33. Schematic diagram of typical PVC calendering system showing the principal elements that are required to convert powdered polymer into finished sheet [1.10]. (Reprinted from Watkins, 1976, by courtesy of *Plastics Engineering*.)

mechanical properties of fibres. Characteristic fibre properties are achieved through the development of an intermolecular chain organization that can generally be described as highly oriented and semicrystalline.

Fibres, natural and man-made, have one particular structural

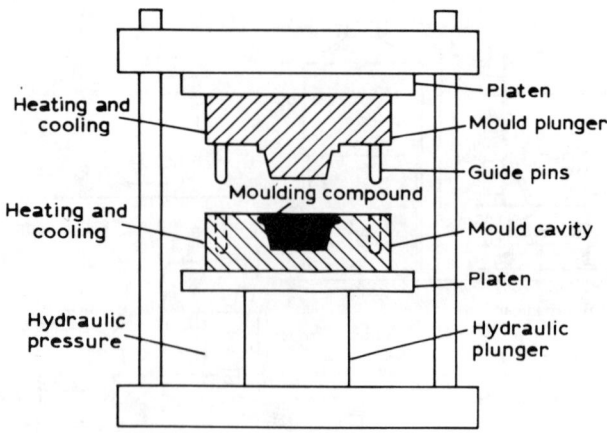

Fig. 1.34. Diagram of compression-moulding press and mould [1.10]. (Reprinted with permission from *Textbook of Polymer Science*, 3rd edn., 1984, by F. W. Billmeyer Jr. Copyright John Wiley.)

feature in common; a preferential orientation of their elemental units with respect to the fibre axis.

In the case of man-made fibres, the preferential orientation is achieved through mechanical drawing operations, during which the filament is extended, immediately after extrusion, to several times its initial length. Drawing has the dual purpose of orienting structural elements with respect to the fibre axis in order to bring these elements into optimal stress-bearing positions and to allow the development of a three-dimensional structural regularity.

Some properties of important textile fibres are given in Table 1.17.

The macromolecular compounds used for making fibres are similar to those used as plastics, but for fibres, the processing must produce an essentially infinite length to diameter ratio. The conversion of bulk polymer to fibre is accomplished by spinning. In most cases, spinning processes require solution or melting of the polymer. The most applied procedures for fibre fabrication are based on:

—melting spinning;
—dry spinning; or
—wet spinning.

If a polymer can be melted under reasonable thermal conditions (to

Table 1.17.

Chemical and physical properties of important textile fibres [1.38]

Fibre	Tenacity at break (g/denier)		Extension at break (%)		Elastic modulus (g/denier)		Density	Moisture regain at 65% RH (%)	Approximate volume swelling in water (%)	Thermal stability
	Dry	Wet	Dry	Wet	Dry	Wet				
Cotton	3–5	3–6	5–12	40–90	30–60	30–60	1·54	7·0–8·0	40	Decomposes at 150°C
Wool	1–1·7	0·8–1·6	25–35	25–50	25–35	20–30	1·32	15·0–17·0	35	Decomposes at 130°C
Silk	3–5	3–4	20–25	30	80–120	80–120	1·25	11·0	30–40	Decomposes at 130°C
Rayon:										
regular	1·5–4·0	0·8–2·5	10–30	15–40	40–70	15–35	1·52	11–13	45–85	Decomposes at 150°C
polynosic	3·0–4·0	2·5–3·5	7–10	9–12	45–70	35–50	1·54	10–12	40	Decomposes at 150°C
Acetate:										
secondary	1·0–1·5	0·7–1·1	25–40	30–45	25–40	20–35	1·33	6·4	10–30	Softens at 165°C
triacetate	1·2–2·1	0·8–1·4	20–28	30–40	35–40	25–35	1·32	4·5	5–15	Softens at 235°C
Acrylic	2·5–5·0	2·5–4·5	15–30	20–35	40–70	30–60	1·17	1·6–2·5	2–5	Softens at 235°C
Nylon	4·8–8·0	3·5–7·0	15–35	20–45	30–50	20–40	1·14	4·0–4·5	2–10	Melts at 220–260°C
Polyester	3·5–6·0	3·5–6·0	15–40	15–40	90	90	1·38	0·4	None	Melts at 250°C
Polypropylene	5·0–9·0	5·0–9·0	15–30	15–30	30–45	30–45	0·90	0	None	Softens at 140°C
Spandex	0·9	0·9	600	600	0·05	0·05	1·21	1·3		Softens at 225–270°C

Reprinted with permission from *Polymer Science and Materials* by A. V. Tobolsky and H. F. Mark (Eds), 1971. Copyright John Wiley.

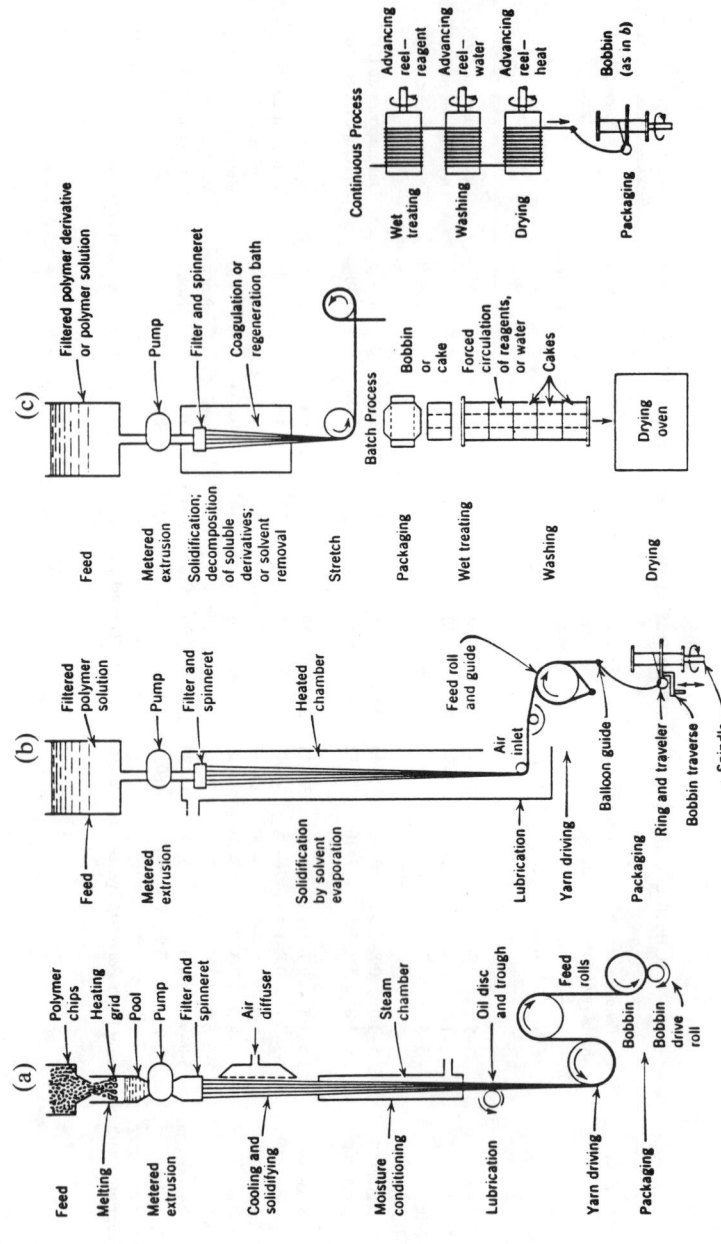

Fig. 1.35. Schematic diagrams of the three principal methods of spinning fibres. (a) Melt spinning, (b) dry spinning, (c) wet spinning. (Riley, 1956) (1.10.)

avoid degradation), the production of a fibre by melt spinning is preferred over solution processes. When melt spinning cannot be carried out, a distinction as to type of process is made depending upon whether the solvent is removed by evaporation (dry spinning) or by leaching out into another liquid which is miscible with the spinning solvent but is not itself a solvent for the polymer (wet spinning). Dry spinning has some advantages over the wet procedure. The three technologies have many features in common [1.10].

Melt spinning is basically an extrusion process. The macromolecular compound is plasticized by melting and prepared through the spinnerette. The fibres are usually solidified by a cross current blast of air as they proceed to the drawing rolls. The drawing operation stretches the fibres, orienting the macromolecules and inducing a high degree of crystallinity, an important condition for the formation of fibres with good properties.

Polyamide and polyester fibres are made by melt spinning.

In dry spinning, a solution of the polymer is forced through the spinnerette. As the fibres proceed downward to the drawing rolls, a counter current stream of warm air evaporates the solvent. PAN fibres are produced by dry spinning.

Wet spinning is similar to the former technology. Here, the solution strands pass directly into a liquid bath (precipitation bath). The liquid might be a non-solvent for the polymer, precipitating it from solution as the solvent diffuses outward into the non-solvent. The bath might also contain a substance that precipitates a polymer fibre by chemically reacting with the dissolved component. Viscose fibre made of cellulose is an example of a wetspun fibre [1.39].

Figure 1.35 gives a schematic of the already mentioned technologies.

REFERENCES

[1.1]   Meyer, K. H. & Mark, H., *Makromolekulare Chemie*, 3rd edn. Akademische Verlagesellschaft Geest and Pasting K. G., Leipzig, 1953.
[1.2]   Bikales, N. M. (Ed.), *Characterization of Polymers.* Encyclopedia Reprints, Wiley-Interscience, New York, 1971.
[1.3]   Moore, G. R. & Kline, D. E., *Properties and Processing of Polymers for Engineering.* Prentice Hall, London, 1984.
[1.4]   Rodriquez, F., *Principles of Polymer Systems,* 2nd edn. Hemisphere Publ. Co., Washington, 1982.

[1.5] Boenig, H. V., *Structure and Properties of Polymers*. Georg Thieme Publishers, Stuttgart, 1973.
[1.6] Odian, G., *Principles of Polymerization*, 2nd edn. Wiley-Interscience, New York, 1981.
[1.7] Simionescu, C. I. & Feldman, D., *Tratat de Chimie Macromoleculara*, Vol. 2. Didactica Bucuresti, 1974.
[1.8] Hiemenz, P. C., *Polymer Chemistry. The Basic Concepts*. M. Dekker, New York, 1984.
[1.9] Collins, E. A., Bares, J. & Billmeyer, F. W. Jr, *Experiments in Polymer Science*. Wiley-Interscience, New York, 1973.
[1.10] Billmeyer, F. W. Jr, *Text Book of Polymer Science*, 3rd edn. Wiley-Interscience, New York, 1984.
[1.11] Hearle, J. W. S., *Polymers and their Properties, Vol. 1*. Ellis Horwood Ltd, Chichester, 1982.
[1.12] Ward, I. M. (Ed.), *Structure and Properties of Oriented Polymers*. Applied Science Publishers, London, 1975.
[1.13] Samuels, R. J., *Structured Polymer Properties*, John Wiley, New York, 1974.
[1.14] Allcock, H. R. & Lampe, F. W., *Contemporary Polymer Chemistry*. Prentice Hall, Englewood Cliffs, New Jersey, 1981.
[1.15] Ritchie, P. D. (Ed.), *Plasticizers, Stabilizers and Fillers*. Iliffe Books Ltd, London, 1972.
[1.16] Seymour, R. B. & Carraher, C. E. *Structure–Property Relationships in Polymers*. Plenum Press, New York, 1984.
[1.17] Powell, P. C., *Engineering with Polymers*. Chapman and Hall, London and New York, 1983.
[1.18] Davis, A. & Sims, D., *Weathering of Polymers*. Applied Science Publishers, London and New York, 1983.
[1.19] Stivala, S. S. & Reich, L., *Polym. Engng Sci.*, **20**(10) (1980) 654–60.
[1.20] Cope, R. & Revirand, G., *Durab Bldg Mat.*, **1** (1983) 225–40.
[1.21] Struik, L. C. E., *Polym. Engng Sci.*, **17**(3) (1977) 165–73.
[1.22] Shalaby, S. W., *J. Polym. Sci. Macromol. Rev.*, **14** (1979) 419–48.
[1.23] Egusa, S., Ishigure, K. & Tabata, Y. *Macromolecules*, **12** (1979) 939–44.
[1.24] Egusa, S., Ishigure, K. & Tabata, Y., *Macromolecules*, **13** (1980) 171–6.
[1.25] Tsuda, M. & Oikawa, S., *J. Polym. Sci. Polym. Chem.*, **17** (1979) 3759–73.
[1.26] Ungar, G., *Polymer*, **21** (1980) 1277–82.
[1.27] Wilson, J. E., *Radiation Chemistry of Monomers, Polymers and Plastics*. M. Dekker, New York, 1974.
[1.28] Brydson, J. A., *Plastics Materials*, 4th edn. Butterworths Scientific, London, 1982.
[1.29] Cullis, C. F., *Brit. Polym. J.*, **16** (Dec. 1984) 253–7.
[1.30] Mascia, L., *The Role of Additives in Plastics*. Edward Arnold, London, 1974.
[1.31] Feldman, D., *Technologia Compusilor Macromoleculari*. Technica, Bucuresti, 1974.

[1.32] Titow, W. V., *PVC Technology,* 4th edn. Elsevier Applied Science, London and New York, 1984.

[1.33] Owen, E. D. (Ed.), *Degradation and Stabilization of PVC.* Elsevier Applied Science, London and New York, 1984.

[1.34] Fried, J. R., *Plastics Engineering,* (Sept. 1983) 37–42.

[1.35] Mathews, G., *Vinyl and Allied Polymers,* 2nd edn. Iliffe Books, London, 1972.

[1.36] DuBois, J. H. & John, F. W., *Plastics,* 6th edn. Van Nostrand Reinhold Co., New York, 1981.

[1.37] Budinski, K., *Engineering Materials; Properties and Selection,* 2nd edn. Reston Publ. Co. Inc., Reston, VA, 1983.

[1.38] Tobolsky, A. V. & Mark, H. F. (Ed.), *Polymer Science and Materials,* Wiley-Interscience, New York, 1971.

[1.39] Rosen, S. L., *Fundamental Principles of Polymeric Materials.* John Wiley, New York, 1982.

Chapter 2

Composites

A number of multiphase systems are known to have a useful combination of properties.

Composite materials are usually a combination of a matrix or binder material, and some kind of reinforcement. The effective method to increase the strength and to improve overall properties is to incorporate dispersed phases into the matrix, which can be an engineering material such as ceramic, metal or polymer.

A solution to the problem of reinforcing a brittle cement matrix with fibres was obtained in the late 19th century, by inclusion of asbestos fibres, and large scale production of flat and corrugated sheets was made possible by the invention of a manufacturing process which was patented at the beginning of the century.

Composites, as complex engineering materials, can be considered to be composed of either:

—one continuous phase, the matrix, and one or more dispersed phases; or,
—two or more continuous phases, plus potentially one or more dispersed phases (reinforcing agent or a gas) in each continuous phase.

The modern polymer–matrix composites industry can be said to have started in 1940 with the realization of glass–phenolic polymer structures. The era from 1940 to 1960 was a period when fabrication methods were the dominant consideration. The majority of present techniques were all brought to being during that period. From 1960 to 1970 was perhaps the era of properties. Here the mechanics of the material were carefully examined and various combinations of many new high modulus reinforcements in matrices were tried.

The era of reinforced thermoplastics, from 1970 to the present, is concentrating on combining the reinforcing properties of fibres with the formability of thermoplastics.

It is generally recognized that the incorporation of an elastomer (rubber-like material) in glassy polymers leads to an increase in their impact strength (i.e. toughening). A notable example of a rubber-modified glassy polymer is HIPS (high impact polystyrene). In such polymers, an increase in impact strength is obtained without adversely affecting the modulus and softening temperature of the polymer.

According to Richardson [2.1], a composite material may be defined as any substance which is made by physically combining two or more existing materials to produce a multiphase system with different physical properties from the starting materials, but in which the constituents retain their identity. If one or more of any dispersed phases are distinguishable extensively and macroscopically, i.e. are larger than 10^{-6} m (1 μm) across their widest or longest axis, or there is more than one continuous phase present, then the material may be considered a *macrocomposite*. If, however, all the individual dispersed phases are between 10^{-8} and 10^{-6} m (10–1000 nm) across their widest or longest axis, and there is only one continuous phase present, then the material may be considered a *microcomposite*. On this basis, polymeric composites may be described as macrocomposites in which at least the matrix is a polymer.

Composites are analogous to reinforced concrete, but because of the scale of the components, it more closely compares to ferrocement, a mixture of fine wire and cement grout.

The main features of the composite materials are:

—high fracture energy,
—ease of fabrication,
—potential low cost.

The low cost is particularly true for glass-reinforced polymers which involve little or no strategic materials and low capital equipment costs, compared to metal processing.

Composites can be fabricated from a wide choice of reinforcing agents and matrices; their properties may therefore vary over a very considerable range.

Composites possess a number of advantages with respect to conventional bulk materials, including metals. These are the following:

—They can be made with high strength and high specific strength relative to general materials of construction.
—They can be made with high stiffness and high specific stiffness relative to general materials of construction.
—Density is generally low.
—Strength can be high at elevated temperatures.
—Impact and thermal shock resistance are good.
—Fatigue strength is good, often better than for metals.
—Oxidation and corrosion resistance are particularly good.
—Thermal expansion is low and can be controlled.
—Thermal conductivity and electrical conductivity can be controlled.
—Stress-rupture life is improved relative to many metals.
—Predetermined properties can be produced if required to meet individual needs.
—Fabrication of large components can often be carried out at lower cost than for metals.

The three common groups of composites are:

(a) fibrous composites;
(b) laminar composites, laminates;
(c) particulate composites.

In the first group, fibres are embedded in a continuous polymer matrix; fibre length varies from tenths of a mm to continuous.

The reinforcing agents in the second group are sheets (paper, textile, etc.), bonded together and often impregnated.

The third group consists of particles such as stone aggregate, sand, or different fillers, which are embedded in a continuous matrix.

In each of these, polymers act as the matrix or binder that holds the reinforcing agent in place.

In the case of fibre reinforced polymers especially, one uses, as well as the matrix and the fibre, the two main components, a third one called the coupling agent. Its role is to enhance adhesion between the matrix and the fibre.

In the construction industry, many polymer based composites are used; some of their best known applications are the following:

—as polymer concrete masses, synthetic granite, synthetic marble, etc.;

—as foams for thermal or acoustic insulation;
—as panels (decorative or structural);
—as stacks, ductwork, tanks, scrubbers, etc.

POLYMER COMPOSITES

Thermosetting or thermoplastic polymers may be used as the matrix. In the thermosetting group, unsaturated polyesters, phenol-formaldehyde resins, melamine-formaldehyde polymers, epoxides and silicones are among the most common. The characteristics of these polymers are presented in Table 2.1.

In the case of unsaturated polyesters, due to the use of an unsaturated component (an acid), double bonds at regular spacings are introduced into the chain, which are potential sites for cross-linking by styrene or other similar monomers; as a result a rigid network is produced.

In the case of unsaturated polyester matrices, by correct control of the main synthesis factors such as:

—the nature and ratio of the components,
—catalysts,
—time and temperature,

a whole class of matrices can be produced to suit the reinforcing agent.

The largest single outlet for polyester–glass composites is in sheeting for roofing and building insulation, and accounts for about one-third of the polymer produced.

Because of their favourable price, polyesters are preferred to epoxy polymers. Generally speaking, the mechanical properties of epoxy composites are better than those of polyesters, although the latter are cheaper. Composites based on epoxy polymers exhibit better water and chemical resistance.

Phenol-formaldehyde polymers are among the cheapest of the used thermosettings. Applications of phenolic-polymer composites include:

—high voltage insulation,
—gear wheels,
—the core of decorative table top laminates,
—insulating foams, etc.

Phenolic foams are more expensive than other foams, but have the

Table 2.1.
Characteristics and uses of reinforced thermosetting resins [2.2]

Resins	Characteristics	Uses	Limitations
Diallyl phthalate polymer	Good electrical properties; dimensional stability; chemical and heat resistance	Prepregs, ducting, radomes, aircraft and missiles	
Epoxy	Good electrical properties; chemical resistance; high strength	Printed-circuit-board tooling, filament winding	Require heat curing for maximum performance
Melamine formaldehyde	Good electrical properties; chemical and heat resistance	Decorative, electrical (arc and track resistance), circuit breakers	
Phenolic	Low cost; chemical resistance; good electrical properties; heat resistance; non-flammable; can be used to 177–204°C (350–400°F)	General—diverse mechanical and electrical applications	Dissolve in caustic unless specially treated
Polyester	Good all-around properties; ease of fabrication; low cost; versatile	Corrugated sheeting, seating, boats, automotive, tanks and piping, aircraft, tote boxes	Degraded by strong oxidizers, aromatic solvents, concentrated caustic
Silicone	Heat resistance; good electrical properties	Electrical, aerospace	

Reprinted with permission from *Encyclopedia of Polymer Science and Technology*, Vol. 12, Copyright John Wiley, New York, 1970, p. 5.

advantage of being self-extinguishing, and of lower toxicity when subject to heat and flames.

Urea-formaldehyde polymers are cheaper, lighter in colour, compared to phenolics, but are inferior in heat and water resistance. Melamine-formaldehyde polymers are generally superior in performance to phenolic and urea polymers, but are more expensive. They manifest a low water absorption, staining and heat resistance, hardness and sustained electrical insulation in damp conditions.

Silicone polymer and polyimide matrices have good thermostability.

As can be seen in Table 2.2, a lot of thermoplastics are used in composite manufacture. They may be grouped together as follows:

—polyolefines: polyethylene, polypropylene;
—vinylic polymers: PVC, PS, PTFE;
—polyamides;
—polyacetals;
—polysulphones;
—polycarbonates;
—polyphenylenes;
—polyimides.

Polyethylene (PE) microstructure may be written as: $[-CH_2-CH_2-]_n$. For every 1000 C atoms, low density polyethylenes (LDPE) may have 20–30 short branches, while high density polyethylenes (HDPE) may have less than 5 short branches, and are therefore nearly linear in structure. The last one has a high degree of crystallinity (85%), is stronger, more rigid, has a higher, more definite melting point, and is hence more likely to be used in engineering applications. There are four quite distinct routes to the fabrication of PE:

—high pressure processes (LDPE);
—Ziegler processes (HDPE);
—Phillips process (HDPE);
—Standard Oil process (HDPE) [2.3].

Low density types have a specific gravity of $0·910–0·925$ g/cm^3, and high density types of $0·941–0·965$. Their densities have a significant effect on the properties (Table 2.3).

PE properties are strongly dependent upon average molecular mass, polydispersity, and degree of crystallinity. Outdoor applications are limited by the susceptibility to ultra-violet light, necessitating the use of stabilizers (carbon black or others); oxidative degradation at high

Table 2.2.

Selected examples of thermoplastic and/or uncross-linked polymers used as continuous phases in various types of composites [2.1]

Thermoplastic continuous phase (matrix)	Typical secondary continuous or dispersed phase(s)	Typical properties modified (amongst others)
Polyamides (e.g. Nylon 6,6)	20–40% glass fibre	Increased tensile strength, modulus of elasticity, hardness, creep resistance, heat deflection temperature
Polystyrene	Up to 40% glass spheres	Increased compressive strength
	Butadiene rubber (plus ABS materials), glass fibre, wood flour	Increased toughness
PVC (Poly(vinyl chloride))	Calcium carbonate, talc silicates, etc.	Cost reduction and/or general reinforcement
	China clay	Electrical insulation
	Asbestos	Wear resistance
	ABS (acrylonitrile–butadiene–styrene) terpolymer	Impact resistance
	MBS (methacrylate–butadiene–styrene) terpolymer	
PE (Polyethylene)	Butyl rubber	Improved environmental stress cracking resistance
	Polyisobutylene	Limited reinforcement in cross-linkable PE
	Carbon black	Stiffness
	Glass fibre	Tensile strength
		Toughness
		Creep resistance
Polypropylene	Butyl (and other) rubbers	Reduces brittleness (now usually achieved by copolymerization with PE, etc.)

Material	Fibre/filler	Properties
Polyacetals	Molybdenum disulphide PTFE Glass fibre	Wear resistance Creep resistance, stiffness, coefficient of expansion (reduced impact strength) Wear (grinding wheels)
Polysulphone	Diamond Glass fibre (typically 40%)	Creep resistance, coefficient of thermal expansion, tensile strength and modulus N.B. Impact strength decreases
PTFE (Polytetra-fluoroethylene)	Glass and asbestos fibres Alumina, silica, lithia	Creep resistance Dimensional stability, electrical insulation
Polycarbonate	Glass fibre	Hardness Flexural strength Tensile modulus Fatigue strength Thermal expansion
Polyphenylene	Glass fibres Carbon fibres Asbestos	Tensile strength Tensile modulus
Polyimide N.B. Thermo-plastic/thermo-setting definitions are difficult to apply here	Glass fibre Carbon fibre Graphite (typically 15%) Diamond	Flexural modulus, etc. N.B. Tensile strength and hardness decreases in some cases Abrasion and wear resistance

Table 2.3.

Relationship of several physical properties to the density of polyethylene [2.4]

Property	Type of polyethylene		
	Low-density	Medium-density	High-density
Specific gravity	0·910–0·925	0·926–0·940	0·941–0·965
Tensile strength, MPa (lb/in^2)	6·89–15·9 (1000–2300)	8·2–24·1 (1200–3500)	21·3–37·9 (3100–5500)
Elongation (%)	90–800	50–600	15–100
Impact strength, Izod test, J/m (ft-lb/in)	No break	26·6–853 (0·5–16)	53–1066 (1–20)
Flexural modulus, MPa (lb/in^2)	55–413 (8000–6 × 10^4)	413–792 (6 × 10^4 – 1.15 × 10^5)	689–1792 (1 × 10^5 – 2·6 × 10^5)
Hardness, Shore D	41–46	50–60	60–70
Resistance to heat (continuous) (°F)	180–212	220–250	250

temperatures and susceptibility to creep, even at room temperature, are further limitations. Although HDPE is theoretically more resistant to oxidation than low density types, in practice they behave similarly.

Polyethylene used on a large scale for drinking water pipes is normally inert to microbiological attack, but unless specially compounded, may not inhibit surface growth, and the relatively low hardness renders it liable to attack by insects and rodents.

Exposure to high energy radiation produces cross-linking with a consequent improvement in certain mechanical properties such as tensile strength and elastic modulus. Transmission of gases and vapours decreases with increase of crystallinity.

Glass fibres are used to improve tensile strength and rigidity, and clay (ceramics in general) is used to improve rigidity in HDPE. Various films, water distribution and sewer pipe, effluent outfalls up to 1·22 m (48 in) in diameter are becoming commonplace applications.

Polypropylene (PP) is a linear hydrocarbon which corresponds to the following microstructure $[-CH_2-CH-]_n$; in several of its charac-
$\qquad\qquad\qquad\qquad\qquad\qquad\qquad\quad CH_3$

teristics it resembles PE; however the methyl groups $[-CH_3]$ attached to alternate carbons in the macromolecular chain, modify the properties causing stiffening, which raises the melting point, rendering it less stable than PE with regard to oxidation. The methyl groups lead to asymmetry and the possibility of obtaining products of different tacticity. The microstructure of PP is arranged to encourage crystal formation, which leads to a good balance of chemical resistance, heat stability and electrical properties, influenced, as in other thermoplastics, by molecular mass and polydispersity. For example, tensile strength and impact resistance increase with molecular mass. According to the crystalline form, the density ranges from 0·87 to 0·95 g/cm^3, and the melting point from 107 to 141°C [2.5]. Some mechanical and thermal properties of PP are presented in Table 2.4.

Because of the tertiary carbon atoms, PP is more susceptible to oxidation than PE, and requires incorporation of appropriate antioxidants. These are especially important when the polymer is exposed to the degradation action of sunlight.

Unlike PE, PP is not subject to environmental stress cracking. It is, however, subject to creep, and its use under continuous loading at high stress is not recommended.

PP is a versatile polymer, a cross-section of end uses being as

Table 2.4.
Some mechanical and thermal properties of commercial polypropylenes [2.3]

Property	Test method	Homopolymers			Copolymers	
		3·0	0·7	0·2	3·0	0·2
Melt flow index	(a)	3·0	0·7	0·2	3·0	0·2
Tensile strength						
lb/in^2	(b)	5 000	4 400	4 200	4 200	3 700
MPa		34	30	29	29	25
Elongation at break (%)	(b)	350	115	175	40	240
Flexural modulus						
lb/in^2	—	190 000	170 000	160 000	187 000	150 000
MPa		1 310	1 170	1 100	1 290	1 030
Brittleness temperature (°C)	ICI/ASTM D.476	+15	0	0	−15	−20
Vicat softening point (°C)	BS 2782	145–150	148	148	148	147
Rockwell hardness (R-scale)	—	95	90	90	95	88·5
Impact strength (ft lb)	(c)	10	25	34	34	42·5
(J)		13·5	34	46	46	57·5

(a) Standard polyethylene grader: load 2·16 kg at 230°C
(b) Straining rate 18 in/min.
(c) Falling weight test on 14 in diameter moulded bowls at 20°C.
Reprinted with permission from *Plastics Materials*, 4th edn by J. A. Brydson, 1982. Copyright J. A. Brydson.

follows: domestic appliances, soil pipe systems, footwear, air ducts, large capacity tanks for chemical plants, pipes for hot water, monofills and multifills for ropes, cordage, carpets, etc.

Poly(vinyl chloride) (PVC) is obtained generally by emulsion or suspension polymerization, and corresponds to the following formula:

$$[-CH_2-CH-]_n$$
$$\underset{Cl}{|}$$

Emulsion polymerizations have complex mechanisms that are affected by many factors, including the numerous free radical reactions (initiation, propagation, termination, chain transfers) and transfer steps in the multiphase system. Key transfer steps are the transfer of the monomer from the dispersed droplets to the dispersed PVC phase, and the transfer of vinyl chloride in the dispersed PVC particles. This latter transfer step involves the diffusion of the vinyl chloride in the pores and through the PVC itself.

The PVCs produced by suspension and bulk polymerizations are generally highly porous, powdery (or granular) products. Since a significant amount of monomer adsorbs in solid PVC, the transfer or diffusion of vinyl chloride in solid PVC is often an important feature of the overall process, and is sometimes the rate-controlling step. The character of the final product (powder) is highly important [2.4].

Commercial PVC polymers and copolymers have molecular weights (MW) that vary from about 50 000 to 150 000. PVC is used as unplasticized or rigid PVC, or flexible (plasticized) PVC, with different properties [2.6–2.10]. For example, the elastic modulus is 280 kgf/mm^2 (397×10^3 psi) for rigid PVC (UPVC), and 0.28–1.29 kgf/mm^2 (397–1833 psi) for plasticized PVC.

For engineering applications, PVC is the most commonly used member of the broad family of thermoplastics. It combines good strength and toughness with excellent resistance to attack from water and many chemicals, good weather resistance, electrical insulating qualities, and resistance to flame spread. It is also economical to process. Additives are almost always used in commercial compounding to achieve the desired performance in end-use application with minimum cost [2.8]. Chemical stability and weatherability are good; it is adversely affected by ultra-violet light (Table 2.5). The stability of plasticized polymer is almost entirely controlled by the nature and amount of additives.

Table 2.5.
Some factors instrumental in the natural weathering of PVC, and examples of their main effects [2.9]

Factor (environmental agent)	Typical action on PVC material	Main observable effects[a]	Remarks
Sunlight	Degradation of PVC polymer by UV component of the radiation Fading of colorants	1, 2, 3, 4, 5, 6, 7, 8	Direct sunlight is also a source of heat and hence can promote temperature effects (see below)
Temperature	Mainly effects of heat and temperature fluctuation, including: (i) Heat degradation (ii) Exudation and volatalisation of components (especially plasticisers) (iii) Physical disruption by local and general stresses caused by temperature changes	1, 2, 3, 4, 5, 6, 8	Temperature effects can promote and enhance those of other agents
Water (including atmospheric precipitation, i.e. rain, snow, hail, vapour and condensate)	(i) Mechanical erosion of surface (especially by wind-borne precipitation) (ii) Leaching out of components (especially plasticisers) (iii) Mechanical disruption (especially of surface) by repeated absorption	1, 3 1, 8, 9 1, 5, 7, 8	Effects aggravated by the action of other factors (e.g. chemical reactions, UV, heat)

Agent	Effect/process	Results[a]	Notes
	and desorption (which may be aggravated by presence of absorbent fillers in the material)		
	(iv) Chemical effects of pollutants (e.g. acids) dissolved in rain	1, 2, 4, 9	
Air	Oxidation of reactive sites in PVC	1, 2, 4, 5, 6	Effects promoted and enhanced by those of UV radiation (instrumental in creating reactive sites, especially double bonds)
Atmospheric pollutants (in vapour, liquid and solid particle form)	(i) Leaching out of components	1, 2, 3, 4, 5, 6, 8, 9, 10	Effects can be accelerated by heat and sunlight
	(ii) Chemical reactions with the PVC polymer and possibly other components of the material		
	(iii) Surface erosion (by wind-borne particulate pollutants)		

[a] 1, dulling, marring and pitting of surface; 2, cracking (in severe or advanced cases); 3, stiffening; 4, discoloration (e.g. yellowing or darkening); 5, reduced strength; 6, reduced extensibility; 7, surface buckling or rippling; 8, distortion (various degrees and kinds); 9, development of microporosity; 10, environmental stress cracking or crazing.
Reprinted with permission from *PVC Technology* by W. V. Titow, 4th ed., 1984. Copyright Elsevier Applied Science Publishers.

Rigid PVC finds its most widespread application as pipes, profiles, rods, tubes, pipe fittings, drain waste, vent pipe in plumbing systems, gas pipe, corrugated roofing, ventilation ducts, wall cladding, house siding, shutters, window and door frames and components [2.10].

Plasticized PVC is used in films, sheet or tube form for flexible hose and ducting, curtaining, cover sheets and tarpaulins, and clothing. It is widely used for wire and cable insulation, footwear, fabric-backed wall covering, vinyl and vinyl-asbestos flooring, etc.

Polystyrene (PS) is produced by free radical polymerization techniques (bulk, solution, emulsion, suspension), is atactic and therefore non-crystalline, and its microstructure may be represented as:

$$[-CH_2-\underset{\underset{C_6H_5}{|}}{CH}-]_n$$

It is a hard, rigid, rather brittle material. It has a relatively low softening point, and does not withstand the temperature of boiling water. PS is highly transparent, transmitting about 90% of visible light; it also has a high refractive index of 1·59, which gives it particular brilliance.

PS is a low cost material and has a good mouldability; these factors, together with its transparency and colorability, are the principal reasons for its widespread application [2.11].

It possesses good thermal stability at processing temperatures, and relatively low shrinkage on cooling. Prolonged exposure to sunlight causes yellowing of clear PS and fading of pigmented types, often with crazing and impairment of mechanical properties.

As one of the main general-purpose thermoplastics, PS is surpassed, in usage, only by PVC and polyolefins.

In construction, PS is used for wall tiles, light-diffusing panels, and electrical appliances.

Expanded PS is extensively used as thermal insulation in buildings, and acoustic tiles for ceilings.

SAN, a non-crystalline copolymer of styrene and acrylonitrile, is hard, rigid, transparent, and tougher than plain PS, but possesses low notch impact strength and elongation. Generally, its mechanical properties and dimensional stability are better than those of PS.

Additives improve outdoor weatherability and ultra-violet stability. Reinforcement with glass fibre improves strength, elastic modulus, impact strength, and heat deflection with loss of transparency. Applications include housings, glazing, containers, etc.

ABS is a very versatile family of thermoplastics produced by using acrylonitrile (A), butadiene (B), and styrene (S). The ratio of those comonomers as well as the microstructure of ABS can be controlled to produce a family of copolymers with a broad range of properties.

ABS copolymers offer a unique combination of toughness, rigidity, chemical resistance, and quality surface appearance, along with the ability to process well into complex parts via almost any thermoplastic processing technology.

For most ABS products, the maximum tensile strength occurs at the yield point, and the tensile strength then drops gradually as the material elongates to failure. In standard products, the highest tensile strength and modulus are available in the medium impact grades [2.12].

Major markets include appliance industries, building and construction, transportation, and business machines. Table 2.6 outlines the 1983 USA sales of ABS in these market areas.

High impact grade ABS has been used to a great extent in pipe applications for drain, waste, and vent purposes, as well as for industrial and chemical uses.

Polytetrafluoroethylene (PTFE) is the most important of polymers containing fluorine; it is a white solid product with a waxy appearance and fill, and has the following microstructure:

$$[-CF_2-CF_2-]_n$$

Table 2.6.
ABS sales in major markets [2.12]

Market area	1983 sales	
	(lb × 10⁶)	*(kg × 10⁶)*
Appliances	188	85
Building and construction	177	80
Transportation	165	75
Business machines and consumer electronics	154	70
	684	310

It is a tough, flexible material of moderate tensile strength, with a tendency to creep under compression. The coefficient of friction is unusually low and is stated to be lower than that of any other solid material. Some PTFE and other fluorine containing polymers are presented in Table 2.7 and Figs. 2.1 and 2.2.

PTFE has high thermal stability and retains its properties over a wide temperature range. The polymer may be used up to about 300°C for long periods without loss of strength, and thin sections remain flexible at temperatures below −100°C. It also has good weather resistance. Its melt temperature (327°C) is extremely high for a polymer [2.13].

PTFE is unaffected by long exposure to exterior weathering and sunlight; it is resistant to micro-organisms and has a high chemical inertness. Resistance to flame spread is good, since its oxygen index is high enough to preclude support of combustion in air.

Typical uses of fluoropolymers include linings for chemical pipe and equipment, high temperature cable coverings, electrical insulators, and non-stick coatings. PTFE is widely used as bridge bearings and as expansion pads to support structural members, steam pipe lines, and similar applications where the low coefficient of friction allows sliding motion.

Polyamides (PA) are well known as nylons, and are defined as polymers which contain recurring amide groups [—CO—NH—] in the main chain.

It may be noted that by far the greater part of the total output of nylons is used for fibre production, but the polyamides have also become of some importance in non-fibrous, particularly engineering, applications. There are also known more complex synthetic aliphatic polyamides which are not fibre-forming, but which are used for adhesives, coatings, etc.

The general formula for aliphatic polyamides obtained by polycondensation is:

$$\{-OC-[CH_2]_m-CO-HN-[CH_2]_n-NH-\}_p$$

The commercial success of nylons (6 and 6,6) is due to the outstanding properties and an economically attractive raw material base. Nylon 6 is the polymer of caprolactam, while nylon 6,6 is produced by melt polycondensation of adipic acid and hexamethylene-diamine. Both these chemicals contain six carbon atoms (nylon 6,6)

Table 2.7.
Properties of PTFE and other fluorine-containing thermoplastics [2.3]

Property	ASTM test	PTFE	PCTFE	PVF[c]	PVDF
Specific gravity	D.792	2·1-2·3	2·1	1·38-1·57[d]	1·76
Tensile strength at 23°C (lbf/in²) (MPa)	D.638	2 500-3 800 / 17-21	4 300-5 700 / 30-39	$9·16-19 \times 10^{3}$ [d] / 66-131	7 000 / 48
Elongation at break at 23°C (%)	D.638	200-300	100-200	110-260[d]	100-300
Izod impact strength at 23°C (ft lbf in⁻¹)	D.256	2·0	1·2-1·3	—	3·5
Deflection temp. under load of 66 lbf/in²(°C)	D.648	121	58	—	150
Water absorption (%)	D.570	0·005	neglig.	<0·5	0·04
Coefficient of friction[a]	—	0·09-0·12	0·4	—	—
Power factor 60 Hz	D.150	<0·003	0·010	0·01(100 c/s)	0·049
Power factor 10⁶ Hz	D.150	<0·0003	0·010	0·08(10⁴ c/s)	0·17
Dielectric constant 60 Hz	D.150	2·1	3·0	6·8-8·5 (10³ c/s)	8·4
Dielectric constant 10⁶ Hz	D.150	2·1	2·5	—	6·6
Volume resistivity (Ω m)	D.257	>10²⁰	10²⁰	10¹⁵-10¹⁶	10¹⁶
Dielectric strength[b] (V/0·001 in)	D.149	400-500	530	3 000-6 000[d]	260

[a] Polymer to metal.
[b] Short time on 0·08 in thick sheet.
[c] Test on Tedlar film. Different test methods involved where marked[d].
PCTFE, poly(chlorotrifluoroethylene); PVF, poly(vinylfluoride); PVDF, poly(vinylidenefluoride).
Reprinted with permission from *Plastics Materials* 4th edn. by J. A. Brydson, 1982. Copyright J. A. Brydson.

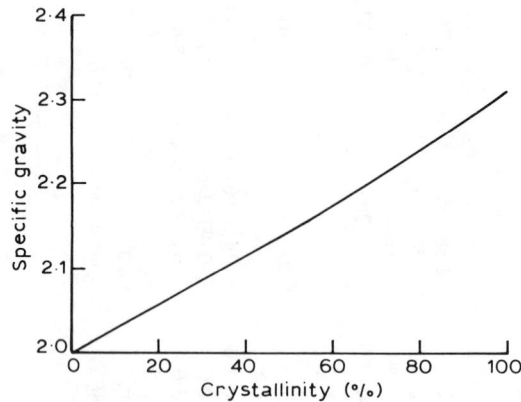

Fig. 2.1. Density as a function of crystallinity in PTFE [2.3]. (After Thomas *et al.*, *Soc. Plast. Engrs J.*, **12**(5) (1956) 89.

and therefore are obtainable from either benzene or other petrochemical products.

Nylon 6,6 is a typical example of the polycondensation of two bifunctional monomers, adipic acid and hexamethylenediamine. This type of process is generally referred to as the AABB reaction. This type of polycondensation has the advantage that the stoichiometry is fixed by salt formation (AH salt). In contrast, AB type polyamides are usually obtained by ring-opening polymerization of lactams.

Two other AABB products, nylon 6,10 and nylon 6,12, are also based on hexamethylenediamine, but use longer chain aliphatic dicarboxylic acids such as sebacic and dodecanoic respectively [2.14].

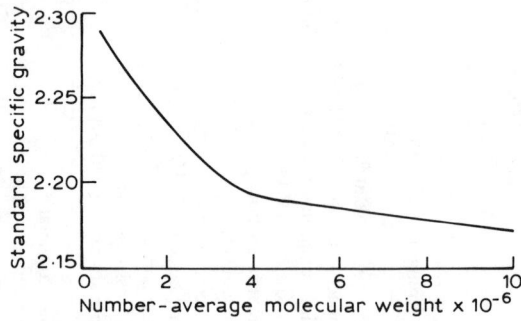

Fig. 2.2. Standard specific gravity of PTFE as a function of MW [2.3]. (After Thomas *et al.*, *Soc. Plast. Engrs J.*, **12**(5) (1956) 89.

Table 2.8.
Melt temperatures of AABB polyamides [2.14]

Polymer	T_m (°C)
6, 6	265
6, 8	240
6, 10	225
6, 12	212

Reprinted with permission from *Introduction to Industrial Polymers* by H. Ulrich, 1982. Copyright C. Hanser Publishers.

As indicated in Table 2.8, the melting point of AABB type linear polyamides differs with the number of carbons in dicarboxylic acids.

Table 2.9 lists the melting points of the AB polyamides. The homopolymers with shorter structural units, like nylon 4, are more hydrophilic, while those with longer units are more hydrophobic.

The absorption of moisture in polyamide moulding compounds results in poor dimensional stability and deterioration of properties (Table 2.10) [2.14].

Polyamides undergo oxidative degradation, with discoloration and loss of strength, on prolonged exposure to sunlight or heat.

Recommended additives include stabilizers to combat degradation and embrittlement at temperatures of 75°C and above, carbon black to

Table 2.9.
Melt temperatures of AB polyamides
[2.14]

Polymer	T_m(°C)
Nylon-4	265
Nylon-6	230
Nylon-7	223
Nylon-11	188
Nylon-12	180

Reprinted with permission from *Introduction to Industrial Polymers* by H. Ulrich, 1982. Copyright C. Hanser Publishers.

Table 2.10.
Absorption of moisture in nylons [2.14]

Polymer	Water absorption[a] at 20°C	Dimensional change[a] (%)
6	2·7	0·7
6·6	2·5	0·6
11	0·8	0·12

[a] At 50% relative humidity.
Reprinted with permission from *Introduction to Industrial Polymers* by H. Ulrich, 1982. Copyright C. Hanser Publishers.

combat sunlight, nucleants to promote crystallization, antifriction agents, fire retardants, etc.

Strength, elastic modulus, and toughness are markedly altered by adding dropped glass fibres, fillers, and fine glass spheres and beads.

Uses are typically for impact resistance and resistance to wear, including cams, gears, containers, seating, hinges and sliders for hardware where loads are moderate, latch parts, wire, connectors, switches and relays, wire coverings, slides, rollers and tracks for cabinets.

Polyacetals are polymers of aldehydes, such as:

—polyformaldehyde $[-CH_2-O-]_n$

—polyacetaldehyde $[-CH-O-]_n$
                                  $|$
                                  CH_3

By definition, polyethers contain the ether linkage —R—O—R—O— as part of their chain structure. By convention, the polymers in which R is the methylene group, —CH₂— (i.e. polyoxymethylenes), are described by the generic terms polyacetals or acetal resins, the name polyethers being applied to polymers in which R is more complex [2.15].

Polyformaldehyde is obtained from pure formaldehyde which is passed over the surface of a stirred solution of catalyst (either a Lewis acid or a base) in a carefully dried inert medium. The polymer is collected by filtration, washed with heptane and acetone, and dried in vacuum at 80°C. In the final stage, the polymer is subjected to esterification to improve its thermal stability; for esterification, acetic

Table 2.11.

Typical values for various properties of formaldehyde homopolymer and copolymer [2.11]

	Homopolymer	Copolymer
Specific gravity	1·425	1·410
Crystalline melting point (°C)	175	163
Tensile strength, lb/in² (MPa)	10 000 (68·9)	8 500 (58·5)
Flexural modulus, lb/in² (MPa)	410 000 (2 825)	360 000 (2480)
Elongation at break (%)	15	60
Impact strength, Izod ft lb/in notch (J/m)	1·4 (74·6)	1·2 (64)
Hardness, Rockwell M	94	80

Reprinted with permission from *Organic Polymer Chemistry* by K. J. Saunders, 1973. Copyright Chapman and Hall.

anhydride is generally preferred. The average molecular weight (M_n) of the homopolymer is in the range 30 000–100 000.

For the synthesis of copolymers, the principal monomer is trioxan, and the second one is a cyclic ether such as ethylene oxide, 1,3-dioxolane or an oxetane; the copolymerization may be carried out in an inert solvent such as hexane at about 60°C, in the presence of boron trifluoride as the most satisfactory initiator. The copolymer is obtained as a slurry and is collected, washed, and dried [2.11]. Some properties of these polymers are presented in Table 2.11.

The homopolymers and copolymers of formaldehyde are rigid materials with broadly similar properties. Commercial polymer is stabilized against thermal degradation, both from oxidative attack on the main chain and from depolymerization, and usually it is suitable for continuous service up to 85°C. It is tough and resilient, even down to low temperatures, and is free from biological attack, but it is susceptible to ultra-violet light and is not recommended for use in exposed situations.

In some applications, polyformaldehyde can advantageously replace zinc die-casting alloys, brass, aluminium, and even cast iron, being appropriate for gear wheels, valve seatings, impellors, and numerous smooth-surface precision-dimensioned mechanical and electrical components.

Polysulphone commercial polymer is obtained through the polycondensation of the disodium salt of 2,2-bis(4-hydroxyphenol) propane (bisphenol A), and 4,4′-dichlorodiphenylsulphone, in a polar aprotic solvent. The polymer, illustrated below, has 50–80 units.

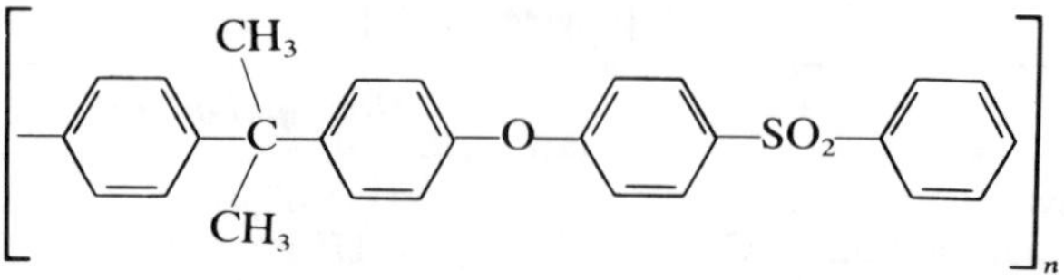

The diphenylene sulphone group confers characteristics such as: thermal stability, oxidative resistance, and rigidity even at elevated temperatures.

Polysulphone absorbs small amounts of moisture, and the equilibrium amounts at 23 and 96°C are shown in Table 2.12.

Polysulphone's limiting oxygen index (ASTM 2683) is 30, and its self-ignition temperature is 621°C.

Typical mechanical properties of polysulphone are given in Table 2.13.

Polysulphone finds uses in applications where heat resistance, dielectric properties, and mechanical strength are required, e.g. housings for electrical and domestic appliances, computer components, switches, etc. Because of its excellent chemical stability, polysulphone offers an attractive alternative to metals in certain fluid-handling applications. It is preferred to PVC, chlorinated PVC (CPVC), and PP in applications where higher temperature resistance

Table 2.12.
Water absorption and diffusion coefficients for polysulphone
at 23 and 96°C [2.16]

Property	Value
23°C	
Water absorption at equilibrium (%)	0·86
Diffusion constant (cm^2/s)	$3·8 \times 10^{-8}$
96°C	
Water absorption at equilibrium (%)	1·00
Diffusion constant (cm^2/s)	$5·9 \times 10^{-7}$

Reprinted with permission from *Engineering Thermoplastics. Properties and Applications* by J. M. Margolis (Ed.), 1985. Copyright Marcel Dekker, New York.

Table 2.13.
Typical mechanical properties of polysulphone [2.16]

Property	ASTM method	At room temperature
Tensile strength, yield, psi (MPa)	D.638	10 200 (70·3)
Tensile modulus of elasticity, psi (MPa)	D.638	360 000 (2 682)
Tensile elongation to yield (%)	D.638	5–6
Tensile elongation to break (%)	D.638	50–100
Flexural strength, yield, psi (MPa)	D.790	15 400 (106)
Flexural modulus of elasticity, psi (MPa)	D.790	390 000 (2 689)
Compressive strength, break, psi (MPa)	D.695	40 000 (276)
Compressive modulus of elasticity, psi (MPa)	D.695	374 000 (2 579)
Shear strength, yield, psi (MPa)	D.732	6 000 (41·4)
Shear strength, ultimate psi (MPa)	D.732	9 000 (61·2)
Poisson's ratio, at 0.5% strain		0.37
Shear modulus (g), psi (MPa)		133 000 (917)
Notched Izod impact strength, 0·25 in. spec., ft-lb/in (J/M)	D.256	1·2 (64)
Notched Izod impact strength, 0·125 in. spec., ft-lb/in (J/m)	D.256	1·3 (69)
Unnotched Izod impact strength, 0·125 in. spec., ft-lb/in (kJ/m)		<60 (no breaks) (<3·2)
Notched Izod impact strength, @ −40°F, 0·125 in spec., ft-lb/in. (J/m)	D.256	1·2 (64)
Tensile impact, short specimen, ft-lb/in² (kJ/m²)	D.1822	200 (420)
Tensile impact, short specimen, ft-lb/in³ (MJ/m³)		430 (35·6)
Falling-dart impact (see note), ft (m)		4–5 (1·2–1·5)
Rockwell hardness	D.785	M69 (R120)

Note: The maximum height required to produce cracks in a 3 in × 4 in × 0·125 in (76·2 mm × 102 mm × 3·2 mm) thick specimen on impact by a falling 10 lb (4·5 kg) cylinder with a 0·5 in (12·7 mm) diameter tip.
Reprinted with permission from *Engineering Thermoplastics. Properties and Applications* by J. M. Margolis. (Ed.), 1985. Copyright Marcel Dekker, New York.

is required. Polysulphone is available as pipe consisting of extruded tubing with a filament-wound fibreglass over-wrap [2.16].

Polycarbonate (PC) is produced by reaction of polyhydroxy compounds with a carbonic acid derivative, which produces a series of polymers with carbonate —O—CO—O— linkages.

The most common polymer of this group is obtained by a polycondensation process of bisphenol A and phosgene. The general representation of the polymer is:

The main differences between the polycarbonate commercial grades are due to: differences in MW, differences in additives, the presence or otherwise of a second polyhydroxy compound.

Polycarbonates possess exceptionally high impact strength, combined with thermal stability and creep resistance. Table 2.14 presents some mechanical properties of commercial polycarbonates.

Polycarbonate has outstanding rigidity and toughness, these properties are retained at both low and elevated temperatures; the maximum possible service temperature is about 135°C. The resistance of the polycarbonate to deformation under load, and the dimensional stability in humid atmospheres are also outstanding. Because of the above mentioned characteristics, it was initially thought that PC would become an important engineering material. This hope had not been fully realized because of the liability of this macromolecular compound to craze or crack under strain or on ageing.

The faint amber colour increases in intensity on exposure to strong light, and for outdoor applications necessitates the use of a stabilizer (carbon black), or lacquering with an ultra-violet absorber.

Relative to other thermoplastics, PC shows good resistance to ionizing radiation; some discoloration and embrittlement may occur, but tensile and electrical properties are maintained.

PC is suitable for gears and bearings that are not heavily loaded.

Polyphenylene, commercial grade, is obtained through two different procedures; the first one is based upon the reaction of aromatic

Table 2.14.
Comparison of mechanical properties of typical commercial bis-phenol A polycarbonates [2.3]

	Units	Test method	Unfilled grades		Glass-filled grades	
			Low	Medium-high		
MW-range			Low	Medium-high		
% glass fibre			Nil	Nil	20	35
Specific gravity		DIN 53479	1·2	1·2	1·33	1·44
Tensile strength −30°C	lbf/in²	DIN 53455	12 000–13 000	12 000–13 000	—	—
	(MPa)		83–90	83–90	—	—
+23°C	lbf/in²	ASTM D638	9 500	9 500	16 000	18 500
	(MPa)		65	65	110	127
+100°C	lbf/in²		5 000	5 000	—	—
	(MPa)	35			—	—
Elongation at yield	%	DIN 53455	6	6–7	4·5	—
at break	%	DIN 53455	110–120	80–120	8·5	2·7
Modulus of elasticity (tensile)	lbf/in²	DIN 53455	345 000	345 000	860 000	1 350 000
	(MPa)		2 400	2 400	5 900	10 000
Flexural strength at yield	lbf/in²	DIN 53452	13 500	13 500	—	—
	(MPa)		93	93	—	—
at break	lbf/in²		—		21 000	30 000
	(MPa)		—		150	210
Izod impact strength	ft lbf/in notch	ASTM D256–56 0·5 in × 0·125 in bar	12–16	15–18	ca 2·5	ca 2·5
Unnotched impact strength	kgf cm⁻²	DIN 53453 (22°)	no fail	no fail	—	50

Reprinted with permission from *Plastics Materials*, 4th edn by J. A. Brydson, 1982. Copyright J. A. Brydson.

disulphonyl chlorides with bi- or terphenyls:

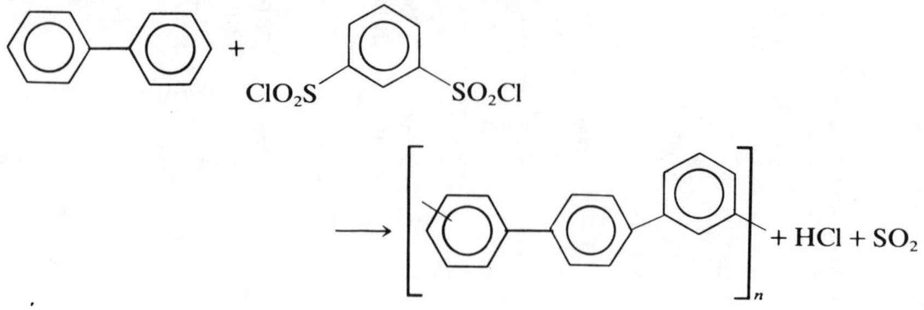

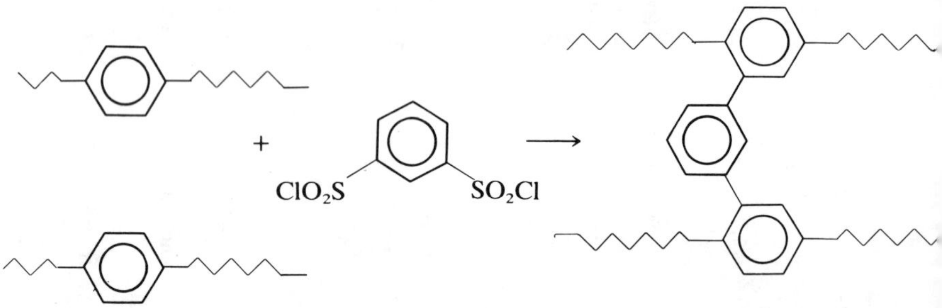

The second procedure produces an acetylenic terminated oligophenylene with the following microstructure:

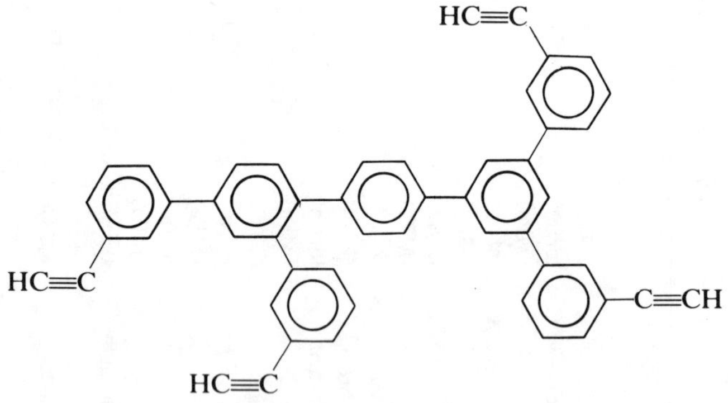

At high temperature the trimerization of the acetylenic groups leads to the formation of phenylene rings [2.17]. After curing, the polymer has the properties presented in Table 2.15.

Table 2.15.
Typical properties of cured resin [2.18]

Density, g/ml	1·145
Flexural strength, MPa (psi)	
at 23°C	48–138 (6 960–20 010)
at 360°C	41–55 (5 945–7 975)
Flexural modulus, GPa (psi)	
at 23°C	4·8–8·3 (696–1 203)
at 360°C	3·5–4·8 (507–696)
Volume resistivity (ohm cm)	
at 23°C	5×10^{17}
at 100°C	5×10^{16}
Dielectric constant	
60 Hz at 23°C	3·0
60 Hz at 100°C	3·2
1 MHz at 23°C	3·2
1 MHz at 100°C	3·3
Dissipation factor	
60 Hz at 23°C	0·001 6
60 Hz at 100°C	0·000 8
1 MHz at 23°C	0·002 5
1 MHz at 100°C	0·002 0
Limiting oxygen index	55
Continuous use temperature in air (°C)	215
Short-term use temperature in air (°C)	350

Reprinted with permission from *J. Elast. Plast.*, **6** (1974) 103. Copyright Technomic Publishing Co. Inc.

Poly-*p*-phenylene is of interest because of its outstanding thermal stability.

Thermogravimetry in air and in nitrogen of the poly-*p*-phenylenes derived from cyclohexadiene, benzene and an acetylene terminated oligophenylene (Fig. 2.3) shows that there is little difference in the stability of the polymers in inert medium, but in air the product obtained from the acetylenic precursor is markedly less stable [2.17].

Thermal properties depend to some extent on the method of synthesis, but generally there is little change on heating in air in the range 300–400°C. Since this polymer is highly insoluble and has a high melting point, it is difficult to process. However, asbestos and carbon fibre composites have been realized.

Polyimides involve the reaction of a diamine with the dianhydride

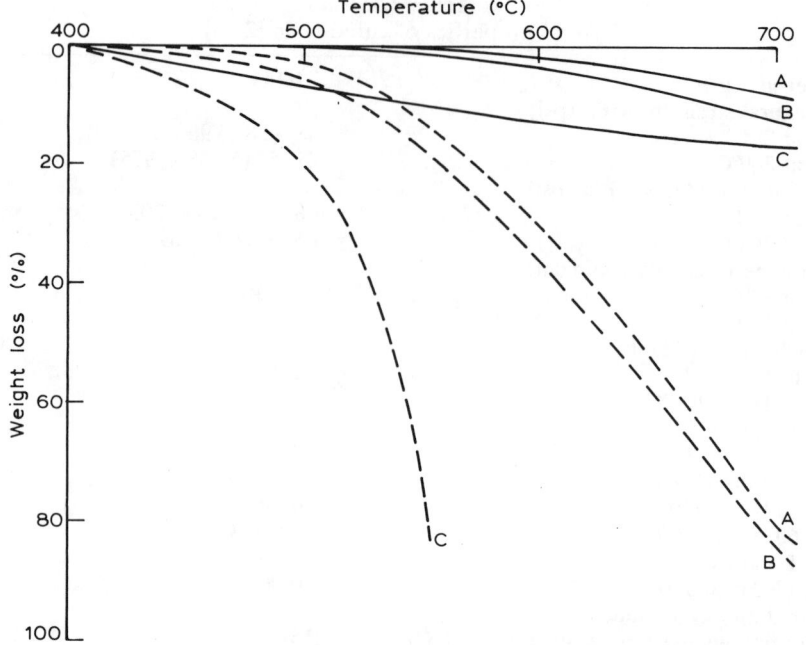

Fig. 2.3. Thermogravimetry of poly-*p*-phenylene (———) in nitrogen, (– – –) in air, (A) from cyclohexadiene, (B) from benzene, (C) from acetylene terminated oligophenylene. Heating rate 10°C/min [2.17]. (Reprinted with permission from *Heat Resistant Polymers* by J. P. Critchley *et al.*, 1983. Copyright Plenum Press.)

of a tetracarboxylic acid in polar solvents, which gives a polyamic acid.

The removal of water from the polyamic acid (by heating or

dehydration) gives the polyimide:

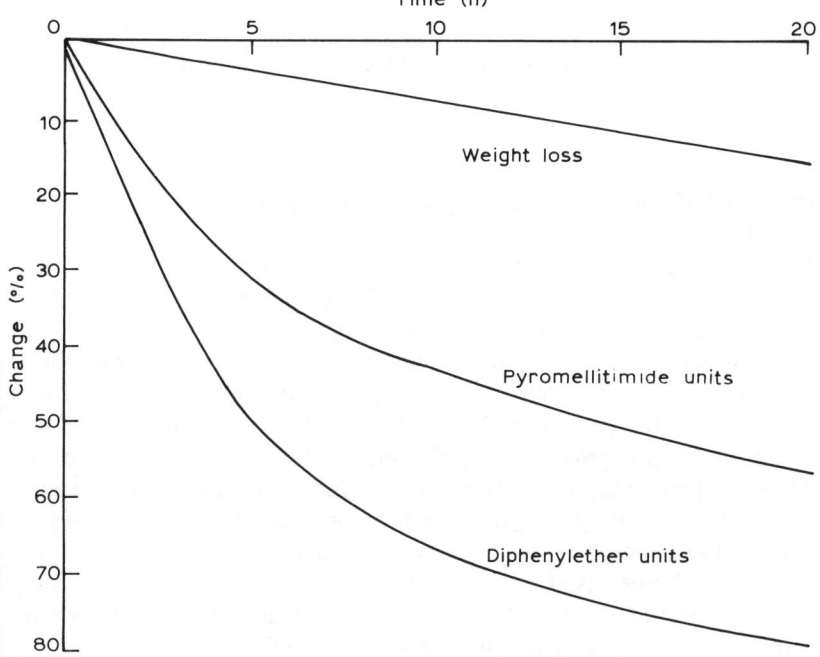

Polyimides have excellent solvent resistance, flame resistance, outstanding abrasion resistance, exceptional heat resistance, excellent resistance to oxidative degradation, most chemicals other than the strong bases, and to high energy radiation.

Figure 2.4 shows the change in composition of a polyimide based on

Fig. 2.4. Change in composition of polyimide based on pyromellitic dianhydride and diphenylether with time of ageing at 400°C in air [2.17, 2.20]. Reprinted with permission from *Heat Resistant Polymers* by J. P. Critchley *et al.*, 1983. Copyright Plenum Press. (Original source: R. A. Dine-Hart *et al.*, *Brit. Polym. J.* **3** (1971) 226.)

Table 2.16.
Some properties of Kapton type H film (1 mil thick) [2.17]

Property	Measured 25°C	200°C
Ultimate tensile strength, MPa (psi)	173 (25 085)	117 (16 965)
Yield point at 3%, MPa (psi)	69 (10 005)	41 (5 945)
Stress to produce 5% elongation, MPa (psi)	90 (13 050)	59 (8 555)
Ultimate elongation (%)	70	90
Tensile modulus, GPa (psi)	2·97 (430 650)	1·79 (259 550)
Dielectric constant at 1 kHz	3·5	3·0
Dissipation factor at 1 kHz	0·003	0·002
Volume resistivity (ohm cm)	10^{18}	10^{14}
Short-term dielectric strength at 60 Hz (V/mil)	7 000	5 600

Reprinted with permission from *Heat Resistant Polymers* by J. P. Critchley *et al.*, 1983. Copyright Plenum Press. (Original source: Dupont Product Bulletin H-1B.)

pyromellitic dianhydride and diphenylether with time of ageing at 400°C in air.

Table 2.16 gives some properties of a polyimide film.

Exposure for 1500 h to a radiation of about 10 rad at 175°C led to embrittlement, but the sample retained shape stability.

Composites produced by impregnation of glass and carbon fibres with polyimides, followed by subsequent pressing, can be used continuously at temperatures up to 250°C, and intermittently to 400°C.

Unsaturated polyesters are polycondensation products of poly-alcohols and dibasic acids. The term includes polyester plasticizers, polyester fibre-forming polymers such as poly(ethylene terephthalate) and polyesters modified by fatty acids and drying oils such as the alkyds (glyphthales) used in the coatings industry.

The unsaturated polyesters are obtained by reacting a glycol with an unsaturated dibasic acid, which is generally maleic or fumaric acid.

Having double bonds, these polyesters can be cross-linked with an unsaturated monomer such as styrene, acrylates, etc.; the main condition is to have reagents with a functionality of 2:3 or more in the polycondensation process.

B. Parkyn *et al.* [2.21] presented the cross-linking of linear chains of poly(ethylene glycol maleate) with styrene in the following way:

$$\cdots O.CH_2.CH_2.O.OC.\overset{\vdots}{\underset{\begin{bmatrix}CH_2\\ \vert\\ CHR\end{bmatrix}_n}{\overset{\vert}{C}H}}.CH.CO.O.CH_2.CH_2.O\cdots$$

$$\cdots O.CH_2.CH_2.O.OC.\underset{\begin{bmatrix}CH_2\\ \vert\\ CHR\end{bmatrix}_m}{\overset{\vert}{C}H}.CH.CO.O.CH_2.CH_2.O\cdots$$

$$\cdots O.CH_2.CH_2.O.OC.\underset{\vdots}{\overset{\vert}{C}H}.CH.CO.O.CH_2.CH_2.O\cdots$$

$$R = C_6H_5 \text{ ring}$$

A general picture of polyester cross-linking and double bonds disappearance can be followed in Fig. 2.5.

Some pertinent properties of a few polyesters of a GP phenolic are compared in Table 2.17.

Madorsky & Straus [2.22] have studied the thermal degradation in vacuum of triallylcyanurate-based polyester at temperatures up to 1200°C, and have compared the results with those of phenolics and an epoxy-novolac (Fig. 2.6). The results have proved that the thermal stability of phenolics is better.

The use of polyesters in composites has led to studies dealing particularly with tensile strength and impact fractures. The effect of temperature has been shown to be a fall in tensile strength as temperature increases. Impact energies of short glass fibre polyester composites have also been studied by summing fracture energy of the matrix, the fibre and the energy necessary to remove the fibres in a crack surface. It was found that the impact fracture energy of the fibres is a controlling factor at ambient temperatures [2.23–2.25].

Cross-linked unsaturated polyesters constitute the basic matrix for sheet moulding compounds (SMC), introduced about 1967, and by 1972 were being produced at the rate of about 20 000 t per year.

Phenol-formaldehyde polymers (phenoplastes) are produced by polycondensation of a phenol or a mixture of phenols with an aldehyde, in most cases formaldehyde.

If an acid catalyst is used and the molar ratio of phenol to

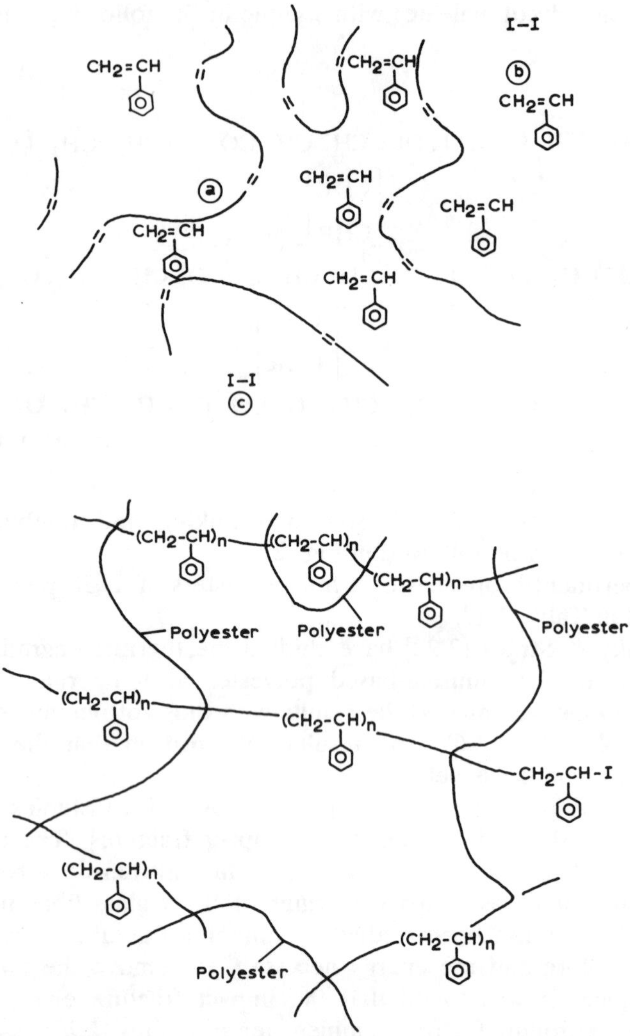

Fig. 2.5. The nature of cured polyester laminating resins. (1) Structures present in polyester ready for laminating: (a) low MW unsaturated resin molecules; (b) reactive diluent (styrene) molecules; (c) initiator (catalyst) molecules. (2) Structures present in cured polyester. Cross-linking via an addition copolymerization reaction. The value of n is 2–3 on average in general purpose resins [2.3]. (Reprinted with permission from *Plastics Materials,* 4th edn, by J. A. Brydson, 1982. Copyright J. A. Brydson.)

Table 2.17.

Properties of thermosetting polyester mouldings [2.3]

	P-FPG	DMC (GP)	Polyester alkyd	DAP alkyd	DAIP alkyd	Units
Moulding temperature*	150–170	140–160	140–165	150–165	150–165	°C
Cure time (cup flow test)*	60–70	25–40	20–30	60–90	60–90	s
Shrinkage	0·007	0·004	0·009	0·009	0·006	cm/cm
Impact strength	0·12–0·2	2·0–4·0	0·13–0·18	0·12–0·18	0·09–0·13	ft lb
	0·016–0·027	0·27–0·54	0·017–0·024	0·016–0·024	0·012–0·017	m kg
Specific gravity	1·35	2·0–2·1	1·7–1·8	1·64	1·8	
Power factor (800 Hz)	0·1–0·4	0·01–0·05	0·01–0·05	0·03–0·05	—	
Power factor (10^6)	0·03–0·05	0·01–0·03	0·02–0·04	0·02–0·04	0·04–0·06	
Dielectric constant (800 Hz)	6–10	5·5–6·5	4·5–5·5	4·0–5·5	—	
Dielectric constant (10^6 Hz)	4·5–5·5	5·0–6·0	4·5–5·0	3·5–5·0	4·0–6·0	
Volume resistivity	10^{12}–10^{14}	$>10^{16}$	$>10^{16}$	$>10^{16}$	$>10^{16}$	Ω m
Dielectric strength (90°)	39–97	78–117	94–135	117–156	117–156	kV/cm
Water absorption	45–65	15–30	40–70	5–15	10–20	mg

Except where marked by an asterisk these results were obtained by test methods as laid down in BS 771.
Reprinted with permission from *Plastics Materials*, 4th edn. by J. A. Brydson, 1982. Copyright J. A. Brydson.

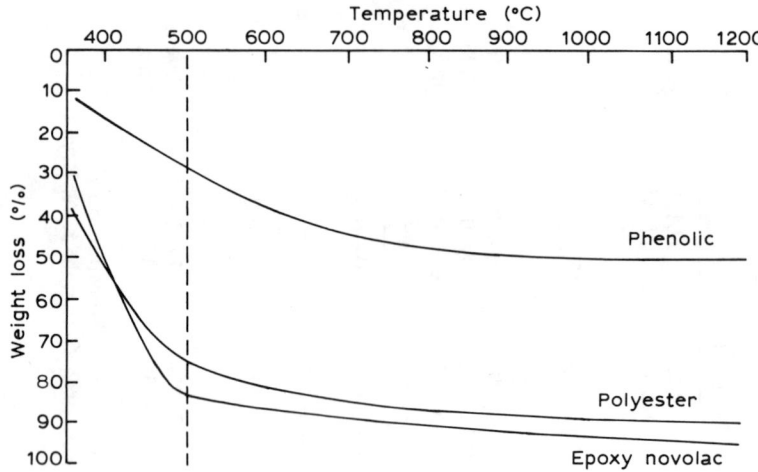

Fig. 2.6. Comparison of thermal stabilities of polyester, phenolic and epoxy novolac resins. Based on weight loss after 30 min at temperatures up to 500°C and 5 min at temperature thereafter [2.22]. (Reprinted with permission from *Modern Plastics,* **38**(6) (1961) 138. Copyright McGraw-Hill Publications Co.)

formaldehyde is greater than 1, the reaction stops at this point. It should be realized that di- and trialcohols are formed and undoubtedly enter into the formation of the straight chains. This structure is that of a typical linear polymer and, as such, has thermoplastic properties in that it is soluble and fusible. Additional formaldehyde or hexa-methylene tetramine is added to form cross-linkages between the chains and thus convert it into a thermosetting three-dimensional structure.

The exact number and positions of the cross-linkages in various grades of phenol-formaldehyde polymers are not known, but it is possible to postulate a very complex highly idealized three-dimensional structure if the ratio of phenol to formaldehyde is correct.

The structure on page 109 is a planar representation of a very complex three-dimensional structure. Ether linkages are also formed, and there is considerably more irregularity than is implied in the diagram.

These resins, when classified according to the nature of the reaction occurring during their production, are of two fundamental types.

One-stage polymers. In these, all the necessary reactants (phenol, formaldehyde, catalyst) required to produce a thermosetting polymer are charged into the polymerization kettle in the proper proportions and react together. An alkaline catalyst is used. The polymer, as discharged from the kettle, is thermosetting or heat-reactive and requires only further heating to complete the reaction to an infusible, insoluble state.

Two-stage polymers. Only part of the necessary formaldehyde is added in the kettle in making these polymers, and an acid catalyst is used. They are permenently fusible or thermoplastic when discharged from the kettle, but will react with additional formaldehyde to produce a thermosetting resin.

This additional formaldehyde is furnished by 'hexa' (hexamethylene tetramine), a white crystalline solid that is added during subsequent processing and breaks down under the influence of heat into formaldehyde and ammonia. The formaldehyde combines with the linear polymer and converts it into a thermosetting product; the ammonia serves as a catalyst for this second stage of the reaction.

Both one- and two-stage polymers are used, separately or in combination, in commercial moulding materials.

The raw materials used in the manufacture of phenol-formaldehyde polymers are phenol and phenol derivatives such as resorcinol or paratertiarybutyl phenol, cresols, formaldehyde, and hexamethylene tetramine.

Since the polymer in phenolic mouldings is cross-linked and highly interlocked, these materials are hard, heat resistant and insoluble.

The properties of phenolic laminates depend on a lot of factors, the following being the most important:

(a) the type of polymer;
(b) the properties of the varnish such as the nature of the solvent, and the viscosity and resin content of the varnish;
(c) the type of reinforcement;
(d) moulding conditions such as time, temperature and pressure.

Epoxy polymers are generally made by the interaction of epichlorohydrin and bisphenol A (diphenylol propane). The useful properties of these polymers only appear after curing. This step transforms the low molecular-weight product to a highly cross-linked space network. The linear product corresponds to the following microstructure:

In commercial epoxies, which are ordinarily mixtures, n is 0–12. The writing of the detailed microstructure, as above, is now known to have a great deal of relevance in correlating molecular motion and mechanical relaxation processes in the glassy solid state of the polymer [2.26]. The diamines $H_2N—R—NH_2$ are the most common cross-linking agents; such a cross-linking process leads to a regular structure of the following network character:

where E represents the difunctional epoxy, and the nitrogen atoms are considered as trifunctional cross-linking units.

Because epoxy polymers are sometimes too viscous for use in certain applications, they are often reduced by adding solvents, plasticizers, modifying agents, and reactive diluents [2.27].

Epoxy polymers have known a wide range of applications mainly because of the versatility of the system. Proper selection of the polymer and its cross-linking agent allows tailoring of properties for the cross-linked products. This versatility has been a major factor in the steady growth rates of this group of polymers. The main characteristics of the properly cured product are the following [2.14]:

—outstanding adhesion to a variety of substrates, especially metals and concrete;
—excellent chemical resistance;
—very low shrinkage on cure;
—high tensile, compressive, and flexural strengths;
—excellent electrical insulation properties;
—corrosion resistance;
—ability to cure over a wide temperature range.

Chu & Seferis [2.28] have done a network study and analysis for relatively complex epoxy–diamine systems of commercial importance as matrix materials for high performance composites. Besides the studies on the microscopic deformation and failure processes of composites by photoelastic studies, Morgan [2.29] correlates the composite matrix performance with measurable, molecular structural parameters of the polymer.

Progressive replacement of amine hardener by a low viscosity flexibilizer will reduce mix viscosity, increase pot life and reduce the heat distortion temperature of the cured system. Higher impact strengths are achieved by using approximately equivalent amounts of hardener and flexibilizer.

By using flexibilizers in addition to the usual amount of hardener, very flexible products may be obtained [2.3].

Table 2.18 shows the influence on epoxy polymer properties of different flexibilizers.

Epoxy polymers are used in a large number of fields, including surface coatings, adhesives, polymer–concrete composites, for laminates in flooring, and to a small extent in road surfacing. The properties of epoxy laminates depend on:

—the type of polymer;
—the type of hardener;
—the amount and type of filler;
—the amount and type of reinforcing agent;
—the curing conditions.

The matrix generally has several functions, of which the most

Table 2.18.
Influence of flexibilizers on epoxy polymer properties [2.3]

	Difunctional amine			Polysulphide			Polyamide	
Flexibilizer	—	25	25	50	25	50	43	100
Epoxy resin	100	100	100	100	100	100	100	100
Amine hardeners	20	13·2	20	20	20	20	—	—
Pot life (1 lb) (min)	20	69	44	76	13	6	150	140
Viscosity (25°C) (cP)	3 700	1 070	870	490	—	—	210 000	210 000
Flexural strength (psi)	16 000	14 400	17 710	—	15 300	—	10 700	11 670
(MPa)	110	99	122	—	105	—	73	80
Compressive yield stress								
(psi)	15 000	13 900	14 330	—	12 300	—	12 800	10 700
(MPa)	103	96	98	—	85	—	88	73
Impact strength (ft lb/0·5 in notch)	0·7	0·82	1·03	8·0	0·5	1·7	0·3	0·32
Heat distortion temperature (°C)	95	44	40	<25	53	32	81	49

Reprinted with permission from *Plastics Materials*, 4th edn by J. A. Brydson, 1982. Copyright J. A. Brydson.

important are:

—it acts as a bridge to hold the fibres in place;
—it protects the filaments from damage by abrasion and chemical attack;
—it transmits stresses to the fibres.

TYPES OF REINFORCING AGENTS

The following types of reinforcing agents are especially used: (a) continuous filaments and fibres, (b) fillers, (c) sheet-like materials. Glass is by far the most common fibre used in polymer–matrix composites. Others are carbon fibre, graphite fibre, boron fibre and steel fibre. Table 2.19 shows tensile strengths and moduli of four typical filaments [2.30].

In composite fabrication, continuous filaments are used in different forms such as: continuous strand mat, twisted (yarns), chopped, wound parallel (rovings), hammer milled (milled fibre) [2.31].

When sheets of various materials are bonded together with a matrix, the end-product is called a laminate, and the reinforcing agent is called a laminating base. The most common laminating bases are woven fabrics, non-woven fabrics and mats, paper and metal foils [2.31].

Fillers are solid, chemically inert substances added to a composite to modify its properties and/or the overall cost. Fillers may be used in the presence or absence of a fibrous reinforcing agent.

Table 2.19.
Tensile strengths and modulus of four typical filaments [2.30]

Material	Diameter		UTS		Modulus	
	in	*cm*	*ksi*	*kN/cm^2*	*10^6psi*	*MN/cm^2*
Glass (type E)	0·000 3	0·000 76	350	241	10·5	7·2
Boron (or tungsten)	0·005	0·012 7	400	278	60	41
Graphite (thermal)	0·000 2	0·000 51	250	172	60	41
			350	241	40	28
Steel (wire)	0·001	0·002 5	500	345	30	21

Reprinted with permission from *Composites: State of the Art* by J. W. Weeton & E. Scala (Eds), 1971. Copyright The Metallurgical Society, 420 Commonwealth Drive, Warrendale, PA 15086, USA.

Table 2.20.
Particle characteristics [2.32]

	Sphere	Cube	Block	Flake	Fibre
Particle class Idealized shape class descriptor[a]	Spheroidal[b]	Cubic[c] Prismatic Rhombohedral	Tabular Prismatic Pinacoid Irregular	Platy[d] Flaky	Acicular Elongated Fibrous
Shape ratios;					
length (L)	1	1^m	1.4–4	1	1
width (W)	1	1^m	1	<1	<0.1
thickness (T)	1	1^m	1–<1	0.25–0.01	<0.1
Sedimentation diameter[e]	1	esd	esd	esd^f	esd^f
Surface area equivalence[g]	1	1.24^h	$1.26–1.5^i$	$1.5–9.9^j$	1.87 for 0.1 2.3 for 0.05^k
Examples	Glass Spheres Microspheres	Calcite[l] Feldspar	Calcite Feldspar Silica Barite Nephelite	Kaolin Mica Talc Graphite Hydrous alumina	Wollastonite Tremolite Wood flour

[a] Preferred to particle class since this is based on relative surface area. First descriptor is preferred.

[b] In the sense that a spheroid approaches a true sphere.

[c] Generally distorted cubes; more nearly prismatic.

[d] Generally having the nature of hexagonal platelets, as illustrated.

[e] According to Stokes's law, esd = equivalent spherical diameter or the diameter of a sphere having the same volume as that of the particle.

[f] Must be modified for disymmetry of greater than 4–1, maximum to minimum particle dimensions.

[g] Equivalent to a spherical diameter of 1; an approximation of the area when the particle has a volume equivalent to an esd of 1.

[h] About the same for cubic and prismatic shapes.

[i] For lengths of 1-4–4, respectively.

[j] Based on hexagonal platelets as follows:

Length/thickness	Area factor
4/1	1·47
6/1	1·78
8/1	2·09
10/1	2·34
100/1	9·88

[k] For a square cross-section.

[l] There are more than 300 crystal shapes for calcite alone, but generally it is an irregular, low-surface area.

[m] Approximate values.

Reprinted with permission from *Handbook of Fillers and Reinforcements for Plastics* by H. S. Katz & J. V. Milewski (Eds), 1978. Copyright Litton Educational Publishing.

Table 2.21.

Some fillers and reinforcements—and their contributions to plastics [2.33]

Filler or reinforcement	Properties improved					
	Chemical resistance	Heat resistance	Electrical insulation	Impact strength	Tensile strength	Dimensional stability
Alumina tabular	O	O				O
Alumina trihydrate			O			
Aluminium powder						
Asbestos	O	O	O	O		O
Bronze						
Calcium carbonate[b]		O				O
Calcium metasilicate	O	O				O
Calcium silicate		O				O
Carbon black[c]		O				O
Carbon fibre						
Cellulose				O	O	O
Alpha cellulose			O		O	O
Coal, powdered	O					
Cotton (chopped fibres)			O	O	O	O
Fibrous glass	O	O	O	O	O	O
Fir bark						
Graphite	O				O	O
Jute				O		
Kaolin	O	O				O
Kaolin (calcinated)	O	O	O			O
Mica	O	O	O			O
Molybdenum disulphide						
Nylon (chopped fibres)	O	O	O	O	O	O
Orlon	O	O	O	O	O	O
Rayon			O	O	O	O
Silica, amorphous			O			
Sisal fibres	O			O	O	O
TFE-fluorocarbon						O
Talc						O
Wood flour			O		O	O

The chart does not show differences in degrees of improvement: calcinated kaolin, for example, generally gives much higher electrical resistance than kaolin. Similarly, differences in characteristics of products under one heading, such as talc (which varies greatly from one grade to another and from one type to another), also are not distinguished.
[a] Symbols: P, in thermoplastics only; S, in thermosets only; S/P, in both thermoplastics and thermosets.
[b] In thermosets, calcium carbonate's prime function is to improve moulded appearance.
[c] Prime functions are imparting of UV resistance and colouring; also is used in cross-linked thermoplastics.
Reprinted with permission from *Plastics*, 6th edn by J. H. DuBois & F. W. John, 1981. Copyright Litton Educational Publishing.

Filler selection is primarily determined by:

(1) the particle shape;
(2) the particle size distribution; and as a consequence of both of these.
(3) the manner in which the particles pack together.

			Properties improved				
Stiffness	*Hardness*	*Lubricity*	*Electrical conductivity*	*Thermal conductivity*	*Moisture resistance*	*Processability*	*Recommended for use in*[a]
							S/P
○					○	○	P
			○	○			S
○	○						S/P
○	○		○	○			S
○	○					○	S/P
○	○				○		S
○	○						S
○			○	○		○	S/P
			○	○			S
○	○						S/P
							S
					○		S
○	○						S
○	○				○		S/P
						○	S
○	○	○	○	○			S/P
○							S
○	○	○			○	○	S/P
○	○				○	○	S/P
○	○	○			○		S/P
○	○	○			○		P
○	○	○				○	S/P
○	○				○	○	S/P
○	○						S
					○	○	S/P
○	○				○		S/P
○	○	○					S/P
○	○	○					S/P
							S

A general classification of filler particles is presented in Table 2.20. Ferrigno [2.32] shows that the classes are based on a somewhat arbitrary classification according to surface area. This classification is based on two primary properties of fillers: (1) surface area and (2) particle size, both of which are directly measurable and serve as a basis for systematizing filler functions.

Most fillers are minerals able to reduce cost, provide body, speed the cure or hardening, minimize shrinkage, reduce crazing, improve thermal endurance, add strength, and provide special electrical, mechanical and chemical properties [2.33]. Their contribution to plastics may be seen in Table 2.21.

Table 2.22.

Chemical composition classification of fillers[a] [2.32]

Chemical class	Derivation	Types	Chemical resistance[b]		
			Acid	Alkali	Other
Oxide	Mineral	Alumina	G	G	
		Gibbsite (alumina trihydrate)	G	G	
Salt	Mineral	Calcium carbonate	P	F	Water, sol.
	Synthetic	Barium sulphate, barite	E	E	
	Animal	Aragonite, calcium carbonate (oyster shell)	P	F	Water, sol.
Silicate[c] neso-	Mineral	Zirconium silicate, zircon	E	E	
ino-		Calcium silicate, wollastonite	P	F	Water, sol.
		CaMg silicate, tremolite	F	G	
phyllo-		Aluminosilicate, kaolinite	G	G	
		K aluminosilicate, mica	G	G	
		Mg silicate, serpentine (asbestos)	G	G	
		Aluminosilicate, pyrophyllite	G	G	
tecto-		Silica	E	P	
		Hydrous silica, opal	E	G	
		Na, K aluminosilicate feldspar, nepheline	G	G	
	Synthetic	Glass (microbeads)	G	G	
Elements	Mineral, synthetic	Calcium silicate, precipitated	P	F	
		Crystalline carbon, graphite	E	E	
	Synthetic	Metals	P		E, P for aluminium
Organic		Coal (anthracite)	E	F	Volatiles
	Vegetable	Wood, bark, cork, nutshell flours	P	P	Reactive with acids, salts

[a] Types used for prime purposes other than as fillers are not included.

[b] E = excellent, G = good, F = fair, P = poor.

[c] Prefixes from the Greek characterize the silica tetrahedral arrangements in the crystal lattice; soro for group, neso for island (isolated), cyclo for ring, ino for chain or thread, phyllo for sheet, and tecto for framework. Soro- and cyclosilicates have been included since no commercial fillers exist for either class, but they may be present as impurities in principal types.

Reprinted with permission from *Handbook of Fillers and Reinforcements for Plastics* by H. S. Katz & J. W. Milewski (Eds), 1978

Chemical composition is a primary property of fillers, and is an essential consideration for their use in many systems. Essentially, chemical reactivity is the chief concern of filler users.

The presence of fillers has effects on the economic aspects, on physical, rheological, optical, electrical and thermal properties of composites.

When the use of fillers is first considered, the aim is usually to reduce the cost of a compound.

Fillers modify the modulus, the tensile strength, flexural strength, elongation, tear resistance, impact strength, compressive strength, creep and stress relaxation, hardness, coefficient of friction, abrasion resistance. Coefficient of thermal expansion, thermal conductivity, deflection temperature, fire retardancy, refractive index, dielectric constant, and dielectric strength are modified, depending on the amount and properties of the filler.

The most prominent physical effect of fillers is the stiffening, or modulus increase, which they cause in composites [2.32].

Mineral classification has been systematized and used in Table 2.22.

THE COUPLING AGENT

It has been established that the properties of composites depend, to a very large extent, on the degree of interfacial bonding between the fibres and the matrix.

Applications of a third component for surface modification of filler and reinforcements in composites have generally been directed toward improved mechanical strength and chemical resistance.

The bonding created by an adhesion promoter occurs by virtue of its having functional groups which can react with the substrate and the matrix. The most common case is that of silanes used as coupling agent for composites based on glass fibres [2.34]. Figure 2.7 presents a schematic of interfaces in composites.

In practice, a strong bond between fibres and polymer matrix is difficult to achieve owing to:

(1) Poor wettability of the polymer matrix over the very large surface area of the fibres, especially with the non-polar, high melt viscosity thermoplastics.

(2) The surface of most reinforcing fibrous materials is hydrophilic

Fibreglass reinforced plastics

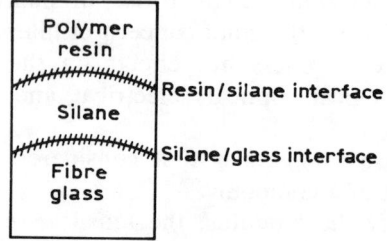

Fig. 2.7. Schematic of interfaces in composites [2.34] (Reprinted with permission from *The Role of Polymeric Matrix in the Processing and Structural Properties of Composite Materials* by J. L. Koenig & C. Chiang, 1983. Copyright Plenum Press.)

and may be coated with multimolecular layers of water which may prevent physical or chemi-adsorption of the matrix molecules. These water layers may be formed on subsequent ageing by diffusion of atmosphere moisture through the matrix.

(3) A weak boundary layer at the interface may be formed by contaminants present on the surface of the fibres (e.g. lubricants, antistatic agents in the case of glass fibres), or by an entropic separation of low molecular mass additives from the matrix.

The interfacial bond may be enhanced and the mechanical performance of composites improved by using suitable 'coupling agents' or 'adhesion promoters'.

Historically, glass fibres have been invariably used as reinforcing fillers, and unsaturated polyesters and epoxide polymers have been used almost exclusively for the matrix phase in laminated composite structures. Consequently, most coupling agents have been developed specifically for these systems.

Five basic types of coupling agents are normally used, which consist of organic functional groupings capable of co-reacting with the matrix. These invariably contain vinyl and allylic groups for polyesters, and amino groups for epoxides, together with inorganic or organo-inorganic groups, which can exert a strong physical-chemical interaction with the hydroxyl groups of the glass surface. Some examples of coupling agents are shown in Table 2.23.

In addition to chemically bridging the matrix to the fibre surface, the function of the coupling agent is to reduce the rate at which water can accumulate at the interface and destroy the bond. To do so, in fact, the water must first hydrolyse the —O—Si—O— bonds at the interface, and displace both resin and coupling agent species.

With short fibre thermoplastic composites (e.g. moulded products), however, the use of conventional coupling agents does not seem to offer great advantages, and the degree of property enhancement depends mainly on the relative natural affinity of the matrix towards the fibres. Thus hydrophilic polymers (e.g. polyamides, polysulphones, phenoxies) give better reinforcement than hydrophobic polymers such as polyolefins, polystyrene, etc.

More successful results have recently been obtained by exploiting the idea of producing a modulus gradient between the fibres and the matrix, by means of cross-linked coatings whose network density gradually decreases towards the bulk of the matrix. Various techniques have been used to obtain a cross-linked coating strongly bonded to both fibres and the bulk of the matrix and, owing to their relatively low sensitivity to temperature changes and time factors, composites with superior properties at high temperatures, and with better creep performance, have been obtained.

Systems which can cross-link the matrix at the interface only have been patented, and include the use of peroxides or chlorinated alicyclic compounds for polyolefin composites [2.35].

Other types of coupling agents, especially metallic complexes, are in the research stage.

Couts & Campbell [2.36] have studied the influence of some coupling agents, such as alkoxides of titanium and silanes, by measuring the fracture energy and flexural strength of the derived composites. The results of investigation confirm that the mechanical performance of wood fibre reinforced Portland cement composite can be altered by the use of coupling agents. Table 2.24 shows the effect of titanate on filled PP.

Boaira & Chffey [2.38] studied the effects of coupling agents on the mechanical and rheological properties of mica-reinforced PP. These authors have established that dimethyl aminoethyl methacrylate is an effective agent for the mentioned system. It increases the flexural strength, above that of samples without coupling agents, almost as much as the widely used silane coupling agent N-(4-vinyl phenyl)methyl-N'-(3-trimethoxysilyl-propyl) ethylenediamine monohydrochloride. The viscosity of the melted system is lowered by the

Table 2.23.

Common coupling agents for glass fibres/thermoset resins laminates [2.35]

Coupling agents	Recommended resin	Reaction with the glass surface
1. Vinyltrichloro-silane $CH_2{=}CH{-}SiCl_3$	Polyesters	 $+ 3HCl$ glass surface
2. Vinyltriethoxy-silane $CH_2{=}CH{-}Si(OEt)_3$	Polyesters	 $+ 3HCl$ glass surface
3. γ-aminopropyl-triethoxysilane $NH_2{-}(CH_2)_3{-}Si(OEt)_3$	Epoxides	 $+ 3EtOH$ glass surface

4. Methacrylo-chromium complex

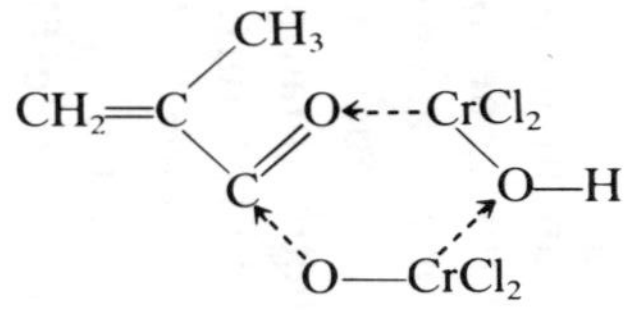

Polyesters

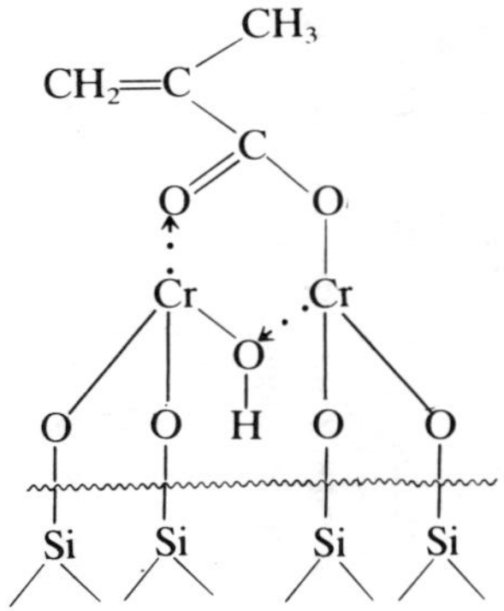

5. Allyltrichloro silane resorcinol

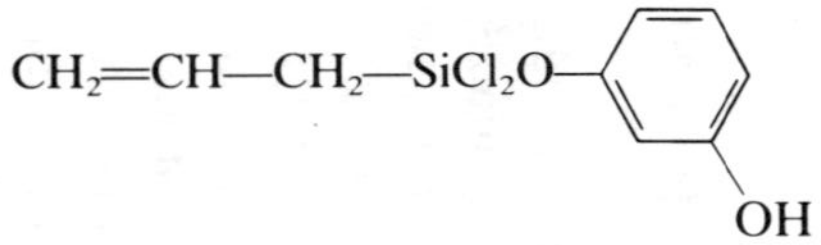

Universal

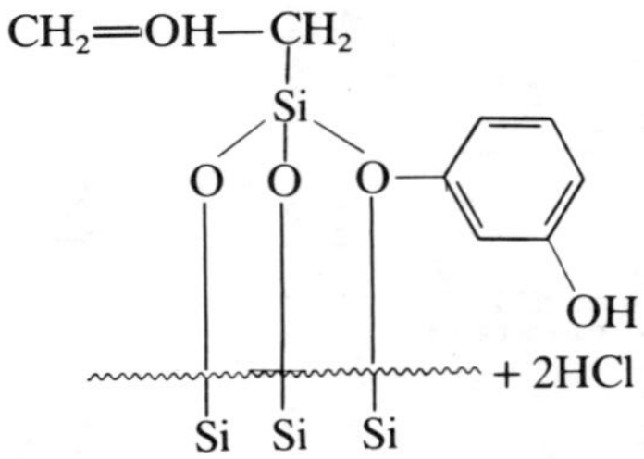

Reprinted with permission from *The Role of Additives in Plastics* by L. Mascia, 1974. Copyright Edward Arnold.

Table 2.24.
Effect of titanate coupling agent on filled polypropylene [2.37]

Material	Ultimate tensile strength, psi (MPa)		Ultimate elongation (%)	Flexural strength, psi (MPa)	
Polypropylene (PP)	4 100	(28·2)	500	7 500	(51·7)
40% Asbestos	5 000	(34·4)	Nil	12 200	(84·1)
40% Asbestos, 1% TTS	5 200	(35·8)	Nil	11 900	(82·0)
40% CaCO$_3$	2 800	(19·3)	350	7 100	(48·9)
40% CaCO$_3$, 1% TTS	3 000	(20·6)	520	5 100	(35·1)
40% Talc	2 500	(17·2)	460	6 500	(44·8)
40% Talc, 1% TTS	2 500	(17·2)	Varies	7 700	(53·0)

Source: S. J. Mente & G. Sugarman. 33rd Annual Tech. Conf. RP/Comp. Inst. SPI Section 2B. Reproduced Courtesy of the Society of Plastics Engineers, 1978.

coupling agent, so that significant reductions in energy needed for processing are possible.

Single crystalline aluminium oxide, sapphire, has been used by Paik *et al.* [2.39] as model surfaces for aluminium, on which the silane coupling agents were adsorbed from aqueous or organic solution. The study shows that the silane film is a polysiloxane network and, in some cases, cross-linking is not complete. With thinner silane films, the cross-link density appears greater than with the thicker films.

FIBRE REINFORCED COMPOSITES (FRC)

A fibre is the basic individual filament of raw material from which threads or fabrics are made. Fibres constitute one of the oldest engineering materials. Jute, hemp, flax, cotton and animal fibres have been used from the earliest days of history. Today, fibres can be organic (plant and animal), mineral, synthetic (man-made from a polymer or ceramic material), or metallic. ASTM further defines a fibre as having a length at least 100 times its diameter, with a minimum length of at least 5 mm. A fibre can be a filament or a staple. Filaments are long, continuous fibres, whereas staple fibres are less than 150 mm in length. Most natural fibres are in staple form, whereas synthetic fibres may exist in both forms.

The whisker is a man-made nearly perfect single crystal with a diameter of about 1 micrometre (μm). Because it contains but few crystalline defects in the direction of the applied load, its strength approaches the theoretical value of more than $6 \cdot 9 \times 10^3 \, \text{MN/m}^2$ (1×10^6 psi). Other materials have strengths well below this value. Man-made whiskers of many ceramic materials such as aluminium oxide (Al_2O_3 or sapphire), silicon carbide (SiC), and silicon nitride (Si_3N_4) with hardness values approaching that of diamond are being used as reinforcements in composite materials.

Continuous fibres are fibres that are essentially infinite in length and extend continuously throughout the matrix.

Discontinuous fibres are fibres less than 30 cm in length that tend to orient themselves in the direction of resin flow [2.40].

On the market we may find at present different kinds of composites known as: FRP, RRIM, SMC, TMC, BMC, XMC, etc. (Fig. 2.8).

Only the FRC which have an organic polymer matrix, known also as fibre reinforced plastics (FRP), will be discussed in the following section. In the case of continuous filaments and fibres, Table 2.25 contains examples of these materials available for reinforcing purposes, including their shape and size.

The most widely applied, so far, is the glass fibre, which paved the way for reinforced structural materials before the beginning of the Second World War. Since that time, steady improvements and continuous market growth have taken place. Glass fibre continues to be the most widely used reinforcement for polymers because of its high strength, ready availability to known specification, low density, and low cost. However, developments in various construction fields have led to a search for better materials for reinforcement, the emphasis being placed on the stiffness–weight ratio, and ability to support higher temperatures than glass. In the late 1950s, ceramic materials were, for the first time, produced in the form of whiskers; these materials (used up to now) are simple compounds such as Al_2O_3 (aluminium oxide), BeO (beryllium oxide), SiC (silicon carbide), SiO_2 (silicon oxide), and B_4C (boron carbide), having stiffness values over five times that of the best glass fibre; moreover, they are light in weight and possess good refractory properties [2.1].

At present, much work aimed at the effective utilization of whiskers in high-performance composite materials is being carried out, but the cost of manufacture is very high.

Carbon is one of the lightest elements and also the most refractory, retaining its strength up to 2000°C and above. Actually, carbon fibre

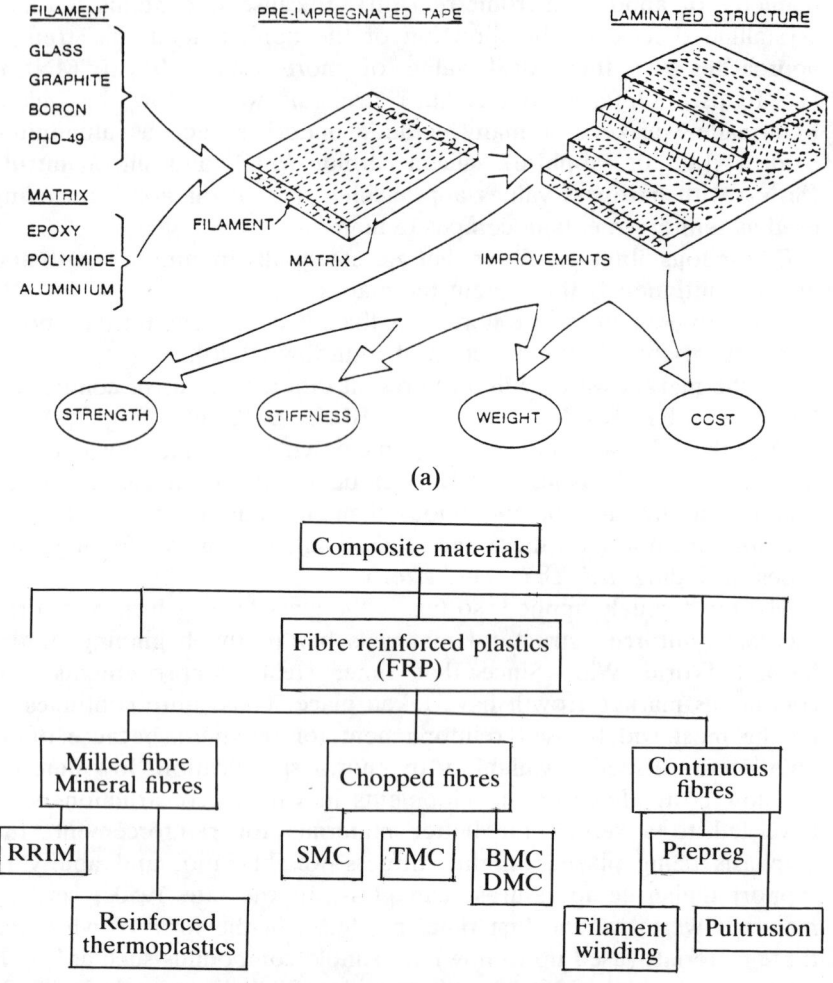

(a)

RRIM: Reinforced reaction injection moulding
SMC: Sheet moulding compound
TMC: Thick moulding compound
BMC: Bulk moulding compound

(b)

Fig. 2.8. (a) Composite materials. (b) Major types of FRP composite materials [2.40]. (Reprinted with permission from *Engineering Materials Technology* by J. A. Jacobs & T. F. Kilduff, 1985. Copyright Prentice-Hall Inc.)

Table 2.25.

Examples of various fibre forming materials [2.1]

Material	Shape and size
Glass fibres	
E-glass; YM-31-A-glass	Continuous length; round cross-sectional shape; 10 μm in diameter
SiO$_2$; S-glass	Continuous length; round cross-sectional shape; 10 μm in diameter
4-HI-glass	Continuous length; round cross-sectional shape; 10–12 μm in diameter
R-108 glass	Continuous length; round cross-sectional shape; 12 μm in diameter
S-1014 glass	Continuous length; round cross-sectional shape; 10–100 μm in diameter
Sil-Temp (98% SiO$_2$)	Continuous length; round cross-sectional shape; 10 μm in diameter
Microquartz (98% SiO$_2$)	Discontinuous length; round cross-sectional shape; 1 μm in diameter
Refrasil (95·5% SiO$_2$)	Continuous or discontinuous length; round cross-sectional shape; 10–12 μm in diameter
Alumina–silica glass	To 3 m in length; round cross-sectional shape; 5–50 μm in diameter
Quartz and silica fibres	
Fused silica	Continuous length; round cross-sectional shape; 35 μm in diameter
Fused silica, carbon coated	Continuous length; round cross-sectional shape; 25 μm in diameter
Fused silica, aluminium coated	Continuous or discontinuous length; round cross-sectional shape
Fused silica-quartz	
Ceramic fibres	
Alumina beryllia fibres National beryllia Atomics Intl. beryllia	3–12 mm in length; 7·5 μm in diameter 7 μm in diameter
Potassium titanate (Tipersul, Dupont)	0·5–1·0 mm in length; 1 μm in diameter
Zirconia fibres	

(continued)

Table 2.25.—*contd.*

Material	Shape and size
	Carbon and graphite fibres
Carbon	Round cross-sectional shape; 15 μm in diameter
Carbon, commercial	Round cross-sectional shape; 5 μm in diameter
Carbon–diamond structure, theoretical	
Graphite	
	Carbon–silica fibres
Carbon–silica	
Silica	
	Metal fibres
Aluminium	Filament staple length; 4–2 μm in diameter
Beryllium	
Boron	
Boron–tungsten filament	
Copper	
Iron	
Iron–nickel wire	Filament staple length; 6–15 μm in diameter
Magnesium	
Molybdenum	
Steel	Up to 1·25 cm in length; 5 μm in diameter
Tantalum	
Tungsten	
	Miscellaneous fibres
Boron nitride (99% boron nitride)	25–37 cm in length; 5–7 μm in diameter
Silicon carbide	

Single-crystal fibres (whiskers)
Hexagonal or rhombohedral in shape; sub-μm in diameter

Alumina
Beryllium oxide
Boron carbide
Chromium
Copper
Graphite
Iron
Nickel
Silicon carbide
SiC   3–25 mm in length; 2–8 μm in diameter
B–SiC   1–10 mm in length; 1–10 μm in diameter

Reprinted with permission from *Polymer Engineering Composites* by M. O. W. Richardson (Ed.), 1977. Copyright Applied Science Publishers.

Table 2.26.
Some new high-modulus fibrous materials [2.1]

Material	Tensile strength GN/m²	Tensile strength (psi × 10³)	Young's modulus GN/m²	Young's modulus (psi × 10⁶)	Density (g/cm³)
Alumina					
single crystal whiskers (A-axis)	10–20	(1450–2900)	700–1500	(101.5–217.5)	4.0
crystals from melt	2	(290)	460	(66.7)	The less compacted forms are of lower density
bulk (microcrystalline)	0.3	(43.5)	350	(50.7)	
sintered fibres	0.2–0.7	(29–101)	140–300	(20.3–43.5)	
Boron					
fibres deposited by thermal decomposition	2.75	(398)	400	(58)	2.3
tungsten core	3.45	(500)	414	(60)	Fibres have higher density depending on proportion of [metal
Boron nitride					
fibres	0.3–1.4	(43.5–203)	28–80	(4–11.6)	2.6
Carbon					
graphite, parallel to planes	—	—	1000	(145)	2.27 (theoretical)
whiskers	—	—	700	(101)	1.8–2.0 (practical values)
fibres (by pyrolysis)	2–3	(290–435)	230–550	(33.3–80)	2.3 (theoretical)
Silicon carbide					
whiskers	9.5	(1377)	480–550	(69.6–80)	1.8–2.0 (fibres only)
fibres deposited by thermal decomposition	1.7–2.5	(246–362)	400–450	(58–65)	3.19; Fibres have higher density depending on proportion of metal core
Silicon nitride					
whiskers	5–7.7	(725–1116)	350–380	(50.7–55.1)	3.2
Aromatic Polyamide fibres (e.g. Kevlar 29 and 49)	up to 3.6	(522)	60–131	(8.7–18.9)	1.44–1.45
N.B. for comparison					
E-glass filaments	2.75–3.45	(400–500)	68.9–72.4	(10–10.5)	2.54–2.55
Stainless-steel wire	4.0	(580)	200	(29)	7.9

rather than carbon in whisker form is considered as an important reinforcing agent of composites. Its organic precursors are basically low in cost and freely available, implying that, for large-scale production, carbon fibre should be competitive in terms of price for a given performance, when compared with other reinforcing materials.

As well as glass, graphite and carbon fibres, other types of fibres are now used for manufacturing composites, such as metals, asbestos, etc.

As can be seen from Table 2.26, fibres have extremely high tensile strengths and moduli.

Since fibres are the highest strength materials, they are the best reinforcements for many applications, ranging from improvement of the strength of low-cost materials or composites to acceptable levels, to the production of superior structures.

Ceramic fibres are made from metal oxides, and feature combinations of properties not previously available. The major advantages are resistance to extremely high temperatures (1370–1650°C), coupled with higher modulus and excellent compressive strength. Their composition gives them exceptionally good chemical resistance, and their small fibre diameter lends flexibility and workability to the fibre [2.1].

GLASS FIBRE MANUFACTURE

There are several basic methods employed for obtaining glass fibres from hot melts to yield products which have some role in reinforcing plastics. These include:

—continuous filament fibres;
—staple fibres;
—blown fibres for mats;
—mechanically drawn fibres for mats.

Continuous filament glass fibres are usually produced by: a marble melt process or a direct melt process.

The marble process is the older one, and in this technology glass is melted in a separate furnace and formed into marbles, which are partially annealed and graded. Then they are fed into heated platinum bushings, where they are remelted and transformed into filaments, being drawn away rapidly at speeds approaching 4000 m/min. There is nothing magical about marbles, except that they are easily formed

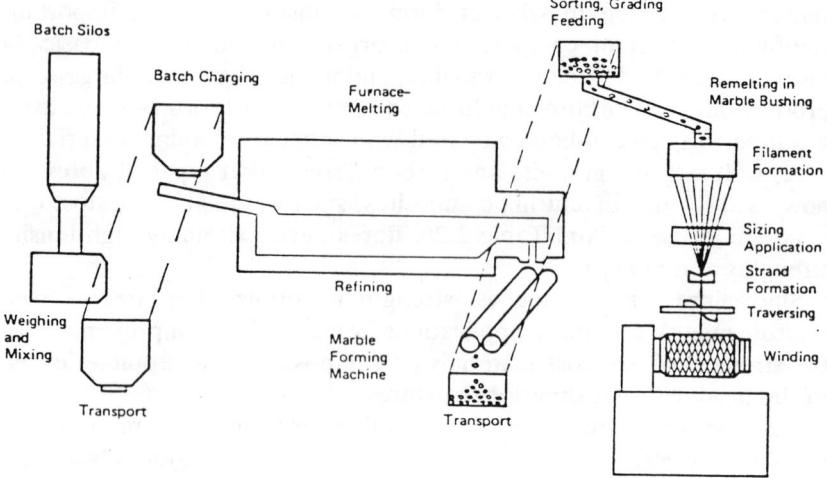

Fig. 2.9. Schematic diagram of marble process [2.41]. (Reprinted with permission from *Handbook of Fillers and Reinforcements for Plastics* by H. S. Katz & J. V. Milewski (Eds), 1978. Copyright Litton Educational Publishers Inc.)

mechanically, and roll by gravity through grading and feeding contrivances.

As technology improved, the remelting step was superseded by installing in-line fibre-producing bushings in the appropriately designed forehearths of a glass tank furnace, and the process known as direct melt glass fibre production evolved. Figures 2.9 and 2.10 illustrate, schematically, the marble melt process and direct melt process, respectively.

Whereas marble bushings were originally capable of 3–4 kg/h, production from larger direct melt bushings may now approach 45 kg/h.

Continuous glass filaments produced commercially range from 2–3 μm to 25 μm in individual fibre diameters. Marble processes are still utilized, being more satisfactory for fabrication of finer filaments. They also permit use of glass marbles, the quality and production efficiency of which have been predetermined, and thus, poor or undesirable glass can be eliminated.

The direct melt process is preferred because it eliminates the remelting step necessary in marble melt, but by comparison, marble

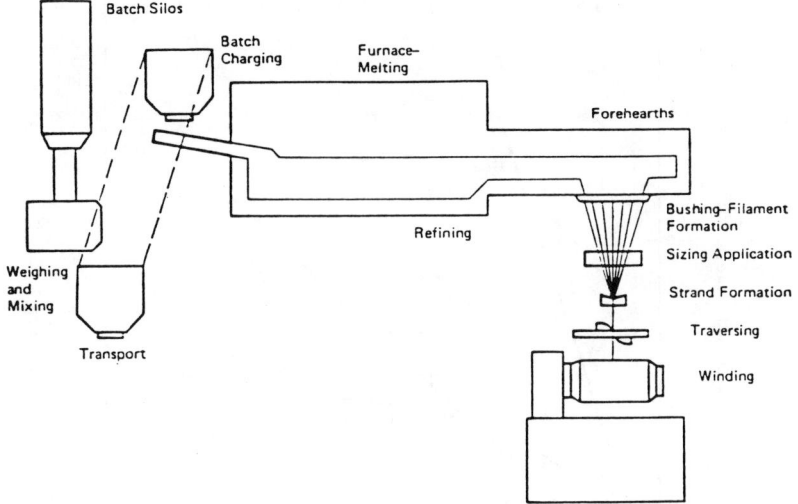

Fig. 2.10. Schematic diagram of direct melt process [2.41]. (Reprinted with permission from *Handbook of Fillers and Reinforcements for Plastics* by H. S. Katz & J. V. Milewski (Eds), 1978. Copyright Litton Educational Publishers Inc.)

bushings obviously favour easy changes to glasses of different composition.

The fineness of a continuous filament is expressed in a special unit known as denier, or strand length to weight relationship, hence filament diameter is determined and influenced by the following:

—glass composition;
—viscosity of the melted glass;
—number of tips or filaments;
—drawing temperature;
—efficacity of cooling;
—winding speed.

For processing composites, glass fibres have to undergo some special treatments; they are treated with two distinct and separate classes of sizing materials:

(a) A starch–oil emulsion type which has humectant properties,

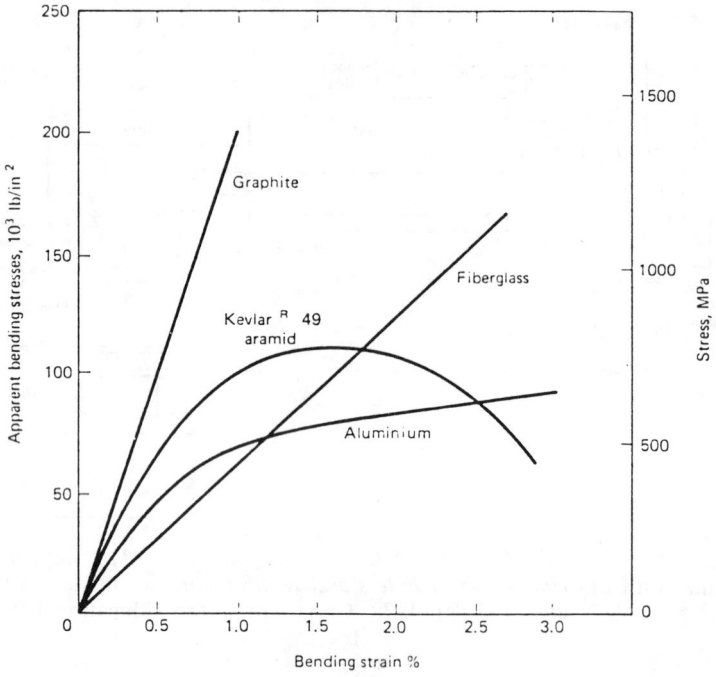

Fig. 2.11. Bending strain [2.40]. (Reprinted with permission from *Engineering Materials Technology* by J. A. Jacobs & T. F. Kilduff, 1985. Copyright Prentice-Hall Inc.)

absorbing water to provide lubrication plus resistance to abrasion in such down-stream processes as twisting, plying weaving, cording, etc.

(b) A plastic emulsion cross-link type of sizing, generally for direct use in composites.

These treatments, including the use of coupling agents, are used to make the reinforcing agent compatible with the matrix [2.41].

Figure 2.11 shows the position of glass fibres in relation to other materials from which reinforcing fibres are being made. It is a plot of bending stress versus bending strain (stress–strain diagram), and compares glass fibres in an epoxy matrix with its competitors [2.40].

One property used to show the effectiveness of the strength of a fibre is its specific strength, that means the ratio of tensile strength to

Table 2.27.
Specific strength [2.40]

Material	Weight density, ρ (kN/m³)	Tensile strength, S GN/m² (psi × 10⁻³)	Specific strength, S/ρ km (miles)
S-glass	24·4	4·8 (696)	197 (122)
E-glass	25·0	3·4 (493)	137 (84·9)
Boron	25·2	3·4 (493)	137 (84·9)
Carbon and graphite	13·8	1·7 (246)	123 (76·2)
Beryllium	18·2	1·7 (246)	93 (57·6)
Steel	77	4·1 (594)	54 (33·4)
Titanium	46	1·9 (275)	40 (24·8)
Aluminium	26·2	0·62 (90)	24 (14·9)

Reprinted with permission from *Engineering Materials Technology* by J. A. Jacobs & T. F. Kilduff, 1985. Copyright Prentice Hall Inc.

weight density. Table 2.27 and Fig. 2.12 list the specific strength for some typical fibres used in composites.

Another indicator of the special properties of fibre composites, in particular the effectiveness of a fibre, is specific stiffness. It is a ratio of the modulus of elasticity (or tensile stiffness) to the weight density of the fibre. Table 2.28, using the same materials as in Table 2.27, and Fig. 2.13 show the average specific stiffness of such fibre materials, starting with materials having the highest values. Note that graphite, boron and carbon have values almost six times that of steel. Graphite is known to have six times the strength of steel with six times less weight. These values of both strength and stiffness of materials in fibre form compared to the same materials in bulk form are most significant [2.40].

Unlike metals, glass—as a brittle material—is characterized by a linear plastic, stress–strain relationship to failure, and in fibre form has a very high strength, far higher than any of the traditional constructional materials. E glass, for instance, in fibre form, exhibits an average tensile strength of about 3·45 kN/mm² (500×10^3 psi) (Table 2.29) in the newly drawn and untouched condition, which is several times the strength of the metals that are used structurally.

The modulus of elasticity, on the other hand, is not exceptional, this property for E glass being about 72·4 kN/mm² (105×10^6 psi), a value

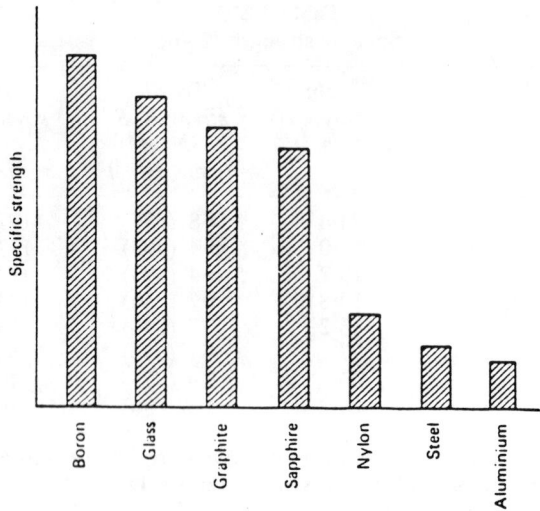

Fig. 2.12. Relative specific strength of typical materials used in composites [2.40]. (Reprinted with permission from *Engineering Materials Technology* by J. A. Jacobs & T. F. Kilduff, 1985. Copyright Prentice Hall Inc.)

Table 2.28.
Specific stiffness [2.40]

Materials	Weight density, ρ (kN/m^3)	Modulus of elasticity, E GN/m^2 $(psi \times 10^{-3})$	Specific stiffness, E/ρ mm $(miles \times 10^3)$
Graphite	13·8	250(36)	18(11·1)
Beryllium	18·2	300(43·5)	16(9·9)
Boron	25·2	400(58)	16(9·9)
Carbon	13·8	190(27·5)	14(8·68)
S-glass	24·4	86(12·5)	3·5(2·17)
E-glass	25·0	72(10·4)	3·0(1·86)
Steel	77	207(30)	2·7(1·67)
Titanium	46	115(16·6)	2·6(1·61)
Aluminium	26·2	73(10·5)	2·8(1·73)

(Reprinted with permission from *Engineering Materials Technology* by J. A. Jacobs & T. F. Kilduff, 1985. Copyright Prentice Hall Inc.)

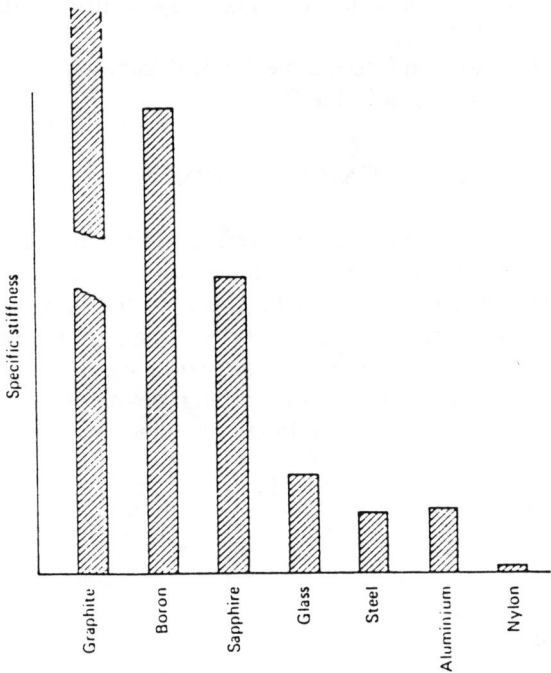

Fig. 2.13. Relative specific stiffness of typical materials used in composites [2.40]. Note: Graphite has a specific stiffness of 5000 versus steel with 25 in units of 10^6 N m/kg (using mass density in kg/m^3). (Reprinted with permission from *Engineering Materials Technology* by J. A. Jacobs & T. F. Kilduff, 1985. Copyright Prentice Hall Inc.)

Table 2.29.
Composition and properties of E and S glass [2.42]

Glass	Composition by wt of major components					Ultimate tensile strength		Modulus of elasticity in tension or compression	
	SiO_2	Al_2O_3	CaO	MgO	B_2O_3	kN/mm^2	$(psi \times 10^{-3})$	kN/mm^2	$(psi \times 10^{-3})$
E glass	54	14	17·5	4·5	10	3·45	(500)	72·4	(10·4)
S glass	65	25	0	10	0	4·60	(657)	85·5	(12·3)

Small quantities of sodium, potassium and iron may be present in these glasses.
Reprinted with permission from *GRP in Structural Engineering* by M. Holmes & D. J. Just, 1983.
Copyright Applied Science Publishers.

similar to that of aluminium. The Poisson's ratio for glass fibres is in the region of 0·2 [2.42.]

The composition and some mechanical characteristics for both E and S glass are shown in Table 2.29.

CARBON FIBRES

Carbon fibre is the most recent application for the reinforcement of polymers or even metals, to yield composite materials of exceptionally high specific stiffness and strength. The fibre for such uses must have greatly increased mechanical properties compared with the grades employed in previous applications. To achieve these properties, special manufacturing processes have been developed.

The raw materials for carbon fibres manufacture are:

—cellulose, cellulose derivatives;
—poly-(acrylonitrile), —PAN;
—pitch;
—asphalt;
—wool;
—lignin, etc.

All organic polymers will yield a carbon end-product after decomposition, but for the production of carbon or graphite fibre, the macromolecular compound must decompose without melting. Carbon fibres have been made successfully from polymers such as PVC, poly(vinyl alcohol) (PVA), polybenzimidazole, or polyimide; in addition, certain kinds of thermosetting resins based on phenolic groups may be used.

Two types of textile fabrics are normally produced, one consisting of carbon filaments, the other of graphite. In the former, the filaments have a fine carbon-type structure; their purity lies in the range of 90–98% carbon. The graphite fibres, on the other hand, have a graphite carbon content of 99·9%.

Cellulose or rayon is the most widely used precursor for making carbon fibres. To convert cellulose to carbon fibre, it is necessary to apply a multi-stage process which needs gradually higher temperatures, up to around 3000°C. The basic stages are the following:

—physical desorption of water;
—dehydration of cellulose units;

—thermal cleavage of the cyclosidic linkage and scission of the other —C—O— bonds, and some $>C\cdot C<$ bonds via a free radical process;
—aromatization.

Cellulose does not melt during decomposition, and yields 15–30% by weight of carbon fibre. The conversion of cellulose in this process implies the transformation of the chains based on glucose units into chains based on aromatic units:

$$\ldots\text{—(glucose)—(glucose)—(glucose)—(glucose)—}\ldots$$
$$\downarrow$$
$$\ldots\text{—AU—AU—AU—AU—AU—AU—AU—}\ldots$$

AU = aromatic unit.

Figure 2.14 shows the mechanism, proposed by Tang and Bacon [2.43], for the conversion of cellulose to graphite, and Fig. 2.15 presents the technological line for the production of carbon-graphite filaments from a cellulose derivative. The main operations carried out in the electrical furnaces are:

—oxidizing (at 260°C),
—carbonizing (at 1000–2000°C),
—graphitizing (at 2500–3000°C).

When PAN is heated in vacuum or air to over 200°C, a number of reactions take place, resulting in the formation of some by-products which are liberated as volatiles. The carbon ring structure produced during the carbonization of PAN fibres under tension has a crystalline form (Fig. 2.16) [2.45]. The dehydrogenation and denitrogenation steps are presented in Fig. 2.17.

A complete flow diagram for the preparation of carbon fibres from PAN is given in Fig. 2.18.

Increased amounts of heat treatment, and in particular graphitization, develop the crystals, reduce fibre porosity, and impart high modulus values. The graphitization stage is necessary to produce high modulus fibre, and temperatures up to 2800°C are involved.

Figure 2.19 indicates that, for the high modulus carbon fibre, the maximum elongation is approximately 0·35%, compared with 1% for the high-strength variety, and 4% for E glass. By comparison, an epoxy polymer will normally elongate up to 5–6% before breaking

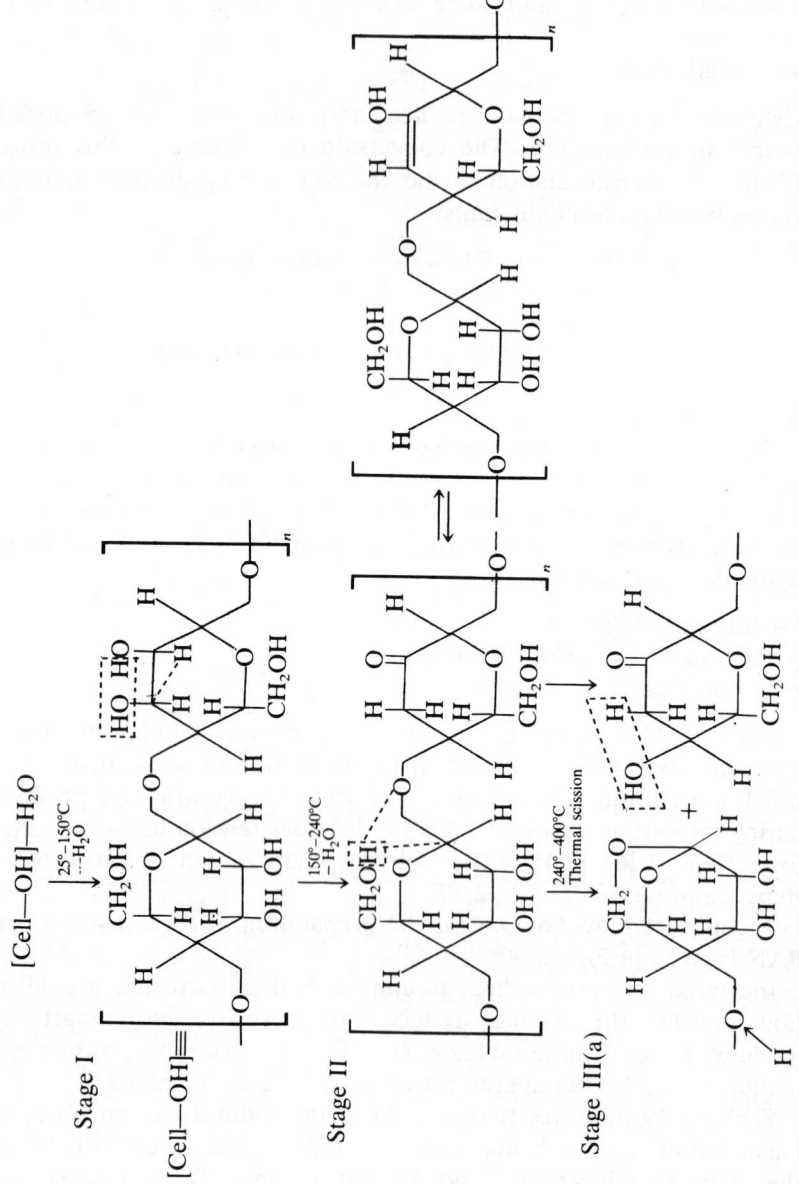

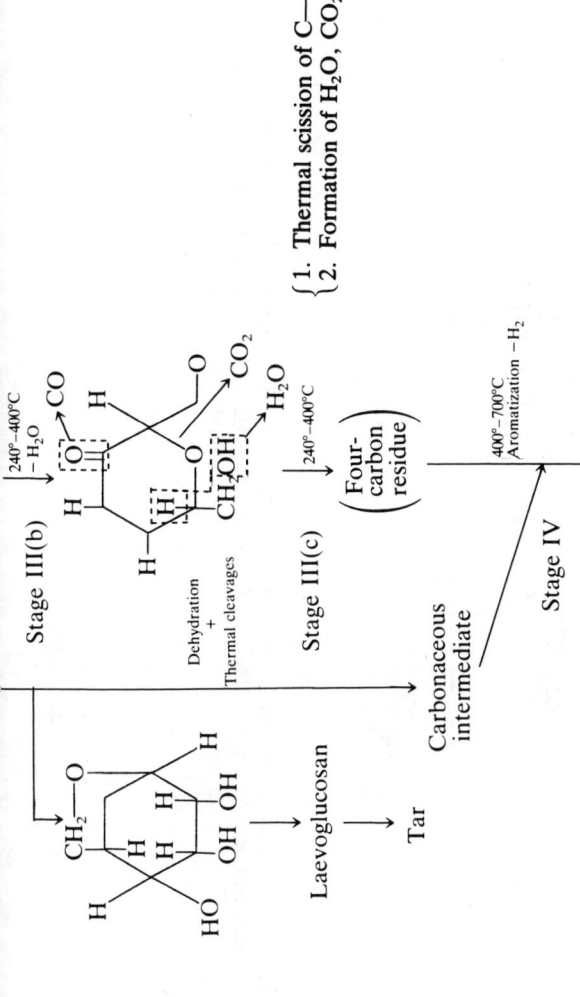

Fig. 2.14. Proposed mechanism for the conversion of cellulose to carbon [2.43]. (From Tang and Bacon, courtesy Maxwell Scientific International Inc.)

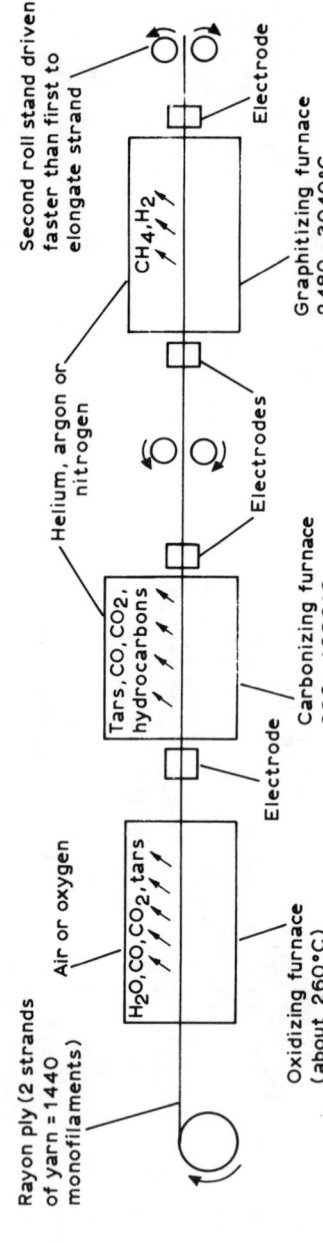

Fig. 2.15. Production line for carbon-graphite filaments from rayon [2.44]. (Reprinted with permission from *Handbook of Fillers and Reinforcements for Plastics* by H. S. Katz & J. V. Milewsky (Eds), 1978. Copyright Litton Educational Publishing Inc.)

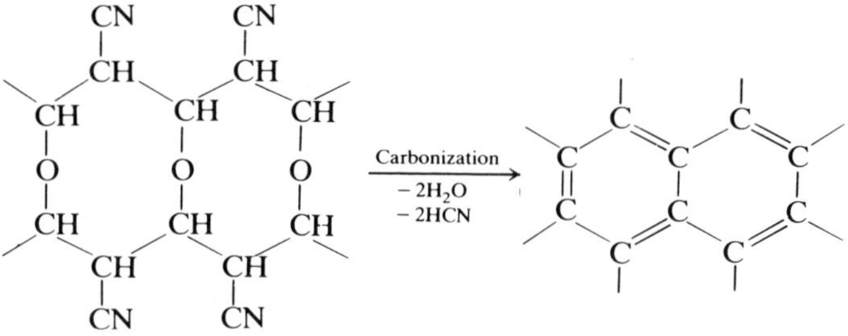

Fig. 2.16. Formation of a carbon ring structure by carbonizing oxidized PAN fibre; the carbon rings will have an orientation which is dependent on that of the original PAN chains and their subsequent oxidized form [2.43]. (Reprinted with permission from *Carbon Fibres in Composite Materials* by R. M. Gill, 1972. Copyright The Plastics Institute.)

occurs. The important advantage of carbon fibres, therefore, is that they can withstand very high stresses without undergoing significant stretching, and hence a matrix material used with carbon is subjected to minimum stresses [2.43].

X-ray and electron diffraction studies of the fibre structure of PAN-based carbon fibres showed that these fibres consisted of long primary units lying parallel to the fibre axis, and bonded together to form a stretched network of branched fibrils that apparently run the full length of the fibre. The fibrils have a width of about 10 nm. A schematic representation of the structure proposed by R. Perret and W. Ruland [2.45] is given in Fig. 2.20.

The future of carbon fibres is under constant evaluation, not only by producers, sellers and users, but also economists, for their strategic importance. The published data have been reappraised many times, oscillating from overenthusiasm to gloom. At present, there is a seeming worldwide frenzy of announcements concerning new openings. It remains to be seen if all these prospects will ever be achieved. The demand for carbon fibres has displayed a remarkable growth of about 50% annually in recent years. The forecasts on market development concerning carbon fibres differ widely from 4000 to 22 000 t/year in 1990 [2.45].

HYBRID SYSTEMS

The most important advantages of carbon-fibre composites over metals or GRP, result from their high values for specific strength and

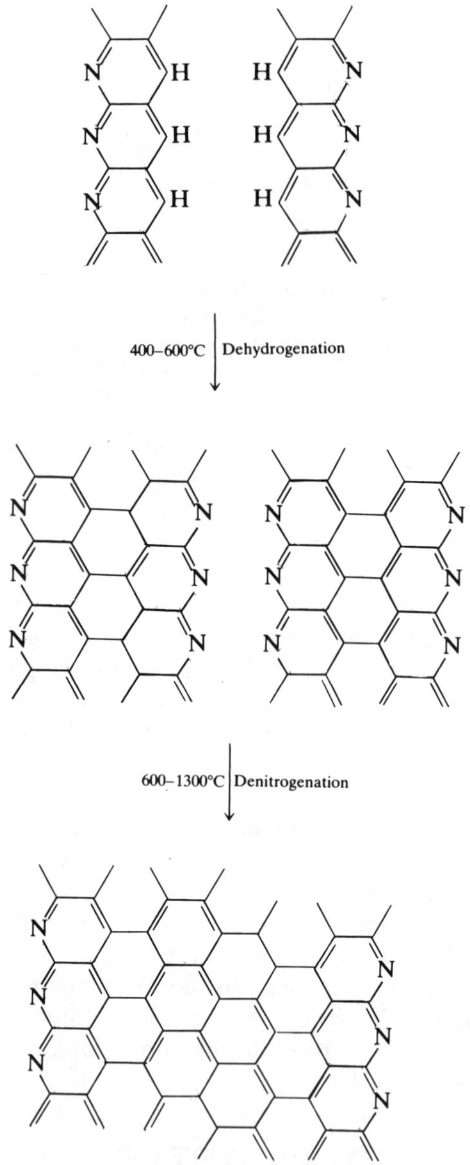

Fig. 2.17. Schematic representation of carbon fibre preparation from PAN. (From Goodhew *et al.*, *Mater. Sci. Engng*, **17** (1975) 3.) [2.45]

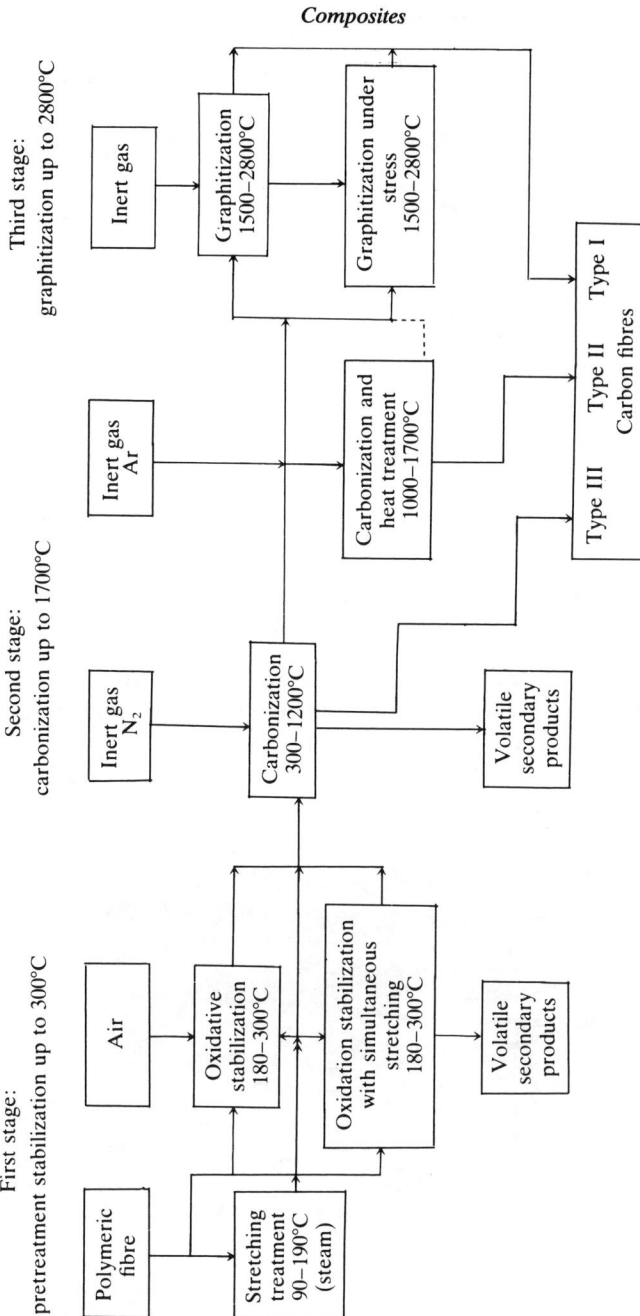

Fig. 2.18. Schematic representation of carbon fibre preparation from PAN. (From E. Fitzer & D. J. Muller, *Chem. Ztg. Chem. Appel,* **96** (1972) 20.) [2.45]

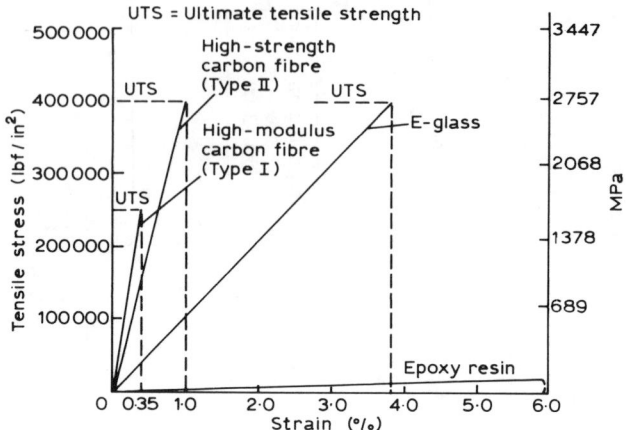

Fig. 2.19. Comparative stress–strain curves for carbon fibres, E-glass and an epoxy resin matrix material; Type I fibre possesses maximum stiffness or modulus, whilst Type II fibre possesses maximum strength [2.43]. (Reprinted with permission from *Carbon Fibres in Composite Materials* by R. M. Gill, 1972. Copyright The Plastics Institute.)

specific stiffness, coupled with very good fatigue strength. At the present time, the cost of carbon fibre is high compared to conventional materials, but in the fabrication of a component it is possible to replace a proportion of the lower-priced material with carbon fibre to obtain improved properties [2.43].

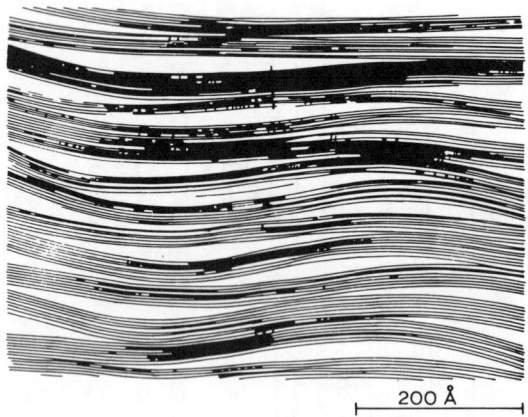

Fig. 2.20. Schematic representation of fibrillar carbon fibre structure. (From R. Perret and W. Ruland, *J. Appl. Cryst.*, **3** (1970) 525.) [2.45]

Table 2.30.
Relative mechanical properties of composites made with E-glass and carbon fibre [2.43]

Materials	Tensile modulus MN/cm^2 $(10^6 lbf/in^2)$	Specific gravity	Specific tensile modulus MN/cm^2 $(10^6 lbf/in^2)$
E-glass fibre composite (glass 50% by volume)	1·4(2·0)	1·66	0·8(1·2)
Carbon fibre tows with parallel alignment (high strength fibre 50% by volume	12·6(18·0)	1·53	8·1(11·8)

Reprinted with permission from *Carbon Fibres in Composite Materials* by R. M. Gill, 1972. Copyright The Plastics and Rubber Institute.

Table 2.30 indicates the relative mechanical properties of a glass-fibre composite in the form of a mat or rovings, compared with the high-strength form of PAN-based carbon fibre aligned tows.

It may be seen that the specific stiffness of the carbon-fibre composite is 9·7 times that of the glass-fibre product. It may be supposed that 453 g of carbon-fibre composite can always replace 4394 g of glass–fibre composite in a component such as a beam.

This is only true if the thickness of the carbon-fibre composite cladding is a small fraction of the total depth of the beam. Where the relative thickness of the carbon-fibre cladding is large, some of the material is used ineffectively, by virtue of the 'depth cube' law for beam strength, and the overall advantage is reduced.

The amount of carbon fibre to be used in any particular application depends on the price that may have to be paid to save weight.

Table 2.31 gives details of the weight and cost of a series of GRP plates clad on both sides with high-strength carbon-fibre reinforced sheets.

As seen in the table, a small addition of carbon fibre gives the greatest benefit. Thus, a 5% addition shows a 22% weight saving; in contrast, 50% addition gives only a 52% weight saving. The reason for this is associated with the 'depth cube' law as applied to the plate. Hence it is possible to estimate the amount of weight saving in a

Table 2.31.

Effect of varying proportions of carbon-fibre cladding to GRP in terms of weight saved and cost [2.43]

Fraction of thickness of plate formed from carbon-fibre composite	Thickness of plate (cm)	Total cost of materials for plate (£/m²)		Weight of plate (kg/m²)	Savings of weight made possible by the use of carbon fibre (%)	Cost of saving 1 kg of wt in the plate (£)	
		Carbon fibre at £100/kg	Carbon fibre at £35/kg			Carbon fibre at £100/kg	Carbon fibre at £35/kg
0	1·00	11	11	16·6	0	0	0
0·05	0·78	49	22	12·9	22	10	3
0·1	0·68	78	30	11·2	32	12	3·5
0·2	0·59	129	46	9·6	42	17	5
0·5	0·50	265	90	8·0	52	30	9
0·7	0·48	359	119	7·5	55	38	12

Notes:
1. The carbon cladding is in the form of sheets containing 60% by volume of fibre.
2. The glass-strand mat consists of randomly oriented fibres of 35% by volume and costing £0·8/kg.
3. In each case the resin matrix is an epoxy.
4. All plates are of constant stiffness and are considered to be bent in one direction only.
Reprinted with permission from *Carbon Fibres in Composite Materials* by R. M. Gill, 1972. Copyright The Plastics and Rubber Institute.

component containing varying amounts of carbon fibre. These data may also be used to ascertain the overall cost of the component and indicate the optimum proportion of carbon fibre to be used [2.43].

Due to its very good properties, in recent years carbon fibre won important applications in civil engineering and a lot of other domains. With respect to thermal behaviour of carbon-fibre reinforced composites, Pilling *et al.* [2.46] studied their thermal conductivity and Loos & Springer [2.47] established how thermal spines affect the moisture absorption and the mechanical properties of different graphite–epoxy composites. The thermal conductivity at very low temperatures was studied by Radcliffe & Rosenberg [2.48].

The transverse properties of the unidirectional carbon-fibre composites are significantly enhanced by the presence of 20% of polymethane in the matrix (epoxy) without, apart from a decrease in the shear modulus, any marked change in the properties. This could prove useful in the applications of carbon-fibre composites. Results for glass-fibre materials are less dramatic, possibly because of poorer adhesion of glass fibre to polyurethane [2.49].

Short & Summerscales [2.50–2.52] have published an interesting review of the literature on carbon fibre and glass fibre hybrid reinforced plastics (CGHRP). The incorporation of two or more fibres within a single matrix is known as hybridization, and the resulting material is generally referred to as 'hybrid' or 'hybrid composite'. The review covers fabrication techniques, design, cost effectiveness, thermal properties, mechanical properties, and applications. The tensile mechanical properties of CGHRP in an epoxy matrix have been evaluated by Manders & Bader [2.53, 2.54] over a range of glass/carbon ratios and states of dispersion of the two phases. Different aspects of carbon-fibre reinforced plastics are discussed in some other papers [2.55–2.64].

Carbon–graphite filaments are relatively inert to most chemicals and resistant to corrosion under ambient conditions. The surface is subject to attack by strong oxidizing agents and halogens, especially at elevated temperatures. This has been a means for surface treatment of the high performance filaments in order to improve the interlaminar shear strength, which was a severe problem with early graphite reinforced composites. In general, carbon and graphite are resistant to common alkaline solutions at all concentrations and temperatures, and to the aqueous solutions of most mineral salts up to their boiling point [2.44].

Asbestos Fibre

Asbestos is the common name given to a number of naturally occurring, hydrated mineral silicates that possess a crystalline structure, and that are incombustible in air and separable into filaments. There are four main types of asbestos, all of which are chemically different and hence have different properties and applications. These are: chrysotile, amosite, crocidolite and anthophyllite. In order to determine the presence and type of asbestos used, it is necessary to examine the bulk sample of the material by microscopy or X-ray analysis.

Asbestos is widely used because it is a relatively cheap material with special chemical and physical properties which make it virtually indestructible, such as: chemical resistance, particularly to acids, fire resistance, mechanical strength, high length to diameter ratio, flexibility, and good friction and wear characteristics.

Among other applications in construction, asbestos fibres are used as a reinforcing agent for cement or polymer matrices.

Chrysotile asbestos occurs in the form of a bundle consisting of fine fibrils characterized by high strength and stiffness. It is soft and flexible, resulting in low wear on processing equipment. Its modulus of elasticity is more than twice that of glass fibres, while the two fibres have similar density and strength properties. Fibre attrition by processing and reprocessing of asbestos filled compounds is low compared with breakage of brittle glass fibres. In addition, chrysotile asbestos imparts fire resistance to polymer matrices, while glass fibres are inert in this respect [2.65].

Kacir & Narkis [2.66] determined mechanical and thermal properties of asbestos-reinforced rigid PVC products. The authors emphasized that the significant advantages of these FRC include higher stiffness, heat distortion, notched Izod impact, and dimensional stability. In addition, the products can be classified as non-burning, non-dripping materials. Table 2.32 contains some properties of chrysotile asbestos-reinforced PVC. 7RF and 4T are grades of asbestos fibres with different lengths. Since all types of asbestos are mineral silicates, they have good heat and weathering resistance. Chrysotile is a highly hydrated magnesium silicate, and its structure actually consists of alternate layers of silica tetrahedra and magnesium hydroxide.

The resistance of asbestos to weathering is well known, and it is one of the reasons for many of its uses in construction. Asbestos fibre is not affected by sunlight, ultra-violet light, oxidation, or ozone. It is often used because of these outstanding characteristics.

Table 2.32.
Properties of chrysotile asbestos-reinforced PVC (injection moulded specimen) [2.66]

Property	Unreinforced	Asbestos-reinforced			
		30 phr[a] (7RF)[b]	50 phr (7RF)	30 phr (4T)[b]	50 phr (4T)
Specific gravity (ASTM D792)	1·387	1·511	1·566	1·517	1·587
Tensile strength (ASTM D638) MN/m² (psi)	45·5 (6 597)	50·3 (7 293)	49·0 (7 105)	67·6 (9 802)	66·2 (9 599)
Tensile modulus (ASTM D638) GN/m² (psi × 10⁻³)	3·0 (435)	5·5 (797)	7·3 (1 058)	8·7 (1 261)	10·1 (1 464)
Tensile elongation (ASTM D638) (%)	—	2·7	1·52	1·79	1·26
Flexural strength (ASTM D790) MN/m² (psi)	7·8 (1 131)	860 (124 700)	830 (120 350)	1 090 (158 050)	1 100 (159 500)
Flexural modulus (ASTM D790) GN/m² (psi × 10⁻³)	3·1 (449)	5·7 (826)	8·0 (1 160)	7·4 (1 073)	8·8 (1 276)
Izod impact strength (ASTM D256) kJ/mm (ft-lb/in)	6·23 (0·33)	6·62 (0·35)	7·37 (0·39)	5·87 (0·31)	6·67 (0·35)
Heat distortion temperature (°C) (ASTM D648) (loading of 1·82 MN/m²)	61	65·5	66	—	—
Flammability (ASTM D635-72)	Non-burning, dripping	Non-burning, non-dripping	Non-burning, non-dripping	Non-burning, non-dripping	Non-burning, non-dripping

[a] phr = parts per hundred.
[b] 7RF and 4T are grades of asbestos fibres having different lengths.
Reproduced from Kacir, L. & Narkis, M., *Composites* 1979, Jan, 31–6, by permission of the publishers, Butterworth & Co. Ltd.

Some of the physical properties of commercial types of asbestos are shown in Table 2.33.

As already mentioned, the strength properties of asbestos such as tensile strength and modulus of elasticity, are very good. Because of the small diameter of the fibrils, it is difficult to get accurate measurements of the strength properties, so there is quite a range reported in the literature. Not being a very hard material, it usually does not cause much wear to processing equipment.

Asbestos is available in many forms, and all of these can be used to reinforce cement or polymers. Much of the literature giving data on asbestos in polymers does not identify the form in which asbestos is used, so the true meaning of the data is obscure [2.65].

Husslein & Fallick [2.67] have realized colloidal asbestos ceraplasts which are particularly useful where transparency or translucency is desired in the composite. Since the diameter of colloidal chrysotile particles, 300 Å, is well under the wavelength of light, transparent polymers, such as polyethylene and polypropylene, can be reinforced with little loss in clarity.

Metal Fibres

Metal fibres as fillers or reinforcements for polymers possess two distinguishing features, namely metal characteristics and a controlled fibre geometry.

Metal fibres are available in a variety of metals and alloys, and in various conditions resulting from heat treatment or working. As a consequence, there is a choice of a wide range of products, and a high degree of flexibility in the application of the product to metal–polymer composites. The disadvantages of metallic fibres are related to their cost and density. Their properties are dependent on both the raw material and the process by which they are manufactured.

There are two known groups of mechanical processes for producing metal fibres:

(1) those which involve attenuation, e.g. wire drawing;
(2) those which rely on the sub-division of a prefabricated metal form, e.g. slitting, broading, shaving, turning, etc.

The chemical, physical and mechanical behaviour of metal fibres as a class of materials will generally approximate the bulk material properties. Exceptions may arise due to the influence of the fiberizing

Table 2.33.
Physical properties of asbestos [2.65]

Property	Chrysotile	Crocidolite	Amosite	Anthophyllite
Colour	White to grey	Blue	Brown	Brown to grey
Tensile strength, psi (GPa)	300 000 (2·06)	500 000 (3·44)	160 000 (1·10)	350 000 (2·41)
Modulus of elasticity, psi (GPa)	$23·2 \times 10^6$ (159·9)	$27·1 \times 10^6$ (186·8)	$23·6 \times 10^6$ (162·6)	$22·5 \times 10^6$ (155·1)
Hardness (Mohs)	2·5–4·0	4·0	5·5–6·0	5·5–6·0
Flexibility	Good	Fair	Poor	Poor
Specific gravity	2·4–2·6	3·2–3·3	3·1–3·2	2·9–3·2
Specific heat, Btu/lb °F (J/kg K)	0·266 (1 113)	0·210 (877)	0·193 (906)	0·210 (877)
pH	10·3	9·1	9·1	9·4
Refractive index	1·5–1·55	1·70	1·64	1·61
Fibril diameter (Å)	160–300	600–900	600–900	600–900
Surface area BET (m²/g)	17–60	9–10·5	8–9	6–7
Coefficient of cubical expansion (°F⁻¹)	5×10^{-5}	—	—	—
Charge in water	Positive	Negative	Negative	Negative
Isoelectric point	11·3–11·8	—	—	—

Reprinted with permission from *Handbook of Fillers and Reinforcements for Plastics* by H. S. Katz & J. V. Milewski (Eds), 1978. Copyright Litton Educational Publishing.

process, or as a result of the diminished cross-section and increased length [2.68].

Deviations from bulk material properties, which occur as a result of the fiberizing process, are summarized in Table 2.34.

A brief listing of some of the properties where physical size may have a major influence is presented in Table 2.35.

In Bigg's paper [2.69], the data which are presented show that the critical volume loading of a metallic filler, needed to induce electrical conductivity in a polymer matrix, can be reduced by adding the metal as randomly dispersed fibres. Composites exhibiting resistivities below 200 Ω cm have been prepared with as little as 7·7 vol. % aluminium fibres. At such low filler loadings, the mechanical properties of the composite are similar to those obtained with an identical loading of milled glass fibres.

The fracture of composites of ductile metal wires in a polymer matrix has been studied by Bowling & Groves [2.70], and a model of the debonding and pull-out processes for individual wires has been found to be applicable.

Aramid Fibres

Aramid fibres are based on aromatic polyamides, and the only commercially available fibre has the trade name Kevlar [2.71]. First, these fibres were prepared from poly(p-benzamide) (PPB), but are now considered to be essentially poly(p-phenylene terephthalamide) (PPT). Another high modulus fibre is based on poly(p-aminobenzhydrazide terephthalamide) (PABH-T), named X-500 (Fig. 2.21).

A high performance fibre can be considered as a fibre having a tensile modulus greater than 40 GN m^{-2} (5·8 × 10^6 psi); this value excludes the conventional fibres including high tenacity polyamides and polyesters (Table 2.36). Following their high Young's modulus values, beside carbon fibres, we find the three types of aramid fibres known as Kevlar 29, Kevlar 49, and X-500 G.

PPT is prepared by the reaction of terephthaloyl chloride and p-phenylene diamine, in a mixture of hexamethyl phosphoramide and N-methyl pyrrolidone, which allows polycondensation to continue to high molecular weights. In strong acids, such as concentrated sulphuric acid, PPT forms an optically anisotropic solution; it is then in a nematic liquid-crystal phase with parallel orientation of the chain molecules, from which it can be spun through a gas phase into a water or dilute sulphuric acid coagulant. There is no subsequent drawing of

Table 2.34.

Property alteration associated with the fiberizing process [2.68]

Fiberizing process	Possible deviation from bulk material properties
Conventional wire drawing	Strongly developed wire texture (preferred orientation of crystallographic directions with respect to wire axis)—particularly in unannealed or hard drawn filament. Such preferred orientation effects can give rise to directionality of the basic properties of the metal in the filament product
Bundle drawing	Preferred orientation effects can be expected with consequent directionality of properties. Surface chemistry alteration can occur and hence changes in initial reactivity. Small cyclical variations in cross-sectional area can be present, resulting in an apparent property change, e.g. in mechanical strength, which is based on the 'weakest link' (smallest cross-sectional area) in the length being measured
Foil shearing or slitting	Fibre properties will reflect the parent foil properties which may or may not show preferred orientation. If preferred orientation is present, the fibre properties will depend on the direction of shear, that is either parallel or perpendicular to the rolling direction in the foil
Shaving and related processes	The occurrence of partial fractures in fibres during processing results in weakened areas along the fibre length which constitute an apparent change in mechanical strength
Melt spin	Surface chemistry change associated with the product of the designed reaction to promote stabilization during fibre forming
Melt extraction	Extreme cooling rates produce fine grain size (particularly on the side of the fibre in contact with the extraction wheel) and can produce amorphous-like structures which will result in major property alteration. This latter phenomenon is exploited to the fullest in a product called 'Metglas' which is not generally categorized as a fibre. The amorphous state is metastable and the unique properties disappear on heating

Reprinted with permission from *Handbook of Fillers and Reinforcements for Plastics* by H. S. Katz & J. V. Milewski (Eds), 1978. Copyright Litton Educational Publishing.

Table 2.35.
Size related property alteration [2.68]

Property	Effect of diminished cross-section and increased length
Corrosion	High surface to volume ratio provides for relatively large reaction surface and hence limited life in corrosive environment
Electrical conductivity	Small cross-section provides efficient use of material for high frequency electrical conduction since the fibre radius can approach the skin depth
Mechanical properties	Sphere of influence of defects (e.g. inclusions in steel) associated with certain fibres approximates to the fibre diameter and results in reduced or degraded mechanical properties for the free fibre
	Whiskers (single crystal fibres) can be essentially dislocation free and approach theoretical strength

Reprinted with permission from *Handbook of Fillers and Reinforcements for Plastics* by H. S. Katz & J. V. Milewski (Eds), 1978. Copyright Litton Educational Publishing.

Fibre B

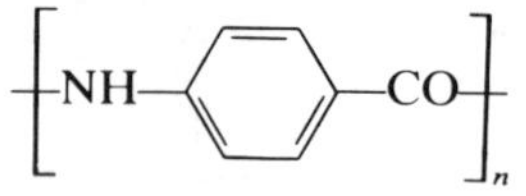

PPB   poly(*p*-benzamide)

Kevlar

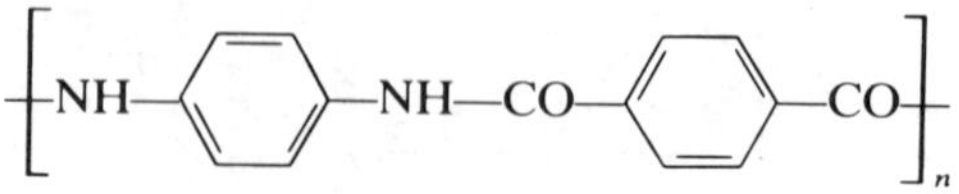

PPT   poly(*p*-phenyleneterephthalamide)

X-500

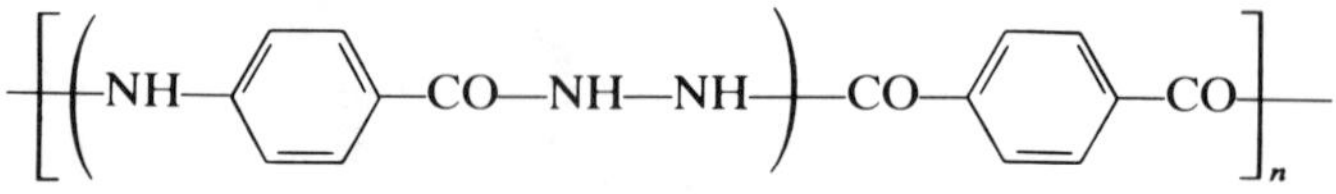

PABH-T   poly-(-*p*-aminobenzhydrazide terephthalamide)

Fig. 2.21. High performance fibres, chemical structures [2.72]. (Reprinted with permission from *Applied Fibre Science, Vol. 3*, 1979 by F. Happey (Ed.). Copyright Academic Press.)

Table 2.36.
Typical physical properties of high performance and other fibres [2.72]

	Young's modulus $GN\,m^{-2}$ (psi × 10^{-6})		Tensile strength $GN\,m^2$ (psi × 10^{-6})		Strain-to-failure (%)	Specific gravity
Polyester HT	12·0	(1·74)	1·0	(0·14)	10·5	1·38
Nylon 6,6 HT	5·0	(0·72)	0·9	(0·02)	13·5	1·14
E glass	70	(10·15)	2·5	(0·36)	4·0	2·6
Steel	200	(29)	4·0	(0·58)	2·0	7·8
Carbon type I	400	(58)	2·0	(0·29)	0·5	1·95
Carbon type II	260	(37·7)	2·6	(0·37)	1·0	1·75
Carbon type A	210	(30·5)	1·9	(0·27)	0·9	1·65
Kevlar 29	59	(8·55)	2·7	(0·39)	4·0	1·44
Kevlar 49	127	(18·4)	2·9	(0·42)	2·6	1·45
X500G	100	(14·5)	2·2	(0·31)	4·0	1·46

HT = High tenacity.
Reprinted with permission from *Applied Fibre Science, Vol. 3*, by F. Hapey, 1979. Copyright Academic Press.

these fibres, thus no draw ratio, only the so-called 'spin-stretch factor' following slight extension on winding up. After washing, a further heat treatment under tension is applied; this, together with the spin-stretch factor, can lead to considerable variation in tensile properties and hence different types of the same basic fibre [2.72].

Sturgeon & Lacy [2.73] show that except for very strong acids and bases, these fibres are relatively immune to a variety of chemical agents (Table 2.37).

In Fig. 2.22, the stress–strain curves of Kevlar fibres are compared to those of galvanized improved, plough steel wire (GIPS wire) and polyester (Dacron).

Kevlar fibre decomposes, without melting, above 500°C [2.74]. Brown & Ennis [2.75] have shown that the final part of the DTA diagram of Kevlar 49 has a melting endotherm at 560°C, and a decomposition endotherm around 590°C (Fig. 2.23).

Milewski [2.76] shows that the structure of aramid fibre is basically a lightly bound bundle of relatively long polymer chains in a semiparallel array, completely different from that of glass or carbon fibres (Fig. 2.24).

Table 2.37.
Stability of Kevlar 29 and Kevlar 49 in chemicals [2.73]

Chemical	Conc. (%)	Temperature (°F)	Time (h)	Strength loss (%)	
				Kevlar 29	Kevlar 49
Hydrochloric acid	37	70	24	NA	0
Hydrochloric acid	37	70	1 000	83	NA
Hydrofluoric acid	10	70	100	12	8
Nitric acid	1	70	100	18	5
Sulphuric acid	10	70	100	14	NA
Sulphuric acid	10	70	1 000	NA	31
Sodium hydroxide	50	70	24	NA	10
Ammonium hydroxide	28	70	1 000	10	NA
Acetone	100	70	24	0	0
Dimethyl formamide	100	70	24	NA	0
Methyl ethyl ketone	100	70	24	NA	0
Trichloroethylene	100	70	24	NA	1·5
Trichloroethylene	100	190	387	7	NA
Ethyl alcohol	100	70	24	0	0
Jet fuel (JP-4)	100	70	300	0	4·5
Jet fuel (JP-4)	100	390	100	4	NA
Brake fluid	100	70	312	2	NA
Brake fluid	100	235	100	33	NA
Transformer oil (Tex No. 55)	100	140	500	4·6	0
Kerosene	100	140	500	9·9	0
Freon 11	100	140	500	0	2·7
Freon 22	100	140	500	0	3·6
Tap water	100	212	100	0	2
Sea water (Ocean City, N.J.)	100	—	1 yr	1·5	1·5
Water (10 000 psi)	100	70	720	0	NA
Water, superheated	100	280	40	9·3	NA
Steam, saturated	100	300	48	28	NA

NA = Not available.
Reprinted with permission from *Handbook of Fillers and Reinforcements for Plastics* by H. S. Katz & J. V. Milewski (Eds), 1978. Copyright Litton Educational Publishing.

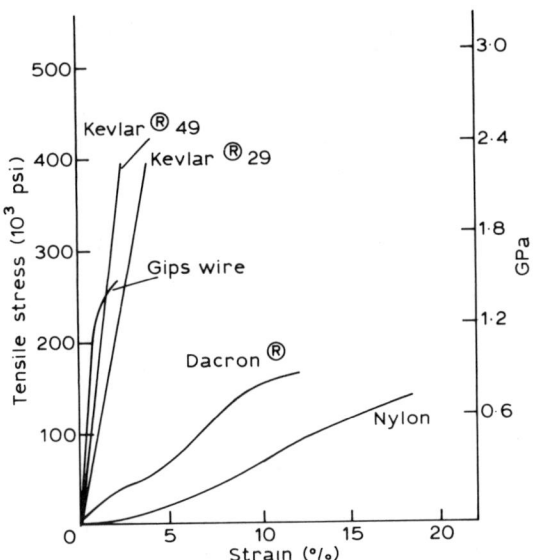

Fig. 2.22. Yarn stress–strain curves for dry twisted yarns, 10 in gauge length, at room temperature [2.73]. (Reprinted with permission from *Handbook of Fillers and Reinforcements for Plastics* by H. S. Katz & J. V. Milewski (Eds), 1978. Copyright Litton Educational Publishing Inc.)

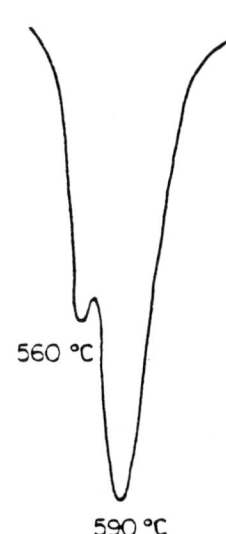

Fig. 2.23. Part of DTA curve for Kevlar 49 showing melting and decomposition point [2.75]. (After Brown & Ennis, 1977. Reprinted with permission from *Textile Research Journal,* **47** (1977) 62. Copyright Textile Research Institute.)

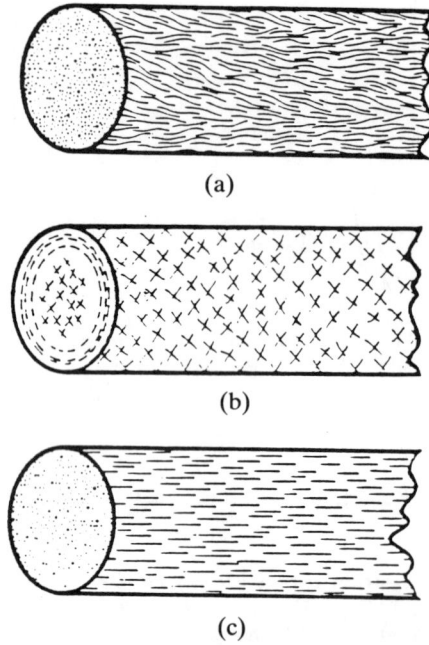

Fig. 2.24. The structures of aramid, glass and carbon fibres. (a) Structure of aramid (Kevlar), a semiloose bundle of relative long polymer chains in semiparallel array. (b) Structure of glass fibre. Internal structure is a random network of amorphous glass. The surface is semioriented, highly stressed in longitudinal compression, which gives it good tensile strength but poor scratch resistance. All strength is associated with the surface. (c) Structure of carbon fibres. Matrix is amorphous carbon filled with acicular microcrystallites of fibrils. The amount of L/D and the degree of alignment determine both the tensile strength and modulus of the fibre [2.76].

COMPOSITE TECHNOLOGY

One of the basic advantages of composites is the inherent ease of forming and manufacturing. *Mouldability* is an important plus factor, which in certain instances may result in savings in tooling, or may allow freedom of structural form not possible or practical with other structural materials.

In principle, the composite manufacture is simple. Each fibre (or other reinforcing agent) should be coated with an adequate layer of polymer, pressed along with a lot of similar fibres (or filler particles) into the required shape and size; then the polymer must be subjected to cross-linking which causes it to become permanently hardened.

However, this apparent simplicity is deceptive and is the cause of many technical difficulties and misunderstandings [2.77].

There are different technologies for the manufacture of polymer composites (Table 2.38), and these may be considered under two main headings:

(1) open mould technology,
(2) closed mould technology.

In the first procedure, during the moulding operation, the material is in contact with the mould on one surface only; this technique is the one generally adopted for civil engineering structural applications.

In the second technology, the composite is formed in a conventional male–female mould; it is employed for the manufacture of small elements which are not necessarily associated with the construction industry.

The most important variations of these technologies are presented in Table 2.39.

Hand Lay-up

This method of composite fabrication is the most versatile manufacturing technique available. In this process, a layer or layers of reinforcement (pre-cut to the desired pattern) are positioned on the mould, and the reinforcement is impregnated with the liquid polymer, either before or after placement. In the latter case, which is more common, the polymer is either brushed, poured or sprayed onto the reinforcing agent as the buildup progresses. In this way, sections can be developed to the desired thickness, and additional amounts of reinforcing agent can be placed as required for local strengthening [2.74].

This technology uses only one part moulds; it is the simplest; because of its cheapness and simplicity it is often chosen for both prototype and production work. The four essential requirements are:

—a mould or form which defines either the inside or the outside contour of the component;
—an ambient temperature setting polymer;
—a lubricant, to avoid the composite sticking to the mould;
—the glass fibre reinforcement, usually in the form of fabric or mat.

Table 2.38.

Reinforced polymer composites moulding methods and typical products [2.77]

Processes	Typical products
A. Room temperature cure	
Hand lay-up	Corrosion equipment, boats, and machine guards
Spray-up, including rigidizing	Tub and shower units
Vacuum bag moulding	Radomes
Tooling	Master moulds, checking fixtures
Cold-press moulding	Motor mounts, hollow ware, finished two sides
Casting, potting, encapsulation	Sinks, synthetic marble, electrical components
B. Intermediate cure	
Architectural panelling	Translucent and opaque, flat and corrugated panels, sandwich panels, foam structures
Centrifugal casting	Pipe and tubing
Pressure bag moulding	Pressure cylinders and tanks
C. High temperature cure (compression moulding)	
Premix and BMC moulding	Distributor caps, electrical gear, shower bases
Preform and SMC moulding	Automotive components, tubs, housings, radomes, aircraft components
D. Throughput processes	
Pultrusion	Rods, tubes, structural beams
Filament winding (including pre-impregnation)	Pipe and tubing
E. Reinforced thermoplastics	
Injection moulding	Gears
Rotational moulding	Hollow structures, tanks
Cold forming	Housing, appliance parts
F. Advanced composites	
Exotic methods and many combinations	Aircraft and aerospace parts

Table 2.39.
Manufacturing routes for fibre-reinforced plastic products [2.78]

Open mould processes
Hand lay-up
Spray up
Vacuum bag, pressure bag, autoclave
    filament winding
Centrifugal casting

Closed mould processes
Hot-press moulding, compression moulding
Injection moulding, transfer moulding
Pultrusion
Cold-press moulding
Resin injection
Reaction injection moulding

Reprinted with permission from *Chemistry & Industry* November 1982, 84. Copyright Society of Chemical Industry, UK.

The layers of reinforcing agent and the polymer loads are worked into intimate contact with the mould surfaces by the use of squeezers or rollers, or by hand. If needed, additional layers are added to build up the proper thickness. Figure 2.25 shows hand lay-up moulding.

Cure is achieved as a result of the cross-linking agent (hardener) or catalyst, that is added to the polymer just prior to its use. Cure can be

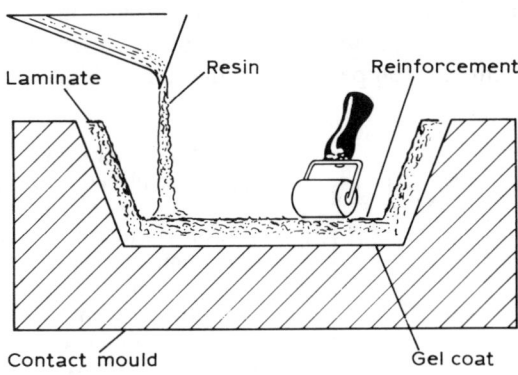

Fig. 2.25. Hand lay-up moulding.

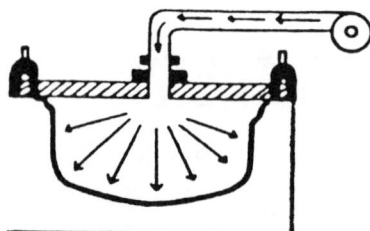

Fig. 2.26. Tailored bag—normally rubber sheeting—is placed against lay-up. Air or steam pressure up to 50 psi (345 kPa) is applied between pressure plate and bag [2.33]. (Reprinted with permission from *Plastics*, 6th edn, by J. H. DuBois & F. W. John, 1981. Copyright Litton Educational Publishing Inc.)

speeded up by heating. This technology can be used for ducts, swimming pools, sheets, tubes and housing. Building and machinery repairs are done by this process also.

Pressure may also be used with open moulds to improve quality, uniformity of density, and better finish on the open side. One procedure makes use of a plastic bag (vinylic or cellophane) that encloses the mould and its contents, as depicted in Fig. 2.26. Air is withdrawn from the bag by a vacuum pump, producing a uniform pressure on the product surface that is not in contact with the mould. The reverse of this process is pressure bag moulding, whereby rubber sheeting is placed against the polymer lay-up, and then exposed to air or steam pressures as high as 3·5 atm, as shown in Fig. 2.27.

This pressure bag type mould may also be inserted, as seen in Fig. 2.28, where steam pressure builds up as high as 7 atm. This process permits maximum loading with glass, as well as greatest product density. The final operation—in all mentioned procedures—is done in order to remove air and matrix excess, and press layers into close contact with each other and with the mould surface. After a convenient time, which depends upon polymerization or cross-linking duration, the moulding is sufficiently rigid to be removed.

In either the ambient temperature or the heat curing systems, it may

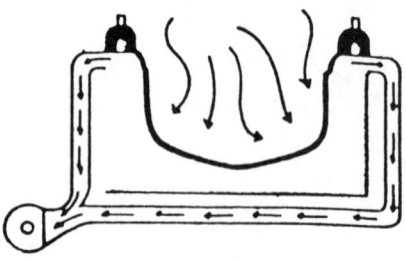

Fig. 2.27. Cellophane or polyvinyl acetate is placed over lay-up. Joints are sealed with plastic; vacuum is drawn. Resultant atmospheric pressure eliminates voids and forces out entrapped air and excess resin [2.33]. (Reprinted with permission from *Plastics*, 6th edn, by J. H. DuBois & F. W. John, 1981. Copyright Litton Educational Publishing Inc.)

Fig. 2.28. Modification of pressure bag method; after lay-up, entire assembly is placed in steam autoclave at 50–100 psi (345–689 kPa). Additional pressure achieves higher glass loadings and improved removal of air [2.33]. (Reprinted with permission from *Plastics*, 6th edn, by J. H. DuBois & F. W. John, 1981. Copyright Litton Educational Publishing Inc.)

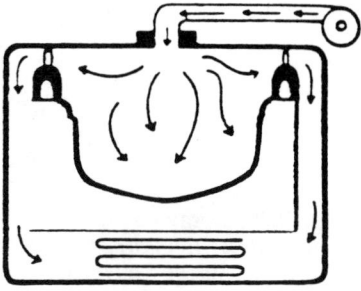

be practical to remove the part from the mould after it has hardened, and then 'post-cure' (post-polymerize) the part in an oven until cure is complete. This step reduces the idle time of the mould and improves the production efficiency [2.33].

Depending on the production schedule, it may be more economical to use elevated temperature cures (heated moulds, autoclave, infra-red units, oven, etc.).

The weatherability of a fibre-reinforced plastic used in construction is dependent upon the quality of the surface which is exposed to the atmosphere, known as the 'gel coat'. Its role is:

(a) to protect the composite from external factors; the most important being moisture penetration between the fibre and matrix, with consequent breakdown of the interface bond;
(b) to provide a smooth finish [2.80].

After the gel coat has become tacky, a coat of polymer is brushed over it and the first layer of reinforcement is applied and consolidated with a brush, roller, etc.

The advantages and disadvantages of hand lay-up are as follows [2.81]:

Advantages
(1) Simple technique.
(2) Low capital costs—cheap moulds.
(3) Large complex shapes can be produced.
(4) No process limitation on mould size.

Disadvantages
(1) High labour costs.
(2) Low production speed.

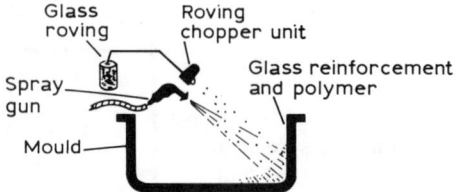

Fig. 2.29. Spray-up moulding technique [2.80]. (Reprinted with permission from *Glass Reinforced Plastics in Construction,* by L. Hollaway, 1978. Copyright Surrey University Press, Blackie Publishing Group, Glasgow, UK.)

(3) Moulding quality is dependent on operator skill.
(4) Mouldings have only one smooth glossy surface.

Spray up [2.79–2.82]

In this technique, glass fibre roving is fed continuously through a chopping unit, which cuts it to a convenient length (e.g. 5 cm); the resulting chopped strands are projected on the mould in conjunction with an atomized polymer jet (Fig. 2.29). Often this method makes use of a multiple-headed gun which blasts chopped glass fibres, polymer and catalyst or hardener simultaneously from one of its three heads. This spray is directed by the moulder to apply a uniform buildup over the entire mould surface. After the desired thickness has been piled up, the exposed product area is hand-rolled, to obtain a smooth surface, to remove air and consolidate the composite. Curing is achieved by one of the aforementioned contact moulding processes.

The spray-up method is a semiautomatic process which proves economical in many manufacturing operations; labour costs are greatly reduced by this process, and it makes use of an inexpensive material, glass roving.

It is a practical method for producing structural building panels, tank linings, large bodies, pools, roofs, etc.

The advantages and disadvantages of spray lay-up are [2.81]:

Advantages
(1) Spray equipment is portable.
(2) Capital outlay is small compared with other mechanized techniques.
(3) Rovings are used, the least expensive form of reinforcement.

(4) Production rates can be higher than with hand lay-up since shorter gel times can be tolerated.
(5) Reduced labour costs are obtained with high volume production.

Disadvantages
(1) Mouldings have only one smooth surface.
(2) Uniformity of lay-up depends, even more than with hand lay-up, on the skill of the operator.
(3) Spray lay-up is uneconomical for small volume production.
(4) An even spray pattern is difficult to achieve on small moulds.

Hand lay-up and spray-up techniques are relatively simple and low tooling cost can result in considerable versatility from the designer point of view. Very large mouldings are done by these methods.

Filament Winding [2.77–2.82]
Filament winding is a highly automated process that can be used to generate extremely efficient structures. This method uses continuous strands of glass fibres or continuous carbon fibres. Two techniques are in common use:

—helical winding, more convenient for long, thin, open-ended shapes such as sections of piping;
—polar winding, used for pressure vessels [2.77].

Such products requiring high mechanical strength in the circumferential and longitudinal directions may be produced by passing continuous glass fibre rovings through a bath of polymer, and winding onto a rotating mandrel. The angle of helix is determined by the relative speeds of the transversing bath and the mandrel (Fig. 2.30). This technology enables a fair degree of automation to be achieved.

When generating large structural shapes, the mandrel rotates about a fixed axis, as the glass fibre feeding mechanism travels the length of the mandrel in a programmed motion. By adjusting the rate of transverse relative to the rate of rotation, the quantity and orientation of the fibre deposition can be varied to best fit the stress levels and directions that will occur in the structure.

Curing can be realized at either ambient or elevated temperatures. As has been discussed, this is a function of the polymer used and the needs of the production schedule.

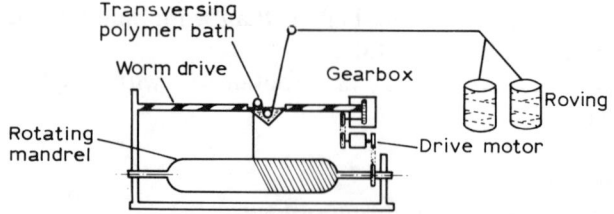

Fig. 2.30. Filament winding technique [2.80]. (Reprinted with permission from *Glass Reinforced Plastics in Construction*, by L. Hollaway, 1978. Copyright Surrey University Press, Blackie Publishing Group, Glasgow, UK.)

Filament winding has several advantages that make it an attractive technique for the manufacture of certain types of structures:

(1) It uses continuous roving, the least expensive form of reinforcement.
(b) The filaments can be oriented to take the stresses efficiently, thereby producing highly efficient composites with resulting savings in material and maximum strength to weight ratio.
(c) Best uniformity of properties.

Highest strength pipe, rocket motor cases, aircraft and missile bodies are produced by this process.

Closed Mould Techniques [2.77–2.83]
Hot and cold moulding systems, pultrusion, and injection moulding are the most important variations of the closed mould technology.
The hot press moulding technique is usually selected as being the most economical, whilst the cold one is an intermediate between the slower open mould systems, which are essentially for large products, and the faster but more expensive hot press moulding method, in which long runs of small to medium composite elements are made.
Closed mould technologies enable a higher fibre glass content to be used in composites with improved mechanical characteristics.
In hot press moulding of glass fibre reinforced unsaturated polyesters, metal moulds are heated to a steady temperature of 90–130°C, which is sufficient to cure the macromolecular matrix in a few minutes, in the presence of a peroxide initiator.
Hot press moulding (Fig. 2.31) is used for the manufacture of: pre-form moulding, sheet moulding compounds, and dough moulding compounds [2.80].

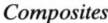

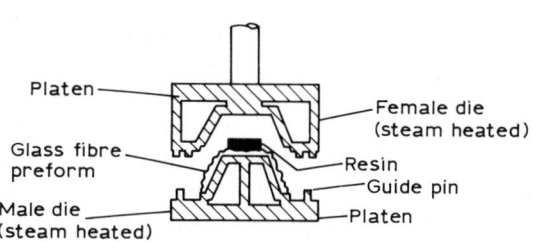

Fig. 2.31. Hot press moulding technique [2.80]. (Reprinted with permission from *Glass Reinforced Plastics in Construction*, by L. Hollaway, 1978. Copyright Surrey University Press, Blackie Publishing Group, Glasgow, UK.)

Pre-form moulding

In this system, chopped rovings are projected onto a rotating fine metal mesh screen which is shaped to the required dimensions. The strands are bound together by spraying the preform with a resinous binder in the form of a powder or an emulsion, and the whole is transferred to an oven at 150°C for 2–3 min, after which time the preform is ready for the press.

Sheet moulding compounds (SMC)

Sheet moulding compound (or prepreg) is a polyester resin based moulding material with E glass fibres, which vary in length and content between 12 and 55 mm and between 20 and 35% respectively. They are produced and supplied in the form of a continuous sheet wound into a roll and protected on both sides by polyethylene film. The latter is removed before loading into the press. The desired composite shape and rapid cure are obtained by the application of heat and pressure in suitable tools and, as the material flows uniformly to produce a homogeneous composite, complex and deep draw mouldings can be produced. It is necessary to comply with the correct conditions for moulding; these include using a suitable press and a mould designed specifically for the material, charging the mould in the correct manner, and using the optimum temperature, pressure and curing time. Moulding pressures of between 3·5 and 14 MN/m^2 (0·5 × 10^3 and 2 × 10^3 psi), moulding temperatures of between 125 and 155°C are generally required; the lower pressures and temperatures are for moulding simple flat shapes and the higher values for the more complex mouldings.

Dough moulding compounds (DMC)

Dough moulding compound (or premix) contains an unsaturated polyester resin, an unsaturated cross-linking monomer such as styrene, suitable mineral fillers and a fibrous reinforcement which is usually chopped strands. The fibres vary in length and content between 3 and 12 mm and between 15 and 20% respectively. Because DMC flow readily, they may be moulded by compression, transfer, or injection, and the pressure required to produce a component is relatively low so that large mouldings can be produced without much difficulty. These compounds do not give such high composite mechanical properties as the SMC due to the geometry of the component parts, but they can be used for intricate mouldings.

Both SMC and DMC have particular uses for appliances and fittings, but have not been extensively used in the structural applications of plastics [2.80].

Pultrusion Moulding

This procedure is a truly continuous process and with an automated cutoff saw, a pultrusion unit can run with virtually no attention, except for occasional checking. Unsaturated polyesters are the most widely used polymers (matrix) as well as other thermosettings such as epoxy which are used primarily in special-purpose applications. Figure 2.32 illustrates the pultrusion technique. Besides the polymer, fillers are used to regulate viscosity and for lowering the cost. Pigments are often added for coloration and light stability.

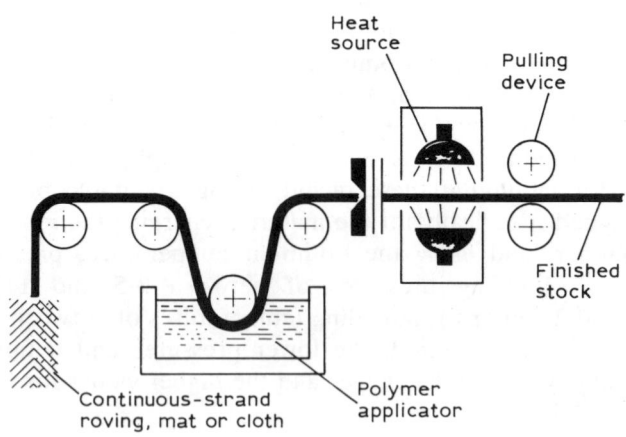

Fig. 2.32. Continuous pultrusion [2.79]. (Reprinted with permission from *Chem. Engng* **86** (2) (1980) 96. Copyright McGraw-Hill Publications.)

In filament winding, much of the wet-out occurs on the mandrel and tensile forces tend to squeeze out air. In pultrusion, as soon as the rovings reach the die cavity, any remaining air is trapped in the laminate, which at this point is cured by the heat source (tunnel oven). The different operations of this technique are: impregnation, shaping, curing and tooling, pulling, and cutting off [2.80].

Surface mats are sometimes added to provide a polymer-rich appearance. Fibreglass surface mats have no strength in the wet state. They should be fed into the structure as close to the curing die as possible. Polyester surface mats (unwoven cloth) have greater strength, but are more expensive.

The hot dies for curing are normally 0·65–1·3 m (2–4 ft) long and should have as high a polish as possible.

The relationship of fibreglass, polymer, filler and stripping-die area are mathematically solvable on a volume basis. According to Rolston [2.79], the total amount of roving can be calculated from the relationship:

$$Y_{tot} = 1 + W_g(\rho_m/\rho_g - 1)/36W_g\rho_m A \qquad (2.1)$$

where

Y_{tot} = total yield of glass rovings in m/kg (yd/lb)
W_g = weight fraction of glass in composite
ρ_m = density of polymer and filler mixture in g/cm^3 (lb/cu. in.)
ρ_g = intrinsic fibreglass density in g/cm^3 (lb/cu. in.)
A = curing die area in cm^2 (sq. in.)

From this, the number of roving packages can be determined from

$$N = Y_g/Y_{tot} \qquad (2.2)$$

where N = number of packages of fibreglass; Y_g = yield of individual package in m/kg (yd/lb).

Pultrusion may be realized also through the polymer injection. In this variation, the reinforcement, accurately positioned and under tension, is drawn through a heated metal die where impregnation of the fibres and cure of the polymer system takes place (Fig. 2.33). Here, by the use of appropriate polymer injection equipment, a short pot-life system can be used [2.81].

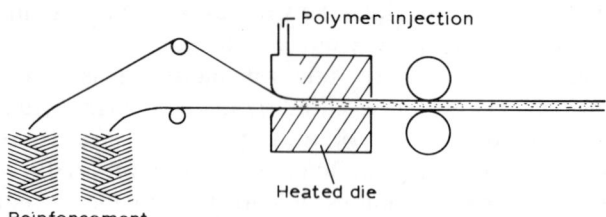

Fig. 2.33. Pultrusion by injection [2.81]. (Reprinted with permission from *FRP Technology, Fibre Reinforced Resin Systems* by R. G. Weatherhead, 1980. Copyright Applied Science Publishers Ltd.)

Injection Moulding

Fibre reinforcement is placed in the bottom mould and allowed to extend beyond its sides; then the top mould is placed over the bottom one, the reinforcement being held firmly in position, whilst the polymer is injected. Due to high mould costs, it is generally suitable for the large-scale production of small-to-medium sized components. Polymers processed in this way are polyester and epoxy DMC, and also phenolics, ureas, melamines and diallyl phthalate moulding compounds.

Darlington & Christie [2.84] carried out an interesting study on time dependent properties of injection moulded composites.

Mechanisms of Reinforcement—Properties

The mechanical and physical properties of the components are controlled by the properties of their constituents. The most important aspect of composite design is anisotropic behaviour, and it is necessary to give special attention to the methods of controlling this property, and its effect on analytical and design procedures.

It has been demonstrated that, with the correct process control and a soundly based material design approach, it is possible to produce composites which can satisfy stringent structural requirements.

The reinforcement of a low modulus matrix with high strength, high modulus fibres uses the polymer flow of the matrix under stress to transfer the load (or stress) to the fibre; this results in a high strength, high modulus composite. The aim of the combination is to produce a complex (two-phase) material in which the primary phase which determines stiffness (i.e. fibres) is well dispersed and bonded by a secondary phase, the polymer matrix. The strength and stiffness of such complex engineering materials depend on the properties of the reinforcing agent, the matrix and interface. Each of these individual phases has to perform certain essential functional requirements based on their mechanical properties so that a system containing them may perform satisfactorily as a composite.

The requirements for fibres are:

—high modulus of elasticity in order to give efficient reinforcement;
—high ultimate strength;
—the variation of strength between individual fibres should be low;
—the fibres should be stable and retain their strength during handling and fabrication;
—the dimensional characteristics (surface and diameter) should be uniform.

The matrix has to fulfil the following functions:

—To bind the fibres together, and protect their surfaces from damage during fabrication in the service life of the composite.
—To keep the fibres dispersed and separated so as to avoid any catastrophic propagation of cracks and subsequent failure.
—To transfer stresses to the fibres efficiently by adhesion and/or friction, when the composite is under load.
—To be thermally compatible with the reinforcement.
—To be chemically compatible with fibres over a long period.

The interface between the fibre and the matrix is an anisotropic transition region which provides adequate chemically and physically stable bonding between the fibres and the polymer [2.80].

In analysing fibre-reinforced matrix materials, the primary aim is to obtain predictions of the average behaviour of the composite from the properties of the components. Of particular interest are the mechanical properties describing strength and stiffness. The factors which influence these characteristics are:

—the fibre volume fraction of the composite;
—the fibre orientation within the matrix;
—the fibre cross-section;
—the mechanical properties of the fibre and matrix;
—the strength of the fibre–matrix bond.

A linear relationship was established between the elastic modulus of the composite and the volume fraction of the fibreglass in an

Polymeric Building Materials

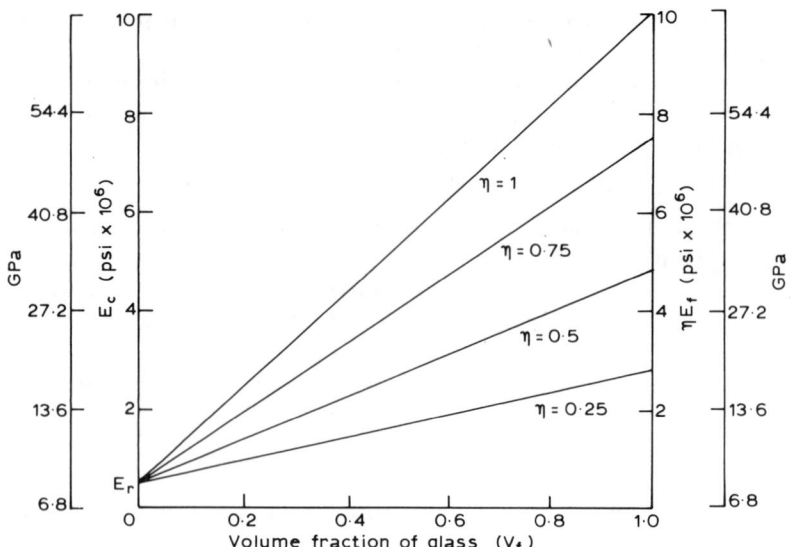

Fig. 2.34. Young's modulus of composites as a function of the volume fraction of the glass-fibre phase V_f and the efficiency of reinforcement η for given Young's moduli of phases E_r and E_f [2.85]. (Reproduced with permission of the publisher, Howard W. Sams & Co., Indianapolis, *Composite Materials for Combined Functions* by E. Scala, copyright, 1973.)

epoxy/fibreglass composite (Fig. 2.34). In the figure, E is the elastic modulus, c refers to the composite, f to the fibre, and r to the polymer.

Plotting the combined modulus follows the weighted average

$$E_c = V_f E_f + V_r E_r \qquad (2.3)$$

The efficiency of the fibres is

$$\eta = \alpha_n \cos 4\phi \qquad (2.4)$$

in which α_n is the volume fraction off angle and ϕ the off angle. The composite modulus then varies with the vol.% of the fibres as follows [2.85]:

$$E_c = \eta(V_f E_f) + V_r E_r \qquad (2.5)$$

As shown in Fig. 2.35, the strength of the finished laminate increases in direct proportion to the amount of glass fibre.

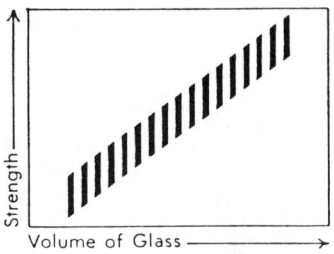

Fig. 2.35. As glass content increases, strength increases [2.83]. (Reprinted with permission from *Reinforced Plastics. Theory and Practice,* 2nd edn. by M. W. Gaylord, 1974. Copyright Cahners Books.)

A part containing 80% fibre and 20% polymer is almost four times stronger than a part containing opposite proportions of these components [2.83].

Milewski [2.76] did a detailed study on the loading levels. Figure 2.36 shows the theoretical and actual effects on composite strength of filler and fibre addition. The theoretical and actual effects for filler additions indicate a steady dilution in strength as filler loading increases. The theoretical curve for fibre addition shows a proportional increase in strength with the increasing amount of fibre; however, the experimental curve for fibre addition does not follow the theoretical curve but instead follows the filler-addition curve at first.

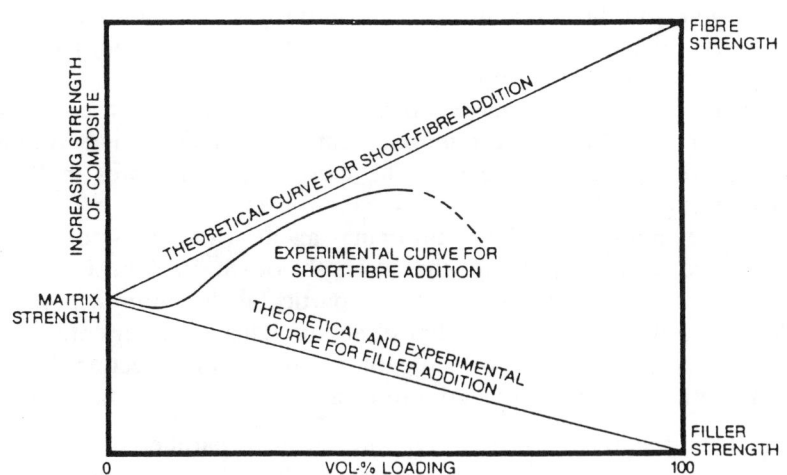

Fig. 2.36. Effect of volume loading of short fibres on composite strength [2.76].

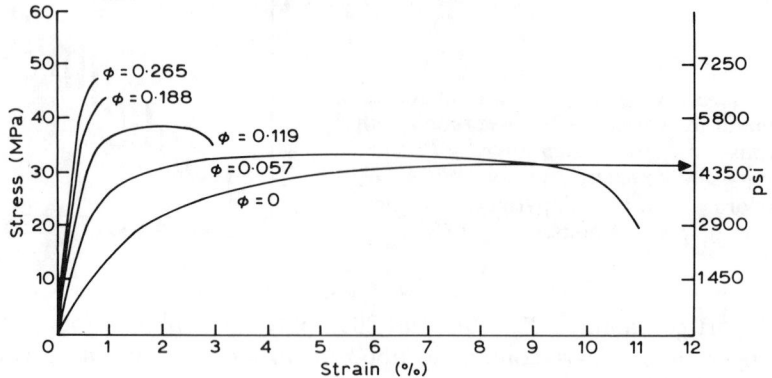

Fig. 2.37. Typical stress–strain curves for carbon fibre reinforced PP at various fibre volume fractions, ϕ. Data obtained from injection moulded tensile bars [2.86]. (After Weiss, 1980.)

The author explains the shape of the experimental curve by the fact that at lower loadings, fibre additions behave as filler additions, diluting the matrix strength and lowering the overall composite strength. When the initial interaction level is reached, the fibres start to reinforce and the curve swings up, approaching the theoretical fibre curve. At higher loadings, the curve dips again, due to packing, distribution and wetting problems.

A number of workers have reported the form of the stress–strain curves for various matrix–reinforcement combinations and volume fractions of the last one [2.86]. The case of carbon fibres in PP is shown in Fig. 2.37.

The orientation of the reinforcing agent (fibres) is of equal importance with respect to the strength of the composite. It is necessary to orient reinforcement in a particular direction in order to achieve maximum strength in that direction. This single orientation is obtained at the expense of strength in the other direction [2.87]. Orientation is generally divided into (Fig. 2.38):

(1) Unidirectional, when all the fibres are laid parallel.
(2) Bidirectional or planar, when half of the strands are laid at right angles to the other half.
(3) Random (isotropic), or multidirectional [2.37, 2.78, 2.83, 2.87].

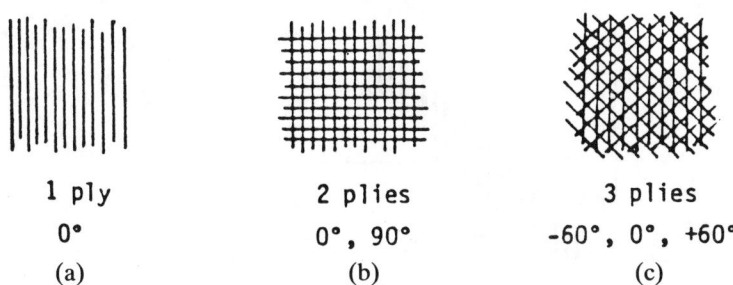

1 ply | 2 plies | 3 plies
0° | 0°, 90° | -60°, 0°, +60°
(a) | (b) | (c)

Fig. 2.38. Different kinds of orientation [2.37]. (Reprinted with permission from *Plastics Product Design Handbook,* Part A, by E. Miller (Ed.), 1981. Copyright Marcel Dekker, New York.)

When all the strands are laid parallel to each other, maximum strength results in one direction. This strength is supplied for end-uses such as solid rods or bars. Product application would include guy strain insulators, golf clubs, etc. The strength perpendicular to the axis of the fibres is dependent on the shear strength of the matrix.

When half of the fibres are laid at right angles to the other half, strength is highest in those two directions. Strength in any one direction is less than with parallel arrangements. This pattern of glass reinforcement is used in structural shapes. When fibres are arranged isotropically, strength is equal in all directions.

There is a relationship between the way fibres are arranged and the amount of fibres that can be loaded into a given object or product. The neater the arrangement or the more precise the placement, the greater the amount of reinforcement that can be placed.

In the case of an all-parallel arrangement, glass loadings from 45 to 90% can be obtained. The maximum theoretical amount is about 92%, while the practical desired maximum is less than 80%. When half the strands are placed at right angles to the other half, glass loadings range from 55 to 75%.

A random arrangement gives glass loading in a range of 15–50%. The relationship of volume glass fibre, strength characteristics, and arrangement of fibres is shown in Fig. 2.39.

Continuous parallel strands give the highest strength range, bidirectional arrangement the next highest, and isotropic the lowest strength range.

Regarding different shapes of glass fibres used for the reinforcing of

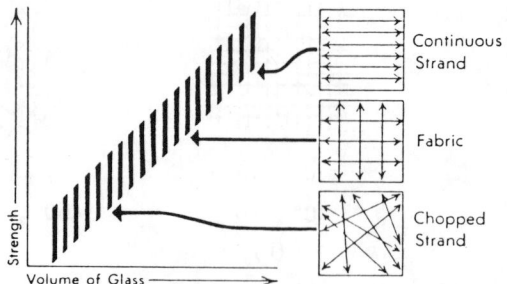

Fig. 2.39. Type of reinforcement determines maximum strength in a given direction [2.83]. (Reprinted with permission from *Reinforced Plastics Theory and Practice,* 2nd edn, by M. W. Gaylord, 1974. Copyright Cahners Books.)

polymer composites, we may mention the following:

—Woven roving gives high strength and is lower in cost than conventional.
—Chopped strand gives random reinforcement.
—Reinforcing mats are lower in cost than fabric and give random reinforcement.
—Fabric essentially reinforces the object in two directions.
—Surfacing mats give virtually no reinforcement but permit a decorative and smoother surface finish [2.83].

Beside the volume and the orientation of the fibres, their length is also an important factor (Fig. 2.40) [2.86].

Taking into account orientation and other main factors, constitutive relationships are described by McCullough *et al.* [2.88] to predict the Young's moduli, shear moduli, Poisson's ratio and coefficients of thermal expansion for two different sheet moulding compounds.

A definite relationship exists between reinforcement, moulding methods and engineering properties. Increased moulding pressure decreases composite thickness and increases reinforcement amount. Strength properties are improved as the quantity of glass reinforcement is increased [2.83].

It may be seen from Fig. 2.41 that the major difference between the three categories of fibre composites (random, bidirectional and unidirectional) is the variation of strength and stiffness with fibre orientation.

Whenever data are published on the strength and stiffness of GRP

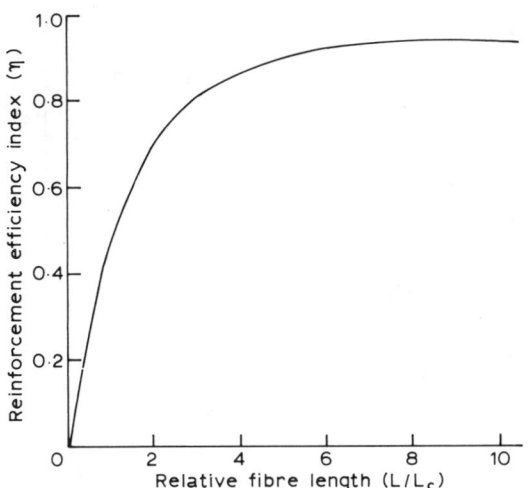

Fig. 2.40. Relationship between relative fibre length and reinforcement efficiency factor η for composite strength. For an ideal composite reinforced with continuous aligned fibres, $\eta = 1$ [2.86]. (After Bader and Boyer, 1973.)

composites, there are two important factors which must be understood when interpreting the information given; these are:

(a) the anisotropic nature of the laminate; the data presented generally relates to the maximum values of the composite;

(b) the method of manufacture of the laminate.

Table 2.40 gives the tensile characteristics of the different types of GRP composites and illustrates the above. It is evident that the tensile stress–strain curves of composites will be a function of the component parts; where the glass content is high the curves established in the direction of fibre alignment will show a close resemblance to the characteristics of glass, and conversely when the glass content is low the matrix characteristics will be reflected [2.80].

The strength-to-density (specific strength) and modulus-to-density (specific modulus) ratios versus temperature behaviours for several of the more important boron and carbon composites are compared in Figs. 2.42 and 2.43. The width of the range of attainable properties with carbon-fibre composites is apparent [2.89].

Table 2.41 gives values of mechanical, physical and electrical properties for pultruded composites averaging 65% fibreglass content.

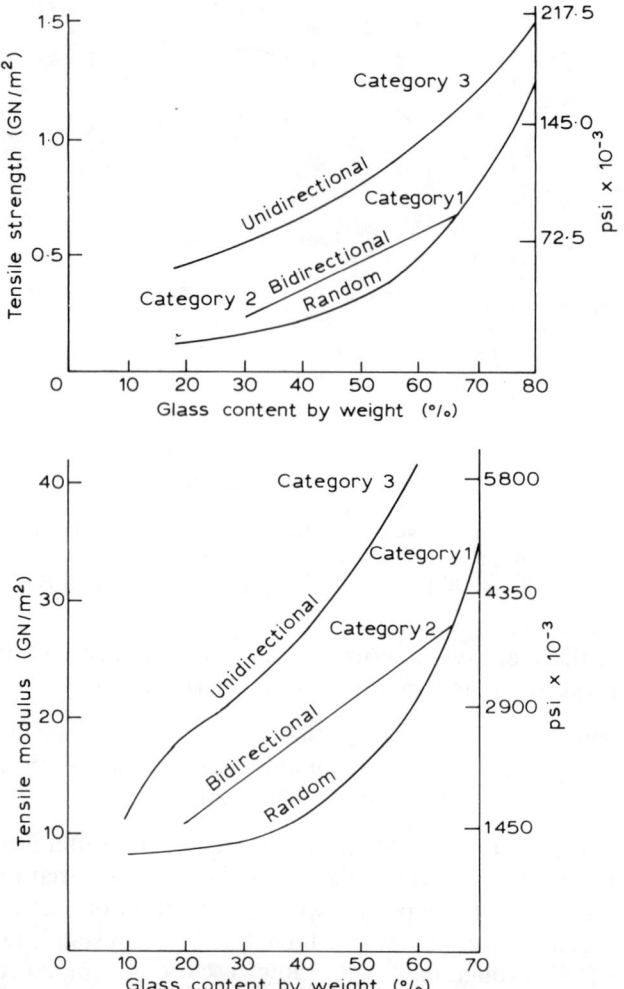

Fig. 2.41. Typical tensile strengths and tensile moduli versus glass content [2.80]. (Reprinted with permission from *Glass Reinforced Plastics in Construction* by L. Hollaway, 1978. Copyright Surrey University Press, Blackie Publishing Group, Glasgow, UK.)

Chemical and weathering characteristics will depend on the type of matrix used [2.82, 2.90].

Typical stress–strain diagrams for three types of fibre orientation are shown in Fig. 2.44. The glass fibre arrangements shown in Fig. 2.44

Table 2.40.
Typical mechanical properties for glass reinforced plastics composites [2.80]

Material	Glass content (% by wt)	Specific gravity	Tensile modulus GN/m^2 ($psi \times 10^{-6}$)	Tensile strength MN/m^2 (psi)
Unidirectional rovings (filament winding or pultrusion)	50–80	1·6–2·0	20–50 (2·9–7·2)	400–1 250 (58 000–181 250)
Hand lay-up with chopped strand mat	25–45	1·4–1·6	6–11 (0·8–1·5)	60–180 (8 700–26 100)
Matched dye moulding with preform	25–50	1·4–1·6	6–12 (0·8–1·7)	60–200 (8 700–29 000)
Hand lay-up with woven rovings	45–62	1·5–1·8	12–24 (1·7–3·4)	200–350 (29 000–50 750)
DMC polyester (filled)	15–20	1·7–2·0	6–8 (0·8–1·1)	40–60 (5 800–8 700)
SMC	20–25	1·75–1·95	9–13 (1·3–1·8)	60–100 (8 700–14 500)

Reprinted with permission from *Glass Reinforced Plastics in Construction* by L. Hollaway, 1978. Copyright Surrey University Press, Blackie Publishing Group, Glasgow, UK.

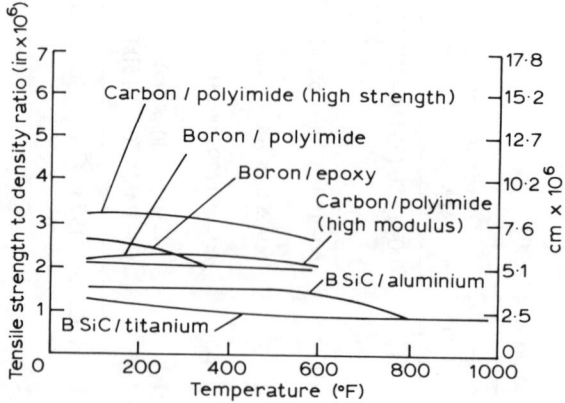

Fig. 2.42. Specific strength of selected composites. BSiC is SiC-coated boron [2.89]. (Reprinted with permission from *Composites, State of the Art* by W. Weeton & E. Scala (Eds), 1974. Copyright The Metallurgical Society, Warrendale, PA.)

are:

   (a) unidirectional, with fibre orientation in line with stress direction;
   (b) cross-ply, i.e. equal amounts of fibre in two perpendicular directions, with stress direction in line with one of the fibre directions of CSM (chopped strand mat), giving an approximately isotropic product [2.42].

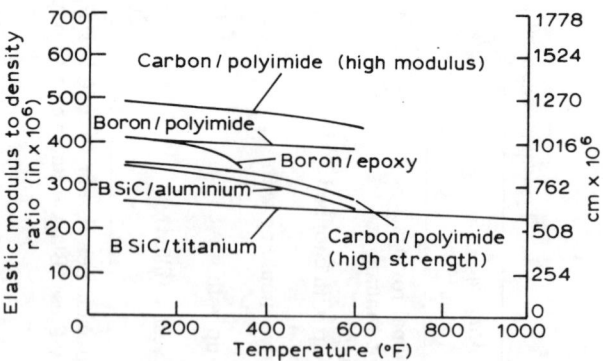

Fig. 2.43. Specific stiffness of selected composites [2.89]. (Reprinted with permission from *Composites, State of the Art* by W. Weeton & E. Scala (Eds), 1974. Copyright The Metallurgical Society, Warrendale, PA.)

Table 2.41.
Technical data for typical pultruded sections [2.82]

Mechanical properties	
Flexural strength	275–700 MN/m^2 (39·8 × 10^3–101 × 10^3 psi)
Flexural modulus	14–34 GN/m^2 (2·3 × 10^6–4·9 × 10^6 psi)
Tensile strength	200–550 MN/m^2 (29 × 10^3–79·7 × 10^3 psi)
Tensile modulus	14–31 GN/m^2 (2·3 × 10^3–4·4 × 10^3 psi)
Compressive strength	96–206 MN/m^2 (13·9 × 10^3–29·8 × 10^3 psi)
Impact strength (Izod)	45–86 J (0·045–0·085 Btu)
Physical properties	
Relative density	1·8
Water absorption	0·5%
Barcol hardness	45–60
Specific heat	942 J/kg K (0·225 Btu/lb °F)
Thermal conductivity	0·37 W/m K (2·56 Btu inch/ft^{2}h°F)
Coefficient of thermal expansion	7 × 10^{-6}/K
Heat distortion temperature	220°C
Electrical properties	
Electric strength at 23°C	5·7 MV/m
Arc resistance	118 arc s
Comparative tracking index	480 V
Surface resistivity	10$^{13·6}$ ohm
Volume resistivity	10$^{14·1}$ ohm cm
Power factor at 1 MHz	0·0127
Permittivity at 1 MHz	3·99

Fracture modes in angle-ply composites have been studied at NASA and the Lewis Research Center where the mechanical behaviour of high modulus graphite fibre/epoxy composites (MODMOR 1-GR/ERLA-4617) was studied by subjecting them to off-axis tensile loading. The most important results obtained by C. C. Chamis and J. H. Sinclair are summarized as follows:

(1) Predominant fracture modes:
   (a) longitudinal tensile (fibre breaks) near 0° load angle;
   (b) intralaminar shear (matrix shear fracture) in the 5–20° angle range;
   (c) transverse tensile (matrix tensile fracture) in the 45–90° load angle range;
   (d) mixed mode (intralaminar shear and transverse tensile) in the 20–45° load angle range.

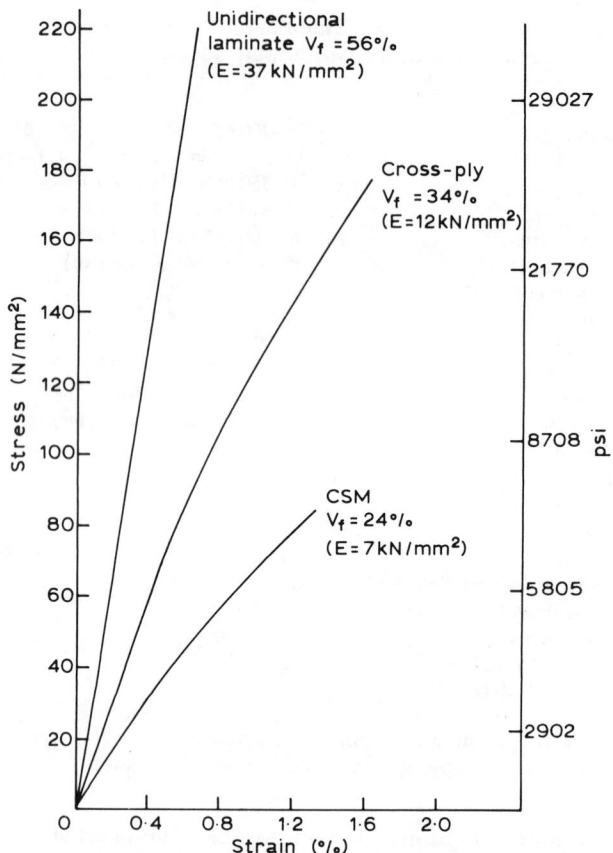

Fig. 2.44. Tensile stress–strain characteristics [2.42]. (Reprinted with permission from *GRP in Structural Engineering* by M. Holmes & D. J. Just, 1983. Copyright Applied Science Publishers.)

(2) Stress–strain curves to fracture are linear (see Fig. 2.45).
(3) Results by linear composite mechanics were in good agreement with measured data for predicting mechanical response to off-axis tensile loads of high modulus fibre/epoxy matrix composites which exhibit linear stress–strain to fracture.
(4) Plotting procedures were developed to identify which stress dominates off-centre tensile fracture [2.91].

Various other properties of composites are discussed in detail in the literature [2.92–2.120].

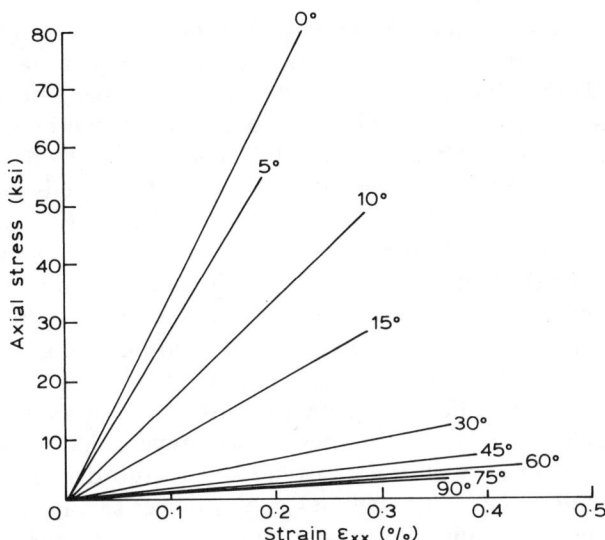

Fig. 2.45. Stress–strain curves for specimens subjected to tensile loading at various angles [2.91]. (Reprinted with permission from *Technology of Carbon and Graphite Fiber Composites* by J. Delmonte, 1981. Copyright Litton Educational Publishing Inc.)

Ideally, designers would like to make monolithic structures, i.e. structures without joints; for many reasons this ideal can never be realized. The designer has three basic types of load carrying joints at his disposal: mechanical joints, adhesive joints (bonded), or joints which are a combination of these two.

The design methods that have been established for joining metals are generally applicable to composites. However, as is to be expected, the physical nature of fibre reinforced composites does introduce problems that are not encountered with metals [2.121].

Holmes & Just [2.42] discuss in detail the three types of structural joints.

Mechanical joints find application where bonding is impractical, uneconomic, or where parts may have to be removed and replaced at some time. Such joints can be made using conventional bolts, screws or rivets, or preferably, the wide range of purpose designed fasteners [2.81].

Godwin & Matthews [2.122] made a review of the published information on mechanically fastened joints; they have drawn the following conclusions. An adequately tightened bolt in a carbon fibre reinforced epoxy can be expected to have a maximum bearing strength between 800 and 930 MN/m² (116×10^3–134×10^3 psi), depending on lay-up. Similar fastening in a glass fibre reinforced epoxy can be expected to have a strength in the region 550–700 MN/m² (79×10^3–101×10^3 psi). With glass fibre reinforced polyester, the strength range is much wider, being between 200 and 600 MN/m² (29×10^3–87×10^3 psi). For pin-loaded or riveted joints, the maximum bearing strength is, at the most, about 60% of the above values. Depending on the lay-up, all materials show sensitivity to the direction of applied loading, a reduction of up to 30% being possible in some cases. Full bearing strength, for bolted joints, is only developed if failure due to tension and shear is suppressed by providing adequate end distance and width (or pitch), and if sufficient restraint is present to prevent local interlaminar failure. A value of 5 diameters would cover both distances for most materials and lay-ups, although in some instances smaller values might be applicable. For pin-loaded and riveted holes, values of about 3 diameters seem adequate.

Adhesive joints should be designed so that forces are transferred by shear, compression or tension; peeling effects should be avoided [2.42].

Configurations of various joints are illustrated in Fig. 2.46, with the efficiencies mentioned in parentheses.

Butt joints afford just a small area of adhesion and are only suitable for resisting compressive forces. Where tensile forces are to be resisted, the area of adhesion can be considerably increased by the use of scarf or lap joints [2.42].

Most aspects of adhesively bonded joints in composites were reviewed by Matthews *et al.* [2.123].

The bonding strength of sheet moulding compounds was also investigated by Feldman *et al.* [2.124, 2.125]. Different types of adhesives and different types of lap and butt joint arrangements were considered. Environmental effects such as humidity, temperature and ultra-violet light on the joint strength were also examined. The results show that epoxy adhesives give good bond strength and the combination of a single lap joint together with adhesive rivets gives strong joints. Environmental cycling of up to 100 cycles in the environmental chamber does not significantly affect the joint strength for the systems considered.

Not all properties of composites will be needed for every application in construction, but some of them will determine the suitability of

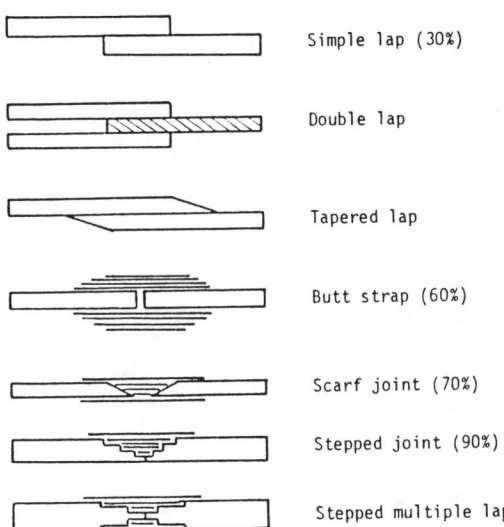

Fig. 2.46. Configuration of adhesive bonded joints [2.81]. (Reprinted with permission from *FRP Technology, Fibre Reinforced Resin Systems* by R. G. Weatherhead, 1980. Copyright Applied Science Publishers Ltd.)

these complex materials for specific applications where other products do not possess the same combination of properties. Properties of interest for building applications compared with properties of some other building materials are listed in Table 2.42.

Reinforced plastics are not surrogates, i.e. substitutes, for other materials. They may be used where traditional materials are used. In many situations they perform as well or better than the traditional materials at the same or lower cost [2.126]. The applications of some reinforced plastics are summarized in Table 2.43.

Composite materials will continue to present the designer with great versatility, opportunity, challenge and cost effectiveness. An increasing array of high stiffness and high strength fibres like silicon carbide, silicon nitride, aluminium oxide and boron nitride, which are under development, show potential as structural reinforcement [2.127].

COMPOSITES IN CONSTRUCTION

In the construction industry, there is a real and fundamental option open to designers; that is, whether to make the materials into their

Table 2.42.
Properties of GRP compared with other materials of construction [2.126]

Property	Units	GRP[a] laminate	Plywood (exterior grade)	Mild steel	Concrete
Specific gravity	—	1·6	0·08	7·8	2·3
Density	kg/m²	1 600	800	7 760	2 310
Actual strength					
Tension	kN/m² × 10⁴	7·6	5·8	55	—
	psi × 10⁴	1·1	0·84	7·9	—
Compression	kN/m² × 10⁴	11·0	2·85	24[c]	0·69[d]
	psi × 10⁴	1·5	0·40	3·4	0·1
Shear	kN/m² × 10⁴	6·9	1·2	38	0·069
	psi × 10⁴	1·0	0·17	5·5	0·01
Specific strength[b]					
Tension	kN/m² × 10⁴	4·7	7·2	7·1	—
	psi × 10⁴	0·68	1·0	1·0	—
Compression	kN/m² × 10⁴	6·9	3·6	3·1[c]	0·30
	psi × 10⁴	1·0	0·52	0·44	0·04
Shear	kN/m² × 10⁴	4·3	1·5	4·9	0·03
	psi × 10⁴	0·62	0·21	0·7	0·004
Actual modulus					
Tension	kN/m² × 10⁴	0·55	0·80	20·7	—
	psi × 10⁴	0·07	0·11	3·0	—

	units				
Shear	kN/m² × 10⁴	0·28	0·034	8·6	—
	psi × 10⁴	0·04	0·004	1·24	—
Specific modulus[b]					
Tension	kN/m² × 10⁷	0·34	1·0	2·65	—
	psi × 10⁷	0·05	0·14	0·38	—
Shear	kN/m² × 10⁷	0·17	0·43	1·1	—
	psi × 10⁷	0·025	0·06	0·14	—
Cost per m³	NP × 10⁴	11·0	0·9	5·4	0·17
Specific cost[b]	NP × 10⁴	6·9	1·1	0·7	0·07
Light transmission	%	85	Opaque	Opaque	Opaque
Maintenance	—	Translucent materials. Maximum useful life of 30 years without maintenance. Opaque materials. Require painting after 10–15 years and thereafter at similar intervals of time	Preservatives, varnishes or paints necessary. Renewed application necessary according to type of finish	Corrosion protection essential and must generally comprise galvanizing or priming followed by decorative finish. Re-painting necessary	None except where decorative finish is required

[a] The figures are minimum design values for laminates made by the hand lay-up or contact moulding process with chopped strand mat reinforcement and resin content about 70%.
[b] Specific properties are determined by dividing the figures for actual properties by the specific gravity.
[c] Yield stress.
[d] Permissible stress.
NP, net price.
Reprinted with permission from *Glass Reinforced Plastics* by B. Parkyn (Ed.), 1970. Copyright The Plastics and Rubber Institute.

Table 2.43.
Principal building applications for GRP [2.126]

Application	Form in which supplied	Advantages over traditional materials	Disadvantages
Rooflights	Corrugated or flat sheet	Stronger than glass, lighter in weight and easier to install	Light transmission less than that of glass. Deterioration of light transmission on ageing
Domelights	One piece components moulded or fabricated	Light in weight, easy to install	Light transmission less than that of glass. Deterioration of light transmission on ageing.
Domes and other roof structures	Modular components, single or double skins	Light in weight, easy to erect	Stiffening with metals or timber may be necessary
Internal partitions	Corrugated or flat sheet	Convenient in use. Special decorative effects can be easily incorporated	Limited use because of cost
Cladding	Flat or profiled unsupported sheet or as surface skin to concrete or asbestos	Lightness, range or decorative effects, versatility for individual designs	Fire performance limitations and the effects of prolonged weathering
Sectional buildings	Modular components, often double-skinned with sandwich construction	Lightness and ease of erection	Fire performance limitations and the effects of prolonged weathering

Bathroom units	Assembled modules	Lightness and ease of erection	No specific disadvantages
Tanks and cisterns	One or twp pieces press mouldings	Lightness, no corrosion, low thermal conductivity	No specific disadvantages
Pipes and ducts	Continuous profiles or as cladding on concrete or PVC pipe	Improves strength of concrete pipes and protects them from chemical attack. Increases temperature range for PVC pipe	No specific disadvantages
Window frames	Assembled press moulded components or as sections cladding timber	Reduces maintenance associated with most other materials	Less suitable than timber for non-standard dimensions
Concrete moulds	Mouldings generally made by hand lay-up, but sometimes by press moulding	Lightness. gives concrete of high quality and excellent finish. Provides a new medium for architectural designs on concrete	Low stiffness means that additional support is often necessary

Reprinted with permission from *Glass Reinforced Plastics* by B. Parkyn (Ed.), 1970. Copyright The Plastics and Rubber Institute.

final shape on site, or previously in factories. The use of composites in construction illustrates these points very clearly. Many processes are used on site and in restoration work are mandatory, but for components the full range of plastics processing techniques have been used to form and fabricate composites.

In many cases, the new composites have to demonstrate their superiority over wood and its derivatives, such as plywood, hardboard and chipboard. On the other hand polymer composites are generally ready for immediate use and do not require a finishing process. Moreover, they usually require very little maintenance after exposure to weather. Although they reduce the labour of finishing on site, they demand more sophisticated processing techniques which require special colour pigments, machinery, etc., in comparison to that which traditional trades utilize.

Speed of construction is perhaps their greatest merit, and in the case of in situ casting or formulation, the rate of polymerization can be controlled and adjusted. For example, this is particularly advantageous in ground reinforcement, or, where polyester and aggregates are used in a mix as grouting cements for bolts, for fast curing floors, or for roadways. Rapid construction is also made possible by the use of prefabricated elements, but these are not, of course, uniquely made of polymers or composites since many traditional materials are offered in prefabricated form for use in the building industry. The special point about polymer composites is that they are generally lightweight and require little labour to install.

A common classification of composites used in construction considers the following two groups:

(a) standard products such as flat sheets, corrugated sheets, or sandwich panels;
(b) structures made to shape for special applications; in this case the civil engineer or the architect must check the characteristics, the behaviour of the composite, and the technology of fabrication before choosing the product for a certain application.

Due to their high coefficient of light transmission (about 85%), to their lightness, to their mechanical properties, and in some cases, to their fire resistance, transparent and translucent sandwich panels have a lot of applications; some special composite panels have been made—in the last few years—for solar collectors.

Composite sheets are used as cladding on structural materials, or take part in panels of structural walls. In the first case, they play a decorative role on brick or concrete structures.

If an opaque composite sheet becomes part of a non-bearing panel structure, it may be used in different manners with other materials for the realization of building cladding. In this case, the composite sheet represents the external membrane. The most used panel is the sandwich panel. The use of these types of prefabricated sandwich panels has proved very satisfactory for the facing of high buildings. These prefabricated panels are light, and as a consequence are easy to install, with very simple tools and very low cost. Walls for mobile houses may also be made with sandwich panels. Easy production of big elements and the simplicity of their joints, with all other properties, recommend the glass fibre reinforced composites for modular construction. These components are prefabricated and assembled very fast on site. They are suitable for construction sites where access is difficult and where the ground cannot support traditional construction without making expensive foundations. These modules or wall components are made by two glass reinforced composite membranes joined as a sandwich by a foam core or by a honeycomb.

Composites are also suitable for prefabricated units or components for bathrooms, for cisterns for cold and hot water, for window frames and for concrete frames [2.128].

Lightness, toughness, strength and mouldability recommend GRP for moulded bath fixtures such as bath tubs and shower stalls. In the case of bath tubs, the surrounding wall surfaces are frequently moulded integrally with the tub, thus avoiding cracks and joints. Increasingly, entire bathrooms, including floors, walls and ceilings, are being moulded of glass fibre reinforced polyester, sometimes with a gel-coated surface, sometimes with a thermoformed thermoplastic sheet surface to the underside of which the fibre reinforcement and the polymer are sprayed.

Because of their resistance to corrosion by many fluids and underground water, glass fibre reinforced polyester or epoxy tanks are employed to store a large variety of fluids such as oil and gasoline. Such tanks are frequently buried in the ground where aggressive water might seriously attack metal tanks [2.129].

Very large enclosures are frequently needed to house activities requiring a great deal of space, or to shelter clusters of activities which would normally require several buildings.

Membranes offer a means of providing such structures with a

minimum of material; they are frequently a composite of fabric, often glass or other high strength fibre such as nylon, coated with a thermoplastic such as PVC.

During the past few years, there has been a great increase in the number and size of structures supported by internal air pressure. These are customarily transparent plastic membranes or plastic-coated glass or polyamide fabrics.

In the case of the house shown in Fig. 2.47, an effort was made to employ structural plastics in their own right, rather than as substitutes for other building materials. To obtain the best compromise between cost and reasonably high elastic modulus, a combination of heavy woven roving reinforcement and polyester was chosen. The resulting composite elastic modulus was 2–3 times higher than random mat, but lower in cost than a finer weave fabric.

The 33 m diameter circular marker building outside Paris (Fig. 2.48) illustrates the use of a curved conoid for a roof structure.

The opaque curved members, parabolic in cross-section, are made of glass fibre reinforced polyester, white gel-coated on the lower surface, and dark on the upper surface. Maximum thickness is approximately 9·5 mm. The conoids are supported on an outdoor peripheral steel tension ring, an inner steel compression ring, and curved pipe ribs along the valley where the conoids are joined to each other on the ribs [2.129]. A lot of other examples may be found in the literature [2.1, 2.80, 2.130, 2.131].

The main corrosion aspects and environmental deterioration of different composites are fully discussed in the literature [2.106, 2.109, 2.132, 2.133].

SANDWICH PANELS

Sandwich panels are generally considered to be a layered construction composed of thin, high modulus facings bonded to a lightweight core. The thin facings carry most of any applied load, and provide the panel with its stiffness and strength characteristics. The core acts to separate the facings and transmits shear forces between them so that they are effective about a common axis.

Most people now realize that composites employing the best features of a number of materials can be made to outperform any

(a)

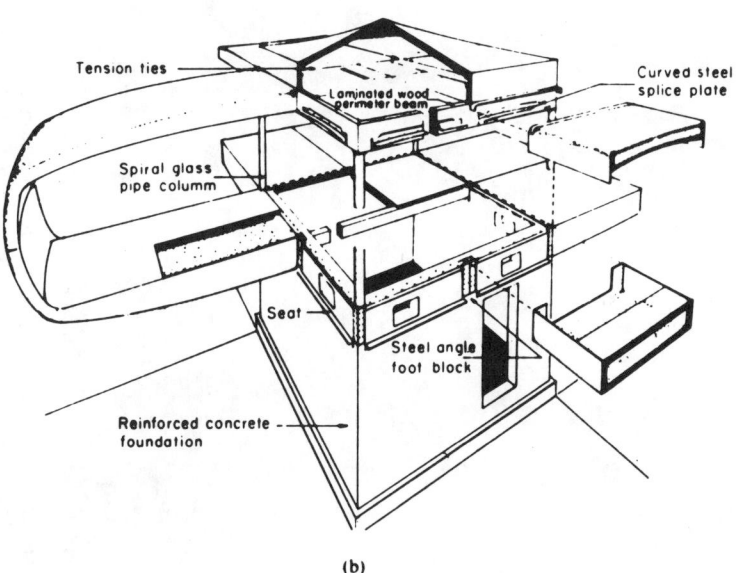

Tension ties

Curved steel
splice plate

Laminated wood
perimeter beam

Spiral glass
pipe columm

Seat

Steel angle
foot block

Reinforced concrete
foundation

(b)

Fig. 2.47. House of the future. (a) External view. (b) Exploded view showing
principal structural elements [2.129]. (Reprinted with permission from
Composite Materials, Vol. 3, by L. Broutman & R. C. Krock, 1974. Copyright
Academic Press.)

Fig. 2.48. Market at Argenteuil, curved conoid roof section being placed [2.129]. (Reprinted with permission from *Composite Materials*, Vol. 3, by L. Broutman & R. C. Krock, 1974. Copyright Academic Press.)

single material. Through proper choice of materials, coupled with proper design, cost effective applications for sandwich panels can be found in many fields.

During the Second World War, substantial amounts of sandwich construction were based on panels with plywood faces and honeycomb cores. Later, new materials such as polymers were being used [2.134].

Sandwich panels can be made entirely of polymers—both faces (skin) and cores—or they can be of a mixed structure in which the core may be made of rigid foams and skins of aluminium, plywood or asbestos–cement plane plates, or paper impregnated with phenoplaste, stainless steel or GFP (glass fibre reinforced plastic) and honeycombs were used, but after a short time they were replaced by polymer foam cores which are lighter and may be produced more economically.

The properties of sandwich panels depend on the selected materials for the faces and core, the ratio between their thickness, the nature of the adhesive used and the technology applied for their production. Figure 2.49 shows a typical sandwich panel.

For building applications, sandwich structures must fulfil certain

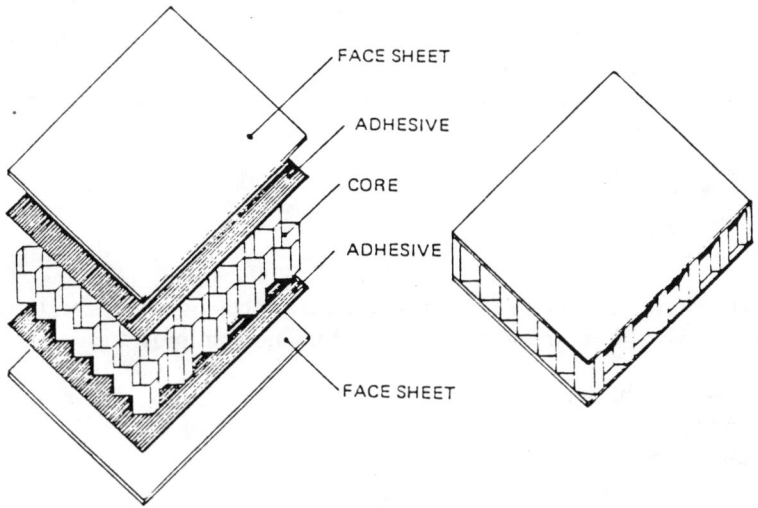

Fig. 2.49. Typical sandwich panel [2.135]. (Reprinted with permission from *Plastic Product Design Handbook,* Part A, by E. Miller (Ed.), 1981. Copyright M. Dekker, New York.)

conditions, such as:

—low thermal conductivity coefficient;
—low specific resistance;
—high permeability for water vapour;
—resistance to steam diffusion;
—low thermal expansion coefficient not affecting the bonds between panel and support elements;
—good performance at high temperatures up to 90°C for thermoplastics and 110°C and over for thermosetting matrix.

Sandwich panels may be classified using various criteria, such as:

—appearance,
—end-use,
—structure.

Panels can be transparent and non-transparent in appearance. In the first case, the core is made of GRP honeycombs or corrugated sheets, and the skins are made of transparent polymers. In the second case, the core is generally a rigid foam and the skins may be metal, plywood, etc. Depending on their application, panels are designed for external or separation walls, roofing, doors, and in two-dimensional and three-dimensional shapes.

However, the more rational classification is perhaps on the basis of core structure, which may be:

—a plastic foam,
—a honeycomb structure,
—a mixed one.

The basic function a sandwich panel faces is to resist tension, compression, bending and shock effects. The skin must:

—resist weathering,
—stop water penetration,
—have good durability,
—be non-flammable,
—have a low thermal conductivity, and
—have low cost of production, mount and maintenance.

Synthetic materials often used for the faces of sandwich panels are

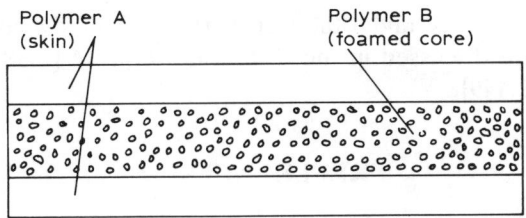

Fig. 2.50. Plan of the cross-section of a sandwiched foam product [2.134].

the following:

—PVC and vinyl chloride copolymers,
—PMMA (poly-(methyl methacrylate)),
—wood–polymer laminates,
—FRP (fibre-reinforced plastics).

The core may consist of materials such as honeycomb, foamed polymers (Fig. 2.50), foamed glass, plywood, and others. Shear stresses in the core itself are usually small, but the rigidity of this element is of significance in determining the structural behaviour of the sandwich panel [2.81, 2.134].

Sandwich panels are usually fastened to an open framework as a transverse web to carry shear loading. This means that sandwich panels are seldom primary structural members, but are attached to the primary structures and carry shear loads. When properly attached to the framework, sandwich panels are capable of withstanding extremely high loadings.

Sandwich panels were initially used in either transportation or architectural applications. The transportation industry utilizes their high strength and low weight, while the building industry is generally concerned with materials that will do an acceptable job at a lower cost [2.135].

Below are some common sandwich panel uses:

Building: prefabricated housing and shelters;
partitions and room dividers;
curtain wall panels;
movable shelters (hospitals);
doors and desk tops.
Aircraft, Aerospace, Transportation fields, etc.

Mechanical aspects and other properties of sandwich panels for construction are discussed in the technical literature [2.38, 2.80, 2.85, 2.128, 2.134–2.149].

REFERENCES

[2.1]   Richardson, M. D. W. (Ed.), In *Polymer Engineering Composites*. Applied Science Publishers, London, 1977, pp. 1–44.
[2.2]   Anon., *Encyclopedia of Polymer Science and Technology*, Vol. 12. Wiley-Interscience, New York, 1970, p. 5.
[2.3]   Brydson, J. A., *Plastic Materials*, 4th edn. Butterworths, London, 1982.
[2.4]   Albright, L. F., *Processes for Major Addition-type Plastics and their Monomers*. R. E. Krieger Publ. Co., Malabar, Florida, 1985.
[2.5]   Roff, W. J., & Scott, J. R., *Fibres, Films, Plastics and Rubbers*. Butterworths, London, 1971.
[2.6]   Nass, L. J., *Encyclopedia of PVC*, Vol. 1. Marcel Dekker, New York, 1976.
[2.7]   Throne, J. L., *Plastics Process Engineering*. Marcel Dekker, New York, 1976.
[2.8]   Ritchie, P. D. (Ed.), *Plasticizers, Stabilizers, and Fillers*. Iliffe Books Ltd, London, 1972.
[2.9]   Titow, W. V., *PVC Technology*, 4th edn. Elsevier Applied Science Publishers, London and New York, 1984.
[2.10] Gomez, I. L. (Ed.), *Engineering with Rigid PVC. Processability and Applications*. Marcel Dekker, New York and Basel, 1984.
[2.11] Saunders, K. J., *Organic Polymer Chemistry*. Chapman and Hall, London, 1973.
[2.12] Hilton, G. B., & Johnson, C. A., In *Engineering Thermoplastics, Properties and Applications*, ed. M. Margolis. Marcel Dekker, New York and Basel, 1985, pp. 355–72.
[2.13] Moore, G. R., & Kline, D. E., In *Properties and Processing of Polymers for Engineers*. Prentice Hall International Inc., London, 1984, p. 59.
[2.14] Ulrich, H., In *Introduction to Industrial Polymers*. Hanser Publishers, Munchen, 1982, pp. 109–17.
[2.15] Billmeyer F. W., Jr, *Textbook of Polymer Science*, 3rd edn. Wiley-Interscience, New York, 1984, p. 415.
[2.16] Harris, J. E., In *Engineering Thermoplastics. Properties and Applications*, ed. M. Margolis. Marcel Dekker, New York and Basel, 1985, pp. 177–200.
[2.17] Critchley, J. P., Knight, G. H., & Wright, W. W., *Heat Resistant Polymers*. Plenum Press, New York and London, 1983.
[2.18] Cessna, L. C., & Jabloner, H., *J. Elast. Plast.*, **6** (1974) 103.
[2.19] Pritchard, G. (Ed.), *Developments in Reinforced Plastics I*. Applied Science Publishers, London, 1980.

[2.20] Dine-Hart, R. A., Parker, D. B. V., & Wright, W. W., *Br. Polym. J.*, **3** (1971) 226.
[2.21] Parkyn, B., Lamb, F., & Clifton, B. V., *Polyesters, Vol. 2.* Iliffe Books Ltd, London, 1967.
[2.22] Madorsky, S. L., & Straus, S., *Mod. Plast.*, **38**(6)/(1961) 134.
[2.23] Miwa, M., Nakayama, A., Oshawa, T., & Hasegawa, A., *J. Appl. Polym. Sci.*, **23** (1979) 2957.
[2.24] Miwa, M., Oshawa, T., & Tahra, K., *J. Appl. Polym. Sci.*, **25** (1980) 795.
[2.25] Miwa, M., Oshawa, T., & Tsuji, N., *J. Appl. Polym. Sci.*, **23** (1979) 1679.
[2.26] Kaelble, D. H., In *Epoxy Resins—Chemistry and Technology*, ed. C. A. May & Y. Tanaka. Marcel Dekker, New York, 1973, p. 328.
[2.27] Mika, T. F., In *Epoxy Resins—Chemistry and Technology*, ed. C. A. May & Y. Tanaka. Marcel Dekker, New York, 1973, p. 298.
[2.28] Chu, H. S., & Seferis, J. C., In *The Role of the Polymeric Matrix in the Processing and Structural Properties of Composite Materials*, ed. J. C. Seferis & L. Nicolais. Plenum Press, New York and London, pp. 53–126.
[2.29] Morgan, R. J., ibid., pp. 207–214.
[2.30] Outwater, J. O., In *Composites: State of the Art*, ed. J. W. Weeton & E. Scala. Proceedings of Sessions, 1971 Fall Meeting, The Metallurgical Society of AIME, Detroit, Michigan, 1974, p. 13.
[2.31] Parker, D. B. V., In *Fillers for Plastics*, ed. W. C. Wake. Iliffe Books Ltd, London, 1971, pp. 129–45.
[2.32] Ferrigno, T. H., In *Handbook of Fillers and Reinforcements for Plastics*, ed. H. S. Katz & J. V. Milewski. Van Nostrand Reinhold Co., New York, 1978, pp. 11–58.
[2.33] Dubois, J. H., & John, F. W., *Plastics*, 6th edn. Van Nostrand Reinhold Co., New York, 1981.
[2.34] Koenig, J. L., & Chiang, C., In *The Role of Polymeric Matrix in the Processing and Structural Properties of Composite Materials*, ed. J. C. Seferis & L. Nicolais. Plenum Press, New York and London, 1983, pp. 503–16.
[2.35] Mascia, L., *The Role of Additives in Plastics*. Edward Arnold, London, 1974.
[2.36] Couts, R. S. P., & Campbell, M. D., *Composites* (Oct., 1979) 228.
[2.37] Richardson, G. C., In *Plastic Product Design Handbook, Part A*, ed. E. Miller. Marcel Dekker Inc., New York and Basel, 1981, pp. 116–118.
[2.38] Boaira, M. S., & Chffey, C. E., *Polym. Eng. Sci.*, **17**(10) (Oct. 1977) 715–18.
[2.39] Paik, C. S., Lee, S. H., & Sung, N. H., In *Adhesion and Absorption of Polymers*, ed. L. H. Lee. Plenum Press, New York and London, 1980, pp. 757–73.
[2.40] Jacobs, J. A., & Kilduff, T. F., *Engineering Materials Technology*. Prentice Hall Inc., Englewood Cliffs, New Jersey, 1985.
[2.41] Mohr, J. G., In *Handbook of Fillers and Reinforcements for Plastics*,

ed. H. S. Katz & J. V. Milewski. Van Nostrand Reinhold Co., New York, 1978, pp. 467–510.

[2.42]   Holmes, M., & Just, D. J., *GRP in Structural Engineering.* Applied Science Publishers, London and New York, 1983.

[2.43]   Gill, R. M., *Carbon Fibres in Composite Materials.* Iliffe Books Ltd, London, 1972.

[2.44]   Katz, H. S., In *Handbook of Fillers and Reinforcements for Plastics,* ed. H. S. Katz & J. V. Milewski. Van Nostrand Reinhold Co., New York, 1978, pp. 562–82.

[2.45]   Donnet, J. B., & Bansal, R. C., *Carbon Fibres.* Marcel Dekker, New York and Basel, 1984.

[2.46]   Pilling, M. W., Yates, B., Black, M. A., & Tattersall, P., *J. Mater. Sci.,* **14** (1979) 1326–38.

[2.47]   Loos, A. C., & Springer, G. S., *J. Comp. Mater.* **13** (Jan. 1979) 17–33.

[2.48]   Radcliffe, D. J., & Rosenberg, H. M., *Cryogenics* (May 1982) 245–9.

[2.49]   Wells, H., & Hancox, N. L., *Polym. Engng Sci.,* **18**(2) (Feb. 1978) 87–96.

[2.50]   Summerscales, J., & Short, D., *Composites* (July 1978) 157–66.

[2.51]   Short, D., & Summerscales, J., *Composites* (Oct. 1979) 215–22.

[2.52]   Short, D., & Summerscales, J., *Composites* (Jan. 1980) 33–8.

[2.53]   Manders, P. W., & Bader, M. G., *J. Mater. Sci.,* **16** (1981) 2233–45.

[2.54]   Manders, P. W., & Bader, M. G., *J. Mater. Sci.,* **16** (1981) 2246–56.

[2.55]   Matthews, F. L., Roshan, A. A., & Phillips, L. N., *Composites* (July 1982) 225–8.

[2.56]   Hancox, N. L., *J. Mater. Sci.,* **16** (1981) 627–32.

[2.57]   Phillips, L. N., *Plastics and Rubber International,* **3**(6) (Nov./Dec. 1978) 239–43.

[2.58]   Mijovic, J. J., & Liang, R. C., *Polym. Engng Sci.,* **24**(1) (Jan. 1984) 57–66.

[2.59]   Philpot, K. A., & Randolph, R. E., *2nd International Conference on Nonmetallic Materials and Composites at Low Temperatures,* Aug. 4–5, Hercules Incorporated, Magna, Utah, USA (1980).

[2.60]   Garg, A., & Ishai, O., *NASA Technical Memorandum No. 84370.*

[2.61]   Garg, A., & Ishai, O., *NASA Technical Memorandum No. 85935.*

[2.62]   Piggott, M. R., & Harris, B., *J. Mater. Sci.,* **16** (1981) 687–93.

[2.63]   Fitzer, E., & Weiss, R., *16th Biennial Conference on Carbon,* University of California, San Diego, July 18–22, 1983.

[2.64]   Lauer, L. D., *SAMPE Quarterly* (Oct. 1983) 31–5.

[2.65]   Axelson, J. W., In *Handbook of Fillers and Reinforcements for Plastics,* ed. H. S. Katz & J. V. Milewski. Van Nostrand Reinhold Co., New York, 1978, pp. 415–28.

[2.66]   Kacir, L., & Narkis, M., *Composites* (Jan. 1979) 31–6.

[2.67]   Husslein, R. W., & Fallick, G. J., In *New Polymeric Materials,* ed. P. F. Bruins. Interscience, New York and London, 1969, pp. 119–35.

[2.68]   Roberts, J. A., In *Handbook of Fillers and Reinforcements for*

Plastics, ed. H. S. Katz & J. V. Milewski. Van Nostrand Reinhold
Co., New York, 1978, pp. 597–616.
[2.69] Bigg, D. M., Composites (Apr. 1979) 95–100.
[2.70] Bowling, J., & Groves, G. M., J. Mater. Sci., 14 (1979) 443–9.
[2.71] Baker, A. A., Metals Forum, 6(2) (1983) 81–101.
[2.72] Johnson, D. J., In Applied Fibre Science, Vol. 3, ed. F. Happey.
Academic Press, London and New York, 1979, pp. 124–62.
[2.73] Sturgeon, D. L. G., & Lacy, R. I., In Handbook of Fillers and
Reinforcements for Plastics, ed. H. S. Katz & J. V. Milewski. Van
Nostrand Reinhold Co., New York, 1978, pp. 511–44.
[2.74] Link, R. S., J. Polym. Sci., Macromolecular Reviews, 13 (1978)
355–87.
[2.75] Brown, J. R., & Ennis, B. C., Text Res. J., 47 (1977) 62.
[2.76] Milewski, J. V., Plastics Compounding (Nov./Dec. 1979) 17–37.
[2.77] Phillips, L. N., In Polymer Science, Vol. 2, ed. A. D. Jenkins.
North-Holland, Amsterdam, 1972, p. 1700.
[2.78] Hull, D., Chem. Ind. (Nov. 1982) 84–9.
[2.79] Rolston, J. A., Chem. Engng, 86(2) (Jan. 1980) 96–110.
[2.80] Hollaway, L., Glass Reinforced Plastics in Construction: Engineering
Aspects. Surrey University Press, Blackie Publishing Group, Glasgow,
UK, 1978.
[2.81] Weatherhead, R. G., FRP Technology, Fibre Reinforced Resin
Systems. Applied Science Publishers Ltd, London, 1980.
[2.82] Spencer, R. A. P., In Developments in GRP Technology, ed. B.
Harris. Applied Science Publishers Ltd, London and New York, 1983,
pp. 1–37.
[2.83] Gaylord, M. W., Reinforced Plastics. Theory and Practice, 2nd edn.
Cahners Books, Boston, 1974.
[2.84] Darlington, M. W., & Christie, M. A., In The Role of the Polymeric
Matrix in the Processing and Structural Properties of Composite
Materials, ed. J. S. Seferis & L. Nicolais. Plenum Press, New York
and London, 1983, pp. 319–55.
[2.85] Scala, E., Composite Materials for Combined Functions. Hayden Book
Co. Inc., Rochelle Park, New Jersey, 1973, Howard W. Sams & Co.,
Indianapolis.
[2.86] Folkes, M. J., Short Fibre Reinforced Thermoplastics. Research
Studies Press, Chichester and New York, 1982.
[2.87] Ogorkiewicz, R. M., In Glass Reinforced Plastics, ed. B. Parkyn. Iliffe
Books Ltd, London, 1970, pp. 192–3.
[2.88] McCullough, R. L., Jarzebski, G. J., & McGee, S. H., In The Role of
the Polymeric Matrix in the Processing and Structural Properties of
Composite Materials, ed. J. C. Seferis & L. Nicolais. Plenum Press,
New York and London, 1981, pp. 261–88.
[2.89] Fleck, J. N., & Hanby, K. R., In Composites: State of the Art, ed. J.
W. Weeton and E. Scala. Proceedings of Sessions 1971 Fall Meeting,
The Metallurgical Society of AIME, Detroit, Michigan, 1974, pp.
25–66.

[2.90]	Butt, L. T., & Wright, D. C., *The Use of Polymers in Chemical Plant Construction*, Applied Science Publishers, London, 1980.
[2.91]	Delmonte, J., *Technology of Carbon and Graphite Fibre Composites*. Van Nostrand Reinhold Co., New York, 1981.
[2.92]	Desvaux, M. P. E., & Smith, G., *Fibre Sci. Technol.* **18** (1983) 53–64.
[2.93]	Springer, G. S., AFWAL-TR-82-4004 (Feb. 1982).
[2.94]	Balow, M. J., & Fuccela, D. C., Thermofil Inc., Engineering Thermoplastics Thermofil, Brighton, Michigan, 1984.
[2.95]	Beckwith, S. W., & Wallace, B. D., *28th National SAMPE Symposium*, April 12–14, 1983.
[2.96]	Bigg, D. M., *Polym. Engng Sci.*, **19**(6) (Dec. 1979) 1188–92.
[2.97]	Humphris, K. J., *Plastics and Rubber International*, **5**(6) (Dec. 1980) 237–42.
[2.98]	Bascom, W. D., Bitner, J. L., Moulton, R. J., & Sieberg, A. R., *Composites* (Jan. 1980) 9–18.
[2.99]	Baggott, R., & Gandhi, D., *J. Mater. Sci.*, **16** (1981) 65–74.
[2.100]	Cunningham, B., Sargent, J. P., & Ashbee, K. H. G., *J. Mater. Sci.*, **16** (1981) 620–6.
[2.101]	Piggott, M. R., & Wilde, P., *J. Mater. Sci.*, **15** (1980) 2811–15.
[2.102]	Mijovic, J., *J. Appl. Polym. Sci.*, **27** (1982) 1149–62.
[2.103]	Kiss, G., Kovacs, A. J., & Wittmann, J. C., *J. Appl. Polym. Sci.*, **26** (1981) 2665–77.
[2.104]	Kadotani, K., *Composites* (Apr. 1980) 87–94.
[2.105]	Hsich, H. S. Y., *J. Mater. Sci.*, **17** (1982) 438–46.
[2.106]	Norris, J. F., Crowder, J. R., & Probert, C., *Composites*, **7**(3) (1976) 165–72.
[2.107]	Boutevin, B., Hervaud, Y., Lafont, J., & Pietrasanta, Y., *Engng Polym. J.*, **20**(9) (1984) 867–73.
[2.108]	Ishida, H., *Polym. Comp.*, **5**(2) (1984) 101–23.
[2.109]	Roylance, D., & Roylance, M., *Polym. Engng. Sci.*, **18**(4) (1978) 249–54.
[2.110]	Runt, J., & Galgoci, E. C., *J. Appl. Polym. Sci.*, **29** (1984) 611–17.
[2.111]	Norwood, L. S., & Millman, A. F., *Composites* (Jan. 1980) 39–45.
[2.112]	Allen, R. C., *Polym. Engng Sci.*, **19**(5) (1979) 329–36.
[2.113]	Blaga, A., & Yamasaki, R. S., *Matériaux et Construction*, **10**(59) (1977) 289–96.
[2.114]	Danusso, F., Tieghi, G., & Lestingi, A., *J. Appl. Polym. Sci.*, **33** (1987) 2137.
[2.115]	Peiffer, D. G., & Nielsen, L. E., *J. Appl. Polym. Sci.*, **24** (1979) 1451–5.
[2.116]	Peiffer, D. G., & Nielsen, L. E., *J. Appl. Polym. Sci.*, **23** (1979) 2253–64.
[2.117]	Douglass, D. C., & McBrierty, V. J., *Polym. Engng Sci.*, **19**(14) (1979) 1054–63.
[2.118]	Marom, G., & Cohn, D., *J. Mater. Sci.*, **15** (1980) 631–4.
[2.119]	Carswell, W. S., & Roberts, R. C., *Composites* **11** (1980) 95–100.
[2.120]	Miwa, M., Ohsawa, T., & Tahara, K., *J. Appl. Polym. Sci.*, **25** (1980) 795–807.

[2.121] Matthews, F. L., In *Developments in GRP Technology*, ed. B. Harris. Applied Science Publishers, London and New York, 1983, pp. 161–89.

[2.122] Godwin, E. W., & Matthews, F. L., *Composites* (July 1980) 155–60.

[2.123] Matthews, F. L., Kilty, P. F., & Godwin, E. W., *Composites* (Jan. 1982) 29–38.

[2.124] Hoa, S. V., & Feldman, D., *Polymer Composites*, **3**(1) (1982) 48–53.

[2.125] Feldman, D., Hoa, S. V., & Coriarty, E., *Polym. Plast. Technol. Engng*, **23**(1) (1984) 99–118.

[2.126] Mitchell, R. G. B., In *Glass Reinforced Plastics*, ed. B. Parkyn. Iliffe Books Ltd, London, 1970, pp. 11–30.

[2.127] Bandyopadhyay, S., In *Recent Advances in Materials Research*, ed. C. M. Srivastava. IBH, Oxford and Bombay, 1984, pp. 343–61.

[2.128] Blaga, A., *Industrialization Forum*, **9**(1) (1978) 24.

[2.129] Dietz, A. G. H., In *Composite Materials, Vol. 3*, ed. L. Broutman & R. C. Krock. Academic Press, New York, 1974, pp. 254–300.

[2.130] Montella, R., *Plastics in Architecture*. Marcel Dekker, New York, 1985.

[2.131] Cheremisinoff, N. P., & Cheremisinoff, P. N., *Fibreglass-Reinforced Plastics Deskbook*. Ann Arbor Science Publishers Inc., Ann Arbor, Michigan, 1978.

[2.132] Hogg, P. J., & Hull, D., In *Developments in GRP Technology—1*, ed. B. Harris. Applied Science Publishers, London and New York, 1983, pp. 37–90.

[2.133] Davis, A., & Sims, D., *Weathering of Polymers*, Applied Science Publishers, London and New York, 1983, pp. 266–85.

[2.134] Fazio, P., & Feldman, D., *Plastics and Rubber International*, **5**(2) (1980) 67–72.

[2.135] Gill, P. C., In *Plastics Product Design Handbook, part A*, ed. E. Miller. Marcel Dekker, New York and Basel, 1981, pp. 355–97.

[2.136] Skeits, J., *Plastics in Building*, Reinhold Publishing Co., New York, 1966.

[2.137] Skeits, J., *The Plastics Manual*, 5th edn, *Applied Plastics*. The Scientific Press Ltd, London, 1971.

[2.138] Ueng, C. E. S., Reprint No. 3695, ASCE Convention, Atlanta, Oct. 23–25, 1979.

[2.139] Platts, R. E., *Forest Products J.*, **12**(9) (Sept. 1962) 429–30.

[2.140] Kourtides, D. A., Gilwee Jr., W. J., & Parker, J. A., *Polym. Eng. and Sci.*, **19**(3) (Feb. 1979) 226–31.

[2.141] Outwater, J. O., In *Composites: State of the Art*, ed. J. W. Weeton & E. Scala. Proceedings of Sessions 1971 Fall Meeting, The Metallurgical Society of AIME, Detroit, Michigan, 1974.

[2.142] Broutman, L. J., & Krock, R. H., *Composite Materials, Vol. 3, Engineering Applications of Composites*. Academic Press, New York, 1974.

[2.143] Theberge, J. E., & Goebel, C. V., 33rd Annual Technical Conference, Reinforced Plastics/Composites Institute, The Society of the Plastics Industry Inc., Section 16E, 1978, pp. 1–15.

[2.144] Belanger, G., *Industrialisation Forum*, **9**(1) (1978) 37.
[2.145] Marsh, C., Criteria for Sandwich Panels in Building, Symposium 1972, Panelized Structural Assemblies, Montreal.
[2.146] Plantema, F. G., *Sandwich Construction*, John Wiley, New York, 1966.
[2.147] Shetty, R., & Han, C. D., *J. Appl. Polym. Sci.*, **22** (1978) 2573–84.
[2.148] Fazio, P., *Journal of the Structural Division, Proceedings of the American Society of Civil Engineers* (May 1972) 1085.
[2.149] Becker, W. E., *NS Sandwich Panel, Manufacturing Marketing Guide*. Technomic Publishing Co., Stamford, 1968.

Chapter 3

Polymer–Concrete Composites

As the world's needs for housing, transportation and industry increase, the consumption of concrete products is expected to increase correspondingly. At the same time, prudent management of energy and natural resources demands ever high levels of performance. Although Portland cement concrete is one of the most remarkable and versatile construction materials, a clear need is perceived for the improvement of properties such as strength, toughness, ductility and durability. One valid approach is to improve the concrete itself; another is to combine technologies in order to make new composites based on cement.

Polymers containing large amounts of filler, such as polymer mortars without cement, and polymer concretes, are increasingly being used in buildings and other structures. Polymer mortars are mainly used as protective coatings on concrete, reinforced concrete, and rarely on steel, while polymer concretes represent a new type of structural material capable of withstanding highly corrosive environments.

The mechanical properties, the corrosion stability and some other useful properties are the reasons for the continuous interest shown in these materials by various design, research and production organizations. However, polymer mortars and polymer concretes have been introduced only recently, and many of their properties are still imperfectly known.

Some polymers are relatively cheap and completely resistant to alkali attack by cement paste. They offer hope in overcoming one of the main problems of fibre reinforced concrete, which is the lack of ductility. The material tends to crack rather than bend under relatively modest loads. Apart from polymers, which are only now being

207

developed, interest is being shown in natural and synthetic fibres, mainly to produce asbestos–cement substitutes.

The development of concrete–polymer composite materials is directed at both improved and new materials by combining the well known technology of hydraulic cement concrete formation with the modern technology of polymers. It should be noted that combinations of siliceous materials with polymers require, in many cases, lower energy inputs per unit of performance than either component alone [3.1].

A wide range of concrete–polymer composites is being investigated, although only some of them are already being applied. The most important are the following:

—polymer impregnated concrete (PIC);
—polymer–cement concrete (PCC);
—polymer concrete (PC);
—fibre-reinforced concrete;
—fibre-reinforced polymer concrete.

PIC is a precast and cured hydrated cement concrete which has been impregnated with a monomer, which is subsequently polymerized *in situ*. This type of cement composite is the most developed of polymer-concrete products.

PCC is a premixture of cement paste and aggregate to which a monomer is added prior to setting.

PC is an aggregate bounded with a polymer binder. This product may be obtained on site. It is called a concrete because according to the general definition, concrete consists of any aggregate bound with a binder.

The last two products are based on natural, metallic or synthetic fibres as reinforcing agents.

POLYMER IMPREGNATED CONCRETE

The largest improvement in structural and durability properties has been obtained with this composite system. In the presence of a high polymer phase, the compressive strength can be increased four times or more, water absorption is reduced by 99%, the freeze–thaw resistance is enormously improved, and in contrast to conventional concrete, polymer impregnated concrete exhibits essentially zero creep properties.

The ability to vary the shape of the stress–strain curve presents some interesting possibilities for tailoring desired properties of concrete for particular structural applications.

Normal concrete consists of particles of a fine and coarse aggregate dispersed in a matrix of hardened cement. Since stone aggregates and sand have very low porosity, the bulk of the porosity of concrete is in the cement phase. Therefore the macromolecular compound must be concentrated in the cement paste phase.

The system becomes more complex when polymer is added to concrete. The high polymer fills the pores of the hardened cement paste phase, creating still another composite, the hardened cement paste–polymer system. The system may be considered to be a composite of coarse and fine aggregate in a matrix of impregnated cement.

The mechanism of polymer–concrete system formation and the physico-chemical phenomena are very complex and not yet very well clarified. The processes which accompany the formation of such types of composites are known only in general features, based on the observations made on the exterior phenomena. One main factor in controlling the properties of the PIC is the porosity, but this cannot account for the different performance recorded with different polymers.

It has for some time been recognized that the pore structure of hardened cement paste plays a decisive role with respect to many important characteristics of concrete. It is expected therefore, that the property improvement due to impregnation is brought about mostly by pore structure modification, and should be a function of this (Fig. 3.1).

Considerable importance has been attached to the role of interphase bonding in determining the role of polymer impregnated composites, though this improvement has not been measured directly.

The chemical materials required for producing polymer–concrete systems are relatively new to the construction industry. For this reason it is very important to have a good knowledge of their properties, their hazards, and the safety precautions required. Safety procedures should be carefully followed whenever different chemicals for polymer–concrete composites are used.

PIC is generally prepared by impregnating dry precast concrete with

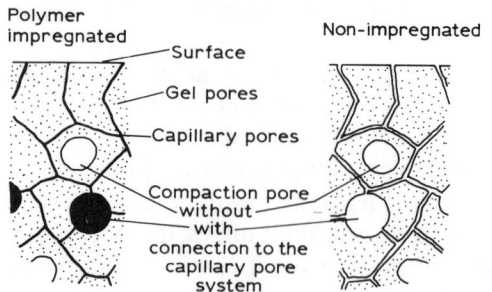

Fig. 3.1. Scheme of pores in hardened cement paste [3.3]. Pore size not drawn to the same scale. (Reprinted with permission from *Polymers in Concrete, 4th International Congress*, by H. Schulz (Ed.), 1984. Copyright Institut für Spanende Technologie und Werzeugmashinen, Technischen Hochschule, Darmstadt.)

a liquid monomer, and polymerizing the monomer *in situ* by thermal, catalytic or radiation methods.

Some of the most widely used monomers for polymer concrete systems include:

—methyl methacrylate (MMA);
—styrene (S);
—butyl acrylate (BA);
—vinyl acetate (VAc);
—acrylonitrile (AN);
—methyl acrylate (MA);
—trimethylpropane trimethylacrylate (TMPTMA).

These monomers may be used alone or in mixtures. Unsaturated polyester–styrene is a very common system for polymer–concrete composites. PIC based on epoxy polymers are more expensive and for this reason less often used, although their properties are superior.

Trimethylpropane trimethylacrylate (TMPTMA) serves as a cross-linking agent used also to decrease the time of polymerization (curing).

Monomers are generally toxic, combustible and volatile liquids. Practice has shown that prolonged stability and safety can be achieved by following recommended practices in storage and handling.

The principal factors that influence the stability of monomers and determine the methods to be used for safe handling are:

—concentration and effectiveness of stabilizers (inhibitors);
—storage conditions (light, moisture, temperature, oxygen, etc.);
—toxicity;
—flammability;
—effect on construction materials with which they are used [3.4];

The role of stabilizers is to prevent premature polymerization; they can act as anti-oxidants and prevent this process by reacting with oxidants that may be formed in the storage tank through contamination or through the formation of a peroxide from the oxygen of the air.

The time required for initial stabilizer concentrations to fall to a critical level varies greatly with storage and handling conditions.

Factors affecting the depletion of stabilizer are water, air, and, most importantly, heat.

The flowchart (Fig. 3.2) of the manufacturing process is generally based on the following main operations:

—preparation of precast concrete;
—drying of the precast concrete;

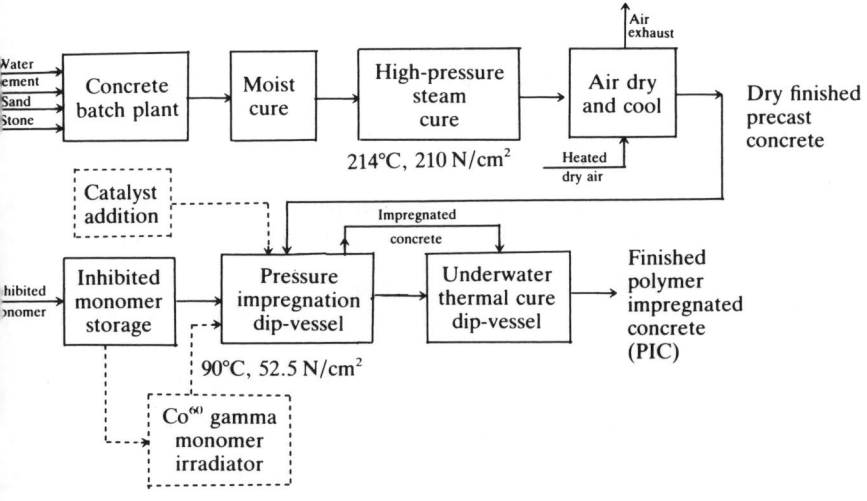

Fig. 3.2. Schematic of PIC process [3.2]. (Reprinted with permission from *Polymers in Concrete*, S.P. 40, 1973. Copyright ACI.)

—impregnation with the monomer;
—polymerization *in situ.*

Kukacka & Romano [3.5] have established that the degree of dryness of concrete strongly affects the amount of monomer absorbed during the impregnation. From their results it appears that a drying temperature of 150°C is required to remove most of the free water without seriously affecting the mechanical properties of the polymer–concrete composite.

Several technologies of concrete curing based on fog, low pressure steam, and high pressure steam, have been studied to determine the effects on the properties of PIC. The data indicate that high pressure steam curing results in specimens with higher polymer loadings, and generally higher strengths than for comparable fog-cured specimens [3.6].

Parameters such as water/cement ratio, entrained air content, and aggregate size and quality, do not greatly affect the properties of such composites. Some studies indicate that the impregnation of low strength concrete produced PIC with essentially the same properties as PIC produced from high strength concrete mixtures.

Property improvement as a result of impregnation is mostly due to pore structure modification and should be a function of the pore structure.

Hastrup *et al.* [3.7] have obtained the pore size distributions of concrete by two different methods:

—analysis of the mercury intrusion data;
—analysis of the adsorption data.

The mercury intrusion data was analysed by assuming cylindrical pores, and relating the intrusion pressure (P) to the pore diameter (d) by the equation:

$$P = \frac{4\gamma \cos \theta}{d} \tag{3.1}$$

where γ is the air/mercury interfacial tension, and θ is the angle of contact between mercury and the pore wall.

Porosity was determined by Feldman & Beaudoin [3.8, 3.9] before and after impregnation by measuring solid volume by helium comparison pycnometry. The apparent volume was established by weighing, in methanol, samples saturated with methanol.

In hydraulic setting engineering materials such as Portland cement and plaster of Paris, a large volume of porosity is inherent in the set structure. This porosity derives mainly from the excess water required to lubricate the powders so that the material may be placed, and it is called capillary porosity. Its amount is dependent upon the original water–cement ratio, and the maturity, or degree of hydration of the cement paste.

For a mature dry paste with a water/cement ratio of 0·5, Auskern & Horn [3.10] have shown that the volume of capillary porosity is about 0·25 cm^3/g. The additional porosity, called 'gel' porosity, is finer than the capillary one and may be present in mature pastes to the extent of about 0·1 cm^3/g. Thus the total cement porosity can be about 0·35 cm^3/g. If all these pores become filled upon monomer impregnation, a weight loading of 6–7% is calculated. Because of compositional variations most normal concretes indicate polymer weight loadings of between 5 and 8% [3.10].

With impregnation by an appropriate monomer, the main effect after polymerization is the sealing of the continuous capillary pore system, resulting in exceptional decreases, to approximately 99%, in the water and salt (such as sulphates and chlorides) permeability. Other effects include increases in the coefficient of thermal expansion (to 30%) and thermal diffusivity (to 13%), and a decrease in specific heat (to 17%). The pore-sealing also minimizes changes in properties, such as dielectric constant and loss, that are sensitive to moisture content [3.1].

Impregnation of the concrete with a liquid monomer occurs by a viscous flow mechanism, and consequently its viscosity is one limiting factor. At the laboratory scale, impregnation can be assisted by applying a vacuum to the concrete to remove entrapped air and after initial vacuum soaking, positive pressure can be applied to the monomer for additional driving force for flow into the pores. However, these approaches are more difficult for field impregnation of large surface areas [3.11].

Monomer diffusion into existing concrete is being carefully studied because special concretes with better corrosion resistance and mechanical properties are required.

Important problems arise in thick sections, which require very low viscosity monomers, but these are inevitably volatile products, and when polymerization is induced, the system must be heated. The polymerization process, being exothermic, evolves large quantities of heat, so that large losses of monomer take place from the outer layers.

The odour, volatility, toxicity and flammability of many monomers restrict their use in composite making, but in cases where penetration of porous structures is called for, then they have obvious advantages, as they leave no residue, nor require evaporation of solvents, water, etc. It will, however, often be found that the viscosity of some monomers is too low, and it is recommended to use oligomers instead.

Volatilization of the monomer during impregnation-polymerization is an important technical and economic factor. To avoid monomer losses, some measures must be taken such as:

—selecting low vapour pressure monomers;
—the use of evaporation barriers or encapsulation methods.

One of the steps in the fabrication of PIC is to wrap the concrete saturated with monomer in aluminium foil or polyethylene film in order to reduce monomer volatilization. In an attempt to further reduce the losses and to eliminate the necessity of wrapping, encapsulation techniques have been introduced, such as:

—replacing the monomer with its oligomers;
—underwater polymerization;
—encapsulation in a form.

Kukacka & Romano [3.5] established that maximum polymer contents in PIC are obtained with the underwater polymerization technique, which consists of a standard vacuum-soak impregnation, placing the specimens in water, and radio-polymerizing underwater.

The mechanical properties of PIC depend to a large measure on the dryness of concrete, some impregnation parameters (vacuum, pressure, soaking time), method of encapsulation, etc.

The type of concrete has a significant effect on the monomer filling rate. Concrete with high air and water content may be saturated with monomer in a shorter time than standard type concrete.

The polymerization of the monomer used for impregnation is realized usually by:

—thermal polymerization; or
—radiopolymerization, especially using gamma rays.

Attention is paid to heating methods such as solar energy, microwave heating, steam, heating blankets, or hot water [3.11].

Radiopolymerization involves some difficult problems with regard to the potential safety hazard in the field.

At present, the use of hot water appears to be the most feasible method of heat application for surface treatments.

From the standpoint of economy, solar energy is a very appealing heat source, but the use of sunlight is limited by variables which cannot be fully controlled. The use of microwave heating for polymerization in the field does not as yet appear feasible.

The initiators most commonly used in polymerization processes for producing concrete–polymer composites are:

—2,2′-azobis-(isobutyronitrile) (AIBN);
—2,2′-azobis (2,4-dimethylvaleronitrile) (AMVN);
—benzoyl peroxide (BP);
—lauroyl peroxide (LP);
—methyl ethyl ketone peroxide (MEKP).

Promoters used besides the initiators have the role to increase the rate of decomposition of the peroxide initiators.

One of the reasons preventing widespread use of PIC is the high cost of monomers. This is why it is very important to gain maximum improvement in the properties of PIC with a minimum consumption of monomer.

Epoxy polymers and polymers obtained through a polycondensation process, such as polyesters, are also used for partial impregnation.

The main consequence of concrete impregnation consists of the sealing of its continuous capillary pore system which leads to an important decrease of the permeability. The improvement depends on the amount of cells that are filled after the polymerization or polycondensation. Auskern & Horn [3.10] show that a reduction of the porosity of the cement phase by 10%, doubles the strength. Kukacka [3.13] mentions some other beneficial effects due to the presence of the polymer such as:

—the polymer acts as a continuous, randomly oriented reinforcing network;
—the polymer increases the cement paste–aggregate bond;
—the polymer absorbs energy during deformation;
—the polymer increases the strength of the aggregate.

To obtain polymer impregnated paste and mortar, Manning & Hope [3.14] have used various monomers such as methyl methacrylate (MMA), styrene (S), vinyl acetate (VAc), methyl acrylate (MA), and acrylonitrile (AN). In particular MA was chosen because its polymer

Polymeric Building Materials

T_g is about 3°C, and, in the bulk state PMA is a rubber-like material at room temperature. AIBN was used as the initiator in all polymerizations in the following amounts: 0·5% in MMA, MA and S, 1·0% in VAc, and 1·5% in S. The polymerization of all these monomers was carried out at 80 ± 5°C.

Ohama [3.15] has established that the polymer loading depends on the polymerization temperature in water according to the following formula:

$$\text{Polymer loading } (\%) = \frac{W_{\text{PIC}} - W_{\text{DC}}}{W_{\text{DC}}} \qquad (3.2)$$

where W_{PIC} = weight of PIC specimen (g) and W_{DC} = weight of dried concrete specimen (g).

Some of the results are as follows:

6·7% PMMA for polymerization at 70°C
6·5% PMMA for polymerization at 80°C
6·1% PMMA for polymerization at 90°C

Tazawa & Kobayashi [3.16] have found that shrinkage occurs through two stages in the production of PIC, namely through initial drying and through polymerization. The second one is peculiar to PIC and could be several times greater than the drying shrinkage as is shown in Fig. 3.3, where such shrinkage is plotted against polymer loading. The measurements were made by the method prescribed in ASTM

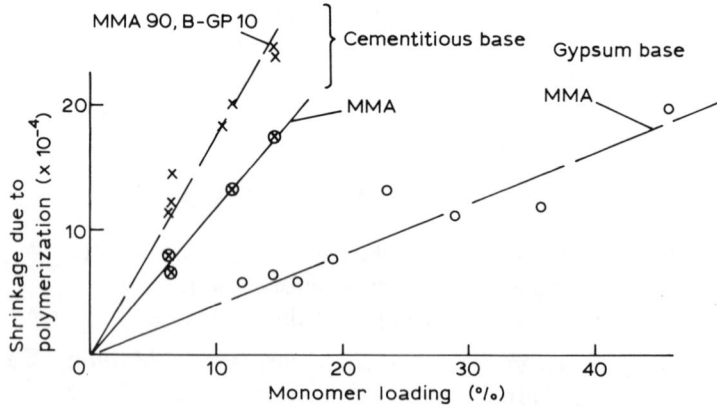

Fig. 3.3. Shrinkage due to polymerization [3.16]. (Reprinted with permission from *Polymers in Concrete*, S.P. 40, 1973. Copyright ACI.)

C-490-65T. Samples were all impregnated completely and polymer loadings in this figure were controlled by varying the amount of water and aggregates. The polymerization was realized at 60°C for 16 h, using 0·5% AIBN.

Godard *et al.* [3.17] have studied the mechanism and kinetics of impregnation. The results as a whole can be summarized by the following empirical equation:

$$D = \frac{D^* \gamma \cos \theta r^3}{\varepsilon \eta} \tag{3.3}$$

where:

D = the coefficient of diffusion
γ = the surface tension
θ = the contact angle
r = the average radius of the pores
ε = the porosity
η = the viscosity of the monomer
D^* = a constant equal to $2 \cdot 00 \times 10^9 \, \text{cm}^{-2}$.

Some experimental results of these authors regarding the kinetics of impregnation of asbestos–cement with different monomers such as: styrene, methyl methacrylate, methyl acrylate, and vinyl acetate, are presented in Fig. 3.4.

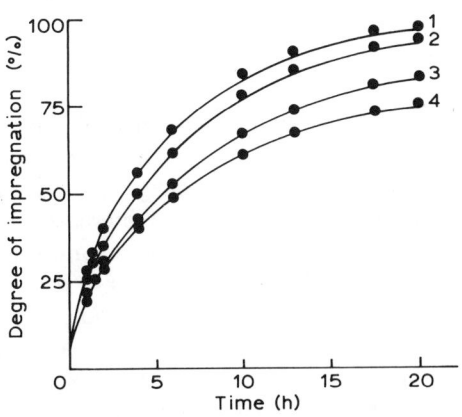

Fig. 3.4. Kinetics of impregnation of asbestos–cement. Porosity, 0·29; average radius, 148 Å; thickness, 4 cm [3.17]. Impregnating liquids: 1, styrene; 2, methylmethacrylate; 3, methyl acrylate; 4, vinyl acetate. (Reprinted with permission from *J. Appl. Polym. Sci.*, **18** (1974) 1477–91. Copyright John Wiley & Sons Inc., New York.)

In another paper, Godard & Mercier [3.18] have investigated the kinetics of polymerization of styrene and of methyl methacrylate in a porous asbestos–cement medium initiated by the thermal decomposition of benzoyl peroxide or of bis(4-tert-butylcyclohexyl) peroxydicarbonate. The results show that the presence of the porous substrate appreciably modifies the rate of polymerization. Two factors have been proposed to explain this effect: a variation of the constant of the thermal decomposition of the initiator resulting from the alkalinity of the cement, and a modification of the rate constant of chain propagation related to the electron-releasing or attracting capacity of the double bond of the monomers. Nevertheless, the polymerization remains solely initiated by radicals resulting from the thermal decomposition of the initiator as the rate varies according to the square root of the initiator concentration and tends towards zero for polymerization containing no peroxide.

When strength is of primary concern, full-impregnation is necessary; this is feasible only with laboratory or plant facilities since the entire specimen must be dried and fully saturated with the monomer. Research has been conducted to develop processes for impregnating concrete to depths of 10–40 mm. Partial depth impregnation, sometimes called surface impregnation, is desirable when durability rather than mechanical properties is of primary concern [3.19].

Manning & Hope [3.14] have found that the vinylic monomers used in their investigations are capable of penetrating the micropores in the cement paste and filling the available pore space to approximately the same degree as methanol.

They have also concluded that the influence of the polymer on the properties of PIC is determined primarily by the change in properties of the cement paste phase and not by changes at the paste–aggregate interface. The macromolecular compound increases the amount of solids per unit volume in the cement paste and simultaneously reduces the effect of stress concentrations from pores and microcracks. The authors consider also that the properties of the polymer in the pores of the cement paste are most likely to be the most important factors in determining the effectiveness of any particular polymer. These properties are not always a reflection of the bulk polymer properties.

In some studies, the polymerization of the impregnated monomer was realized through an irradiation procedure based on gamma rays. Thus, Hastrup & coworkers [3.7] led the polymerization of MMA in an MMA–concrete system under Co^{60} gamma radiation.

Feldman & Beaudoin [3.9] have vacuum saturated, on a laboratory scale, concrete specimens with methyl methacrylate. After saturation the samples were raised above the excess monomer and, in the presence of its vapour, exposed to Co^{60} radiation ≈ 2500 rad/min for 17 h. After the mechanical properties and porosity were measured, the samples were re-impregnated to reduce residual porosity due to shrinkage during polymerization, and re-exposed to the radiation. The authors concluded that large increases in Young's modulus and microhardness of Portland cement concrete can be achieved by impregnation with MMA. Increases in microhardness due to impregnation with this monomer are greater than with sulphur at the same volume fraction.

Madruga *et al.* [3.20] have extracted PMMA and have characterized the PMMA from a PIC product. The polymer isolation was carried out according to the different steps presented in Fig. 3.5.

PMMA in finely powdered PIC samples was first extracted with benzene as solvent. The residual powder was then treated with a saturated salicyclic acid–methanol solution, and a second fraction of PMMA was extracted from the new residue using benzene as solvent. The polymer fractions in benzene solution were precipitated with methanol, filtered and dried to constant weight. The yield of the extracted PMMA was determined gravimetrically. The characterization of the PMMA formed in the capillary pores of concrete shows the possibility of formation of polymer–cement associates which are similar to the intermolecular charge transfer complexes. The analysis of the MW and stereochemical configuration seems to indicate that the kinetics of polymerization are essentially similar to the MMA radical bulk polymerization, although the rate constant can be affected by the cement chemical composition.

For full impregnation of concrete specimens, the following process has been shown to provide high quality PIC using dense concrete specimens with cross-sections up to 300 mm [3.19]:

—Dry to constant weight at 150°C which requires approximately 24 h.
—Maintain vacuum for 30 min.
—Apply monomer under vacuum and then pressurize to 70 kPa. If vacuum is not used, pressure-soak for 60 min at 700 kPa.
—Remove section from monomer and place under water, or drain monomer from impregnation chamber and then fill with water.
—Polymerize monomer *in situ* with radiation, hot water, or steam.
—Remove water and clean specimen.

Several encapsulation techniques have been used to minimize

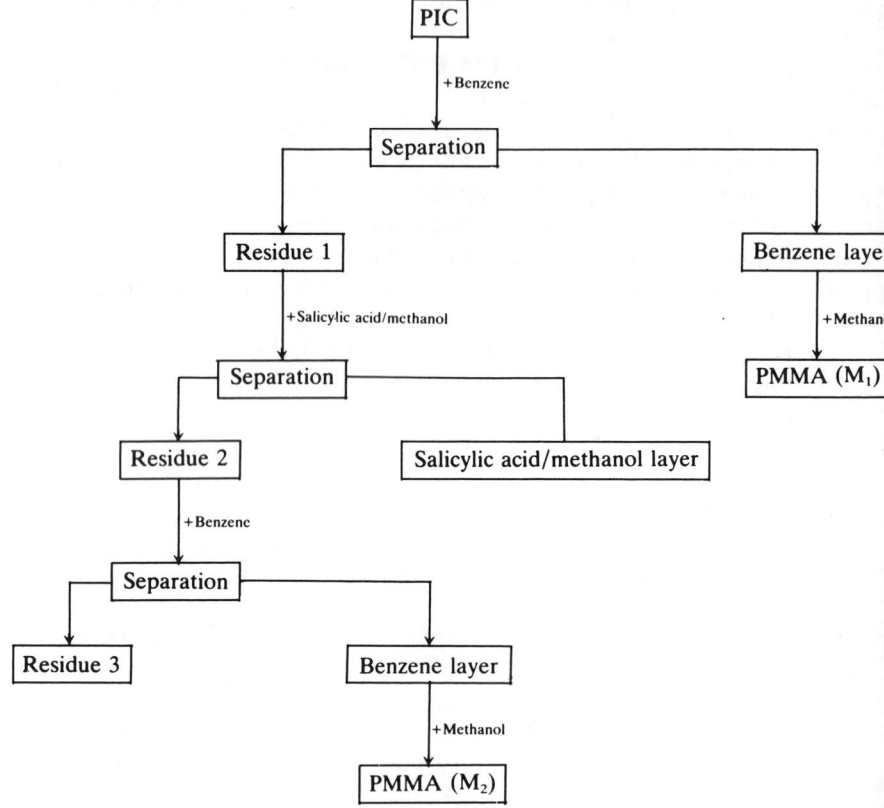

Fig. 3.5. Extraction of PMMA from PIC [3.20]. (Reprinted with permission from *Polym. Engng Sci.*, **19** (1979) 825. Copyright Society of Plastics Engineers.)

monomer evaporation losses during polymerization. Wrapping the specimen in polyethylene film and/or immersing the specimens in water has been found to be effective.

In their study for finding an appropriate process for manufacturing high-strength concrete, Fukuchi & Ohama [3.21] have used, for the impregnation of concrete, a mixture of styrene and trimethylacrylate as comonomers, AIBN as initiator and a silane type coupling agent. A similar formulation based on MMA was applied by Webster *et al.*

[3.22] for partial-depth polymer impregnation of concrete bridge decks.

Blaga & Beaudoin [3.23] have pointed out that the monomers most widely used in the impregnation of concrete are of the vinyl type.

PIC Properties

The incorporation of a relatively small volume of a macromolecular compound in the capillary pores of concrete, results in an important increase in the mechanical properties and durability of this engineering material. In their study, Auskern & Horn [3.10] have established that the main effects of the presence of the polymer in such composites are:

—to improve the strength and increase the modulus of the current phase;
—to improve the cement–aggregate bonding.

A number of workers have developed composite models based on the structure of concrete. One such concrete model can be put into the following form:

$$\frac{E}{E_a} = 1 + \frac{2V_a(r-1)}{(r+1) - V_a(r-1)} \tag{3.4}$$

where:

E = modulus of the composite
E_d = modulus of the aggregate phase
E_m = modulus of the matrix
V_a = volume fraction of the aggregate
r = ratio E_a/E_m

This equation can also be applied to mortars. The aggregate phase now consists of sand particles of modulus approximately 70×10^3 MN/m^2 ($10 \cdot 15 \times 10^6$ psi) and the matrix is common cement of modulus 7×10^3 MN/m^2 ($1 \cdot 01 \times 10^6$ psi) or polymer filled cement of modulus $19 \cdot 6 \times 10^3$ MN/m^2 ($2 \cdot 84 \times 10^6$ psi). The moduli of the two mortars can be calculated for various aggregate loadings. This is illustrated by Auskern & Horn [3.10] in Fig. 3.6, where theoretical curves predicted from the equation are presented along with Young's modulus for a series of normal and polymer filled mortars with sand contents ranging from 40 to 65% by volume. The calculated results are in fairly good agreement with the experimental results although the agreement appears better for the polymer filled series. The results

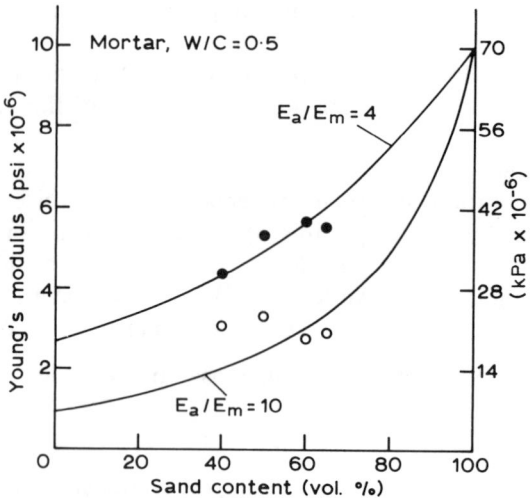

Fig. 3.6. Predicted and experimental variation of Young's modulus of mortar with sand content [3.10]. ●, Polymer filled; ○, unfilled. (Reprinted with permission from *Polymers in Concrete*, S.P. 40, 1973. Copyright ACI.)

indicate that for mortar, the only effect of the polymer is to increase the modulus of the cement phase.

Some concretes were also investigated by adding aggregate to mortar in concentrations of 10–50% by volume. The elastic moduli of the concretes were calculated with the former equation using $56 \times 10^3 \, \text{MN/m}^2 \, (8 \cdot 12 \times 10^6 \, \text{psi})$ as the modulus for the limestone aggregate and the observed modulus of the 40% sand mortar. These results are shown in Fig. 3.7. In this case the experimental results are below the calculated curves, and indicate an extrapolated mortar modulus (0% aggregate) lower than that used for the calculation.

In contrast to conventional concrete, PIC exhibits essentially zero creep properties. By polymer impregnating concrete, conventional concrete is transformed from a plastic material, or to be precise an elasto-plastic body, to essentially an elastic one with an increase of at least two times in the modulus of elasticity. This is reflected by the linearity of the stress–strain curve for PIC in Fig. 3.8, obtained by Steinberg [3.2]. The ability to vary the shape of the stress–strain curve presents, as mentioned previously, some very interesting possibilities for tailoring desired properties of concrete for particular structural applications.

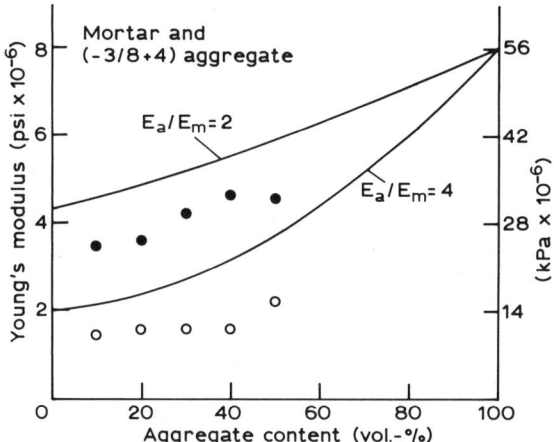

Fig. 3.7. Predicted and experimental variation of Young's modulus of concrete with 3/8 + 4 mesh aggregate content [3.10]. ● Polymer filled; ○ unfilled. (Reprinted with permission from *Polymers in Concrete*, S.P. 40, 1973. Copyright ACI.)

Tazawa & Kobayashi [3.16] have established that the bending strength is not simply related to polymer loading or water/cement ratio, while the compressive strength showed an inverse relation with polymer loading or volumetric concentration of the polymer (Fig. 3.9).

Some typical properties of PIC are presented in Tables 3.1 and 3.2, as discussed by DePuy [3.24].

In some of their investigations, Tazawa & Kobayashi [3.16] have replaced MMA with different amounts of DOP (dioctyl phthalate). Both the bending strength (Fig. 3.10) and the compressive strength (Fig. 3.11) were affected by the quantity of DOP, with the optimum replacement being approximately 5% by weight.

Specimens of polystyrene-impregnated concrete were also tested for compressive strength by Fukuchi & Ohama [3.21]. Some results are shown in Fig. 3.12, where the strength improvement factor (α) is expressed as follows:

$$\alpha = \frac{\sigma_i}{\sigma_o} \qquad (3.5)$$

where

σ_i = strength of PIC
σ_o = strength of unimpregnated concrete.

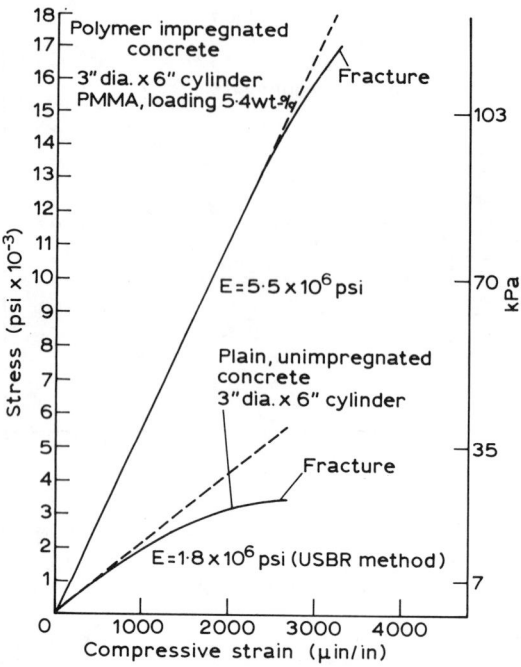

Fig. 3.8. Compressive stress–strain curve for PMMA-impregnated concrete. Impregnated shows elastic behaviour. Unimpregnated shows plastic behaviour [3.2]. (Reprinted with permission from *Polymers in Concrete*, S.P. 40, 1973. Copyright ACI.)

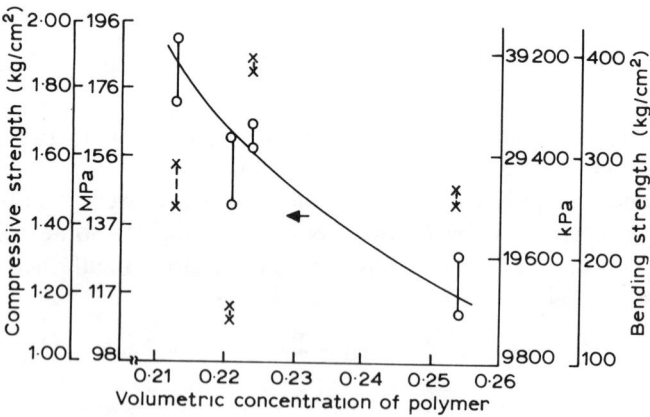

Fig. 3.9. Relationship between volumetric concentration of polymer and mechanical strength [3.16]. ○, Compressive strength, ×, bending strength. (Reprinted with permission from *Polymers in Concrete*, S.P. 40, 1973. Copyright ACI.)

Table 3.1.
Typical properties of polymer-impregnated concrete [3.24]

Property	Unimpregnated	MMA	MMA + 10% TMPTMA
Polymer loading, % by weight	0	4·6–6·7	5·5–7·6
Compressive strength, MPa	34	125	131
(psi)	(4 930)	(18 125)	(16 375)
Modulus of elasticity, 10^3 MPa	19	43	33
(psi)	(2 755)	(6 235)	(4 785)
Direct tensile strength, MPa	2·31	10·4	8·62
(psi)	(335)	(1 508)	(1 249)
Modulus of rupture, MPa	4·34	15·8	—
(psi)	(629)	(2 291)	
Flexural modulus of elasticity, 10^3 MPa	20	49	—
(10^3 psi)	(2 900)	(7 105)	
Abrasion resistance (mm)	1·27	0·38	0·019
Cavitation resistance (in)	8·13	0·51	—
Thermal coefficient of expansion, millionths/°C	7·24	9·45	9·11

Specimens oven-dried at 150°C prior to impregnation.
Thermal-catalytic polymerization.
MMA = methyl methacrylate.
TMPTMA = trimethylolpropane trimethylacrylate.
Reprinted with permission from *Polymers in Concrete, 4th International Congress*, by H. Schulz (Ed.), 1984. Copyright Institut für Spanende Technologie und Werzeugmashinen, Technischen Hochschule, Darmstadt.

Authors have argued that the compressive strength of PS impregnated and unimpregnated autoclaved concretes tends to increase (Fig. 3.12):
—with increasing cement and silica content;
—with decreasing W/C ratio in the concrete batch.

At the same cement amount, α of polystyrene, impregnated autoclaved concrete is much the same regardless of the W/C ratio or silica content. The compressive strength of this type of material is hardly influenced by polymer loading.

Gunduz *et al.* [3.25] have studied the effect of radiation on the physical properties of styrene–acrylonitrile (S–AN) copolymer impregnated concrete. Monomer conversion, compressive and tensile

Table 3.2.
Creep of polymer-impregnated concrete [3.24]

Specimen	Polymer loading (wt. %)	Load (MPa)	Creep $(10^{-6}/MPa)$ 31 days	836 days
Tensile creep[a]				
MMA impregnated	4·6	1·23	7·4[b]	74·7[b]
		2·38	2·6[b]	35·1[b]
Unimpregnated	0	1·23	27·7	23·8
Compressive creep[c]				
MMA impregnated	5·5	4·76	3·0	12·3
	5·4	15·9	3·3	8·1
	5·4	48·3	3·8	7·5
Unimpregnated	0	4·76	19·3	58·2
	0	15·9	28·7	62·0

115 × 130 mm cylindrical specimens oven dried at 150°C prior to impregnation.
Radiation polymerization.
[a] Results for single specimens.
[b] Negative creep/length change is in direction opposing applied load.
[c] Results are averages of two specimens.
MMA = methyl methacrylate.
Reprinted with permission from *Polymers in Concrete, 4th International Congress*, by H. Schulz (Ed.), 1984. Copyright Institut für Spanende Technologie und Werzeugmashinen, Technischen Hochschule, Darmstadt.

strengths, molecular weight and water absorption were determined with changes in polymer loading. The impact strength and acid resistance of S–AN copolymer impregnated specimens were also determined. Polymer loading was increased using vacuum and pressure impregnation techniques.

The changes of compressive and tensile strengths with increased polymer loading by vacuum and pressure impregnation techniques are given in Fig. 3.13.

The impact strength test also shows a great beneficial property of S–AN copolymer impregnated specimens. Acid resistance and water absorption were significantly improved. The decrease in water penetration is explained by the decrease of the volume of pores, as shown in Fig. 3.14.

Fowler & Taylor [3.19] have found that the strength and durability

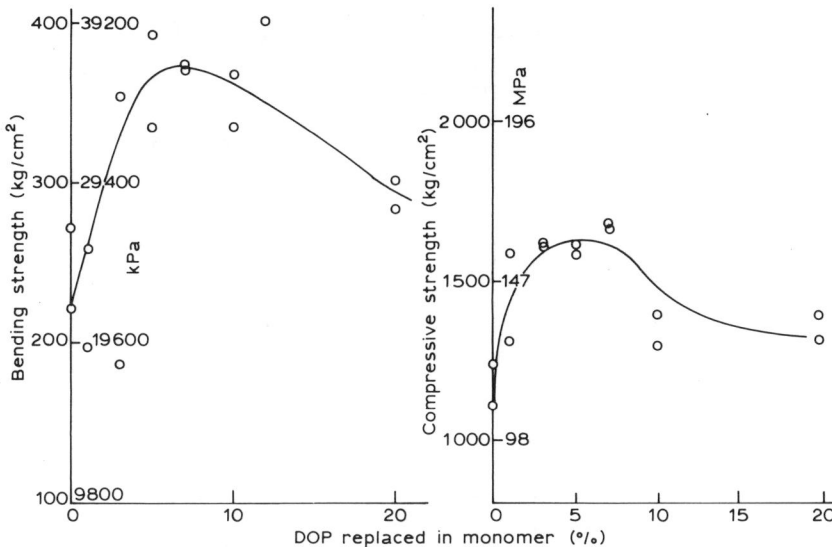

Fig. 3.10. (Left) Effect of DOP dosage on bending strength [3.16]. (Reprinted with permission from *Polymers in Concrete,* S.P. 40, 1973. Copyright ACI.)

Fig. 3.11. (Right) Effect of DOP dosage on compressive strength [3.16]. (Reprinted with permission from *Polymers in Concrete,* S.P. 40, 1973. Copyright ACI.)

properties of PIC, particularly PIC produced by the full impregnation process, are generally much greater than for the plain concrete. The stress–strain relationship for PIC produced with MMA has been shown to be very linear, with practically no ductility. The addition of a monomer which may lead to a copolymer with a lower T_g results in much more ductile behaviour. Figure 3.15 shows the stress–strain curves for PIC made with varying combinations of MMA and butyl acrylate [3.26]. The stress–strain curves were generated from repeated high intensity loading of PIC cylinders. It can be seen that the ultimate compressive stress decreases and the ultimate strain increases with increasing amount of butyl acrylate.

Most of the increases in durability properties can be explained by the fact that the pores are filled, or nearly filled, with the macro-molecular component, which greatly restricts the intrusion of water or acids into the concrete. Both strength and durability are strongly

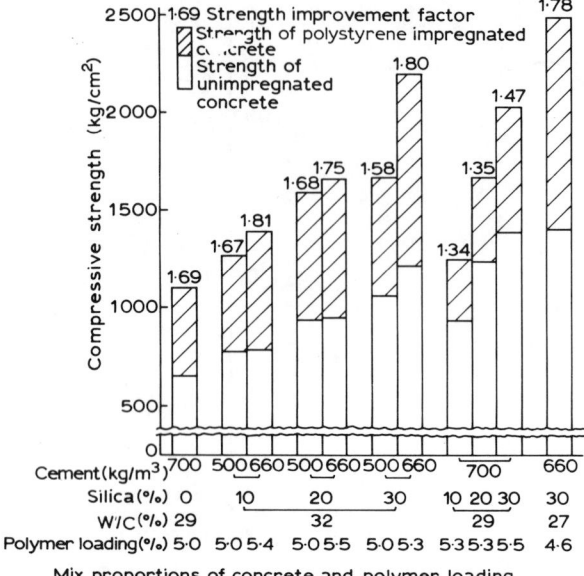

Fig. 3.12. Compressive strength of polystyrene-impregnated and unimpregnated autoclaved concretes vs. mix proportions of concrete and polymer loading [3.21]. (Reprinted with permission from *Polymers in Concrete*, S.P. 58. Copyright ACI.)

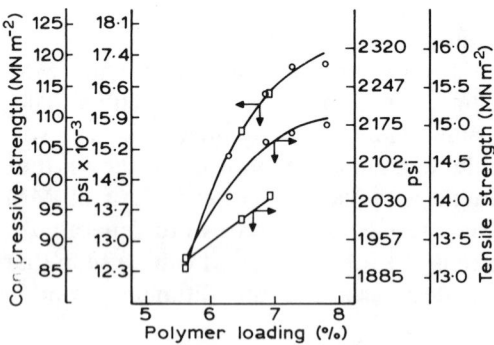

Fig. 3.13. Change of strength with polymer loading [3.25]. ○, Pressure data; □, vacuum data. (Reprinted with permission from *J. Mat. Sci.*, **16** (1981) 221–25. Copyright Chapman and Hall Ltd.)

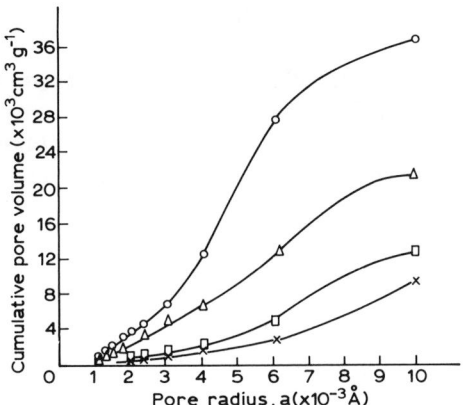

Fig. 3.14. Porosity distribution [3.25]. ○, Unimpregnated control samples (0% polymer); △, samples impregnated at 0·9 atm (5·65% polymer); □, samples impregnated at 10 atm (7·30% polymer); ×, samples impregnated at 5 mm Hg vacuum (6·90% polymer). (Reprinted with permission from *J. Mat. Sci.*, **16** (1981) 221–25. Copyright Chapman and Hall Ltd.)

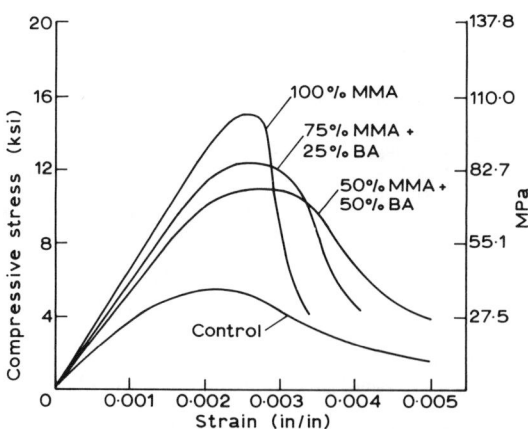

Fig. 3.15. Stress–strain relationship of polymer-impregnated concrete made with MMA and butyl acrylate (BA) [3.26].

dependent upon the fraction of the porosity of the cement phase which is filled with polymer.

Maximov [3.27] has introduced the coefficient K into PIC technology, representing the ratio between increase of concrete strength (R_{PIC}) after its impregnation, and amount of monomer; it has been assumed as an efficiency criterion of monomer used when obtaining PIC.

$$K = \frac{R_{PIC} - R_C}{M} \qquad (3.6)$$

where:

R_C = strength of concrete (MPa)
M = quantity of monomer.

Table 3.3 represents some data concerning the effect of the concrete impregnation method as well as various moulding techniques, composition and curing regime on the strength of PIC.

Table 3.3.
Effect of impregnation method on strength and coefficient K [3.27]

Method of test	W/C	Regime of hardening	Regime of impregnation	Monomer of consumption	R_C (MPa)	R_{PIC} (MPa)	K
Vibration	0·65	Steaming	I	76	24	41	0·22
		Steaming	II	123	24	102	0·53
		Steaming	III	140	24	114	0·64
		Normal	III	129	26	120	0·73
	0·55	Steaming	I	70	33	55	0·32
		Steaming	II	113	33	106	0·65
		Steaming	III	134	33	121	0·66
		Normal	III	123	34	127	0·75
	0·34	Steaming	I	64	51	72	0·32
		Steaming	II	105	51	130	0·75
		Steaming	III	125	51	149	0·78
		Normal	III	121	51	140	0·71
Pressure vacuum treatment	0·40	Steaming	III	77	51	143	1·20
Centrifugation	0·40	Steaming	III	67	81	169	1·31

I. Open immersion in dipping bath.
II. Vacuum treatment down to residual pressure of 0·002 MPa and impregnation at atmospheric pressure.
III. Vacuum treatment down to residual treatment of 0·002 MPa with subsequent impregnation at 0·25 MPa pressure created by means of compressed nitrogen.

Comparing the technological parameters by their influence on the K value, the author concludes that the efficiency of monomer application basically depends on the application regime and moulding method. Alexeev *et al.* [3.28] have applied radiopolymerization technology for the impregnation of concrete and other materials such as asbestos–cement, wood, etc. Experiments showed that electron accelerators are better than isotope sources or nuclear reactors. Technology and necessary equipment were investigated and designed. A plant was built for automatic production of $500\,000\,\mathrm{m}^2$ of composite materials per year for architectural proposals. The durability of such composites is due to their high moisture, bio-, and frost resistance, their ability to resist corrosion, high temperatures, thermal shocks, etc.

DePuy & Dikeou [3.29] have classified PIC under their investigation as follows:

—PIC for normal temperature applications;
—PIC for desalting plant applications;
—partially impregnated concrete;
—surface impregnated concrete.

Durability tests made by these authors have shown highly significant increases of PIC as compared with conventional concrete. The testing programme includes freeze–thaw, sulphate attack, and resistance to acids and bases for PIC designed for normal temperature applications, and exposure to brine and demineralized water at elevated temperatures for PIC designed for desalting plant applications.

PIC Applications
The most important applications of PIC are:

—bridge decking;
—tunnel support-lining systems;
—pipes; hydrostatic tests show that PIC supported about twice as much hydrostatic pressure as the unimpregnated pipe;
—desalting plants;
—beams; ordinary reinforced beams and post-tensioned beams;
—underwater habitats;
—dam outlets, off-shore structures, underwater oil storage vessels, ocean thermal energy plants, etc.

PIC has been also evaluated for use in multistage flash distillation

Table 3.4.
Acid resistance of PIC [3.31]

Corrosive medium		No. of cycles						
		0	*100*			*100*		
		Compressive strength (MPa)	Compressive strength (MPa)	Degree of change (%)	Change in appearance[a]	Compressive strength (MPa)	Degree of change (%)	Change in appearance
Sulphuric acid 5%	U	39·2	21·1	−46·2		6·6	−83·1	×
	I	71·7	46·6	−35·0		18·3	−74·5	×
Sulphuric acid 15%	U	39·2	5·3	−86·5	×	10·0	−86·1	×
	I	71·7	32·0	−55·3				
Hydrochloric acid 15%	U	39·2	6·7	−83·0	×			
	I	71·7	62·9	−12·3		32·8	−54·1	
Acid–alkali alternation	U	52·3	32·2 (15 cyc.)	−31·0				
(H$_2$SO$_4$ 5% and NaOH 5%)	I	79·4	53·9 (50 cyc.)	−32·0		26·9	−66·2	×

Data in above table are results of rapid cycle tests.
[a] Corrosion (scaling and loosening).
× Serious corrosion (serious scaling or loosening).
U Unimpregnated concrete.
I Impregnated concrete.
Reprinted with permission from *Polymers in Concrete, 4th International Congress*, by H. Schulz (Ed.), 1984. Copyright Institut für Spanende Technologie und Werzeugmashinen, Technischen Hochschule, Darmstadt.

vessels for desalting plants. It was concluded that prestressed PIC vessels would be cost-competitive with more conventional types [3.19].

Since PIC has been proved as a new composite, more research is needed to determine its structural behaviour. The 300% or more increase in compressible and tensile strengths, and up to 100% increase in stiffness, indicate excellent potential for PIC structural beams, columns, floor and roof joists, and piling, especially in severe environments where the excellent durability properties can be utilized [3.30].

In view of PIC applications in chemical fertilizer plants, Chen & Xu [3.31] conducted detailed research on PIC chemical stability. Some of the results are presented in Table 3.4.

When PIC is compared with unimpregnated concrete, the acid resistance of PIC is improved, the degree of improvement varying with different acid solutions used (Table 3.4), but as the base material is alkaline, acid resistance is still relatively poor on the whole. When specimens were immersed in a 15% solution of hydrochloric acid, after undergoing 40 dry–wet cycles, the compressive strength of PIC was reduced by about 12%, that of plain concrete by about 83%; whereas in a 15% solution of sulphuric acid, the compressible strength of PIC was reduced by about 55%, and that of concrete by about 86%. The mode of damage PIC suffers in the acid solution is not the same during exposure. For instance, hydrochloric acid corrodes in layers from the surface to the interior, leaving behind a loose configuration that still appears intact, though the strength has decreased. Sulphuric acid scales the concrete in layers also from the surface to the interior, making pebbles show, but the portion that is not corroded still possesses very high strength.

PIC must be considered a new complex material with specific characteristics, which place it in a position, from the viewpoint of quality and cost, between traditional concrete and other groups of engineering materials such as metals and ceramics.

POLYMER CEMENT CONCRETE

The results obtained by mixing various organic compounds with concrete are relatively modest in improvement of strength and durability. Common monomers such as MMA or S either interfere with the hydration process of the cement paste or are degraded because of the alkalinity of some of the cement components. Polyester–styrene systems or epoxies can be effective, though fairly high proportions are usually required if mechanical properties are to be improved. Attention has been given to the use of latices based on polymers. In the case of latices, the physical process of film formation is required rather than the chemical process of polymerization. Since latices are aqueous emulsions, less water is usually needed for workability [3.1, 3.23].

Under the best conditions, compressive strength improvements over conventional concrete of about 50% are obtained with relatively high polymer concentrations of approximately 30%.

In the production of polymer cement concrete (PCC), monomers such as vinylidene chloride or furan, prepolymer–monomer complex systems such as unsaturated polyester–styrene or epoxy–styrene, and latices such as acrylics, styrene–budadiene copolymers, poly(vinylidene chloride), poly(vinyl-esters), etc, have been used with limited success.

The relatively low improvement of the properties in the case of PCC is explained by Steinberg [3.2] by the fact that organic substances are incompatible with aqueous mineral systems, and in many cases interfere with the alkaline cement hydration process.

The incentive to attain improved premix concrete materials is that they can be cast in place for field applications whereas PIC requires a precast structure.

A schematic for PCC in comparison with other systems is presented in Fig. 3.16.

Maslow [3.32] underlines the principal characteristics of epoxy polymers as used with concrete:

—high strength adhesion to most building materials;
—very low shrinkage during and after curing;
—outstanding dimensional stability;
—void filling qualities;
—thermosetting-resistance to softening;
—optimum chemical resistance;
—fatigue resistance;
—creep resistance;
—ability to withstand thermocycling;
—good electrical insulation.

Popovics's [3.33] test results demonstrate the extent of quality improvement produced by proper epoxy modification. The author

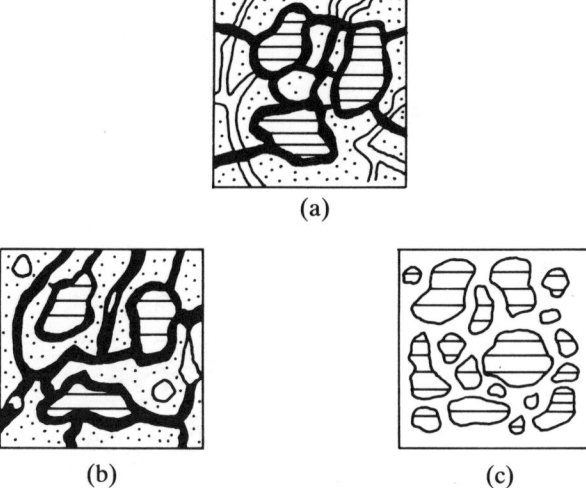

(a)

(b)   (c)

Fig. 3.16. Schemata for (a) polymer-Portland cement-concrete, (b) polymer impregnated concrete, and (c) polymer-concrete (not to scale). Crosshatched areas represent aggregate particles; open and polymer filled capillary pores are shown in (a) and (b); dots represent gel pores, some of which are polymer-filled in polymer impregnated concrete [3.1]. (Reprinted with permission from *Applications of Polymer Concrete*, S.P. 69, 1981. Copyright ACI.)

concluded that not only can the compressive strength of an epoxy modified concrete (PCC) be greater than the strength of a comparable non-modified concrete, but it is also considerably greater than the strength of a comparable latex modified concrete.

Using chemical admixtures, silica fume and different mixing methods (called S, W, X, Y, or Z), Popovics has obtained results presented in Tables 3.5 and 3.6.

In addition, the strength loss caused by the presence of water in concrete is less in epoxy modified concretes than in latex modified concrete. The quality improvement is attributed to the following factors:

—reduction of the volume of air or water filled pores;
—modification of the crack propagation in the concrete under load.

The author considers that the good properties of the epoxy modified concretes can be further improved by the simultaneous use of suitable chemical admixtures. For instance, the addition of a superplasticizer produced a much higher 7-day strength than the control strength, and

Table 3.5.
Test results of epoxy modified concretes with chemical admixtures [3.33]

Mix no.	Unit weight kg/m³ (lb/ft³)	Compressive strength, MPa (psi)				Splitting strength MPa (psi)		Mixing method
		1 day	3 days	7 days	28 days	7 days	28 days	
AC	2470 (154·1)	13·82 (2 003)	34·90 (5 058)	42·26 (6 125)	58·31 (8 450)	—	—	S
E-1	2458 (153·7)	—	—	51·65 (7 492)	65·27 (9 467)	3·41 (495)	5·00 (726)	S
ME	2522 (157·7)	—	—	65·26 (9 467)	87·44 (12 683)	4·63 (671)	5·72 (829)	S
SE	2468 (154·3)	—	—	59·70 (8 658)	73·43 (10 650)	3·80 (551)	5·04 (731)	W
F-1	2464 (154·0)	14·57 (2 112)	35·31 (5 117)	51·25 (7 433)	75·21 (10 908)	—	—	S
F-2	2458 (153·4)	21·89 (3 173)	44·62 (6 466)	60·60 (8 783)	71·93 (10 425)	—	—	S
F-2 (R)	2492 (155·6)	21·67 (3 140)	46·46 (6 733)	60·32 (8 742)	72·97 (10 517)	—	—	S
FC	2461 (153·6)	—	—	43·99 (6 375)	62·04 (8 992)	3·70 (536)	5·20 (754)	W
FC-1	2465 (153·8)	—	—	42·15 (6 108)	57·09 (8 275)	3·33 (483)	4·24 (615)	W
FN	2468 (153·9)	—	—	51·87 (7 517)	66·70 (9 667)	3·88 (562)	5·03 (729)	W
FN-1	2479 (154·7)	—	—	57·73 (8 367)	72·05 (10 442)	4·65 (674)	5·78 (837)	X
FN-2	2471 (154·2)	—	—	59·34 (8 600)	73·04 (10 858)	4·19 (607)	5·87 (851)	Y
FN-3	2475 (154·5)	—	—	60·94 (8 833)	76·07 (11 025)	4·70 (681)	5·84 (847)	Z

(R) Indicates repetition.
Reprinted with permission from *Polymers in Concrete*, *4th International Congress*, by H. Schulz (Ed.), 1984. Copyright Institut für Spanende Technologie und Werzeugmashinen, Technischen Hochschule, Darmstadt.

Table 3.6.
Test results of epoxy modified concretes with silica fume [3.33]

Mix no.	Unit weight kg/m³ (lb/ft³)	Compressive strength, MPa (psi)				Splitting strength MPa (psi)		Mixing method
		1 day	3 days	7 days	28 days	7 days	28 days	
AC	2 470 (154·1)	13·82 (2 003)	34·90 (5 058)	42·26 (6 125)	58·31 (8 450)	—	—	S
E1	2 458 (153·7)	—	—	51·65 (7 492)	65·27 (9 467)	3·41 (495)	5·00 (726)	—
SE	2 468 (154·3)	—	—	59·70 (8 658)	73·43 (10 650)	3·80 (551)	5·04 (731)	W
S	2 409 (150·3)	11·97 (1 735)	21·56 (3 125)	28·17 (4 083)	36·14 (5 238)	—	—	S
S-1	2 306 (143·8)	8·74 (1 266)	18·82 (2 727)	25·03 (3 628)	30·81 (4 465)	—	—	W
S-2	2 345 (146·3)	12·77 (1 850)	25·43 (3 685)	28·79 (4 172)	34·07 (4 938)	—	—	W
S-3	2 341 (146·1)	7·70 (1 116)	30·35 (4 398)	37·32 (5 408)	48·93 (7 092)	—	—	—

Reprinted with permission from *Polymers in Concrete, 4th International Congress*, by H. Schulz (Ed.), 1984. Copyright Institut für Spanende Technologie und Werzeugmashinen, Technischen Hochschule, Darmstadt.

higher than the strength of the comparable epoxy modified concrete without the superplasticizer.

Furan polymers have excellent stability to acidic or alkaline aqueous media as well as to strong polar solvents such as ketones, aromatics and halogenated compounds. Mortars and grouts can be readily formulated with these resins for use with chemically resistant brick when installing floors or linings, which are resistant to strong non-oxidizing acids, hot basic cleaning solutions, and most organic solvents. Such composites also exhibit excellent resistance to elevated temperatures and to extreme thermal shock [3.34].

Latices for PCC

Most of the PCC composites are based on different kinds of latices obtained especially by emulsion polymerization. In this process the system consists of the monomer, water containing emulsifier, and a water soluble initiator. In batch operation, polymerization was carried out in an autoclave whose design depended mainly on monomer properties. For instance, in the case of vinyl chloride, the autoclave is designed for operation at its vapour pressure (40–60°C), up to about 1 MPa. In the presence of the emulsifier(s), agitation of the charge in the autoclave disperses the monomer into very fine droplets (down to about $0.1\,\mu$m). The initiator (commonly potassium or ammonium persulphate alone or with a reducing agent, or a more complex redox system, e.g. $H_2O_2/FeSO_4$/ascorbic acid) produces free radicals in the aqueous phase, where initiation takes place at the boundary with the monomer phase. The conversion is normally about 90%, the reaction being terminated by venting off excess monomer.

Considerable advances were made in the area of applications of emulsion polymerization during the Second World War, when continuous processes for synthetic rubber production were developed for styrene–butadiene rubber (SBR) and acrylonitrile-butadiene rubber (NBR) copolymers to alleviate the shortage of natural rubbers. Since then, many other types have come on to the market, with polyvinyl acetate and copolymers, acrylics and carboxylic-SBR types being the major products.

Emulsion polymerization has certain advantages over other industrial methods such as bulk or solution polymerization, including:

—it produces very high molecular weight polymers;
—as invariably an aqueous medium is used, there are no problems with heat dissipation;

—use of water minimizes cost in comparison with the use of solvents in solution polymerization;

—no particular fire or toxicity hazards;

—ease of control of each stage of the process;

—results in a latex which can be used in many applications;

—emulsion polymerization processes are easily adaptable to continuous running conditions [3.35].

The above-mentioned main components of an emulsion polymerization are used in the following proportions:

	Parts by mass
Monomer(s)	100
Emulsifier (surfactant)	up to 7
Initiator(s)	up to 1
Water	up to 200

A latex may be defined as a stable dispersion of fine polymer particles in water, also containing some non-polymeric constituents.

The properties of the latex formed and of the polymer or copolymer are very dependent on how the various constituents are put together.

One of the early formulations used for the production of SBR (or GR–S rubber) is presented in Table 3.7; it is typical of all emulsion polymerization systems.

The action of the emulsifier is due to its molecules having both hydrophilic and hydrophobic parts. Various other components may

Table 3.7.

Composition of a GR–S recipe for emulsion polymerization of styrene–butadiene [3.36]

Component	Parts by weight
Styrene	25
Butadiene	75
Water	180
Emulsifier (Dresinate 731)	5
n-Dodecyl mercaptan	0·5
NaOH	0·061
Cumene hydroperoxide	0·17
$FeSO_4$	0·017
$Na_4P_2O_7 \cdot 10H_2O$	1·5
Fructose	0·5

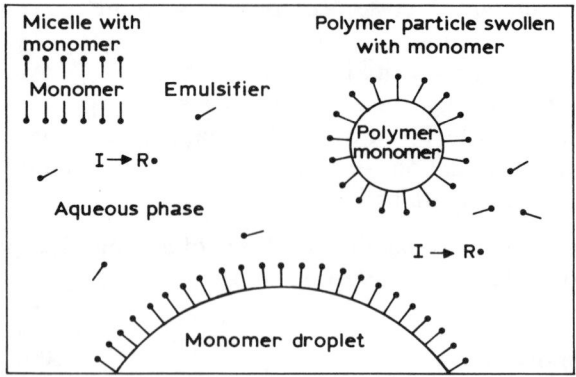

Fig. 3.17. Simplified representation of an emulsion polymerization system
[3.36]. (Reprinted with permission from *Principles of Polymerization,* 2nd
edn., 1981, by G. Odian. Copyright John Wiley, New York.)

also be present in the system; a mercaptan (*n*-dodecyl mercaptan) is
used as a chain transfer agent to control the MW of the polymer. In
the recipe for the above copolymerization, the initiator is the
hydroperoxide-ferrous ion redox system, and the fructose role is to
generate ferrous ions by reducing the ferric ion produced during the
initiation of the process. The sodium pyrophosphate acts to solubilize
the ion salts in the strongly alkaline reaction medium.

Emulsion polymerization takes place almost exclusively in the
interior of emulsifier molecule aggregates called micelles. These
aggregates act as a meeting place for the monomer and the water
soluble initiator. As polymerization proceeds, the micelles grow by the
addition of monomer from the aqueous solution whose concentration
is replenished by dissolution of monomer from the monomer droplets.
In Fig. 3.17, a schematic representation of an emulsion polymerization
system is shown.

The system consists of three types of particles: monomer droplets,
inactive micelles in which polymerization is not occurring, and active
micelles in which polymerization is occurring. The latter are no longer
considered as micelles but are referred to as polymer particles. An
emulsifier molecule is represented as •— to indicate one end; (•) is
ionic or polar and the other end (—) non-polar [3.30]. I represents the
molecule of the initiator and R· the radicals obtained after the
decomposition of the initiator.

Table 3.8.

General properties	Specific properties
Total solids content	Soap coverage
pH	Soap/rubber (resin) ratio
Viscosity	Chemical stability
Surface tension	Minimum film forming temperature
Particle size	Glass transition temperature
Particle size distribution	Freeze–thaw stability
Coagulum content	Gel content (isolated polymer)
Mechanical stability	Mooney viscosity (isolated
Residual monomer content	polymer)

The latices obtained through emulsion polymerization contain small polymer particles of 0·05–5 μm.

A list of latex properties is given in Table 3.8. Some properties apply to all latices, but others only apply to latices for use in specific end applications. The list is divided into general and specific properties.

To obtain the required properties for any given latex, there are a large number of contributing factors that must be taken into account. These may be broadly divided into two groups, chemical and operational (Table 3.9).

Commercially, polymerization techniques fall into three categories, namely batch, semi-continuous and continuous.

Table 3.9.

Chemical	Operational
Nature and amount of emulsifier	Reactor capacity, profile
Nature and amount of initiator	Agitator-stirrer type and speed
Nature and amount of modifier	Temperature, pH, pressure
Nature and amount of electrolyte	Mode of addition of emulsifier, mono-
Phase ratio (monomers/water)	mers, etc (i.e. batch, incremental,
Relative water solubility of monomers	or continuous)
Relative reactivities of monomers in copolymerization (R_1, R_2)	

Styrene–Butadiene Copolymer Latex (SBR, GRS)

Before 1950, copolymerization was usually carried out at about 50°C using a water-soluble initiator such as potassium persulphate with the MW being controlled largely by chain transfer agents such a *t*-dodecyl-mercaptan. Conversion to copolymers was taken to about 72%, at which value a polymerization stopper was added (e.g. hydroquinone).

The second generation of styrene–butadiene copolymers appeared about 1950. The copolymerization was done at about 5°C using a redox initiation system. Many of the additives used in this 'cold' process are multifunctional. For example, the surfactant serves to emulsify the comonomers, solubilize them in the micelles, and to stabilize the latex particles formed. The mercaptan is not only a chain transfer agent, but also a promoter of polymerization in the case of the 'hot' process. The 'hot' copolymers are more extensively branched than the 'cold' ones, and contain much microgel, which arises through cross-linking within the copolymerization locus (micelle or droplet).

Styrene and butadiene may be copolymerized in any desired ratio from 100% polybutadiene to 100% polystyrene. While both *cis*- and *trans*-polybutadienes have low T_gs of about $-100°C$, polystyrene has much higher T_g of about $+90°C$. The T_g of butadiene–styrene closely follows a linear interpolation between the two so that a 50:50 copolymer would be expected to have a T_g of about $-5°C$. SBRs have a styrene content of about 23·5%, this being generally believed to give the best balance of properties [3.37]. Unreacted butadiene is removed by flash distillation, and unreacted styrene by steam distillation.

Latex prepared by emulsion copolymerization has a solids content of 30–35%, and a particle size of 65–75 nm, with a monodisperse distribution. Use of this latex is limited, mainly due to solids content which can be increased by evaporation, but at solids values in excess of 40%, the viscosity becomes too high [3.35].

Acrylonitrile–Butadiene Copolymer (NBR)

A typical recipe for a medium acrylonitrile content NBR is given in Table 3.10.

The emulsion polymerization is realized at a constant temperature typically above 30°C for so-called 'hot' polymerization, and 5–30°C for 'cold' polymerization.

In a typical 'hot' copolymerized NBR containing 33% acrylonitrile there is 60% 1,4-butadiene and 7% 1,2-butadiene. The vinyl group

Table 3.10.

	Parts by weight
Butadiene	67·00
Acrylonitrile	33·00
Water	200·00
Emulsifier	3·50
Electrolytes	0·30
Initiator	0·10
Activator	0·05
Modifier	0·50
Stopper	0·10
Stabilizer	1·25

resulting from 1,2-addition is readily attacked by free radicals and hence is prone to oxidative attack, and may also be the site of branched chain growth during copolymerization [3.38].

Acrylonitrile reacts with the butadiene radical, in a growing chain, more rapidly than butadiene monomer ($r_{AN} = 0·03$, $r_B = 0·30$). However, a change of B/AN of mole ratio 1·67 (azeotropic composition will provide the same copolymer composition as the charge composition.

AN levels may vary in the range 10–45%, with an average of about 33%. Nitriles are commonly referred to as having a low, medium or high AN content, indicating levels of 25, 33 or 45% respectively. The monomer ratio is probably the most important variable, with the polarity of the copolymer increasing with increasing AN content, affecting a series of important properties such as: flexibility, oil resistance, adhesion to polar substrates. Many NBR latices are of the modified type, which contain other reactive groupings (up to 10%), and which may give self-cross-linking properties, or allow the polymer to cross-link with other polymers or additives. The main groups used are carboxyl, amide and substituted amide.

Acroylonitrile–butadiene copolymer latices are used in applicational areas where oil and abrasion resistance, and high binding power to polar substrates are required. The main applications are in the textile and paper industries, surface coatings and adhesives [3.35].

Polychloroprene Latex

A typical recipe for emulsion polymerization of chloroprene might be as in Table 3.11.

Table 3.11.

	Parts by weight
Chloroprene	100·0
Water	150·0
Resin (stabilizer)	4·0
Sodium hydroxide	0·8
Emulsifier	0·7
(Methylene bis(naphthalene-sulphonic acid) sodium salt)	
Sulphur (modifier)	0·6
Potassium persulphate (initiator)	0·1–0·2

Polymerization is carried out at 40°C and is allowed to continue until a 90% conversion is reached [3.39]. At this value the conversion is short-stopped by the addition of thiuram disulphide. During an ageing period at ambient temperature, the sulphide linkages are cleaved by the thiuram disulphide, which thereby reduces the MW.

The solids contents for these types of latices is in the range 35–60%, with a high pH, and particle size in the range 50–190 nm. By variation of polymerization conditions such as temperature, emulsifier, initiator and modifier, different types of latices are produced.

Polychloroprene latices may be classified as general purpose or speciality. General purpose grades are homopolymers, anionic, and of the 'gel' type.

Polychloroprene latices are used for foams, adhesives, binders for non-woven fabrics, paper, textiles, and modifiers for concrete and asphalt [3.40].

Polyvinyl Acetate (PVAc) Latices

Vinyl acetate monomer differs from the majority of common vinylic monomers in that it is appreciably soluble in water, having a solubility of 2·0–2·5% by mass at ambient temperatures, and in contrast to other water-soluble monomers, vinyl acetate monomer and poly(vinyl acetate) are completely miscible.

In a typical emulsion polymerization, approximately equal quantities of vinyl acetate and water are stirred together in the presence of poly(vinyl alcohol) as emulsifier, together with sodium lauryl sulphate, and potassium persulphate as a water-soluble initiator. Polymerization takes place over a period of about 4 h at 70°C. The polymerization is exothermic. In order to achieve better control of the process, and to obtain particles with a smaller particle size, part of the monomer is first polymerized, and the rest, with some of the initiator, is then steadily added over a period of 3–4 h. To minimize the hydrolysis of the monomer during the polymerization, it is necessary to control the pH throughout the reaction. For this purpose, sodium acetate is used as a buffer [3.41].

Typically, latices have solids contents up to 55%, and an average particle size, dependent on the end use, in the range 100–1000 nm.

The monomers used in the copolymerization processes of vinyl acetate are the vinyl esters of long chain fatty acids, and trialkyl acetic acids, esters of acrylic, maleic and fumaric acids, and ethylene. Vinyl acetate copolymers are available at solids contents up to 55%, and average particle sizes up to 1500 nm. The higher solids and larger particle size latices are normally used in adhesive applications. Latices for paint, textile and paper coating applications generally have solids contents up to 50%, with an average particle size of 100–200 nm.

In many applications, vinyl acetate polymers and copolymers are preferred to those containing butadiene as they exhibit superior resistance to oxidation and UV light. Their stability, however, may decrease on storage due to a decrease in pH arising from hydrolysis of residual monomer [3.35]. The introduction of the highly polar carboxyl groups on maleic acid during vinyl acetate emulsion polymerization, improves the adhesion characteristics of the poly(vinyl acetate). Typical physical properties of a poly(vinyl acetate) copolymer latex suitable for use with Portland cement are given in Table 3.12 [3.32]:

This kind of emulsion leaves a dried film which is flexible without the necessity of adding plasticizer. The PVAc film has excellent water resistance, light stability, good ageing characteristics, and good film

Table 3.12

Solids (%)	52
pH	4·10–6·5
Average particle size	<2 μm
Free monomer %	<1
Odour	slight
Borax compatible	no
Particle charge	negative

consolidation. Its particular characteristics of compatibility with cement have led to wide uses of this latex as an adhesive and a main component in polymer–mortar and polymer–concrete composites.

Poly(Vinyl Chloride) (PVC) Latices

PVC latices may be classified according to the nature of the polymer of the particulate phase, i.e. vinyl chloride homopolymer or copolymer latices, and the presence or absence of external plasticizer, i.e. plasticized or unplasticized latices.

PVC latices provide an advantageous alternative to the other two liquid PVC systems—pastes and solutions—for a number of applications. In many they are the only type of liquid system suitable in practice. The main areas of use are the bonding and coating of certain textiles, some paper treatments, leather finishes, and heat-activated adhesive applications [3.42].

Acrylic Latices

Acrylics generally are polymers and copolymers of the esters of acrylic and methacrylic acids. Typical physical properties of an acrylic latex suitable for use with Portland cement are:

Solids (%)	45
pH	9·4–9·9
Specific gravity	1·054

The uses made of this latex as an admixture in cementitious compounds are similar to those made with SBR and PVAc latices. Among these, we can include patching and resurfacing work, floor underlays, terrazzo tile flooring, spray coat and fill coat applications, cement plaster and stucco, tile grouts, crack fillers and precast panel surfacing. The addition rates of acrylic latices are comparable to those of the SBR and PVAc latices, ranging from 0·10–0·20 part latex solids: 1 part Portland cement. It is also possible to add acrylic latex at a ratio as low as 0·05.

The generally recommended addition rates fall in the range of 0·1–0·2. Above this level, some properties actually begin to decrease. Below this level, many properties are still more than adequately higher than for unmodified mortars. Many commercial formulations are based on an addition rate at this level for price reasons [3.32].

Table 3.13.
Typical properties of latices for PCC [3.43]

Polymer type	Polyvinyl acetate	Styrene– butadiene	Acrylic	Neoprene
Solids (%)	50	48	46	42
Stabilizer type	Non-ionic	Non-ionic	Non-ionic	Anionic
Specific gravity at 25°C	1·09	1·01	1·05	1·10
Weight/m³ (kg at 25°C)	1 102	1 006	1 054	1 114
pH	2·5	10·5	9·5	9·0
Particle size (Å)	NA	2 000[a]	NA	NA
Surface tension (dynes/cm² at 25°C)	NA	34[b]	40	40
Shelf life	NA	>2 years	excellent	NA
Freeze–thaw stability (−15°C to 25°C)	NA	5 cycles	5 cycles	NA
Viscosity (cP at 20°C)	17	24	250	10

lb/gal = 119·8 kg/m³.
[a] Å = 0·1 nm.
[b] dynes/cm² = 0·1 Pa.
NA, not available.
Reprinted with permission from *Plastics in Material and Structural Engineering* by R. Bares (Ed.), 1982. Copyright Elsevier Science Publishing Co.

Some general characteristics of latices for PCC are shown in Table 3.13.

Selection of the proper latex depends in a high measure on the service life requirements. For instance, SBR is excellent for exterior exposure or environments where moisture is present.

Although not as widely used as latex-modified concrete, latices of lower MW thermosetting polymers (predominantly epoxies) and their curing agents are also used commercially in PCC [3.32, 3.43–3.45].

In all cases, the operations for preparing PCC are similar to those used for common concrete and mortar. Curing, however, is different. After only 1 day of moist cure the surface of PCC work can be uncovered. The film already formed on the surface retains the necessary amount of moisture for the full hydration of the cement. In principle, after only a few days of air cure at around the normal ambient temperature (20°C), the PCC can be put in service.

Kukacka [3.43] emphasizes the following requirements for the polymer:

—the latex must be able to form a film under ambient conditions, coat cement grains and aggregate particles, and form a strong bond between the cement particles and aggregates;
—the polymer network must possess the capacity to intercept a growing microcrack and, by dissipating energy through microfibril formation, hinder crack propagation.

PCC Properties

Latex type PCC possesses excellent bonding to steel reinforcement and to old concrete, good ductility, resistance to penetration by water and salt, and excellent durability under freeze–thaw cycling. The properties of PCC do indeed depend on the type of polymer and/or its amount.

Bhargava & Rehnstrom [3.46] studied the dynamic properties of PCC; in some cases they used specimens which were reinforced by polypropylene (PP) fibres. The dynamic strength of obtained concrete was 40–45% higher than the static strength. Compared to plain concrete, PCC showed 30–35% higher dynamic strength, and significantly higher energy transmission capacity.

From the amplitude of incident and transmitted pulses, the stresses in concrete shown in Fig. 3.18 were calculated by these authors. As the amplitude of loading pulse was gradually increased, the transmitted stresses reached an optimum value and then decreased. This optimum value was considered to be the dynamic strength of the material.

The value of the stress due to the transmitted wave, $(\sigma_t)_{max}$, obtained when the specimens were subjected to maximum amplitude pulse directly (marked by ∇ in Fig. 3.18) was approximately the same as that obtained by the increasing amplitude method. Thus the complete σ_t–σ_i (transmitted stress–incidental stress) curve AB, including the horizontal extrusion, gives the stress amplitude envelope for the particular concrete quality.

The fact that the slope of the σ_t/σ_i curve was the same for all the concretes indicated that the viscoelastic character of concrete was not changed by the polymer and fibre admixtures. The authors considered that the increase in impact strength was rather due to the improvement in deformation capacity.

Figure 3.19 shows the relative magnitudes of W_t, W_r and W_a

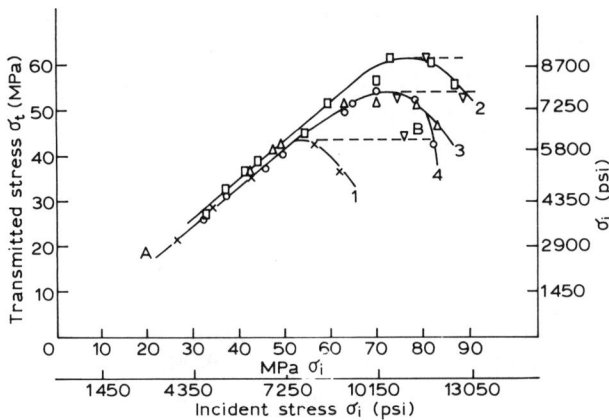

Fig. 3.18. Reaction between the incident and transmitted pulses. (1) Plain concrete, (2) polymer cement concrete, (3) fibre-reinforced concrete, and (4) fibre-reinforced polymer concrete [3.46]. (Reprinted with permission from *Polymers in Concrete*, S.P. 58, 1978. Copyright ACI.)

(transmitted and reflected wave, and the energy attenuated in the specimen), with increasing incident energy levels for different concretes.

From this figure it can be observed that:

(1) Even at early stages of loading, about 10–20% of the energy was attenuated in the specimens, which resulted in some stable crack formation.

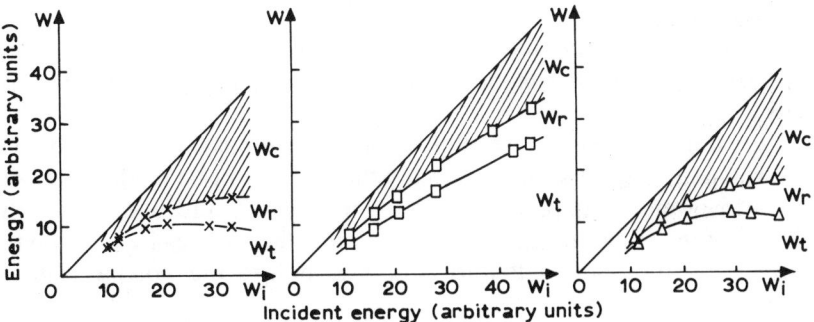

Fig. 3.19. Energy transmission and attenuation for different concretes [3.46]. (a) Plain concrete; (b) polymer concrete; (c) fibre-reinforced concrete. (Reprinted with permission from *Polymers in Concrete*, S.P. 58, 1978. Copyright ACI.)

(2) PCC showed a much higher impact endurance (lower energy attenuation) and improved energy transmission capacity, than the corresponding plain concrete.
(3) Reflex was rather large in PCC. This was probably due to the large amount of air entrapped during mixing.

POLYMER CONCRETE

PC may be considered as an aggregate filled with a polymeric matrix. The main technique in producing polymer concrete (PC) is to minimize void volume in the aggregate mass so as to reduce the quantity of the relatively expensive polymer necessary for binding the aggregate.

A wide variety of monomers, prepolymers, and aggregates have been used to realize PC. The list includes epoxy polymer, polyester–styrene system, methyl methacrylate (MMA) and furane derivatives, usually in conjunction with cross-linking agents.

With polymer as the matrix, some of the drawbacks of conventional concrete which can be overcome include:

—formation of internal voids when alkaline Portland cement is used;
—on freezing, can readily crack due to water being entrapped;
—alkaline cement can be chemically attacked by acidic substances and deteriorated.

Curing times and the duration for development of a high proportion of maximum strength can be readily varied from a few minutes to hours.

A great number of macromolecular compounds are hydrophobic and provide PC resistant to chemical attack, and can be made compact with a low amount of voids. In order to improve the bond strength between the macromolecular matrix and the aggregate, a silane coupling agent is added to the monomer before the polymerization process.

Bond strengths to substrates are usually high. In spite of high cost, PC is particularly useful for maintenance and repairs, especially when delay and inconvenience are important. Thus the cost/benefit ratio is favourable [3.43].

Table 3.14.

Typical properties of polymer concrete composites [3.1]

	PCC	PIC	PIC(A)^a	PC	PPCC
Compressive strength, MPa	34·47	137·88	268·86	130·98	37·91
psi	5 000	20 000	39 000	19 000	5 500
Tensile strength, MPa	2·41	10·34	14·47	9·65	5·51
psi	350	1 500	2 100	1 400	800
Young's modulus					
MPa × 10^{-3}	24·12	41·36	48·25	34·47	13·78
psi × 10^{-6}	3·5	6·0	7·0	5·0	2·0
Shear strength,					
kPa	861	—	—	≥4 481	≥4 481
psi	125			≥650	≥650
% water adsorption	5·5	0·6	≤0·6	0·6	—
Freeze–thaw resistance	700/25	3 500/2	—	1 600/0	—
Acid resistance^b	—	10×	≥10×	>20×	4×
Benefit/cost	1	2	3	4	—

^{a} Based on autoclaved concrete.
^{b} Improvement factor.
Reprinted with permission from *Applications of Polymer Concrete*, Publication SP-69, 1981.
Copyright ACI.

PC Properties

By carefully grading the aggregate, it is possible to wet the aggregate and fill the voids by the use of as little as 7–8 wt% (~14–16%) polymer. With high degrees of packing, high compressive strength can be obtained (Table 3.14).

Flexural strengths, though much higher than for plain concrete, are limited by the aggregate–matrix bond strength and by disparities of the aggregate surfaces, which can introduce stress concentrations. As was previously mentioned, the use of a coupling agent has been shown to improve compressive strength, presumably by improving bond strength [3.1].

Steinberg [3.2] has shown that the main PC problems arise from the viscoelastic properties of the polymer. Usually the macromolecular compounds have a low modulus of elasticity, which means that they are flexible and manifest creep properties. This explains why high polymers can't be used alone in structural units. Using a polymer as a matrix, with aggregates as the second component, some of these difficulties are overcome. To obtain the best chemical resistance, complete curing of the polymer is necessary. This usually is achieved by careful heating, or by using an appropriate cross-linking agent.

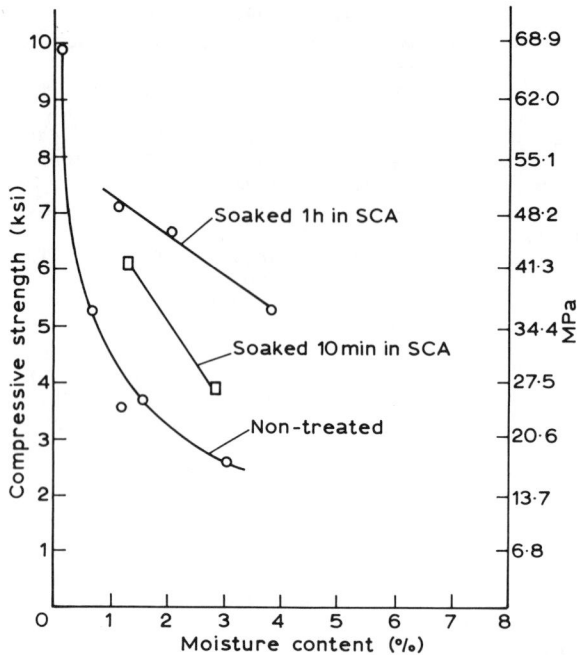

Fig. 3.20. Effect of silane coupling agent (SCA) coated aggregate on compressive strength of polymer-concrete made with wet aggregate [3.47]. (Reprinted with permission from *Applications of Polymer Concrete*, S.P. 69, 1981. Copyright ACI.)

Potential solutions for improving the strength of PC made with wet aggregate have been investigated by Fowler *et al.* [3.47]. Chemical additives, aggregate treatment and addition of fibres have been evaluated. Moisture contents used in research ranged up to 7%, with the maximum values usually 4–5%. No chemical additives for monomer were identified which provide significant strength increases. Aggregate treatments consisted of coated aggregate and moisture additives for wet aggregate. Aggregates treated with a silane coupling agent (SCA) provided very good strength, even when the moisture content was in excess of 4% (Fig. 3.20). Several types of fibres were investigated; hooked steel fibres were found to provide the greatest strength. It was found that the addition of approximately 5 wt% steel fibres provided good strength increases and ductility.

The importance of the nature of the reinforcing agent on the

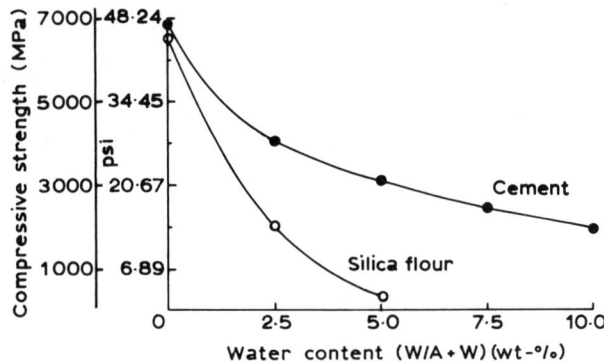

Fig. 3.21. Compressive strength vs. water content of aggregate for cement and silica flour-filled UP polymer mortar at a curing age of 1 h at 24°C [3.48]. (Reprinted with permission from *Polymers in Concrete, 4th International Congress,* by H. Schulz (Ed.), 1984. Copyright Institut für Spanende Technologie und Werzeugmashinen, Technische Hochschule, Darmstadt.)

hydrothermal stability of PC composites has been shown. While aggregates such as quartz, silica, fly-ash and Portland cement give products serviceable up to ~220°C, a combination of silica sand with Portland cement is required for use at high temperatures.

The PC systems based on unsaturated polyester and wet aggregates result in significant strength improvements (Fig. 3.21). Martinez *et al.* [3.48] have studied different methods for treatment of a siliceous sandstone with tris-(2-methoxy ethoxy)-vinylsilane and its influence on the compressive strength of polymer concretes obtained with unsaturated polyester.

On the basis of results obtained from an IR spectroscopy study, Sugama *et al.* [3.49] gave a theoretical presentation of the mechanism of chemical bonding with the aggregate due to the formation of Ca-unsaturated polyester complexed ionomer (Fig. 3.22).

In the first stage, the strongly nucleophilic Ca^{2+} ions are released rapidly from the surface of the cement grains during mixing of the Portland cement, water and unsaturated polyester–styrene system. The ester groups existing in the main chains of the polyester are easily converted into carboxylate ions ($-COO^-$) brought about by a hydrolytic reaction. Ethylene glycol is simultaneously produced as a by-product. Ca^{2+} cations produced in an aqueous medium, complex up

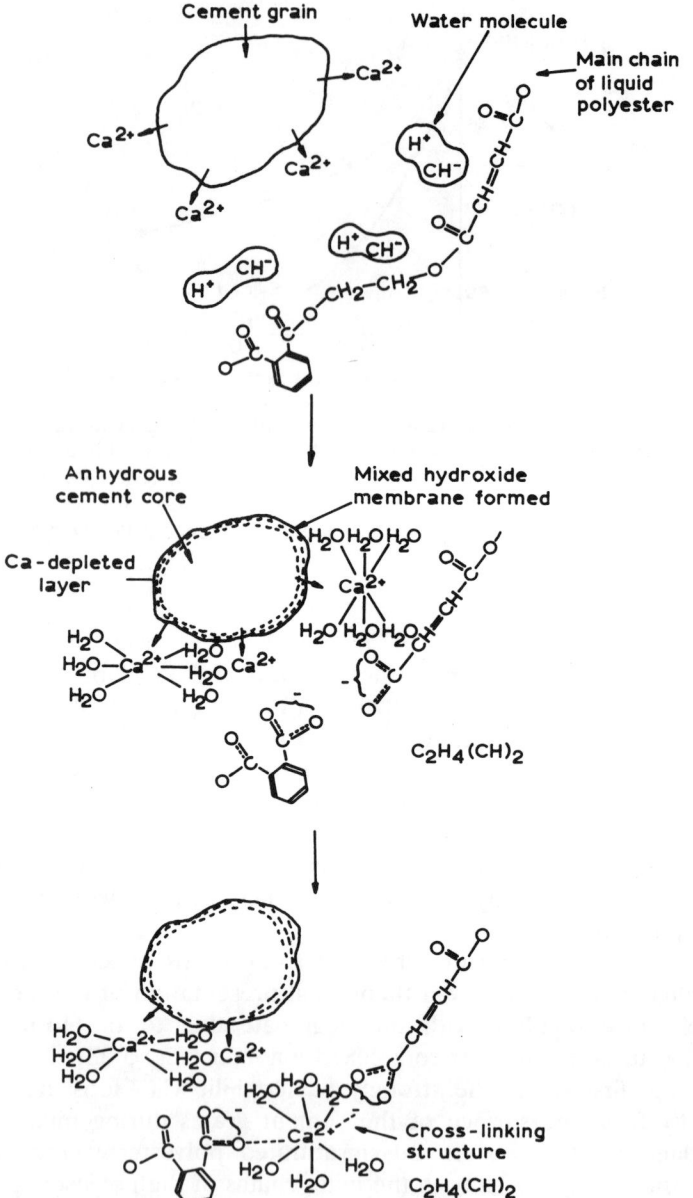

Fig. 3.22. Interaction mechanism occurring between Ca^{2+} cations of Portland cement particles and carboxylate anions of unsaturated polyester in an aqueous medium [3.49]. (Reprinted with permission from Sugama, T. *et al.*, *Cem. Concr. Res.*, **11** (1981) 429–42. Copyright Pergamon Journals Ltd.)

to six molecules of water in the form of an octahedral structure. The neutral water molecules coordinated to Ca^{2+} cations are stable enough to remove water vapour from ordinary air at ambient temperatures.

In the final stage, the Ca-unsaturated polyester complex consisting of the Ca^{2+} salt bridge structure is synthesized by the formation of an ionic bond between the Ca^{2+} ions having six coordinated H_2O ligands and the carboxylated anions (COO^-) from the polyester. The chemical effect of Ca^{2+} may be presumed to be intermolecular cross-linking, acting to connect the two COO^- groups formed in the main chains of unsaturated polyester molecules. It is further assumed that the Ca-unsaturated polyester complex contributes to the prevention of the scission of the macromolecular chains caused by the hydrolysis of polyesters.

Zeldin *et al.* [3.50] have established that the greatest hydrothermal stability has been obtained using mineral-organic type monomers such as organosiloxane in conjunction with a filler composed of silica sand and Portland cement. Such monomers with —Si—O—Si— bonds provide PC with a greater thermal stability than the —C—C— bond characteristics for the vinyl monomers.

An investigation of various formulations for PC was undertaken by Bourleson *et al.* [3.51] in order to determine possible applications of these composites in building construction. A number of polyester–styrene–methyl methacrylate systems were mixed with combinations of different types of fillers such as sand, gypsum, gravel, clay and chopped glass fibres, then cast and polymerized (no Portland cement was used). The resulting products were then tested. Table 3.15 presents some formulations used in this study.

Based on this investigation, the authors have drawn some conclusions such as:

Maximum individual values of strength obtained for the PC composites were as follows:

Compressive strength	$145 \cdot 6 \, N/mm^2$ (21 112 psi)
Flexural tensile strength	$41 \cdot 4 \, N/mm^2$ (6003 psi)
Splitting tensile strength	$15 \cdot 2 \, N/mm^2$ (2204 psi)
Modulus of elasticity	$10 \cdot 1 – 22 \cdot 8 \, N/mm^2$ (1464–3306 psi)
Density	$1 \cdot 97 – 2 \cdot 35 \, kg/m^3$

Figure 3.23 presents the stress–strain diagrams for flexure for PC type A, PC type B, and plain concrete.

Table 3.15.

Resin formulations and aggregate systems tested (all quantities are parts by weight) [3.51]

Resin Mix A		Resin Mix B	
Polyester	40	Polyester	20
Styrene	40	Styrene	30
Methyl methacrylate	20	Methyl methacrylate	50
Hexachloro-1,3-butadiene	2	Hexachloro-1,3-butadiene	2
Divinyl benzene	0·6	Divinyl benzene	0·6
Methyl ethyl		Methyl ethyl	
ketone peroxide	1·6	ketone peroxide	1·5
2,2'-azobisisobutyro-		2,2'-azobisisobutyro-	
nitrile	0·04	nitrile	0·1
Resin Mix C		Resin Mix D	
Polyester	45	Polyester	20
Styrene	25	Styrene	60
Methyl methacrylate	30	Methyl methacrylate	20
Methyl ethyl		Hexachloro-1,3-butadiene	5
ketone peroxide	1·4	Divinyl benzene	1·5
2,2'-azobisisobutyro-		Methyl ethyl	
nitrile	0·6	ketone peroxide	2·4
		2,2'-azobisisobutyro-	
		nitrile	0·4

System 1		System 2	
Resin mix	15–20	Resin mix	20–25
Concrete sand	67	Sea sand	67
Moulding plaster	33	Moulding plaster	33

Reprinted with permission from *Polymers in Concrete, International Symposium*, Publication SP-58, 1978. Copyright ACI.

PC type A is based on:

	Parts by weight
Polymer system	166
Fine sand	556
Gypsum plaster	278

PC type B is based on:

	Parts by weight
Polymer system	263
Fine sand	465
Gypsum plaster	233
Chopped glass fibre	39

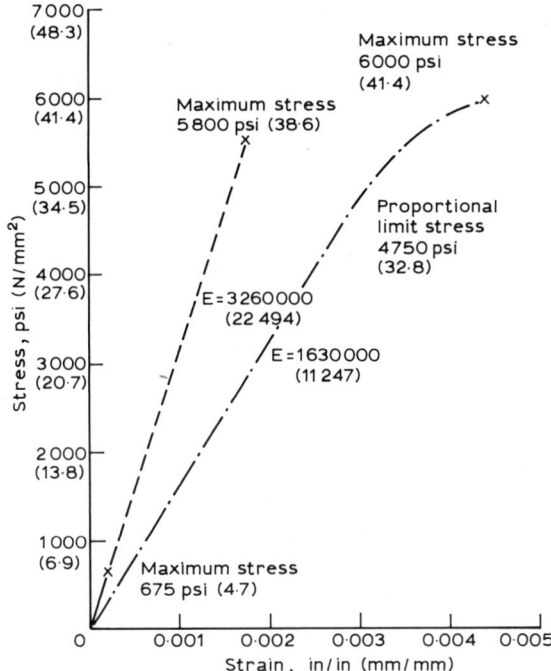

Fig. 3.23. Stress–strain diagrams for flexure of polymer concretes A (– – –) and B (–·–·–) and Portland cement concrete (——) [3.51]. (Reprinted with permission from *Polymers in Concrete*, S.P. 58, 1978. Copyright ACI.)

Rapid curing at ambient temperatures was possible. The addition of chlorendic anhydride has improved the polymerization process at both ambient and elevated temperatures.

The addition of MMA to a polyester–styrene system provides a hard, clear mirror finish, improves the workability without reducing the strength, and enhances the durability.

Fine sands produced a more workable mix which may be cast in relatively thin sections.

PCs investigated cost more than conventional concretes for equal volumes of material, but compare well with conventional concretes on a cost-to-strength basis.

Ohama [3.52] has established the working life of fresh PC depending on the polymer content. The properties of the used polyester, fillers and aggregates are presented in Tables 3.16 and 3.17.

Table 3.16.
Properties of unsaturated polyester resins for testing [3.52]

Type of resin	Acid value	Specific gravity (20°C)	Viscosity (20°C, cP)	Styrene content (%)
UP-1	19·6	1·050	340	38·3
UP-2	17·0	1·120	445	39·3
UP-3	25·4	1·122	430	38·0
UP-4	19·5	1·171	48	50·0

Reprinted with permission from *Polymers in Concrete, International Symposium,* Publication SP-58, 1978. Copyright ACI.

The fresh PC mixtures were tested for working life in accordance with the penetration, pull-out resistance and finger-touching methods. The mix proportions are given in Table 3.18.

In another paper, Ohama & Demura [3.53] have evaluated the performance of mould-releasing agents for PC based on polyesters. Jokiel *et al.* [3.54] experimented with some similar types of PC in view of their application in mining construction. Another study [3.55] has been done on the action of a new combination of mechanical low-frequency vibration in the ultra-sound range on a PC made of an epoxy matrix. The authors have established that the ultra-sound treatment of a highly filled PC speeds curing and leads to the strengthening of the structure of the specimens.

Table 3.17.
Properties of filler and aggregate for testing [3.52]

Type of filler and aggregage	Grading (mm)	Specific gravity	Water content	Organic impurities
Calcium carbonate, heavy	<2·5 × 10⁻³	2·70	0·08	No
Isawa crushed	10–20	2·58	0·03	No
Andesite	5–10	2·58	0·04	No
Fujigawa	1·2–5	2·56	0·06	No
River sand	<1·2	2·55	0·08	No

Reprinted with permission from *Polymers in Concrete, International Symposium,* Publication SP-58, 1978. Copyright ACI.

Table 3.18.
Mix proportions of polyester resin concrete [3.52]

Materials	*Per cent weight (%wt)*		
Binder, unsaturated polyester resin	9·00	11·25	13·00
Filler, calcium carbonate	9·00	11·25	13·00
crushed size, 10–20 (mm)	15·39	14·55	13·89
Aggregate Ardesite (mm) 5–10	15·39	14·55	13·89
River sand size, 1·2–5	10·16	9·60	9·17
(mm) <1·2	41·06	38·80	37·05

Reprinted with permission from *Polymers in Concrete, International Symposium,* Publication SP-58, 1978. Copyright ACI.

PC mass production requires sophisticated manufacturing plants in which the PC parts are produced continuously so that short curing times can be taken advantage of. Due to special processing techniques and moulds, it is also possible to cast equipment, such as wash basins, bath tubs, etc., plain or with a marble or onyx effect. Figure 3.24 presents a flow chart of PC based on polyester manufacture [3.56].

The scheme presents the plant designed for slabs of 20 mm thickness, with many different surface effects, even marble with two or more colours. These are injected in the casting machine and they can be adjusted in their quantities in very wide ranges so that all kinds of appearances are possible. The thickness required depends on the size and the application of the slabs. With 20 mm as an average, the slabs can not only be made thicker, but a lot thinner as well. The main operations in this technique are:

—mixing of the raw materials (polymer and fillers);
—casting the mixture in the moulds;
—curing.

Depending on the designs and the ratio between the components, the slabs stay in the moulds up to 30 min. On average, this time is about 6–12 min, and by this time the slabs already have about 50% of their final strength. After another 3 h, the curing rate has advanced to about 90%, and curing is finished after approximately 24 h.

Kapasny [3.57] gives information about the grading of aggregates for PC systems based on epoxy polymer.

Hop & Miodynski [3.58] have obtained a PC system, a phenolic mortar with very good stability to water (including hot and boiling),

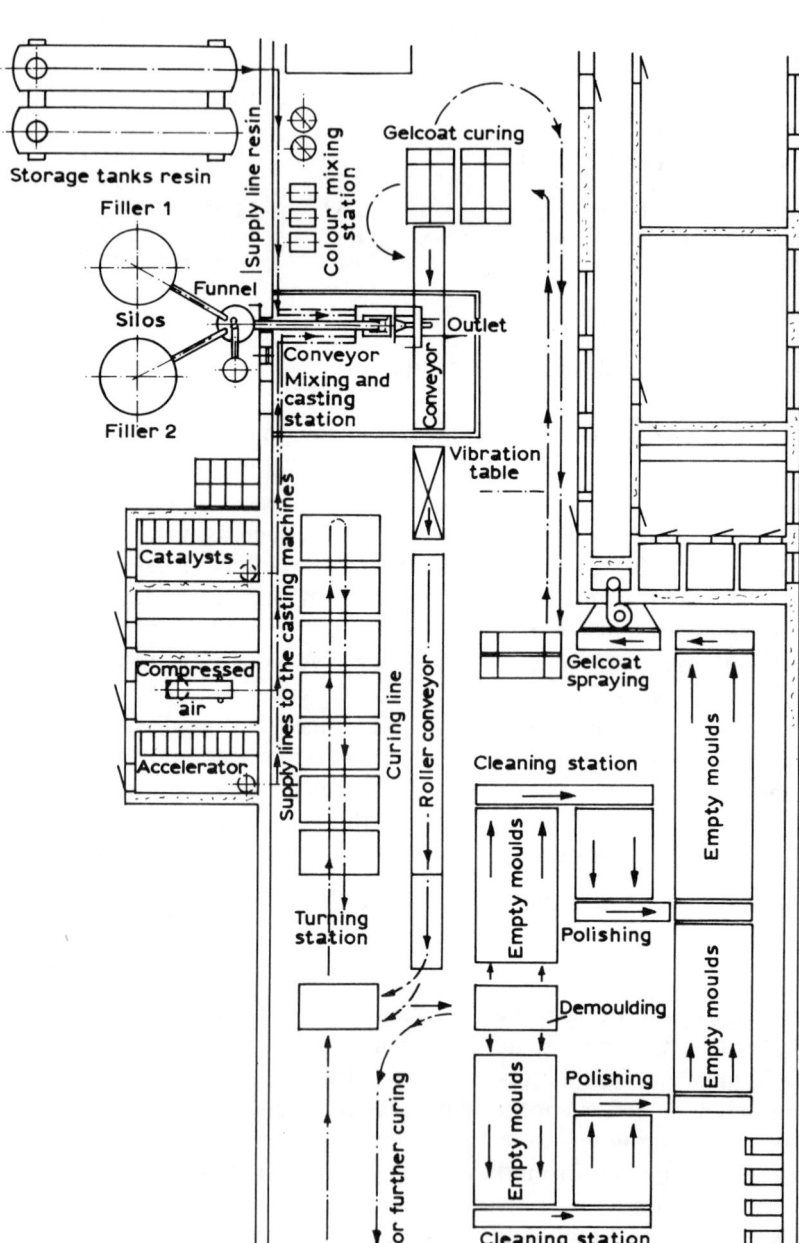

Fig. 3.24. Layout of a continuous production plant for polyester marble design Respecta [3.56]. (Reprinted with permission from *Plastics in Material and Structural Engineering* by K. A. Bares (Ed.), 1982. Copyright Elsevier Science Publishing Co.)

benzene, transformer oil and sulphuric acid (30% conc.). It satisfactorily endures the influence of hydrochloric and nitric acids (10%). Its resistance is less in the case of 3% sulphuric acid solution. Affected by sodium hydroxide (in the form of 3 and 10% solutions), the phenolic mortar is destroyed in a relatively short time. Epoxy mortars have been used in oil wells [3.59], or as flooring materials [3.32].

Ohama *et al.* [3.60] studied polyester concrete using various amounts of styrene, shrinkage reducing agent, catalyst and accelerator, and its early age length change to 24 h after mixing. The conclusions drawn were:

—the setting shrinkage of polyester concrete increases with a rise in content of catalyst and accelerator, and is markedly affected by catalyst content rather than accelerator content;
—the shrinkage of PC based on polyester decreases with increasing styrene content contained in the matrix;
—the setting shrinkage decreases with increasing content of shrinkage-reducing agent, and the product expands at high content of shrinkage-reducing agent.

Staynes [3.61] shows that substantially increased stiffness has been achieved by pre-packing the aggregate into a mould and then impregnating the voids with an epoxide resin system. The pre-packing technique results in a significant change in internal stresses induced by applied loads, and increased thermal conductivity. As a result, a considerable improvement in the fatigue performance has been achieved, particularly at the high frequencies used in the electricity generation industry.

Hirano *et al.* [3.62] have reported that glycerol methacrylate/styrene polymer concrete and mortar show high mechanical strength, good wear resistance, good adhesive strength to cement concrete, less permeability, and skid resistance as good as that of cement or asphalt concrete.

Recent papers [3.63–3.66] discuss in detail some properties and new applications of PC, such as beams [3.63], cladding panels reinforced with glass fibre [3.64], industrial buildings [3.65], and in bridge and tunnel rehabilitation [3.66].

Although the density of PC is similar to that of the cement bonded material, the finished products of PC are much lighter because they have thinner walls. Less weight means less cost for transport and handling, and less static load.

The excellent resistance to chemicals allows many applications in the construction of sewer systems, sewer treatment plants, animal stables, oil separations, chlorine electrolysis, machine foundations, high resistance floors, etc. The versatility in formulation and in processing has led to many applications such as flooring, cast articles of various kinds, patching, overlays for highway bridge decks, panels for facades, marine applications, and in electrotechnical industries as bearing carrying materials.

FIBRE REINFORCED CONCRETE

The use of fibrous reinforcement to improve the strength and deformation properties of concrete is now well established. The concept of fibre reinforcement is to use the deformation of the matrix under stress to transfer load to the fibre. Substantial improvements in static and dynamic strength properties could then be achieved if the fibres were strong and stiff, and loaded to fracture, provided there is, of course, a minimum fibre volume fraction.

The first solution to the problem of reinforcing a brittle cement matrix with fibres was obtained in the late 19th century by the inclusion of asbestos fibres. The success of this material is probably unparalleled in the field of fibre composites. The world production of asbestos–cement in recent years has been in excess of 20×10^6 t/annum. However, this market is likely to decline due to health problems, and it has also been forecast that the supply of asbestos may be exhausted by the year 2000 [3.67].

Fibres currently being used in fibre–concrete composites can be broadly classified into two types:

(a) low modulus, high elongation fibres;
(b) high strength, high modulus fibres.

In the first group we have polyamide, polypropylene and polyethylene, which are capable of large energy absorption characteristics; they do not lead to strength improvement, rather they impart toughness and resistance to impact and explosive loading.

The second group contains fibres such as steel, glass, asbestos, carbon and graphite, which are able to produce strong fibre–concrete composites. They primarily impart characteristics of strength and stiffness to the composite, and to varying degrees, dynamic properties [3.68].

Some important characteristics of the fibrous reinforcing agent are:

—fibre geometry;
—length/diameter ratio;
—fibre volume;
—fibre orientation;
—fabrication technique.

All of these characteristics profoundly influence the properties and mode of failure of the fibrous composite.

The role of fibres is essentially to arrest any advancing cracks by applying pinching forces at the crack tips, thus delaying their propagation across the matrix, and creating a distinct slow crack propagation stage. The ultimate cracking strain of the composite is thus increased to many times greater than that of the unreinforced concrete. The synthetic fibres already mentioned have, without exception, a lower Young's modulus than the Portland cement or concrete matrix in which they must be incorporated. Properties of various types of fibres used in fibre reinforced concrete are discussed in an ACI report [3.69].

Among the synthetic fibres, polypropylene is one of the cheapest; polypropylene fibres are readily available in a great variety at prices lower than those of polyamides, polyesters, polyacrylonitriles, etc.

The potential of polypropylene as a substitute of mineral asbestos fibre in concrete has not as yet been fully explored, although it seems to be very promising. This may be due to the differences in properties between asbestos and polypropylene fibres. Asbestos fibres have a high tensile strength, a high Young's modulus, and good compatibility with Portland cement, which means a high affinity for Portland cement particles. On the other hand, polypropylene fibres have a low Young's modulus, are hydrophobic, and are incompatible with Portland cement powder. A good fibre–matrix bond is a prime requisite for any type of fibre–concrete. The hydrophobic polypropylene' fibres cannot ensure the same type of bonding as exists between cement and mineral fibres, such as glass fibres, asbestos, etc.

As Zonsveld [3.70] reported in the early 1960s, stretched polypropylene (PP) films with longitudinal fibrillation were developed for twines used in agricultural and rope making applications. This fibrillated PP film applied in cement technology allows the fluid water–cement paste to penetrate between the fibrils, and to achieve a mechanical bond within the hardened (cured) matrix, in place of the physical adhesion.

The stretching of PP films during extrusion induces orientation and crystallization, and provides a strong film in the processing direction that is prone to splitting over short distances.

PP exhibits fairly good mechanical properties, a relatively high melting point (165°C), a low density (0·91 kg/dm³), and good chemical stability. In the case of a chemical attack on the PP reinforced concrete composite, the matrix will be destroyed long before PP is affected.

Fibrillated films occur either naturally fibrillated, as on rubbing or twisting of the film material, or mechanically fibrillated to a regular pattern, as obtained after pin-rolling of the film.

Fibrillated PP networks can be incorporated in a concrete matrix in a multiplicity of layers at percentages of over 5% by volume, and the resulting sheet material can meet the requirements for asbestos–cement sheeting. The continuous PP network in cured cement produces a highly ductile and impact resistant sheet material at a cost which is comparable to that of asbestos–cement.

PP–cement composite has potential for many of the end-use applications presently served by asbestos–cement, such as:

—planking, now served by wood;
—cladding and light walling, where aluminium or PVC are most popular at present;
—rainwater goods;
—non-pressure pipes in use for drainage and sewage;
—pressure pipes;
—permanent shuttering, etc.

Explosive testings of fibre-reinforced cement composites were made by Raouf *et al.* [3.71].

Tests were performed on square-section bars $25 \times 25 \times 210$ mm, reinforced with different types of fibres such as PP, E-glass, steel and carbon.

In their conclusions, the authors have established that the effectiveness of the fibres in these tests can be listed in order of significance as follows:

—E-glass fibre;
—polypropylene;
—steel;
—carbon.

The greater resistance of E-glass fibre composites may be due to the uniform distribution of the fibres throughout the mix, and to the fact that these fibres have the highest tensile strength of the four fibres used in this study. In the case of PP fibres, no spalling of the composite occurred because of the presence of the fibres across the crack, acting as crack stoppers. In applying theoretical considerations to fibre–cement composites, it is assumed that the material is elastic, having no damping, and having the same modulus in tension as in compression.

Tables 3.19 and 3.20 show typical properties of fibres and matrices used in fibre–cement composites.

It is interesting to know that in the past, PP fibres have not been regarded as potentially satisfactory for reinforcing materials in direct tension or flexure, because of their low elastic modulus, high Poisson's ratio and poor bonding with cement.

The patent of Hannant *et al.* [3.73] lies in the use of continuous networks of fibrillated PP film to produce adequate bonding and thus to utilize the full strength of the polymer. The inclusion of PP mesh in a cement mortar mix more than doubles the load capacity of the beams, and can produce closely spaced multiple cracking at bond stresses well within the measured range of PP and cement. The material has the additional advantage of high energy absorption under impact.

PP films were chosen for their low cost, strength and chemical inertness, to the extent that the cement matrix would be the first to deteriorate should there be contact with aggressive chemicals.

Figure 3.25 represents the stress–strain curve in tension for a composite containing 2·3% by volume of PP film networks.

The drawbacks of this reinforcing agent also have to be taken into account, namely: combustibility, low modulus of elasticity, poor bond with the matrix, sensitivity to oxygen and sunlight. The low Young's modulus means that the inclusion of fibres reduces the cracking strength of the composite and results in very large strains before multiple cracking is complete. It is necessary to emphasize the fact that the surrounding concrete in the PP–concrete composites protects the PP fibres so well that the sensitivity to light and oxygen is removed altogether.

PP fibres have been used with good results to increase the toughness of concrete subject to impact loading. The high impact strength of PP–cement systems is partly due to the large amount of energy

Table 3.19.
Typical fibre properties [3.72]

Fibre	Diameter (μm)	Length (mm)	Density (kg/m³ × 10³)	Young's modulus (GN/m²)(psi × 10⁻⁶)[a]
Asbestos				
Chrysotile (white)	0·02–30	<40	2·55	164 (23·78)
Crocidolite (blue)	0·1–20	—	3·37	196 (28·42)
Carbon				
Type 1 (high modulus)	8	10	1·90	380 (55·1)
Type 2 (high modulus)	9	cont.[c]	1·90	230 (33·35)
Cellulose	—	—	1·2	10 (1·45)
Glass Cem-fil filament	12·5	10–50		80 (11·6)
E	8–10		2·54	72 (10·44)
204 filament strand	110 × 650		2·7	70 (10·15)
Kevlar		6–65		
PRD 49	10		1·45	133 (19·28)
PRD 29	12		1·44	69 (10·00)
Nylon				
Type 242	>4	5–50	1·14	up to 4[b]
Monofilament	100–200	5–50	0·9	up to 5[b]
Polypropylene				
Fibrillated	500–4 000	20–75	0·9	up to 8
High tensile	100–600			200 (29)
Steel		10–60	7·86	
Stainless	10–330			160 (23·2)

[a] Note: 1% elongation = 10 000 × 10⁻⁶ strain.
[b] Rate dependent.
[c] Continuous.
Reprinted with permission from *Fibre Cements and Fibre Concretes* by D. J. Hanant, 1978. Copyright John Wiley, New York.

absorbed in debonding, stretching, and pulling out of the fibres, which occurs after the matrix has cracked.

Comparative impact tests between the PP and steel fibres have shown that PP can absorb as much energy as some steel fibres for the same fibre volume.

The area under the load–deflection curve in slow flexure is another measure of the ability to absorb flexural impact energy. Figure 3.26 presents such a diagram for a beam with short chopped fibres at 1·2% by volume. It can be seen that compared with the plain matrix, the work to fracture is greatly increased.

With regard to the durability, there was little modification in modulus of rupture or impact strength with time after a severe test. Results indicated that the high impact strength derived from PP will remain stable over very long periods of time in normal use.

Poisson's ratio	Tensile strength MN/m^2 (psi × 10^{-3})		Elongation[a] at break (%)	Typical volume in composites
0·3	200–1 800	(29–261)	2–3	10
—	3 500	(507)	2–3	—
0·35	1 800	(261)	9·5	2–12
0·35	2 600	(377)	1·0	
	300–500	(43·5–72·5)		10–20
0·22	2 500	(362·5)	3·6	2–8
0·25	3 500	(507)	4·8	
—	1 250	(181·2)	—	
				<2
0·32	2 900	(420·5)	2·1	
—	2 900	(420·5)	4·0	
0·40	750–900	(108·7–130·5)	3·5	0·1–6
—	400	(58)	18	0·1–6
0·29	400	(58)	8	0·2–1·2
0·46	700–2 000	(101·5–290)	3·5	
0·28				0·5–2
	2 100	(304·5)	3	

Regarding the flammability test, the presence of PP fibres made no difference to the behaviour of concrete. During the test, the fibres had melted, and at higher temperatures the products of their degradation had volatilized, leaving fine channels and an additional porosity in the composite [3.73].

In a more recent paper, Hibbert & Hannant [3.74] have reviewed the methods used to measure the toughness or impact resistance of different cement and concrete composites with PP fibres, steel wire, glass strand or Kevlar 49 as reinforcing agent. They have commented that in fibre cements and concretes there are essentially two modes of failure: one involving fibre pull-out, and the other involving multiple cracking of the matrix. In the first case, the energy is often absorbed in a small volume of the composite, either side of the crack across which the fibres are pulling out. Even if many cracks develop in the initial

Table 3.20.
Typical properties of matrix [3.72]

Matrix	Density (kg/m^3)	Young's modulus GN/m^2 $(psi \times 10^{-6})$	Tensile strength MN/m^2 (psi)	Strain at failure $\times 10^{-6}$
Ordinary portland cement paste	2 000–2 200	10–25 (1·45–3·62)	3–6 (435–870)	100–500
High alumina cement paste	2 100–2 300	10–25 (1·45–3·62)	3–7 (435–1 015)	100–500
OPC mortar	2 200–2 300	25–35 (3·62–5·07)	2–4 (290–580)	50–150
OPC concrete	2 300–2 450	30–40 (4·35–5·8)	1–4 (145–580)	50–150

Reprinted with permission from *Fibre Cements and Fibre Concretes* by D. J. Hannant, 1978. Copyright John Wiley, New York.

stages of loading, failure usually occurs due to the propagation of one crack along a particular plane of weakness.

However, in the case of multiple cracking followed by fibre fracture, rather than fibre pull-out, the energy is absorbed throughout the volume of material which is stressed above the matrix cracking stress.

The amount by volume of PP used for fibre–cement or fibre–

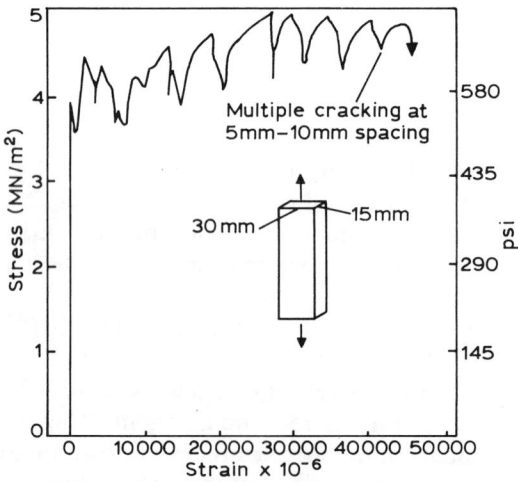

Fig. 3.25. Stress–strain curve in tension for a composite containing 2·3% by volume of opened PP film networks [3.73]. (Reproduced from Hannant, D. J. *et al. Composites,* April 1978, 83, by permission of Butterworth & Co. Publishers Ltd.)

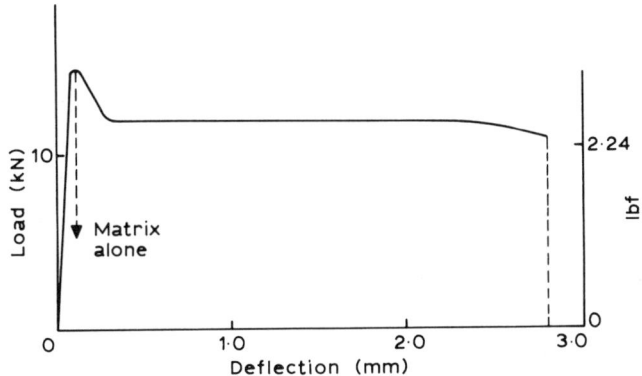

Fig. 3.26. Load–deflection curve for 100 × 100 × 500 mm beam containing PP chopped fibres (1·2% by volume of fibrillated PP, 700 m/kg, length 75 mm) [3.73]. (Reproduced from Hannant *et al. Composites,* April 1978, 83, by permission of Butterworth & Co. (Publishers) Ltd.)

concrete composites might be 5% or more [3.67]. At amounts less than 2%, PP–cement composites have a better impact strength and a lower Young's modulus than composites with other fibres. Table 3.21 lists results obtained by Wells [3.75].

Carbon fibres were also used by Aly *et al.* [3.76] in carbon fibre–cement composites; they have used different fibre orientations. Some of their results are presented in Table 3.22, where it can be observed that 3% by volume of random fibre mat produces about a 70% increase in tensile properties, but the impact strength is decreased. However, with 3·7% by volume of continuous aligned fibres, a five-fold increase in tensile strength can be achieved.

Figure 3.27 shows that the random mat composite has very little postcracking ductility.

Most fibrous composites are based on chopped fibres in random array, and the initial work on carbon fibre–cement composites similarly have used chopped fibre. Early studies of possible structural units made with carbon fibre–cement have shown that the directional properties of a continuous fibre system had a marked cost advantage over the uniform properties of a dropped fibre system [3.77].

Aly *et al.* [3.76] have established the good durability of carbon fibre–cement composites by measuring the strength retention of

Table 3.21.

Influence of the type and amount of fibres on the properties of concrete composites [3.75]

Product	% by wt. of fibre	Dry Charpy impact kJ/m² (ft lb/ft²)	Modulus of rupture MPa (psi)				Combustibility (BS 476, Part 4)
			Dry	Wet	Wet aged at 20°C	Wet aged at 50°C	
Asbestos/cement	10	2–4 (137–274)	30 (4 350)	25 (3 625)	—	—	Pass
Polypropylene monofilament-reinforced alternative	0·5–2	6 (411)	16–17·5 (2 320–2 537)	16–17·5 (2 320–2 537)	—	—	Fail 70°C temp. rise 600 s flaming
Carbon fibre reinforced-alternative	1·0–2·0	2·5–4·5 (171–308)	18–22 (2 610–3 190)	17–19 (2 465–2 755)	17·5 (2 537) (182 days)	16·0 (2 320) (122 days)	Pass
Refined cellulose reinforced-alternative	3–5	1·5–2·5 (102–171)	15–18 (2 175–2 610)	13–15 (1 885–2 175)	12–14 (1 740–2 030) (84 days)	11·0 (1 595) (84 days)	Pass

Reproduced from R. W. Wells, *Composites*, 1982, 169–172, by permission of Butterworth & Co. (Publishers) Ltd.

Table 3.22.
Mechanical properties of high modulus carbon fibre cement composites [3.76]

Fibre volume	Fibre orientation	Young's modulus GN/m^2 $(psi \times 10^{-6})$		Stress MN/m^2 (psi)		Strain $\times 10^{-6}$	Impact strength kJ/m^2 $(ft\,lb/ft^2)$
0	—	13·8	(2·0)	5·52	(800)	300–400	2 (137)
3·0	Random in plane chopped fibre mat	18·2	(2·6)	9·6	(1 444)	570	1·4–1·8 (95–123)
3·7	Continuous aligned	26·1	(3·7)	26·6	(3 857)	2 160	3·6–4·5 (246–308)

Reprinted with permission from *Cem. Concr. Res.*, **2**(21) (1972) 201–212, by M. H. Aly *et al.* Copyright Pergamon Journals Ltd.

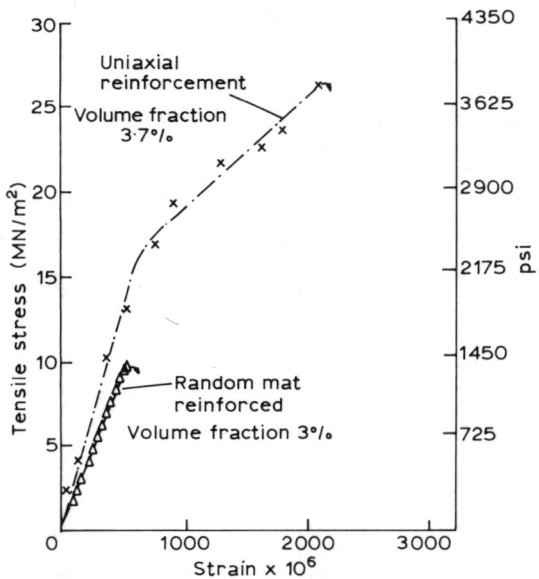

Fig. 3.27. Stress–strain curves for carbon fibre-reinforced cement composites in tension [3.76]. (Reprinted with permission from Aly, M. A., *et al.*, *Cem. Concr. Res.*, **2**(2) (1972), 201–12. Copyright Pergamon Journals Ltd.)

samples which were kept under water at 18°C and 50°C for up to 1 year.

With 2% by volume carbon fibre, typical mechanical properties measured at ambient temperatures were:

Ultimate tensile strength   16 MN/m² (2320 psi)
Modulus of rupture   44 MN/m² (6380 psi)
Impact strength   17 kJ/m² (1165 ft lb/ft²)

These values are not likely to be affected by age. The composite showed excellent fatigue resistance at stresses well above the elastic limit. No failures were recorded for samples stressed below the elastic limit of 15 MN/m² (2175 psi) after 10^6 cycles [3.78].

In a report presented in 1975, Lankard [3.79] mentioned that the existing and potential applications for fibre-reinforced concrete can be conveniently placed into two main groups:

(1) Mass concrete applications—in which the final product is realized at the construction or building site.

(2) Precast concrete applications—in which the product is realized in a plant and shipped elsewhere for use.

Table 3.23.
Mass concrete application areas for FR concrete for which field trials have been performed [3.79]

Application area	Fibre types used in concretes evaluated in the field	Countries in which significant field work has been done
Bridge decks, overlays, and construction	Steel	USA
Highway, street, and airfield pavement overlays and construction	Steel, glass	USA, UK, Canada
New pavement construction	Steel	USA
Mass concrete maintenance and repairs (dams, slabs, pavements, bridges, culverts, etc.)	Steel	USA, UK, Western Europe
Mining and tunnelling	Steel	USA, UK
Rock slope stabilization	Steel	USA
Industrial floors	Steel, glass	USA, Italy, UK
Refractory applications	Steel	USA

Reprinted with permission from *Fibre Reinforced Cement and Concrete*, RILEM Symposium, by A. Neville (Ed.), 1975. Copyright Longman Group Ltd.

Table 3.23 shows the mass concrete application areas studied and the types of fibres that have been used for these applications.

The list of precast product applications for fibre-reinforced concrete is more comprehensive from the point of view of the type of fibres; many applications are based on PP fibres and some, as shown in Table 3.24, are based on metallic fibres.

Table 3.24.
Precast concrete product applications for FR concrete [3.79]

Product application	Type of fibre evaluated	Countries in which significant field work has been done
Car part deck slabs	Steel	UK
Concrete pipe	Steel, glass	USA, UK
Concrete piling	Polypropylene	UK
Ceramic tooling	Steel	USA
Floating pontoon units	Steel, glass	UK
Dolosse (break-waters)	Steel	USA
Boat hulls	Steel, glass	USA, UK
Burial vaults	Steel	USA
Concrete steps	Glass	USA
Decorative garden units	Glass	USA
Utility poles	Steel	Canada
Decorative building panels	Polypropylene	UK
Structural units	Steel	USA
Manhole assembly	Steel	UK
Weight coatings for gas and oil transmission line	Steel, polypropylene	USA, UK
Pile tips	Steel, glass	USA
Machine pads	Steel	USA
Machine frames	Steel	USA
Precast refractory shapes	Steel (stainless and carbon)	USA
Underground utility vaults	Steel	USA

Reprinted with permission from *Fibre Reinforced Cement and Concrete* by A. Neville (Ed.), 1975. Copyright John Wiley, New York.

Table 3.25.
Tensile strength of various composites [3.80]

No.	First fibre				Glass wt.	Curing	Tensile strength MN/m² (psi × 10⁻³)			
	Type	Form	Length	wt.			7 days	28 days	180 days	1 year
Vk	Poly-propylene	Tape 1000d	19 mm	1·0%	5·0%	Air	11·7 (1·69)			10·3 (1·49)
						Water		10·7 (1·55)		8·8 (1·27)
						NW[a]				9·2 (1·33)
						60°C				2·1 (0·30)
Yf	Poly-propylene	Fib. 1000d	51 mm	1·0%	4·3%	Air		5·7 (0·82)	11·1 (1·60)	11·1 (1·60)
						Water		13·2 (1·91)	7·1 (1·02)	8·8 (1·27)
						NW		12·7 (1·84)	9·6 (1·39)	9·9 (1·43)
						60°C			2·3 (0·33)	3·1 (0·44)
Yc	Poly-propylene	Mono 170d	51 mm	1·2%	4·3%	Air		7·5 (1·08)	12·4 (1·79)	11·7 (1·69)
						Water		13·4 (1·94)	9·3 (1·34)	8·8 (1·27)
						NW		11·9 (1·72)	9·9 (1·43)	8·8 (1·27)
						60°C			3·3 (0·47)	2·9 (0·42)
Yd[b]	Poly-propylene	Mono 170d	51 mm	1·1%	4·3%	Air		5·2 (0·75)	10·2 (1·47)	9·9 (1·43)
						Water		11·1 (1·60)	8·5 (1·23)	9·7 (1·40)
						NW		9·0 (1·30)	8·6 (1·24)	9·0 (1·30)
						60°C		9·4 (1·36)	3·4 (0·49)	3·4 (0·49)
Yb	Poly-propylene	Mono 170d	51 mm	2·3%	4·0%	Air		5·5 (0·79)	9·2 (1·33)	10·2 (1·47)
						Water		10·3 (1·47)	7·6 (1·10)	7·3 (1·05)
						NW		9·1 (1·31)	9·2 (1·33)	8·4 (1·21)
						60°C		8·9 (1·29)	4·4 (0·63)	6·0 (0·87)
								5·6 (0·81)		

Mix	Fibre	Matrix	Fibre length	Vol %	Vol %	Storage	(1)	(2)	(3)	(4)
Xy	Poly-propylene	Mono	51 mm	2·3%	—	Water	4·4 (0·63)			
Ye	Poly-propylene	Mono 170d	51 mm	2·8%	—	Air		5·7 (0·82)		
						Water		5·6 (0·81)		
						60°C		5·9 (0·85)		
Yg 1[c]	Poly-propylene	Mono 170d	51 mm	1·0%	5·0%	Air		9·9 (1·43)	8·8 (1·27)	9·2 (1·33)
						Water		9·3 (1·34)	6·4 (0·92)	7·0 (1·01)
						NW				7·8 (1·13)
						60°C		3·8 (0·55)		2·8 (0·40)
Yg 2[b]	Poly-propylene	Mono 170d	51 mm	2·3%	—	Air		4·7 (0·68)	2·9 (0·42)	
						Water		6·7 (0·97)		
						60°C		4·4 (0·63)		
Yz	Nylon	Mono 50d	51 mm	0·2%	4·0%	Air		11·7 (1·69)		
						Water		10·8 (1·49)		
						60°C		3·9 (0·56)		
We	Carbon		11 mm	1·3%	—	Water	11·4 (1·65)			10·8 (1·56)
						NW				10·3 (1·49)

[a] NW = natural weathering.
[b] Matrix contains 40% pfa (pulverized fly ash).
[c] Matrix contains 33% sand.

Reproduced from Walton & Majumdar, *Composites*, 1975, Sept., 209–216, by permission of Butterworth & Co. (Publishers) Ltd.

Table 3.26.

Properties of composites reinforced with nylon fibres [3.85]

No.	Nylon fibre monofilament		Wt.	Fabrication method	Curing	MOR^a MN/m^2 $(psi \times 10^{-3})$		IS^b kJ/m^2 $(ft\,lb/ft^2)$	
	Denier	Length				7 days	60 days	7 days	60 days
Va	24	25 mm	0·5%	Pre-mix and spray	Water	7·1 (1·02)		7·4 (507)	
					60°C				
H'	24	25 mm	2·0%	Pre-mix and spray	Water	11·2 (1·62)		21·8 (1 494)	
					60°C		9·1 (1·31)		18·9 (1 295)
J'	24	25 mm	4·0%	Pre-mix	Water	9·0 (1·30)		29·3 (2 008)	
					60°C		10·5 (1·52)		29·8 (2 042)
K'	24	25 mm	3·3%	Pre-mix	Water	9·1 (1·31)		23·9 (1 638)	
					60°C		9·5 (1·37)		27·7 (1 898)
F'	24	13 mm	1·0%	Pre-mix	Water	11·2 (1·62)		7·9 (541)	
					60°C		9·7 (1·40)		5·2 (356)
Zi	50	51 mm	0·9%	Spray-up	Air	7·5 (1·08)	6·9[c] (1·00)	13·5 (925)	8·4[c] (575)
					Water		10·5[c] (1·52)		9·6[c] (657)
					60°C		8·2[c] (1·18)		10·9[c] (747)

[a] Modulus of rupture.
[b] Impact strength.
[c] Curing 28 days.

Organic polymer fibres such as polyamide and polypropylene, when added to cement or concrete, even in very small proportions, substantially improve the impact resistance of the matrix but have very little effect on its tensile or bending strength. Walton & Majumdar [3.80] have obtained composites produced using a mixture of organic and inorganic fibres which exhibit the advantages of both.

Table 3.25 presents their results regarding the tensile strength of various composites obtained with different mixtures of fibres; Table 3.26 shows the values of the modulus of rupture (MOR) and impact strength of composites reinforced with polyamide fibres.

The accelerated testing results indicated that the high impact strength derived from synthetic fibres such as PP and polyamide, will remain stable over very long periods of time in normal use. Improved behaviour in bending may be obtainable with such synthetic fibres and by using higher volume fractions.

The strength of the bond between synthetic fibres and cement or concrete is likely to be poor. This poor interfacial bond is largely responsible for the excellent impact strength of the PP fibre cement composites. For these composites to be of practical value, however, their tensile and bending properties must be improved. This can easily be accomplished by using a second fibre such as glass or asbestos.

Some studies deal with Kevlar type polyamide–cement and concrete composites.

The tensile stress–strain diagram of the Kevlar–cement system is shown in Fig. 3.28, along with some other composite systems studied by Majumdar & Laws [3.81].

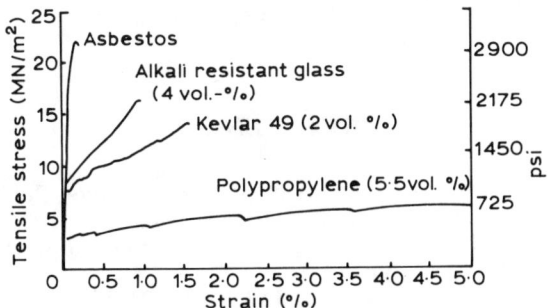

Fig. 3.28. Tensile properties of some typical fibre cement composites [3.81]. (Reprinted from Majumdar, H. J. & Laws, V., *Composites*, **10** (1979) 17, by permission of Butterworth & Co. (Publishers) Ltd.)

In the case of Kevlar–cement composites, as shown in Fig. 3.28, cracking begins at more or less the same stress as in glass-reinforced composites, but the ultimate failure strain is much higher in the case of composites containing Kevlar, reflecting perhaps higher effective tensile strength of this fibre, and relatively poor interfacial bond strength as in the case of PP fibres.

Since Kevlar fibres have a better thermal stability than PP or nylon type polyamide containing cement composites, the fire resistance of Kevlar cement is expected to be appreciably better than the former.

For cement boards, in which chopped fibres are distributed randomly in two dimensions, typical mechanical properties obtained with 1·9 vol% fibre addition are as follows:

Ultimate tensile strength	$16 \, MN/m^2$ (2320 psi)
Modulus of rupture	$44 \, MN/m^2$ (6380 psi)
Impact strength	$17 \, kJ/m^2$ (1165 ft lb/ft^2)

The durability of these composites is likely to be very good in most environments [3.82].

Nylon type polyamide fibre was one of the first synthetic fibres to be included in cement and concrete, but due to its relatively high cost compared with that of PP, its commercial potential may be limited. Walton & Majumdar [3.83] achieved moduli of rupture of up to $11 \, MN/m^2$ using 4% by weight of 25 mm long nylon monofilaments, and large increases in impact strength were observed which were not diminished by ageing. Both pre-mix and spray-up techniques were successfully used.

Other information about fibre–cement or fibre–concrete composites may be found in the various literature [3.84–3.95].

REFERENCES

[3.1]   Manson, J. A., Applications of polymer concrete. Publication SP-69, Americian Concrete Institute, Detroit, 1981, pp. 1–20.
[3.2]   Steinberg, M., *Polymers in Concrete, International Symposium*, SP-40, ACI, Detroit, 1973, pp. 1–3.
[3.3]   Swamy, R. N., *J. Mater. Sci.*, **14** (1979) 1521–53.
[3.4]   Fowler, D. W. *et al.*, *Polymers in Concrete, International Symposium*, Publication SP-58, ACI, 1978, pp. 123–37.
[3.5]   Kukacka, L. E. & Romano, A. J., *Polymers in Concrete, International Symposium*, SP-40, ACI, Detroit, 1973, pp. 15–32.

[3.6] Kukacka, L. E., *Polymers in Concrete,* The Construction Press, Hornby, UK, 1976, p. 26.

[3.7] Hastrup, K., Radjy, F. & Bach, L., ibid., p. 43.

[3.8] Feldman, R. F. & Beaudoin, J. J., *Cem. Concr. Res., 7* (1977) 19.

[3.9] Feldman, R. F. & Beaudoin, J. J., *Cem. Concr. Res., 8* (1978) 425–32.

[3.10] Auskern, A. & Horn, W., *Polymers in Concrete, International Symposium,* SP-40, ACI, Detroit, 1973, pp. 223–46.

[3.11] Fowler, D. W., Houston, J. T. & Paul, D. R., ibid., pp. 93–118.

[3.12] Horn, W., *Cem. Concr. Res., 4* (1974) 785.

[3.13] Kukacka, L. E., *Plastics in Material and Structural Engineering,* ed. R. A. Bares. Elsevier Science Publishing Co., Amsterdam, 1982, p. 815.

[3.14] Manning, D. G. & Hope, B. B., *Proceedings of the 1st Congress on Polymer Concrete,* May 1975, Concrete Construction Publications, p. 37.

[3.15] Ohama, Y., ibid., p. 43.

[3.16] Tazawa, E. & Kobayashi, S., *Polymers in Concrete, International Symposium,* SP-40, ACI, Detroit, 1973, pp. 57–92.

[3.17] Godard, P., *J. Appl. Polym. Sci., 18* (1974) 1477–91.

[3.18] Godard, P. & Mercier, J. P., *J. Appl. Polym. Sci., 18* (1974) 1493–504.

[3.19] Fowler, D. W. & Taylor, T. K., 1981 Plastics Seminar, NACE, Texas, October 27–29, 1981, Reprint No. 13.

[3.20] Madruga, E. L. *et al., Polym. Engng Sci., 19*(12) (1979) 825–8.

[3.21] Fukuchi, T. & Ohama, Y., *Polymers in Concrete, International Symposium,* SP-58, ACI, Detroit, 1978, pp. 215–24.

[3.22] Webster, R. P., Fowler, D. W. & Paul, D. R., *Polymers in Concrete, International Symposium,* SP-58, ACI, Detroit, 1978, pp. 249–66.

[3.23] Blaga, A. & Beaudoin, J. J., *Canadian Building Digest,* DBR, Ottawa, CBD-241, 1985, p. 2.

[3.24] DePuy, G. W., *Polymers in Concrete, 4th International Congress,* ed. H. Schulz, Sept. 1984, Darmstadt, pp. 79–83.

[3.25] Gunduz, G. *et al., J. Mater. Sci., 16* (1981) 221–5.

[3.26] Limsuwan, E. *et al.,* Research Report 114-6, Center of Highway Research, The University of Texas at Austin, June 1978.

[3.27] Maximov, Y. V., *Plastics in Material and Structural Engineering.* Elsevier Science Publishing Co., Amsterdam, 1982, pp. 279–92.

[3.28] Alexeev, A. S. *et al.,* ibid., pp. 275–8.

[3.29] DePuy, G. W. & Dikeou, J. T., *Polymers in Concrete, International Symposium,* SP-40, ACI, Detroit, 1973, pp. 33–56.

[3.30] Fowler, D. W. *et al., Polymers in Concrete, International Symposium,* SP-58, ACI, Detroit, 1978, pp. 187–203.

[3.31] Chen, H. & Xu, Y., *Polymers in Concrete, 4th International Congress,* Sept. 1984, Darmstadt, pp. 109–18.

[3.32] Maslow, P., *Plastics, Mortars, Sealants and Caulking Compounds,* ed. R. B. Seymour. ACS Symposium Series 113, Washington, D.C., 1979, pp. 39–60 and 79–100.

[3.33] Popovics, S., *Polymers in Concrete, 4th International Congress,* Sept. 1984, Darmstadt, pp. 369–72.

[3.34] Leitheiser, R. H., Londrigan, M. E. & Rude, C. A., *Plastics, Mortars,*

Sealants and Caulking Compounds, ed. R. B. Seymour. ACS Symposium Series 113, Washington, D.C., 1979, pp. 7–26.

[3.35] Calvert, K. O., *Polymer Latices and Their Applications.* Applied Science Publishers Ltd, London, 1982.

[3.36] Odian, G., *Principles of Polymerization,* 2nd edn. John Wiley, New York, 1981.

[3.37] Brydson, J. A., *Rubber Chemistry.* Applied Science Publishers Ltd, London, 1978.

[3.38] Dunn, J. R., *NBR Chemistry and Markets.* New York Rubber Group, Education Series, March 1982.

[3.39] Saunders, K. J., *Organic Polymer Chemistry.* Chapman and Hall, London, 1973.

[3.40] Roff, W. J. & Scott, J. R., *Fibres, Films, Plastics and Rubbers.* Butterworths, London, 1971.

[3.41] Brydson, J. A., *Plastics Materials,* 4th edn. Butterworths, London, 1982.

[3.42] Titow, W. V., *PVC Technology.* Elsevier Applied Science, London and New York, 1984.

[3.43] Kukacka, L. E., *Plastics in Material and Structural Engineering,* ed. R. A. Bares. Elsevier Science Publishing Co., Amsterdam, 1984, pp. 815–26.

[3.44] Popovics, S. & Tamas, F., *Polymers in Concrete, International Symposium,* ACI Publication, SP-58, Detroit, 1978, pp. 357–66.

[3.45] Nawy, E. G., Ukadike, M. M. & Sauer, J. A., ibid., pp. 329–55.

[3.46] Bhargava, J. & Rehnstrom, A., ibid., pp. 313–28.

[3.47] Fowler, D. W., Meyer, A. H. & Paul, D. R., *Applications of Polymer Concrete,* ACI Publication SP-69, Detroit, 1981, pp. 107–22.

[3.48] Martinez, A., Salla, J. M., Aros, M., Saura, P. & DeCea, A., *Polymers in Concrete, 4th International Congress,* Sept. 1984, Darmstadt, pp. 219–21.

[3.49] Sugama, T., Kukacka, L. E. & Horn, W., *Cem. Concr. Res.,* **11** (1981) 429–42.

[3.50] Zeldin, A. N., Fontana, J., Kukacka, L. E. & Carciello, N., *Int. J. Cem. Comp.,* **1** (1979) 1.

[3.51] Bourleson, J. D., Long, C. G. Jr, Armeniades, C. D. & Krahl, N. W., *Polymers in Concrete, International Symposium,* ACI Publication SP-58, Detroit, 1978, pp. 1–20.

[3.52] Ohama, Y., *Polymers in Concrete, International Symposium,* ACI Publication SP-58, Detroit, 1978, pp. 31–40.

[3.53] Ohama, Y. & Demura, K., *Applications of Polymer Concrete,* ACI Publication SP-69, Detroit, 1981, pp. 179–88.

[3.54] Jokiel, M., Kachlicki, B., Rozicki, J. & Spychala, A., *Plastics in Material and Structural Engineering,* ed. R. A. Bares. Elsevier Science Publishing Co., Amsterdam, 1982, pp. 301–6.

[3.55] Uriev, N. B. & Chernomaz, V. E., ibid., pp. 341–4.

[3.56] Schwabe, A., ibid., pp. 325–32.

[3.57] Kapasny, L., ibid., pp. 307–10.

[3.58] Hop, T. & Miodynski, Z., ibid., pp. 297–300.

[3.59] Sosa, J. M., *Plastics, Mortars, Sealants and Caulking Compounds,* ed. R. B. Seymour. ACS Symposium Series 113, Washington, D.C., 1979, p. 34.

[3.60] Ohama, Y., Demura, K. & Komiyama, M., ibid., p. 67.

[3.61] Staynes, B. W., *Polymers in Concrete, 4th International Congress,* Sept. 1984, Darmstadt, pp. 313–17.

[3.62] Hirano, T., Nagano, N. & Katsuse, K., ibid., pp. 333–8.

[3.63] Fowler, D., ibid., pp. 159–64.

[3.64] Van Den Berg, R. P., ibid., pp. 185–8.

[3.65] Paturoev, V. V., ibid., pp. 215–18.

[3.66] Kane, J. F., ibid., pp. 447–50.

[3.67] Hannant, D. J. & Zonsveld, J. J., Royal Society Discussion Meeting, New Fibres and Their Composites, 18–19 May, 1978, University of Surrey, Brighton, pp. 2–17.

[3.68] Swamy, R. N., *Polymers in Concrete,* ACI Committee 548 Report, ACI, Detroit, 1977.

[3.69] Hoff, G. C. (Ed.), *State of the Art Report on Fibre Reinforced Concrete, International Symposium,* ACI SP-81, Detroit, 1984, pp. 411–32.

[3.70] Zonsveld, J. J., *9th International Congress of the Precast Concrete Industry,* Vienna, 9–13 Oct., 1978, pp. 120–7.

[3.71] Raouf, Z. A., Al-Hassani, S. T. S. & Simpson, J. W., *Concrete,* **10** (1976) 28–30.

[3.72] Hannant, D. J., *Fibre Cements and Fibre Concretes,* John Wiley, Chichester, 1978.

[3.73] Hannant, D. J., Zonsveld, J. J. & Hughes, D. C., *Composites,* April 1978, 83.

[3.74] Hibbert, A. P. & Hannant, D. J., *Composites,* April 1982, 105–11.

[3.75] Wells, R. A., *Composites,* April 1982, 169–72.

[3.76] Aly, M. H. *et al., Cem. Concr. Res.,* **2**(21) (1972) 201–12.

[3.77] Waller, J. A., *Fiber Reinforced Concrete, International Symposium,* ACI, SP-44, Detroit, 1974, p. 143.

[3.78] Aveston, J., Mercer, R. A. & Sillwood, J. M., *Proceedings of the National Physical Laboratory Conference,* April 1974, pp. 93–103.

[3.79] Lankard, D. R., *Fibre Reinforced Cement and Concrete, RILEM Symposium,* ed. A. Neville. Construction Press Ltd, 1975, pp. 3–19.

[3.80] Walton, P. L. & Majumdar, A. J., *Composites,* Sept. 1975, 209–16.

[3.81] Majumdar, A. J. & Laws, V., *Composites,* **10** (1979) 17–27.

[3.82] Walton, P. L. & Majumdar, A. J., *J. Mater. Sci.,* **13** (1978) 1075.

[3.83] Diamond, S., *Proceedings of Durability of Glass Fibre Reinforced Concrete Symposium,* PCI, November 12–15, Chicago, 1985.

[3.84] Suaris, W. & Shah, S. P., *Fiber Reinforced Concrete, International Symposium,* ACI, SP-81, Detroit, 1984, pp. 247–66.

[3.85] Brandt, A. M., ibid., pp. 267–86.

[3.86] Lankard, D. R. & Newell, J. K., ibid., pp. 287–306.

[3.87] Majumdar, A. J., *Composites,* Jan. 1975, 7–16.

[3.88] Nair, N. G., *Fibre Reinforced Cement and Concrete, RILEM Symposium,* ed. A. Neville. Construction Press Ltd, 1975, pp. 81–93.

Polymeric Building Materials

[3.89] West, J. M., Majumdar, A. J. & DeVekey, R. C., *Composites*, Jan. 1980, 19–24.
[3.90] Moore, J. F. A., Building Research Establishment, I.P. 5/84, Garston, UK, March 1984, pp. 1–3.
[3.91] Morgan, D. R. & Mowat, D. N., *Fiber Reinforced Concrete, International Symposium*, ACI, SP-81, Detroit, 1984, pp. 307–24.
[3.92] Ibukiyama, S., Seto, K. & Kokubu, K., ibid., pp. 351–74.
[3.93] Naaman, A. E., Shah, S. P. & Thorne, J. L., ibid., pp. 375–96.
[3.94] Zollo, R. F., ibid., pp. 397–410.
[3.95] Johnson, C. D., *Composites*, April 1982, 113–21.

Chapter 4

Polymer Foams

INTRODUCTION

Polymer foams, known also as cellular polymers, cellular plastics or expanded polymers, are multiphase material systems (composites) that consist of a polymer matrix and a fluid phase, the fluid phase usually being a gas. Foams is a general term that refers to any degree of communication between the voids. The word cellular is used as a general term whereby the cells may have any degree of interconnection; expanded polymers are composites with closed cells.

Most polymers can be expanded into cellular products, but only a small number have been exploited commercially. A summary of US foamed polymer markets is presented in Table 4.1. As we may see from Table 4.2, in terms of volume, polystyrene, PVC and polyurethane have dominated the US markets.

In Europe, the four major contenders for insulation are polystyrene, polyurethane, rigid-PVC and phenol-formaldehyde foams. The last two are becoming more interesting because they are naturally flame retardant. The size of the Western European polymeric foam market in 1990 is likely to be in the order of 1 400 000 t [4.2]. However, over the last few years there has been increasing utilization of engineering structural foams for load-bearing applications.

The polymers used include polyolefins, modified polyphenylene oxide, polycarbonate and ABS. In some cases, depending on the uses, additional solid phases may be present in the cellular system. Examples include fibre-reinforced foams and syntactic foams (which are composite materials consisting of hollow glass, ceramic or plastic microspheres dispersed throughout a polymer matrix) [4.3–4.5].

Table 4.1.
Summary of US foamed plastics markets [4.1]

	Millions of pounds (kg)				
	1972	1975	1976	1977	1980
Furniture	452(204)	617(279)	659(298)	670(303)	907(412
Transportation	329(149)	436(197)	470(213)	479(217)	680(308
Bedding	110(50)	142(64)	153(69)	160(72)	207(94)
Carpet underlay and flooring	154(70)	213(96)	235(106)	251(114)	310(140
Textile	19(9)	76(34)	88(40)	95(43)	122(55)
Packaging	214(97)	267(121)	274(124)	296(134)	353(160
Construction	143(65)	198(90)	221(100)	241(109)	291(132
Tanks/pipe	32(14)	44(20)	48(26)	53(24)	70(32)
Appliances	51(23)	70(32)	77(35)	84(38)	113(51)
Marine	22(10)	29(13)	31(14)	34(15)	41(19)
Miscellaneous	52(24)	100(45)	88(40)	116(53)	132(60)
Total	1 578(715)	2 192(993)	2 344(1 062)	2 479(1 123)	3 226(1 46

Reprinted with permission from *US Foamed Plastics Markets & Directory*, 1977. Copyri
Technomic Publishing Co., Westport, Conn.

because of their complex nature, polymer foams have been classified
following different criteria such as:

—the cellular morphology;
—the mechanical behaviour;
—the composition; or
—the density.

Table 4.2.
Summary of US foam markets [4.1]

	Millions of pounds (kg)				
	1972	1975	1976	1977	1980
Rigid urethane	250(113)	340(154)	373(169)	409(185)	519(235)
Flexible urethane	810(367)	1 044(473)	1 126(510)	1 138(515)	1 627(737)
Polystyrene	285(129)	363(164)	380(172)	410(186)	490(222)
PVC	304(138)	515(233)	560(254)	603(273)	690(313)
Total	1 649(747)	2 262(1 025)	2 439(1 104)	2 560(1 160)	3 326(1 507)

Reprinted with permission from *US Foamed Plastics Markets & Directory*, 1977.
Copyright Technomic Publishing Co., Westport, Conn.

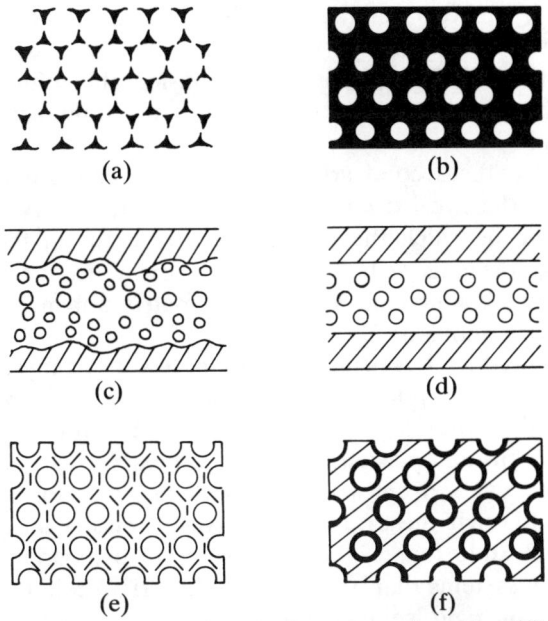

Fig. 4.1. Schematic representations of sections through different types of cellular polymer. (a) Low density open cell foam, (b) high density closed-cell foam, (c) single component structural foam with cellular core and integral solid skin, (d) multicomponent structural foam, (e) fibre-reinforced closed cell foam, and (f) syntactic foam [4.6]. (Reprinted with permission from *Mechanics of Cellular Plastics* by N. C. Hilyard (Ed.), 1982. Copyright Macmillan Publishing Co. Inc.)

From the morphological point of view we may use open cell foams or closed cell foams. With closed cell polymers, the second phase, i.e. the gas, is dispersed in the form of discrete gas bubbles and the polymer matrix forms a continuous phase. In open cell foams, the voids coalesce so that both the solid and the fluid phases are continuous [4.6]. Schematic representations of the different physical forms of cellular polymers are given in Fig. 4.1.

With open cell structures, as depicted in Fig. 4.1(a), the gas phase is able to flow through the polymer matrix under the action of some driving potential, whereas in closed cell structures (Fig. 4.1(b)), gas transport takes place by diffusion through the cell walls. The ease of movement of the fluid phase through the matrix is one of the factors governing the physical and mechanical properties of polymer foams.

In practice, the two cellular morphologies can co-exist so that a polymer foam is not always completely open or closed cell. The volume fraction of closed cells has a considerable influence on the mechanical behaviour of these systems so it is an important structural characteristic.

According to the second criterion, the mechanical behaviour, i.e. their stiffness, the two extremes are *rigid* and *flexible*. A rigid foam is defined as one in which the polymer matrix exists in the crystalline state, or, if amorphous, is below its T_g. Following from this, a flexible cellular polymer is a system in which the matrix polymer is above the crystalline T_m or above its T_g.

According to this classification, most polyolefin, polystyrene, phenolic, polycarbonate, polyphenylene oxide, and some polyurethane foams are rigid, whereas rubber foams, elastomeric polyurethanes, certain polyolefins and plasticized PVC are flexible [4.6]. Intermediate between these two extremes is a class of polymer foams known as *semi-rigid*. Although these materials have an elastic modulus higher than that of flexible foams, their stress–strain behaviour is closer to that of flexible systems than that exhibited by rigid cellular polymers.

The group of rigid cellular polymers can be further subdivided according to whether they are used for:

(a) *non-load bearing* applications, such as thermal insulation; or as
(b) *load bearing* structural materials, which require high stiffness, strength and impact resistance.

A particular class of flexible foam is known as *high resiliency* (also termed cold-cured polyurethane foam). This classification refers to the mechanical hysteresis exhibited by the material when taken through a compression–deformation cycle. High resiliency in compression is intermediate between that of conventional polyurethane flexible foam and rubber latex foam. The mechanical hysteresis of this material is much smaller than that of open cell flexible polyurethane foam and is close to that of rubber latex foam.

Polymer foams exist in a wide range of different structural forms. These may be homogeneous with a uniform cellular morphology throughout or they may be structurally anisotropic. They may have an integral solid polymer skin or they may be multicomponent in which the polymer skin is of different composition to the polymeric cellular core.

Table 4.3.

Cellular forms in which plastics are normally encountered [4.7]

Flexible		Rigid	
Closed cell	*Open cell*	*Closed cell*	*Open cell*
Polyvinyl chloride[a,d]	Polyvinyl chloride[f]	Polyvinyl chloride[a,d]	PVC membranes[e]
	Polyurethane[b,d]	Polystyrene[a,c,d]	Polyethylene sponge[e]
	Polyvinyl-formal[f]	Polyurethane[b,d]	Polyethylene porous
	(wet)		membranes[e]
Silicone rubber[a,b]	Rubber latex[f]	Cellulose acetate[d]	Polyethylene porous
Grafted cross-linked	Polyvinyl-formal[f]	Polyethylene[a,c,d]	Gelatin[f] (dry)
polyolefins[a]			
Gelatin[f] (wet)		Polymethyl meth-	Urea-formaldehyde[f]
		acrylate[b,d]	Phenol-formaldehyde[b]
		Urea-formaldehyde[b]	Polyvinyl-formal[f] (dry)
		Silicones[a]	
		Ebonite[c]	
		Epoxides[a,g]	

Note: Superscripts indicate the principal method(s) by which the materials are produced.
[a] Thermal decomposition of a chemical blowing agent.
[b] Blowing by in situ chemical reaction.
[c] Low-pressure release of dissolved gas.
[d] Blowing by vapour from a volatile liquid.
[e] Extraction of a temporaty filler.
[f] Mechanical entrainment of gas.
[g] The use of microspheres.
Reprinted with permission from *Plastic Foams*; *the Physics and Chemistry of Product Performance and Process Technology*, Vol. II, by C. J. Benning, 1969. Copyright John Wiley, New York.

Some of the cellular foams in which plastics are normally encountered are presented in Table 4.3.

The description of cellular foams as low, medium or high density, very common in practice, is not exact, as the different density ranges which correspond to each of these terms are not strictly defined. The following figures can, however, serve as a rough general guide [4.8]:

	kg/m^3	lb/ft^3
Low density range	10–50	0·6–3
Medium density range	50–350	3–21
High density range	350–900	21–54

FOAMING (BLOWING) AGENTS

Blowing agents make up a group of additives used in the manufacture of polymer foams. These materials are capable of producing pores, or cells, throughout the polymer mass. The resultant cellular body is a solid–gas composite consisting of a continuous polymer phase and a gas phase, either *continuous* or *discrete,* created by the foaming agent. Depending on the nature of the cell-forming process, that is, whether it is a physical change of state or a chemical composition, blowing agents are classified as *physical* or *chemical.*

Foaming promoters are adjuncts that facilitate the formation of uniform cells and can increase the stability of the foam.

Physical blowing agents undergo a phase change during foaming. For example, compressed gases dissolved under pressure in a plastisol (polymeric gel) can develop a cellular structure on release of the internal pressure. Volatile liquids incorporated in a polymer mix are capable of producing a foam by passing from a liquid to a gaseous state.

Physical Foaming Agents (gaseous or liquid)

Nitrogen, air and carbon dioxide are prime examples of *gaseous blowing agents* used in preparing cellular polymers. The first two are used in the high-pressure 'gassing' process, while CO_2 itself is used in a low-pressure adsorption method.

Many useful physical blowing agents are found among *volatile liquids* with boiling points not exceeding 110°C. These compounds are usually selected from odourless, non-toxic, non-corrosive, and non-flammable liquids possessing good thermal stability in the gaseous state. Since the efficiency of the liquid blowing agents is directly related to the ratio of specific volume of vapour to the volume of liquid, products with high specific gravity combined with low molecular mass are most effective. Fluorinated aliphatic hydrocarbons possess all these desirable properties and therefore are ideal physical blowing agents (Table 4.4) [4.9].

Chemical Foaming Agents

Mineral and organic compounds that liberate large volumes of gas, as a result of thermal decomposition at elevated temperatures, compose a group of additives known as chemical blowing agents. These are usually solid materials with good thermal stability at ambient temperatures. At relatively elevated temperatures, these agents undergo a rapid decomposition in well-defined temperature intervals. The most commonly used foaming agents liberate, in addition to nitrogen, other non-condensable gases such as carbon

Table 4.4.
Properties of some blowing agents [4.9]

Blowing agent	Molecular weight	Density at 25°C (g/cm^3)	(lb/ft^3)	Boiling point (°C)
Cyclohexane	84·00	0·774	48·37	80·8
Trichloroethylene	131·40	1·466	91·62	87·2
1,2-Dichloroethane	98·97	1·245	77·81	83·5
1,1,2-Trichlorotri-fluoroethane	187·39	1·565	97·80	47·6
Acetone	58·08	0·785	49·0	56·2

Reprinted with permission from *Encyclopedia of PVC, Vol. 2*, by L. J. Nass (Ed.), 1976. Copyright Marcel Dekker.

dioxide, carbon monoxide and hydrogen. On decomposition of the agent, the residue becomes a component of the polymer matrix and therefore should not detract from any of the valuable properties of the polymer compound.

Mineral blowing agents, mostly alkali salts of weak acids, can liberate gas either by thermal dissociation or in the presence of activators by chemical decomposition. The most important of these agents are listed in Table 4.5.

Among the mineral salts, ammonium bicarbonate is of some interest because it leaves no residue on decomposing:

$$(NH_4)HCO_3 \rightarrow NH_3 + CO_2 + H_2O$$

Organic compounds that release nitrogen as the main component of

Table 4.5.
Inorganic blowing agents [4.9]

Agent	Decomposition temperature (°C)	Gas yield (ml(STP)/g)	Main decomposition products
Ammonium bicarbonate	60	850	NH_3, CO_2, H_2O
Sodium bicarbonate	100–140	267	CO_2, H_2O
Sodium borohydride ($NaBH_4$)	300	2 370	H_2

Reprinted with permission from *Encyclopedia of PVC, Vol. 2*, by L. J. Nass (Ed.), 1976. Copyright Marcel Dekker.

Table 4.6.
Characteristic functional groups of organic blowing agents [4.9]

—N=N—	Azo
>N—NO	N-nitroso
—SO₂—NH—NH—	Sulphohydrazo

—N⟨ (N triangle structure)   Azido

—N with N≡N azido ring structure

| structure | Azido |

Reprinted with permission from *Encyclopedia of
PVC, Vol. 2,* by L. J. Nass (Ed.), 1976. Copyright Marcel Dekker.

liberated gas are the most important foaming agents for cellular
plastics. The thermal decomposition of organic foaming agents is an
irreversible exothermic reaction independent of external pressure.

Those blowing agents that have achieved commercial significance
possess good storage stability and an excellent dispersibility. Their
decomposition occurs in well-defined, usually narrow temperature
ranges.

Chemically, organic foaming agents are characterized by one of the
functional groups shown in Table 4.6. With the exception of azides, all
functional types of organic foaming agents are represented among
commercial products.

CELLULAR POLYMER MANUFACTURE

Cellular polymer products are made, in principle, by generating gas
bubbles in a polymer and then stabilizing the expanded structure. Such
processes are used to make polyurethane foams, polystyrene foams,
beverage cups and furniture, PVC fabric coatings, ordinary sponge
rubber and epoxy flotation devices.

In a very simple way the cellular expansion process is illustrated in
Fig. 4.2. Three methods are available for the formation of gas bubbles:

(1) Latex foam rubber is made by mechanically induced frothing of
    a latex or a liquid rubber, followed by cross-linking the polymer
    in the expanded state.

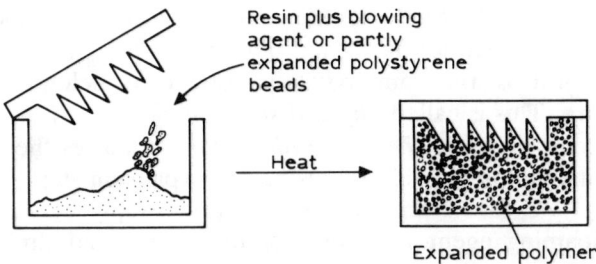

Resin plus blowing
agent or partly
expanded polystyrene
beads

Heat

Expanded polymer

Fig. 4.2. Preparation of moulded objects from expanded cellular polymers [4.10]. (Reprinted with permission from *Contemporary Polymer Chemistry* by H. P. Allcock & F. W. Lampe, 1981. Copyright Prentice Hall Inc.)

(2) The polymer suspension (or liquid monomer) is mixed with a chemical foaming agent which liberates a gas when heated (e.g. AIBN [azo-bis-isobutyronitrile] evolves nitrogen gas when heated; sodium bicarbonate liberates carbon dioxide).

(3) A low boiling point liquid or gas is dissolved in the polymer suspension under pressure. Heating of the suspension causes boiling of the liquid to generate bubbles. Pentane, hexane or halocarbons are commonly used expansion agents.

A simple flowchart of the process is shown below:

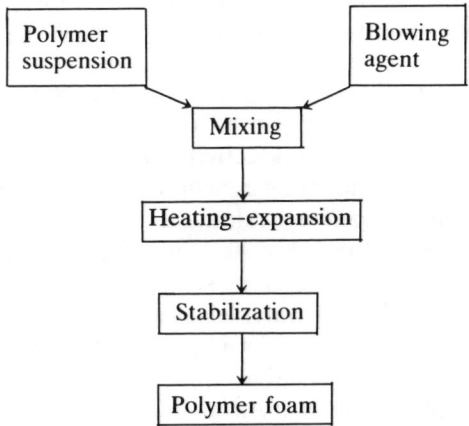

Once the polymer has been expanded, the cellular structure must be stabilized rapidly; otherwise it would collapse. Two stabilization methods are used. First, if the macromolecular compound is a thermoplastic, expansion is carried out above the T_g or melting point, and the form is then immediately cooled to below the melting temperature. This is called *physical stabilization.*

The second method, *chemical stabilization,* requires the polymer to be cross-linked immediately following the expansion step.

PS and PVC foams are usually stabilized simply by cooling. Such polymer-foaming agent mixtures are often extruded through a slit. Expansion and simultaneous cooling occur as the polymer is extruded. Polyurethane foams (PU) are expanded by CO_2 bubbles generated from the reaction of excess isocyanate with water or carboxylic acids in the system, and by the expansion of volatile organic expansion agents. Cross-linking occurs during foaming. Epoxy foams are stabilized by cross-linking, as are those from phenolic polymers. Silicone foams are expanded by chemical blowing agents, and are stabilized by the reaction of a cross-linking catalyst [4.10].

Foams differ in the process conditions under which they are generated. A material that requires a high temperature or an external pressure for foam formation cannot be easily *foamed in place,* as could a system that can be mixed and poured at room temperature. And of course, the polymer from which the foam is desired can be, as already mentioned, a thermoplastic or a thermoset, rubbery or glassy.

For some PU foam formulations, the blowing agent is a fluorinated hydrocarbon which increases the total volume of the product without itself participating in the chemical reactions that are occurring. Concern for environmental air pollution by fluorocarbon vapour is an incentive to replace this particular ingredient with other volatile liquids. For thermal insulating purposes fluorocarbons are the best gas because of their low thermal conductivity, as we may see in Table 4.7.

In the case of PU the foaming reaction takes place so rapidly that large-scale productions demand automatic mixing equipment. In one arrangement, mixing and reaction start in a travelling head that cycles back and forth to make a continuous slab (Fig. 4.3).

The PU foam is a typical polymer system in that the number of variables that can significantly affect important properties is very large. The optimization of the relative quantities of only the specific components in Table 4.8 is in itself a time consuming and expensive exercise even with the aid of a computer and multiple-regression analysis. We must add to that problem the possibility of varying each ingredient chemically.

Table 4.7.
Properties of gases at 1 atm and 273 K [4.11]

Gas	Boiling temperature		Thermal conductivity	
	(°C)	(°F)	$(10^{-5}cal/cm\,s\,°C)$	$(10^{5}J/mm\,s\,°C)$
Hydrogen	−253	−423	40·3	16 872
Nitrogen	−196	−320	5·72	2 394
Oxygen	−183	−297	5·78	2 419
Carbon dioxide	−78	−108	3·56	1 490
Freon (CCl_2F_2)-12	−30	−21·6	2·68[a]	1 122[a]

[a] Established at 423°K.
Reprinted with permission from *Handbook of Heat Transfer*, 1973. Copyright McGraw-Hill.

Rigid foams result when the polyol is short-chained and highly functional so that the distance between cross-links is small, or when the polyol backbone contains bulky groups that raise the T_g. Fillers such as fibres or finely divided silica can be used to stiffen a foam, but they are seldom used because they increase the density of the cellular material. Components (polyisocyanates) of high functionality can be employed to increase the cross-link density. Some of these are condensation products of the commonly used toluene diisocyanate.

The structural foam process (Fig. 4.4) gets its name from the application for its product rather than the mechanics of the process itself. In a manner directly opposite to the vented extruder, a blowing agent, often nitrogen, is injected into the melt in the extruder.

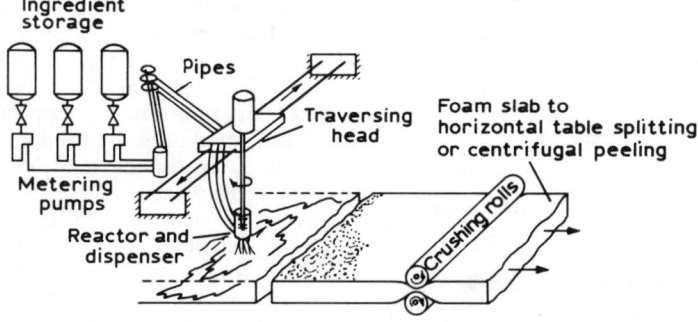

Fig. 4.3. Continuous polyurethane foam slab production [4.10]. (Reprinted with permission from *Principles of Polymer Systems*, 2nd edn, by F. Rodriguez, 1982. Copyright Hemisphere Publishing Co., 1982.)

Table 4.8.
Polyurethane one-shot foam system [4.12]

Ingredient		Parts by weight
Poly(propylene oxide) $M_n = 2000$, 2 OH/molecule	$HO—(—CH_2\overset{\displaystyle CH_3}{\underset{\displaystyle \mid}{C}HO—)—H}$	35·5
Poly(propylene oxide) started on trifunctional alcohol, $M_n = 3\,000$, 3 OH/molecule	$HO—\underset{\displaystyle \underset{\displaystyle OH}{\mid}}{R}—OH$	35·5
Toluene diisocyanate (80% 2,4 and 20% 2,6-substituted)	$(C_6H_3)(CH_3)(NCO)_2$	26·0
Dibutyl tin dilaurate	$(C_4H_9)_2Sn(O_2C_{12}H_{23})_2$	0·3
Triethyl amine	$(—CH_3—CH_2—)_3—N$	0·05
Water		1·85
Surfactant (silicone)		0·60
Trichloromonofluoromethane, CCl_3F		12
Final density = $1·4\,lb/ft^3 (2·0\,lb/ft^3$ if CCl_3F omitted)		

Reprinted with permission from *Principles of Polymer Systems*, 2nd edn, by F. Rodriguez, 1982. Copyright Hemisphere Publishing Co.

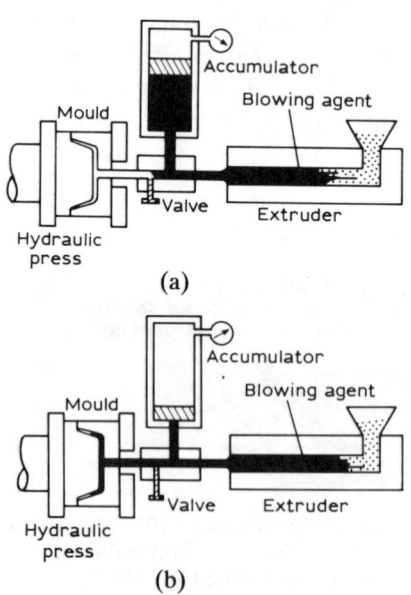

(a)

(b)

Fig. 4.4. Structural foam process. (a) Filling the accumulator. The blowing agent—usually nitrogen—is injected into the melt in the extruder. The melt is passed into the accumulator, where it is maintained at a pressure and temperature high enough to prevent foaming until the predetermined charge is collected. (b) Filling the mould. The accumulator ram then injects the charge into the mould, where the reduced pressure allows the gas to foam the resin and drive it throughout the mould [4.10]. (Reprinted with permission from *Principles of Polymer Systems*, 2nd edn, by F. Rodriguez, 1982. Copyright Hemisphere Publishing Co.)

Polymer melt that has already been injected with gas goes to the accumulator. When a sufficient charge has accumulated it is transferred into the mould. The melt foams and fills the mould at a relatively low pressure (1·3–2·6 MPa) compared to the much higher pressure in the accumulator. The lower pressures in the moulds make the moulds less expensive than those used for conventional injection moulding. However, the cycle times are longer because the foam, being a good insulator, takes longer to cool. Structural foams can also be made using a chemical blowing agent rather than an inert gas. A change in pressure or temperature on entering the mould triggers gas formation.

Today it is believed that structural foam injection moulding is a very fast growing polymer processing technique that can be used to modify the properties of thermoplastics to suit specific applications. With this technique, a product with a cellular core and a solid skin can be moulded in a single operation.

Villamizer & Han [4.13] carried out an experimental study to gain a better understanding of the dynamic behaviour of gas bubbles during the structural foam injection moulding operation. Polymers used were PS, HDPE and polycarbonate. As chemical blowing agent, sodium bicarbonate, a proprietary hydrazide and 5-phenyl tetrazole, both the latter generating nitrogen, were used.

Injection pressure, injection melt temperature, and mould temperature were raised to investigate the kinetics of bubble growth (and collapse) during the foam injection moulding operation.

It was found that the processing parameters such as pressure, concentration of blowing agent and temperature have a profound influence on the nucleation and growth rates of gas bubbles during mould filling.

Figure 4.5 gives a schematic diagram, describing the bubble growth during the mould filling step under isothermal conditions. There are two different mechanisms of bubble growth; one mechanism is due to a sudden decrease in internal (melt) pressure during the mould-filling and the other is due to diffusion of gas from the polymer melt.

Some specific observations made from the same study are the following:

(1) An increase in melt temperature, mould temperature and blowing agent concentration brings about an increase in bubble growth but more non-uniform cell size and its distribution.

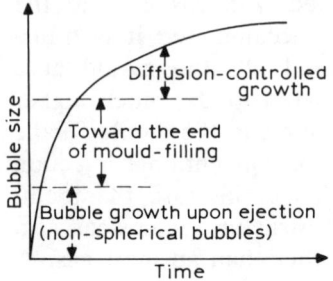

Fig. 4.5. Schematic representing various steps of bubble growth during the foam injection moulding operation of gas-charged molten polymer [4.15]. (Reprinted with permission from *Polym. Engng Sci.*, **18**(9) (1978) 639. Copyright Society of Plastics Engineers.)

(2) An increase in injection pressure brings about a decrease in bubble growth but more uniform cell size and its distribution.

In the context of the author's observations of bubble growth under different processing conditions, a general trend of bubble growth rate as affected by processing variables is summarized schematically in Fig. 4.6.

SPECIAL FOAMS

In a foam the gas is distributed in cells, whereas the solid matrices enclose these voids to form the cell walls. A foam may have a predominantly 'closed cell' structure or an 'open cell' structure.

Most rigid plastic foams, however, are neither completely open, nor completely closed cells but are characterized by a fraction of open or closed voids.

The arrangement of the gas phase and solid phase depends especially upon the forces which are involved during the expansion of the melted polymer. These are:

(a) the gas pressure which causes the flow of the melted polymer as the volume of the cells increases;

(b) the reaction of viscoelastic forces of the polymer, forces which resist its flow;

(c) the surface tension forces which cause the flow of the polymer from cell walls to the point at which they intersect.

The most favoured cell structure is one resulting in a minimum surface tension for the expanding high polymer.

Density, when applied to a rigid polymer foam, refers to its bulk density, defined by the ratio of total weight/total volume of the

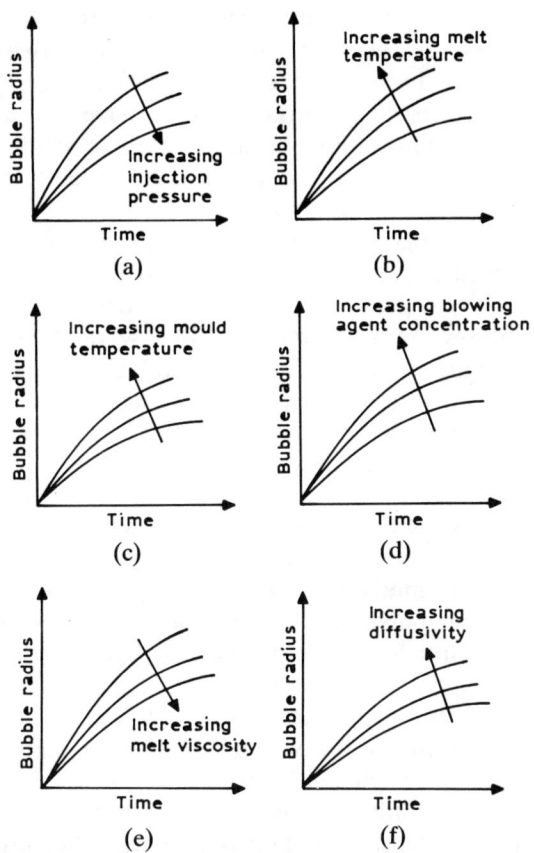

Fig. 4.6. Schematic showing the effects of various processing variables on the growth of gas bubbles during mould filling: (a) the effect of injection pressure; (b) the effect of injection melt temperature; (c) the effect of mould temperature; (d) the effect of initial blowing agent concentration; (e) the effect of melt viscosity; (f) the effect of diffusivity [4.13]. (Reprinted with permission from *Polym. Engng Sci.*, **18**(9) (1978) 639. Copyright Society of Plastics Engineers.)

polymer and gaseous component. Obviously the gas phase contributes considerably to the volume of the end-product, while the solid component contributes almost to the entire weight [4.14].

Some special types of foams are:

—structural foam;
—syntactic foams and multifoams;
—reinforced foams.

The term, 'structural foam', designates components possessing full-density skins and cellular cores, similar to structural sandwich constructions, or to human bones, whose surfaces are solid, but cores are cellular. For structural purposes, they have favourable strength and stiffness-to-weight ratios, because of their sandwich type configuration. Frequently, they can provide enhanced structural performance at reduced cost of materials.

Structural foams may be manufactured by high-pressure processes or by low-pressure processes. The first one may provide denser, smoother skins with greater fidelity to fine detail in the mould than may be true of low pressure processes. Fine wood detail, for example, is used for simulated wood furniture and simulated wood beams. Surfaces made by low pressure processes may show swirl or other textures, not necessarily detracting from their usefulness. Almost any thermoplastic or thermosetting polymer can be formulated into a structural foam.

In the case of syntactic foams, instead of employing a blowing agent to form bubbles in the polymer mass, preformed bubbles of glass, ceramics, or other macromolecular compounds are embedded in a matrix of unblown polymer. In multifoams such preformed bubbles are combined with a foamed polymer to provide both kinds of cells. Reduction in weight is an obvious objective. However, this may be accompanied by other properties.

Synthetic 'wood' for instance, is provided by a mixture of polyester and small hollow glass spheres (microspheres).

In contrast to syntactic foams in which the microspheres are inclusions within a continuous polymer matrix, the three-phase composite is an aggregate of microspheres bonded together by a small amount of polymer which may not form a continuous matrix. The cell space is obtained by limiting the amount of polymer which is mixed with the microspheres [4.15]. The microspheres are made of phenolformaldehyde polymers or of glass.

Figure 4.7 is a graphical representation of foam structures in which the microspheres are dispersed (a) randomly, and (b) uniformly in close packing. In both structures, the two components fill completely the whole volume (no dispersed voids) and the density of the products is thus calculated from the relative proportions of the two. Measured density values often differ from the calculated ones due to the existence of some isolated or interconnected irregularly shaped voids as shown in (c). The voids are usually an incidental part of the

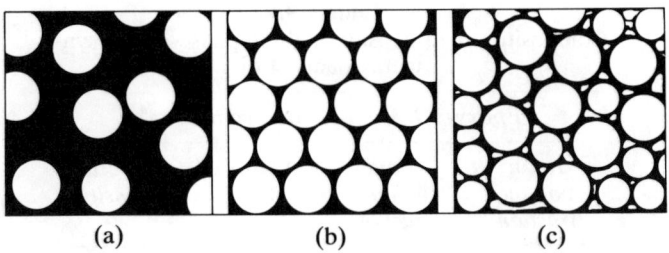

Fig. 4.7. Graphical representations of syntactic foam structures. (a) Random dispersion of spheres. Two-phase composite. (b) Hexagonal closed packed structure of uniform sized spheres. Two-phase composite. (c) Three-phase composite containing packed microspheres, dispersed voids and binding resin [4.4]. (Reprinted with permission from *J. Cell. Plast.*, July/Aug 1980, 223. Copyright Technomic Publishing Co.)

composite, as it is not easy to avoid their formation. Nevertheless, voids are often introduced intentionally to reduce the density below the minimum possible in a closed-packed system two-phase structure. In these three-phase systems the resin is mainly a binding material holding the structure of the microspheres together.

The composition of three-phase systems (polymer, hollow microspheres and voids) can be represented by a ternary phase diagram, as in Fig. 4.8. Point A on the diagram denotes a composite of the

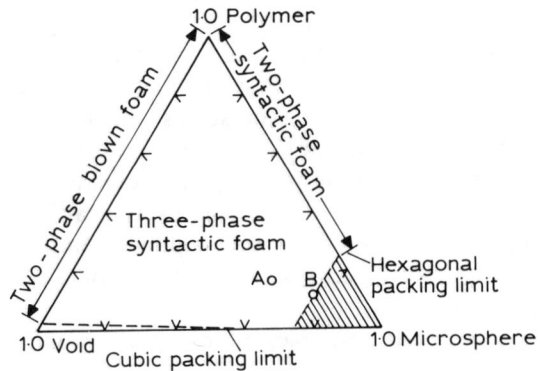

Fig. 4.8. Phase diagram of three-component syntactic foams [4.15]. (Reprinted with permission from *J. Comp. Mater.*, **10** (1976) 314. Copyright Technomic Publishing Co.)

Table 4.9.
Composition and density of three-phase epoxy syntactic foams [4.4]

Epoxy/VT Microballoon composite[a]			
Resin volume fraction	*Void volume fraction*	*Measured density*	*Expected density*
0·016	0·384	0·19	0·182
0·035	0·365	0·21	0·205
0·060	0·340	0·22	0·234
0·094	0·306	0·26	0·273
0·141	0·259	0·33	0·328
0·212	0·188	0·42	0·410

[a] Microballoon volume fraction = 0·60; bulk density = 0·16 g/cm^3.
Reprinted with permission from *J. Cell. Plast.*, July/Aug (1980) 223. Copyright Technomic Publishing Co.

following volume fractions:

resin	0·15
microspheres	0·60
voids	0·25

A limiting case is point B which represents a composition of microspheres 0·74, polymer 0·11 and voids 0·15. The microspheres in this case are arranged in hexagonal close packing.

Syntactic foams exhibit their best mechanical behaviour in the compressive mode. The spheres themselves are an extremely strong structure and hence can withstand such stresses very well.

Table 4.9 presents the compositions and densities of some epoxy syntactic foams and Table 4.10 summarizes typical results for compressive properties of epoxy–glass microsphere syntactic foams as a function of density showing the strength and modulus increase with polymer concentration [4.4].

In Table 4.11 selected examples of composites containing microspheres are presented [4.16].

The fairly dense varieties of polymer foams may be reinforced, usually with short glass fibres, and also other fibres such as asbestos or metal, and other reinforcements such as carbon black. The reinforcing

Table 4.10.
Compressive properties of epoxy syntactic foams [4.4]

Density		Compressive strength		Compressive modulus	
(g/cm³)	*(lb/ft³)*	*(kg/cm²)*	*(kPa)*	*(kg/cm²)*	*(MPa)*
0·16	10·0	6	588	340	33·3
0·19	11·8	17	1 666	1 700	166·6
0·22	13·75	30	2 940	2 600	254·8
0·28	17·5	59	5 782	4 600	450·8
0·32	20·0	96	9 408	7 000	686·0
0·42	26·2	140	13 720	7 500	735·0

agent is generally introduced into the basic components and is blown along with them, to form part of and to reinforce the walls of the cells. When this is done, it is not unusual to obtain increases in mechanical properties, especially in thermosettings of 400–500% with fibre glass content up to 50% by weight. With liquid foams such as some of the thermosetting ones, reinforced fibres may be introduced into the mould itself before or after foaming, but increases in mechanical properties are likely to be much lower than those above.

The principal advantages of reinforcement, in addition to increased strength and stiffness, are:

—improved dimensional stability;
—improved resistance to extremes of temperature;
—improved resistance to creep.

Reinforced foams are heavier, may be more abrasive to moulds and machinery, and are likely to be more costly than plain foams [4.17].

The free-rise process and the restrained-rise process for the manufacture of glass-reinforced foam laminates are presented in Figs. 4.9 and 4.10 respectively.

Various types of facing sheets may be used, such as aluminium foil for building insulation products, asphalt-saturated felts for roof insulation or any other material, e.g. paper, plastic films, etc., which is processable in web form and is otherwise suitable for use in the process.

Table 4.11.

Selected examples of composites containing microspheres [4.16]

Generic description of constituent materials	Summary of salient design properties	Optimum design application areas	Limitations
Hollow glass microspheres and carbon particles in resin	Pack-in-place absorbers of graded dielectric constant for 'electrically tapered' energy dissipators in microwave transmission lines	As indicated in summary of properties	—
Hollow glass microspheres in epoxy resin	Low viscosity, 5000 cP; low weight, specific gravity 0·72; flexural strength, 4000 psi; unicellular; negligible moisture absorption; combines insulation properties of glass and air; strength with chemical and moisture resistance of resin	Potting compound for high density electronic modules and other units likely to encounter hydrostatic pressures	—
Hollow glass microspheres and powdered aluminium in resin	One-part artificial dielectric foam; silica bubbles provide low weight; resin gives strength, and aluminium has controllable electrical properties	Core material for sandwich construction of radomes	Not completely unicellular and must be contained or sealed
Hollow glass microspheres in resin	Pack-in-place foam; bulk density 20 lb/ft^3; cell size, 30–50 µm diam.; withstands 300°F continuously; room temperature cure	Electrical potting adhesive or caulking compound; for sandwich structures, thermal insulation	Requires surface sealant since structure is not completely unicellular
Hollow glass microspheres in aluminium matrix	Density, 1·5 g/cm^2; tensile strength, 2 700 psi; compressive strength, 22 000 psi; thermal conductivity, 280 Btu/h/ft^2/°F/in; volume resistivity, 10^{-4} ohm-cm; water absorption, nil; high machinability	Aerospace and extreme hydrostatic pressure (oceanographic) applications in view of low weight and high compressive strength; core materials in sandwich	Cost may be a factor
Epoxy–silica micro-balloon syntactic foam	Easy to machine; dimensionally stable; weighs approximately 20 lb/ft^3; one-component system based on powdered epoxy, with unlimited pot life	Used to make component holder for electronic hardware where weight is a consideration	Must be encapsulated in a shell of impervious material because of porous nature

Reprinted with permission from *Composite Materials for Combined Functions* by E. Scala, 1973. Copyright Hayden Book Co.

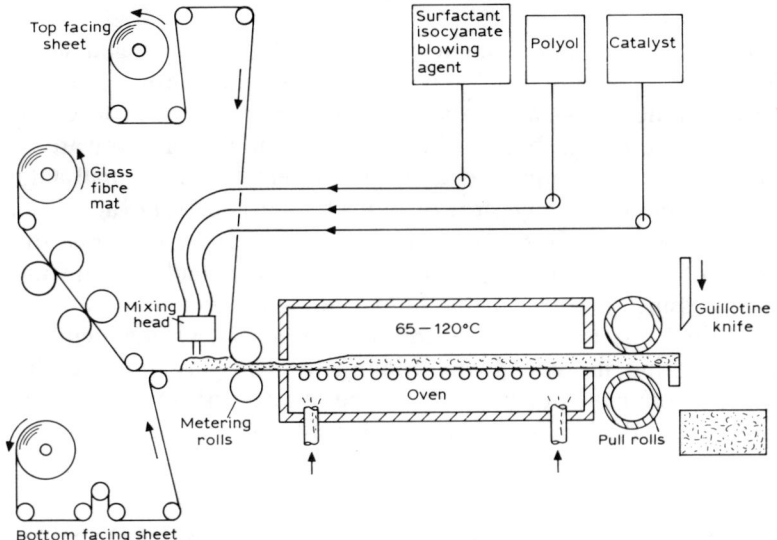

Fig. 4.9. Free-rise process for manufacture of glass-reinforced foam laminates [4.17]. (Reprinted with permission from *Advances in Urethane Science and Technology,* Vol. 7, by K. C. Frisch & D. Klempner (Eds), 1980. Copyright Technomic Publishing Co.)

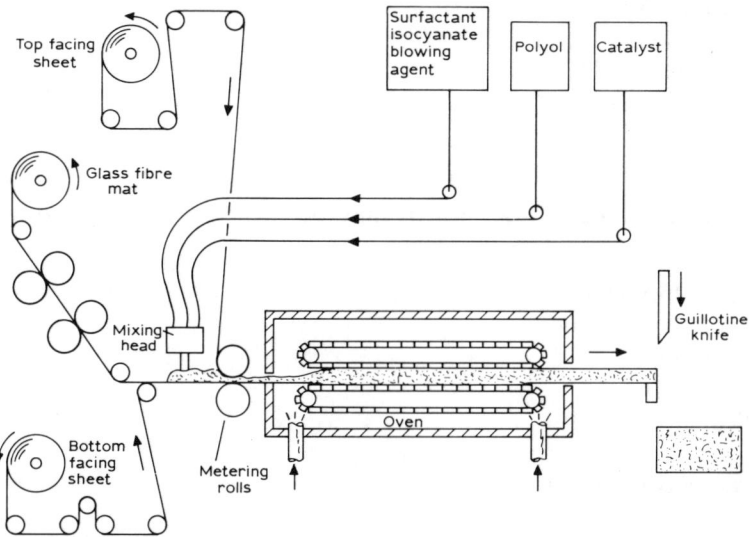

Fig. 4.10. Restrained-rise process for manufacture of glass-reinforced foam laminates [4.17]. (Reprinted with permission from *Advances in Urethane Science and Technology,* Vol. 7, by K. C. Frisch & D. Klempner (Eds), 1980. Copyright Technomic Publishing Co.)

The glass fibre reinforcement is a thin (0·25–1·25 mm) mat supplied in roll form. It is substantially incompressible, and consists of layers of relatively long (1·5–4 m) glass fibres, the plane of each layer being parallel to the facing sheets. The glass fibres in a given layer are at an acute angle to the fibres in each next adjacent layer. A small amount of a silane-modified polyester, or other binder is present, at a level of 2–10% by weight. This type of glass mat is relatively porous to the passage of liquids, and sufficiently dimensionally stable for unwinding and processing, yet it is capable of expanding within a mixture of rising foam chemicals, to provide a uniform, three-dimensional reinforcing network within the final foamed laminate. The glass fibre reinforcement is functionally effective when used at levels of from 4 to 24 g per board foot of the laminate foam product [4.17].

THERMOPLASTIC AND THERMOSETTING FOAMS

Foams can be made with both thermoplastic and thermosetting plastics. The well known commercial thermoplastic foams are:

—polystyrene (PS),
—polyvinyl chloride (PVC),
—polyethylene (PE),
—polypropylene (PP),
—ABS,
—cellulose acetate.

The thermosetting plastics which may be mentioned, among others, are:

—phenol-formaldehyde,
—urea-formaldehyde,
—polyurethane,
—epoxy polymers,
—silicone polymers.

Thermoplastic Foams

Polystyrene (PS)
Polystyrene (PS) foams are the most used from the first group mentioned above. Cellular polymers based on PS are generally closed

cell, rigid foams, that can be manufactured in densities ranging between 16 and 480 kg/m^3 (1–30 lb/ft^3); most are in the 16–80 kg/m^3 (1–5 lb/ft^3) density range. For the main types of construction PS foams the following technologies are used:

—moulding bead foam,
—extrusion.

Although there are applications in which either type may be employed, the two materials should not be considered readily inter-changeable. The first type is produced from expandable PS beads, by using a blowing agent such as hydrocarbons, halocarbons and/or mixtures of both. The expandable beads are converted to foam in the following manner: they are first matured and conditioned by subjecting them to heating by steam, hot water or hot air to yield pre-expanded beads. The pre-expanded beads are again heated so that they undergo additional expansion, flow to fill the spaces between particles and fuse. This produces an integral moulded item. Low density foam structures are usually made by placing pre-expanded PS beads in a mould and steaming the particles to complete the expansion process and to obtain a good bead-to-bead consolidation [4.7].

To produce extruded foam, a molten PS-based compound containing a blowing agent is extruded at a certain temperature range and pressure through a slit orifice to atmospheric pressure; in this condition, the mass expands to about 40 times its pre-extrusion volume. It is extruded in board form with a continuous surface skin or in large billets that can be cut into standard board or fabricated into desired shapes. Extruded foam has a simple, more regular structure than moulded bead foam, and also has better strength properties and higher water resistance.

The flow chart of the technology for producing PS foams is presented in Fig. 4.11.

The suspension in water obtained after the polymerization of styrene in the reactor (1) is sent through a filter (2) to a vessel (3) for partial gasification. After another filtering (4) beads are brought into the autoclave (5) with an equal amount of hot water. At this point, various additives and foaming agents are added to the mixture. At predetermined conditions of temperature and pressure, the blowing agent is absorbed in the PS beads; after a washing operation in a centrifuge (6) the beads are transported pneumatically to an ageing vessel (7), after which they are pre-expanded with steam (8) and

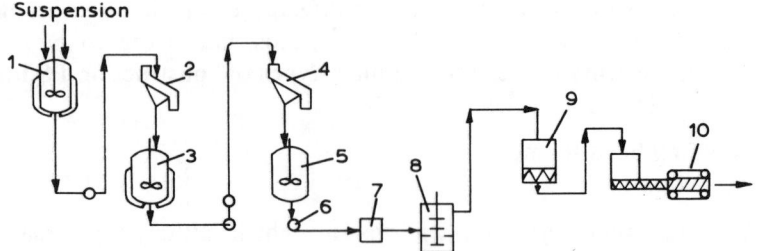

Fig. 4.11. Polystyrene foam technology (Feldman [4.18]).

further aged in another vessel (9), prior to the addition of flame retarders. The beads are then sent to the extrusion machine (10) for the fabrication of blocks of varying cross-section [4.18].

PS foams have poor outdoor weathering resistance; they resist moisture well but deteriorate when exposed to direct sunlight for long periods of time as evidenced by a characteristic yellowing of the foam matrix. Multiple coats of water dispersed exterior paints will provide protection against weathering; Portland cement plaster, latex modified plasters and asphaltic emulsions are often used to provide protection against physical damage.

As with other thermoplastic foams, PS foams undergo slight deterioration of their mechanical properties when the temperature is raised to the T_g, which means that different grades of PS are affected by temperatures in the range of 71–77°C. GPPS (general purpose polystyrene) foams are flammable; under larger fluctuating temperature and moisture gradients some PS foams can absorb a high proportion of water.

PS foams are used in construction as:

—perimeter insulation,
—roof insulation,
—masonry wall insulation.

The requirements for perimeter wall insulation, applied below ground level along the edges of a concrete foundation, are relatively high thermal resistance for a given thickness, good moisture resistance, good compressive strength.

The product for roof insulation should have good dimensional stability and high flexural and compressive strength, and should

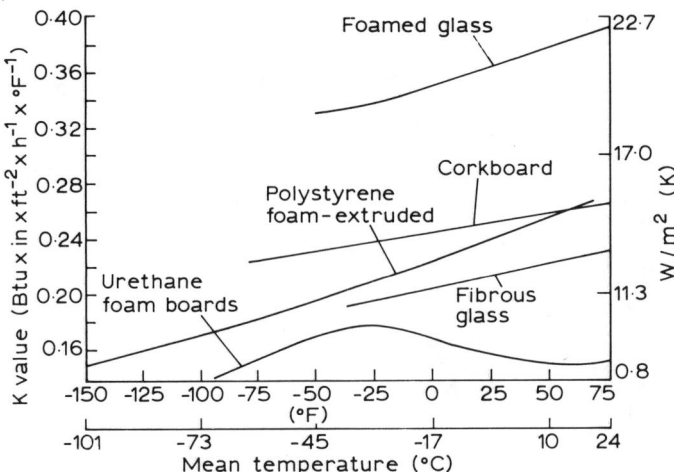

Fig. 4.12. Thermal conductivity (*K* value) variation with temperature [4.19]. (Reprinted with permission from *Plastics in Building*, by I. Skeist, 1966. Copyright Reinhold Publishing Co.)

preferably be of fire retardant grade. It must also be protected from overheating and melting when hot asphalt or coal tar pitch is used to adhere PS foam to the roof deck.

Masonry buildings can readily be insulated by placing foam board between exterior and interior walls or by bonding the foam directly to the wall.

Due to its relatively good thermal resistance, stability to water and water vapour, ease of fabrication and ease of bonding to metals, wood, concrete or plaster faces, PS foams are widely used as a core material for sandwich panels. The most important drawback with respect to this application is that it undergoes heat distortion. In Fig. 4.12 the thermal conductivity of PS and polyurethane (PU) foams is compared with those of other cellular materials.

PS foams are also suitable for applications in other areas [4.20].

Chemical cross-linking through the use of difunctional comonomers is an attractive way for modifying the structure and properties of polystyrene containing blowing agents. Small percentages of tung oil (0·025–0·7%) or divinyl benzene (DVB; 0·025–0·14%) have an important effect upon the foaming behaviour and mechanical properties of the cross-linked polymer [4.7].

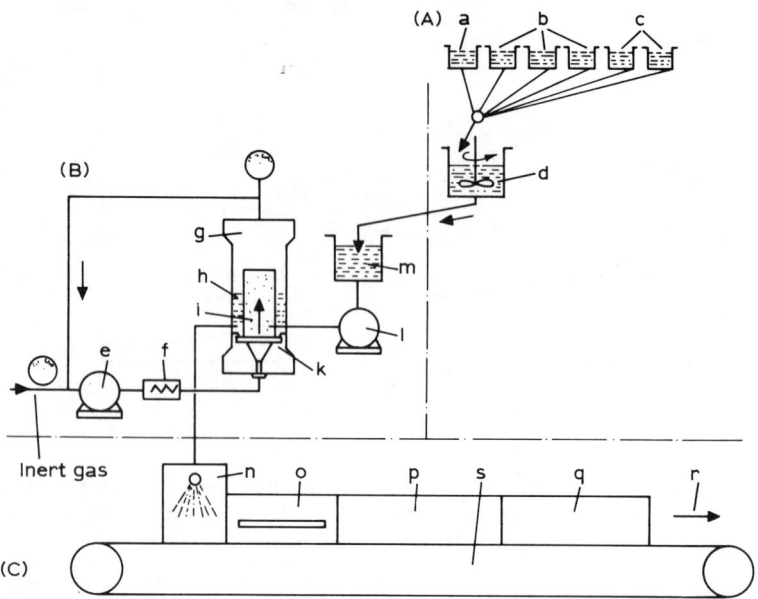

Fig. 4.13. Schematic representation of the Trovipor process. (A) Paste preparation: a, PVC; b, plasticizer; c, additives; d, mixer. (B) Gasification plant: e, gas circulation pump; f, cooler; g, autoclave; h, paste; i, gas stream; k, perforated plate; l, paste feed pump; m, paste feed container. (C) Spray and fusion plant: n, spray tower; o, hf heating; p, after heating; q, cooling tunnel; r, to transverse or longitudinal cutters; s, conveyor belt [4.8]. (Reprinted with permission from *PVC Technology*, 4th edn, by W. V. Titow, 1984. Copyright Elsevier Applied Science Ltd).

Polyvinyl chloride (PVC)

The greatest interest in rigid PVC foam is in applications where low flammability requirements prevail. It can be produced either by a mechanical blowing process or by one of several chemical blowing techniques. It has an almost completely closed cell structure and therefore low water absorption. PVC foams are produced in rigid or flexible forms. The PVC foam manufactured by the German Trovipor process (Fig. 4.13) is flexible, mainly open cell (about 90% of the cells are intercommunicating) and at low to medium density (60–270 kg/m³, 3·75–16·87 lb/ft³). It is normally produced in the form of continuous sheet [4.8].

Flexible PVC foam is commonly made by dispersing PVC in a

plasticizer to form a plastisol, adding AIBN blowing agent, and oven-heating to produce foaming and fusion [4.7, 4.21].

While the rigid PVC foam is used as the cellular layer of some sandwich and multi-layer panels, the flexible PVC foam is used widely as the foam layer in coated-fabric flooring. Its low vapour transmission is an advantage when condensation might be a problem. Because of its rigidity, it is often used in sandwich construction to increase the stiffness of the composite.

Some of the most important properties of PVC rigid foams are:

—high tensile, shear and compressive strength;
—does not crumble under impact or vibration;
—low thermal conductivity;
—low water vapour permeability;
—resistance to termites and bacterial growth;
—good chemical stability [4.14].

Typical mechanical properties for four standard grade PVC foams are given in Table 4.12.

A comparison between PVC and PU foams shows that structures with a high mass concentration in the cell walls are stiffer and

Table 4.12.
Typical mechanical properties of standard grade PVC foams [4.14]

Properties	Standard grades			
Density kg/m^3 (lb/ft^3)	40(2·5)	55(3·43)	75	100(6·25)
Compressive strength MN/m^2 (psi)	0·34(49·3)	0·48(69·6)	0·93(134·9)	1·38(200)
Compressive modulus MN/m^2 (psi)	10·34(1 500)	17·23(2 498)	19·30(2 798)	27·58(4 000)
Shear strength MN/m^2 (psi)	0·34(49·3)	0·48(69·6)	0·93(134·9)	1·14(165·3)
Shear modulus MN/m^2 (psi)	6·89(1 000)	8·27(1 149)	20·68(2 998)	27·58(3 999)
Flexural strength MN/m^2 (psi)	0·38(55·7)	0·55(79·7)	1·03(149)	1·38(200)
Tensile strength MN/m^2 (psi)	0·48(69·6)	0·55(79·8)	1·00(145)	1·72(249)

Reprinted with permission from *Glass Reinforced Plastics in Construction*, by L. Hollaway, 1978. Copyright Surrey University Press, Blackie Publishing Group, Glasgow, UK.

therefore more appropriate for construction utilizing this aspect of material efficiency [4.22].

Polyolefins

Polyolefin foams can be produced with closely controlled density and cell structure. Table 4.13 presents some relationships between the properties of foam based on polyolefins of different structure. The data show that a decrease in the modulus, density or crystallinity of the polymer leads to a corresponding increase in the flexibility of the foam [4.7]. Cross-linking improves foam stability and polymer properties.

The production volume of polyolefin foams is growing rapidly, but not as fast as that of polystyrene or polyurethane foams. This is due to the higher cost of production and some technical difficulties in the production of polypropylene foams [4.23].

Generally the mechanical properties of polyolefins lie between those of a rigid and a flexible foam (Fig. 4.14). polyolefin foams have a very good chemical and abrasion resistance as well as good thermal insulation properties.

A variety of systems can be produced from various types of polyethylenes and cross-linked systems having a very wide range of physical properties, and foams can be tailor-made to a specific application. High modulus products can be made by combining rubber-like modifiers and other types of low modulus additives.

Han *et al.* [4.24] carried out a study with a view to determining the effect of processing variables on the quality of the polyethylene foam produced. Figure 4.15 shows the relationship between the tensile strength and the specific gravity of the LDPE foam. Interpreting this relationship, one can say that the tensile strength increases with specific gravity.

PP has a higher thermostability than PE and can be cross-linked. From a knowledge of the behaviour of the PP at elevated temperature it is anticipated that the PP foam would exhibit desirable behaviour under loads for extended periods of time at elevated temperature [4.7].

ABS

The tensile properties of a general purpose ABS copolymer are reduced as a result of foaming and are dependent on the specific amount of density reduction (Figs. 4.16 and 4.17). Improved tensile,

Table 4.13.
Physical properties of foams based on polyolefins of different structure [4.7]

Property	96D-5MI[a]	0.95D-5MI[b]	-Resin-0.1MI[c]	LDPE[d]	LDPE/PE-PVAc[e]
Gel (%)	52	61	61	74	72
Density, g/cm³ (lb/ft³)	0·13 (8·12)	0·13 (8·12)	0·13 (8·12)	0·135 (8·43)	0·13 (8·12)
Cell size, mil[f]	3–4	2–3	4–6	2–3	3–4
Flexural modulus, psi (MPa)	800 (46·8)	6 500 (44·8)	1 967 (13·5)	830 (5·7)	580 (3·9)
Compressive load, psi (kPa)					
at 5% defl.	108 (745)	101 (696)	26 (179)	24 (165)	18 (124)
at 10% defl.	122 (841)	116 (800)	40 (276)	37 (255)	25 (172)
at 25% defl.	120 (827)	122 (841)	48 (331)	46 (317)	31 (214)
Tensile modulus, psi (MPa)	6 900 (47·5)	5 700 (39·2)	1 941 (13·3)	698 (4·8)	567 (3·9)
Tensile strength, psi (MPa)	250 (1·7)	260 (1·7)	193 (1·3)	195 (1·3)	165 (1·1)
Elongation at failure (%)	178	180	263	162	174

[a] Linear PE homopolymer, 5 melt index, 0·958 g/cm³ (59·4 lb/ft³).

[b] Ethylene/butene copolymer (5–7 C's/1000 C's = branching), 5 melt index, 0·950 g/cm³ (59·4 lb/ft³).

[c] Grafted copolymer of ethylene–maleate ester, 1 melt index, 0·93 g/cm³.

[d] LDPE, 1·9 melt index, 0·918 g/cm³.

[e] Blend of LDPE–Elvax (70/30 PE/VAc).

[f] One mil = 25·4 μm.

Reprinted with permission from *Plastics Foams; the Physics and Chemistry of Product Performance and Process Technology*, Vol. II, by C. J. Benning, 1969. Copyright John Wiley, New York.

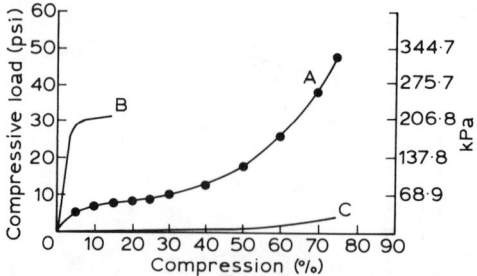

Fig. 4.14. Stress–strain curves for PE foams compared to latex rubber and PS foams [4.23]. A, Polyethylene $2.0\,\text{lb/ft}^3$ ($32 \times 10^{-3}\,\text{g/cm}^3$); B, polystyrene $2.0\,\text{lb/ft}^3$ ($32 \times 10^{-3}\,\text{g/cm}^3$); C, latex rubber foam. (Reprinted with permission from *Mechanical Properties of Polymeric Foams* by E. A. Meinecke & R. C. Clark, 1973. Copyright Technomic Publishing Co.)

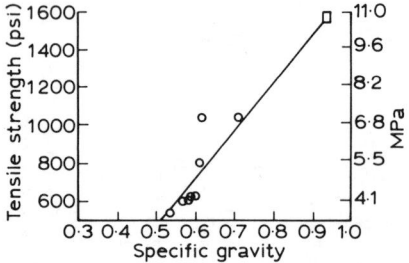

Fig. 4.15. Tensile strength versus specific gravity for low density PE foam without (□) and with (○) $0.5\,\text{wt}\%$ blowing agent (azodicarboamide). (Reprinted with permission from *J. Appl. Polym. Sci.*, **20** (1976) 1593. Copyright John Wiley, New York.)

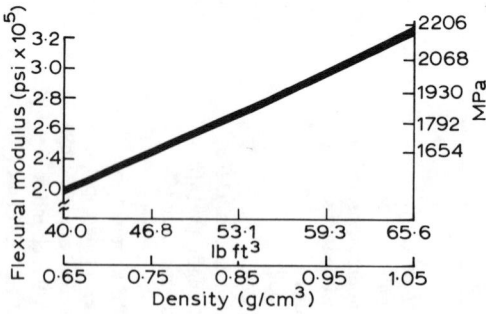

Fig. 4.16. Flexural modulus of ABS versus density reduction [4.25]. (Reprinted with permission from *J. Elast. Plast.*, **11** (1979) 133–9. Copyright Technomic Publishing Co.)

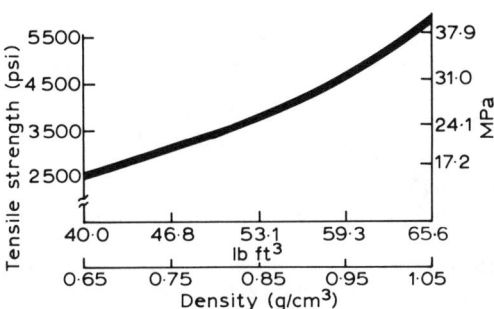

Fig. 4.17. Tensile strength of ABS versus density reduction [4.25]. (Reprinted with permission from *J. Elast. Plast.*, **11** (1979) 133–9. Copyright Technomic Publishing Co.)

flexural and heat distortion properties occur with the addition of glass fibres (Tables 4.14 and 4.15) [4.25].

Cellulose Acetate

Cellulose acetate foams have good mechanical properties, solvent resistance, high thermal stability and wide service temperature range (−57 to +177°C).

Thermosetting Foams

In the production of *thermosetting* foams, foaming takes place at the same time as the polymer is built up by the chemical reaction between

Table 4.14.
Properties of glass reinforced ABS foam [4.25]

	Non-reinforced		10% glass fibre reinforced	
Tensile strength, psi (MPa)	4 100	(28·2)	5 400	(37·2)
Tensile modulus, psi (MPa)	290 000	(2 000)	430 000	(2 964)
Elongation (%)	3·1		1·6	
Flexural strength, psi (MPa)	9 000	(62·0)	9 700	(66·8)
Flexural modulus, psi (MPa)	320 000	(2 206)	440 000	(3 033)
Heat distortion (°F) at 264 psi (1·8 MPa)	177		190	
	10% density reduction			

Reprinted with permission from *J. Elast. Plast.*, **11** (1979) 133. Copyright Technomic Publishing Co.

Table 4.15.
Properties of glass reinforced foam [4.25]

	5% glass reinforced		15% glass fibre reinforced	
Tensile strength, psi (MPa)	5 000	(34·4)	6 600	(45·5)
Tensile modulus, psi (MPa)	340 000	(2 344)	510 000	(3 516)
Elongation (%)	2·9		1·2	
Flexural strength, psi (MPa)	9 200	(63·4)	12 900	(88·9)
Flexural modulus, psi (Mpa)	330 000	(2 275)	470 000	(3 240)
Heat distortion (°F) at 264 psi (1·8 MPa)	187		198	

Reprinted with permission from *J. Elast. Plast.*, **11** (1979) 133. Copyright Technomic Publishing Co.

the starting materials. The most common thermosetting foams are made of:

—polyurethane (PU),
—phenol-formaldehyde,
—epoxy,
—silicone polymers.

Of these four polymers, polyurethane is the most versatile.

Polyurethane (PU)
PU foam is available in flexible and rigid forms, closed and open cell, as a liquid, etc. The characteristics of PU foam depend on:

—the nature of the starting components,
—the type of foaming agent,
—the technology used to produce it.

The most used techniques for producing rigid PU foams include:

—foam-in-place,
—spraying,
—continuous slabbing.

Foaming-in-place is a useful technology convenient especially for filling irregular voids or cavities; it is especially suited to high-rise applications and gives good uniformity in density and foam structure.

The spraying technique permits thin layers of PU foam to be built up on large surface areas and additional layers can be applied almost immediately in consecutive passes.

The slab produced by the slabbing process can be cut after curing, or formed to a specific shape and size.

Chemically, rigid PU foam may be considered the most complex of all polymer foams. The complexity comes from the fact that besides the main initial components, a considerable number of additives are used, such as blowing agents, catalysts, surfactants, etc.

The main advantages of PU over other foams lie in its:

—low thermal conductivity (0·02 W/m °C);
—good thermal resistance (up to 120°C);
—low vapour permeability;
—light weight and strength;
—in-situ foamability.

The behaviour of the foam in fire is not good, although flame retardant grades are available by adding halogenated compounds at the time of preparation.

Owing to a combination of interesting properties PU foam is the most widely used of all thermosetting foams; it is more expensive than other polymer foams. Some properties for two different types of PU foam are given in Table 4.16.

Rigid PU foams are used as thermal insulation over a wide range of temperatures (−88 to +127°C). Applications include use as perimeter insulation, wall and roof insulation, curtain wall panels, low temperature insulation and insulation of industrial pipe and storage tanks. PU foam used in roof deck insulation has an additional advantage over PS foam, in that it can be hot mopped with bitumen without damage to its structure. It is also a very good core material for sandwich panels and some structural components.

The recently introduced poly-isocyanurate foams, closely related to the PU foams, have higher thermal stability and inherently lower flammability characteristics than PU [4.26].

The chemistry of PU foams is relatively complex and involves reactions taking place simultaneously and then gradually terminating.

$$HO-R-OH + {}_nO=C=N-R'-N=C=O \longrightarrow$$

(polyol) (diisocyanate)

$$HO-R-O-\left[-\underset{\underset{O}{\|}}{C}-NH-R'-NH-\underset{\underset{O}{\|}}{C}-O-R-O-\right]_{n-1}-\underset{\underset{O}{\|}}{C}-NH-R'-N=C=O$$

Table 4.16.
Typical mechanical properties of standard and nilflame polyurethane foams
[4.14]

	Standard foam	Nilflame isocyanurate foam
Density, kg/m³ (lb/ft³)	29(1·81)	32(2·0)
Compressive strength in direction of rise, kN/m² (psi)	172(24·9)	172(24·9)
Compressive strength across rise, kN/m² (psi)	124(17·9)	96(13·9)
Tensile strength, kN/m² (psi)	276(40·0)	206(29·8)
Shear strength, kN/m² (psi)	172(24·9)	138(20·0)
Temperature limit (°C)	110	150

Reprinted with permission from *Glass Reinforced Plastics in Construction* by L. Hollaway, 1978. Copyright Surrey University Press, Blackie Publishing Group, Glasgow, UK.

The isocyanate group reacts with water:

$$R—C{=}N{=}O + H_2O \longrightarrow [R—NH—\overset{\overset{\displaystyle O}{\|}}{C}—OH] \longrightarrow R—NH_2 + CO_2$$
$$\text{carbamic acid}$$

This reaction releases CO_2 gas which expands the liquid into a foam. The amine then reacts with additional isocyanate to form a substituted urea.

$$R—NH_2 + R—N{=}C{=}O \longrightarrow R—NH—\underset{\underset{\displaystyle O}{\|}}{C}—NH—R$$

Although water to a greater or lesser degree is used in practically all polyurethane foams, relatively inert low boiling point liquids such as fluorocarbons and chlorocarbons (methylene dichloride) are used as auxiliary agents to expand the foam. The reaction between isocyanate and water is affected by temperature, pressure, mixing efficiency and other variables that are somewhat decreased with the use of an inert gas blowing agent.

Besides the main components, the polyol (polyester or polyether), also involved in producing foam are catalysts, antioxidants, surfactants such as organosilicones, auxiliary blowing agents, fillers, dyes and plasticizers. Although polyesters represented the polyol base of early polyurethane foam production, polyethers, more chemically stable, have taken over practically the entire market in the past few years.

A common formulation for a slab PU foam consists of [4.24]:

Polyol (MW = 3000)	100·0
Silicone	1·0
N-Ethylmorpholine	0·4
Triethylene diamine	0·05
Stannous octoate	0·45
Water	3·30
Toluene diisocyanate (TDI)	42·50
Freon 11	varied
Methylene chloride	varied

The chemistry of rigid foam is essentially the same as that used for making flexible foams, and the equipment is also very similar. It may be processed in slab form, sprayed, and moulded into all types of shapes.

The thermal insulation value of rigid PU foam is excellent, and it is used for making low density insulation panels for buildings. The rigid PU foam can be sandwiched between two or more metal slabs to make structural members.

The future of rigid PU foams looks extremely good because of increasingly higher costs of making wood counterparts. Also, rigid PU foam can be compounded to make it relatively non-flammable; and this factor is becoming increasingly important [4.27].

Rigid foam is generally anisotropic because of the generation of flow lines in the foam during its formulation. In this spread of flow lines it is similar to timber and has an identifiable 'grain' direction. Some physical properties depend on the grain direction, which may vary from place to place in the product. Other properties are independent of grain direction and may be regarded as bulk properties. The most important grain-dependent properties are strength characteristics and thermal conductivity. As for timber, the compression and tensile properties are greatest parallel to the grain direction, and least perpendicular to it.

Rigid polyurethane foam as normally used for insulation has a density of 30–35 kg/m^3. Thus the product is about 97% gas, which is contained in non-interconnecting cells with diameters in the range of

0·2–1·0 mm. The physical properties therefore depend on the interaction and separate contributions of the gaseous and solid phases. Strength characteristics depend mainly on the polymeric phase, thermal conductivity depends on the gas phase and dimensional stability depends on both phases. The gas in the cells is usually Fluorocarbon 11 ($CFCl_3$) with perhaps a little air and CO_2.

Polyurethane and polyisocyanurate rigid foams are usually anisotropic in their strength characteristics. The anisotropy arises from the fact that the major cell diameters align themselves along the direction of flow of the material (grain direction) during the formation of the foam. In sealed cavities (i.e. by panel injection) it is possible to pressure the foam so that the diameters are made nearly equal and the anisotropy is reduced. Thus it is clear that the exact strength characteristics at any given density depend mainly on the method of manufacture and only to a secondary degree on the chemical formulation. However, with any given manufacturing process, consistency of strength characteristics can be maintained without difficulty.

Strength is dependent on density. An empirical relationship has been found: $S = kD^n$, where S is the strength parameter, k is a constant, D is density and n is a power index close to 2.

Dransfield [4.28] has investigated the creep behaviour at 60°C of rigid foams in compression. He showed that under constant load the strain in the material could be represented to a high degree of approximation by an equation of the form:

$$E = E_0 + kt^{1/6}$$

where E is the strain at time t, E_0 is the original strain and k is a constant. A typical curve illustrating his results is given in Fig. 4.18.

Dimensional stability in the case of closed cell foams is dependent on the ability of the foam to resist atmospheric pressure. If the cell pressure is too low, the foam may shrink unless the polymer network can withstand the inward pressure and, conversely, if the cell pressure is above atmospheric, swelling takes place unless the network can withstand these forces also.

When the foam is newly made the cell pressure is below atmospheric, but on exposure to the atmosphere this can increase to a value above atmospheric due to inward diffusion of air. The final value is shown in Fig. 4.19. Changes in the internal cell pressure are due to the fact that the initial gas in the cells ($CFCl_3$) does not diffuse out quickly. The foam, to be stable, must resist the differential pressure. Properly

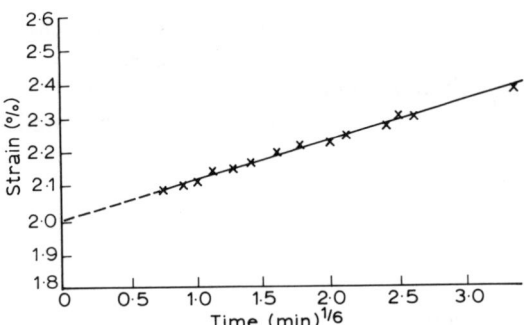

Fig. 4.18. Creep behaviour of rigid polyurethane foam at 60°C and stress 55 kN/m² (7·97 psi) [4.28]. (Reprinted with permission from *Developments in Polyurethane—1,* by J. M. Buist (Ed.), 1978. Copyright Applied Science Publishers.)

manufactured foams will do this, but obviously as density is reduced the strength of the network becomes too low to resist the pressure differentials, and this is the factor which determines the lower density limits. At present this figure can be as low as 27 kg/m³, although normally it is in the range of 32–35 kg/m³.

Changes in dimensions up to perhaps 0·5% may occur as a result of temperature changes or even ageing.

Thermal conductivity (*K* value) is the most important physical characteristic of polyurethane rigid foam [4.29, 4.30]. It has been shown that heat transfer is predominantly dependent on conduction through the gaseous phase, and the exceptionally low figure for rigid

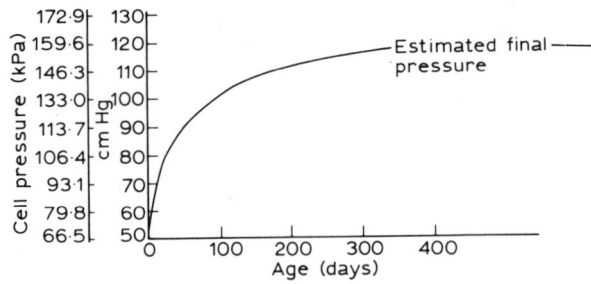

Fig. 4.19. Changes in internal cell pressure at 20°C on ageing [4.28]. (Reprinted with permission from *Developments in Polyurethane—1,* by J. M. Buist (Ed.), 1978. Copyright Applied Science Publishers.)

polyurethane foam is due to the presence of fluorocarbon gas ($CFCl_3$) in the cells. A subsidiary factor is cell size, as transfer of heat by radiation is directly dependent on the cell diameter.

The conduction of heat in a foam may be described by the general equation:

$$K_f = K_g + K_s + K_r + K_c$$

where:

K_f = thermal conductivity

K_g = thermal conductivity of the cell gas

K_s = thermal conductivity of the solid phase

K_r = thermal conductivity associated with radiation across the cells

K_c = thermal conductivity due to convection within the cells

It has been shown that $K_c = 0$ for cell diameters below 10 mm, and K_s is a constant at normal foam density. K_g and K_r are important variables which determine the performance of the foam as an insulator. It has been shown theoretically that the radiation component of conductivity should vary directly with the cell diameter [4.31], so that the larger the diameter the higher the conductivity. Buist *et al.* [4.32] showed in practice that this was indeed the case (Fig. 4.20). This means that, other things being equal, fine celled foams will have a lower conductivity than coarse grained ones.

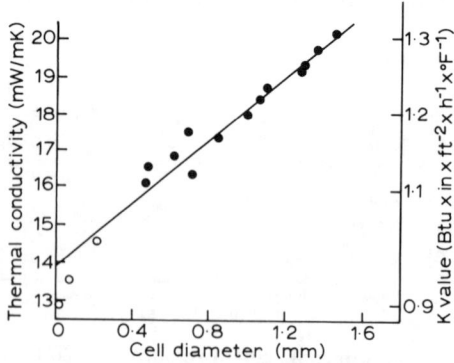

Fig. 4.20. Influence of cell size on K value [4.32]. ●, Values obtained previously; ○, new values for urethane vertical lamination foams.

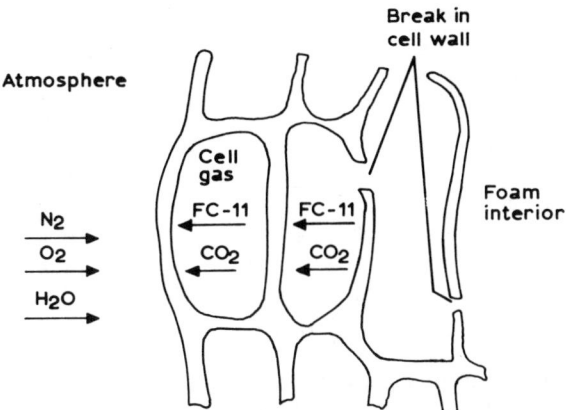

Fig. 4.21. Diffusion of N_2 and O_2 across cell membranes into cells and simultaneous counter diffusion of FC-11 ($CFCl_3$) and CO_2 out of the cells [4.33].

K_g in general does not remain constant during the life of the foam because of diffusion of gases which may take place in and out of the cells through the cell walls. Since most rigid polyurethane foams are blown with $CFCl_3$, loss of this gas to the environment is also of interest. Figure 4.21 shows the diffusion process [4.33].

When the foam is manufactured, the cell gas consists mainly of FC-11 with CO_2 in some cases. The polyurethane cell membranes are permeable to each of the gases present, though to differing extents, as measured by the individual diffusion coefficients. The larger and heavier $CFCl_3$ molecules diffuse much more slowly than the smaller, lighter N_2, O_2, H_2O and CO_2 molecules. The transport of each species occurs by three modes:

—permeation through the cell walls,
—diffusion across the interior of the cells,
—infusion and effusion through breaks and holes in the cell walls.

As time progresses, the concentration of FC-11 and CO_2 in the cell gas diminishes while the O_2 and N_2 concentrations increase. The net effect of the dilution of low thermal conductivity FC-11 with higher thermal conductivity O_2 and N_2 is a gradual increase in the thermal conductivity of the foam. The principal reason for the increase in foam thermal conductivity is the influx of air rather than the efflux of FC-11,

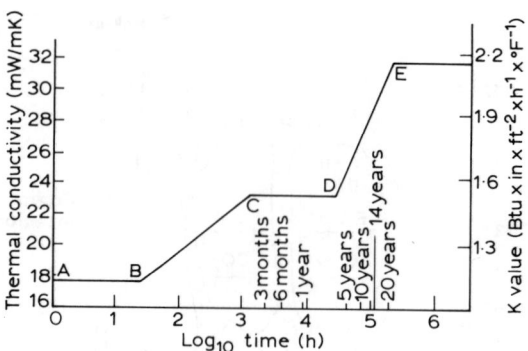

Fig. 4.22. Schematic diagram showing theoretical changes in *K* value with age [4.29]. (Reprinted with permission from *J. Cell. Plast.*, **6**(21) (1970) 66. Copyright Technomic Publishing Co.)

since the change in FC-11 content is small compared to the increase in the concentration of N_2 and O_2 in the cell gas. The above is best illustrated by reference to the schematic diagram used by Ball [4.29] in explaining the effect (Fig. 4.22).

For testing purposes, specimens of standard size ($30 \times 30 \times 4$ cm) are compared, having been aged to equilibrium under standard conditions (23°C, 50% RH). Figure 4.23 is typical of the curve obtained.

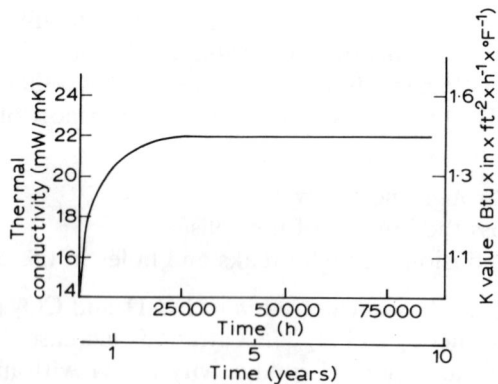

Fig. 4.23. Typical variation of *K* value with age [4.29]. (Reprinted with permission from *J. Cell. Plast.*, **6**(21) (1970) 66. Copyright Technomic Publishing Co.)

Ball *et al.* [4.29] have carried out an extensive analysis of the variation in K value which occurs in practice. Both the difference between different manufacturing techniques and the manufacturing tolerances due to day-to-day variations within each of these techniques were examined.

An important consideration for any product intended for use in the building industry is its ageing and durability characteristics. Rigid polyurethane foam is a relatively new product in this context, but information on its behaviour in service over a period of close to 20 years is now becoming available. In addition, the results of laboratory ageing studies can be used to supplement and confirm what is known from behaviour in the field.

From the earliest days it was known that rigid polyurethane had insufficient resistance to ultra-violet radiation. Thus it has never been used commercially in situations where it was exposed to direct sunlight. Laboratory tests also indicated that it had poor resistance to mineral acids, and moderate resistance to a wide range of organic solvents. The foam will withstand without difficulty both water and a wide range of petroleum products.

Many detailed studies have been carried out to determine the effects of long term laboratory ageing on the K value.

The characterization of the construction materials in terms of their flammability properties is a very complex task because of the fact that these engineering materials can potentially be exposed to a large variety of ignition sources and conditions, and their performance in a fire is influenced by their environment. Small scale laboratory tests have been performed but failed to correlate with actual fire situations [4.34–4.48].

Deleon & Snider [4.49] have used the release rate apparatus for flammability tests. A schematic of the apparatus is shown in Fig. 4.24.

A typical chart for a sample tested in this apparatus is shown in Fig. 4.25. Point C is the maximum rate of heat release (RHR). Slope E is the measure of ease of ignition and Slope B is the measure of flame travel rate. Typically, materials were tested in a flux range of 658–1862 kJ/m² min (58–164 Btu/ft² min). Several foams and building materials were tested in this flux range and the results are summarized in Table 4.17.

Polyisocyanurate foams appeared superior to polyurethane or poly-stryene foams in terms of low RHR and low rate of smoke release, probably due to the higher decomposition temperature of this type of

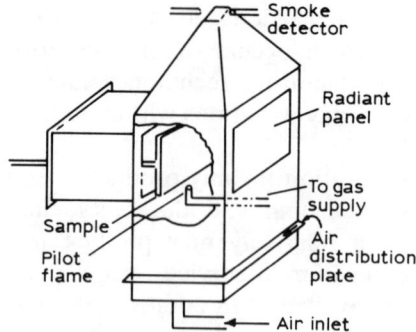

Fig. 4.24. Schematic of combustibility apparatus [4.49]. (Reprinted with permission from *Advances in Urethane Science and Technology*, Vol. 9, by K. C. Frisch & D. Klempner (Eds), 1984. Copyright Technomic Publishing Co.)

foam. Thermogravimetric analysis typical of polyisocyanurate and polyurethane foams used in this study is shown in Fig. 4.26.

Pearce [4.50] obtained improved flame retardation in the case of polystyrene foams, provided thermally stable cross-links could be formed at high temperature between the aromatic rings.

Annamalai & Sibulkin [4.43] conducted flammability experiments on commercial samples of flexible and rigid foams. Samples of prime (pure, without fillers), loaded, bonded and fine retarded PU foams were used. Tests were also done on samples of polyisocyanur-ate, polystyrene, polyethylene, PVC and latex foams. The results for ignition and flame spread times are shown in Table 4.18. As observed,

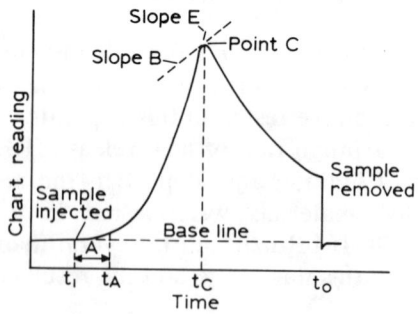

Fig. 4.25. Typical chart [4.49]. (Reprinted with permission from *Advances in Urethane Science and Technology*, Vol. 9, by K. C. Frisch & D. Klempner (Eds), 1984. Copyright Technomic Publishing Co.)

Table 4.17.
Rate of heat release test results [4.49]

| | Unfaced | | | | | | Extruded polystyrene foam | | | Red oak | | | Exterior plywood | | |
	Urethane foam			Isocyanurate foam											
Thickness, in (mm)	1·0 (25·4)	1·0 (25·4)	1·0 (25·4)	1·0 (25·4)	1·0 (25·4)	1·0 (25·4)	1·0 (25·4)	1·0 (25·4)	1·0 (25·4)	0·78 (19·8)	0·78 (19·8)	0·78 (19·8)	0·78 (19·8)	0·78 (19·8)	0·78 (19·8)
Heat flux, Btu/ft²min (W/m²)	58 (30·4)	121 (63·5)	164 (86·1)	58 (30·4)	121 (63·5)	164 (86·1)	58 (30·4)	121 (63·5)	164 (86·1)	58 (30·4)	121 (63·5)	164 (86·1)	58 (30·4)	121 (63·5)	164 (86·1)
Max. heat release (HR), Btu/ft²min (W/m²)	230 (120·7)	650 (341)	680 (357)	<20 (<10·5)	85 (44·6)	110 (57·7)	0	1 040 (546)	1 180 (619)	335 (175)	450 (736)	450 (736)	430 (225)	625 (328)	700 (367)
Cumulative HR, Btu/ft² — 3 min	145	870	850	<25	100	80	0	1 300	1 200	520	950	1 200	690	1 090	1 190
5 min	160	>1 000	>1 000	<40	190	150	0	1 650	1 500	625	1 700	2 130	880	1 510	1 700
10 min	—	—	—	—	—	—	0	1 700	1 500	790	3 700	4 450	1 100	2 820	3 500
Max. smoke release rate (particles/ft²min)	3000	4800	6500	30	2300	2700	0	12 000	15 000	<20	<30	130	75	80	180

These results indicate the need to evaluate each material at more than one flux. For example, polystyrene foam at a heat flux of 58 Btu/ft²min, had zero heat and smoke release because it melts away from the heat flux source. However, at higher heat fluxes this product has very high heat and smoke release.

Reprinted with permission from *Advances in Urethane Science and Technology*, *Vol. 9*, by K. C. Frisch & D. Klempner (Eds), 1984. Copyright Technomic Publishing Co.

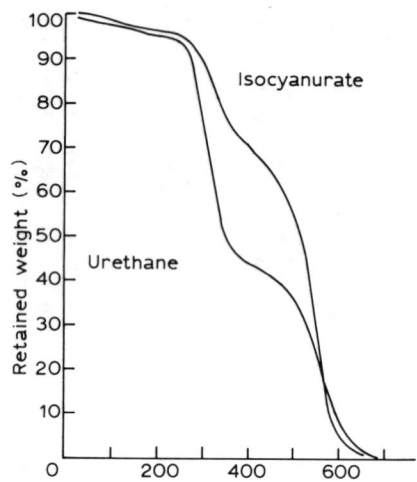

Fig. 4.26. Thermogravimetric analysis [4.49]. (Reprinted with permission from *Advances in Urethane Science and Technology,* Vol. 9, by K. C. Frisch & D. Klempner (Eds), 1984. Copyright Technomic Publishing Co.)

there are generally three classes of foam response based upon ignition and flame spread behaviour:

(1) igniting foams, subdivided into:
    (a) upward flame propagation, and
    (b) downward flame propagation;
(2) non-igniting foams; and
(3) melting foams.

The authors of the study emphasize that these classes are not mutually exclusive since some foams which can melt can also ignite. Except for those foams which melt, all other foams were flammable.

Hurd [4.51] has considered that it is probably unrealistic to expect that massive improvements in the fire resistance of PU or polyiso-cyanurate (PIR) foams are still possible by chemical modification of the polymer. The problem is not so much a question of ignitibility but more one of low density and high insulating power. Indeed, as the following figures show, ease of ignition as expressed by self-ignition temperature (ASTM D1929) is quite favourable compared with other

Table 4.18.
Summary of results of piloted ignition and flame spread tests [4.43]

No.	Material	Density kg/m^3 (lb/ft^3)	Ignition time (s)	Flame spread time (s)	Transient flaming
1.	Polyurethane—flexible	18·6 (1·2)	2·6	6·3	
2.	Polyurethane—flexible	19·2 (1·2)	1·8	5·1	
3.	Polyurethane—flexible	24·6 (1·5)	4·4	7·1	
4.	Polyurethane—flexible	43·5 (2·7)	9·9	31·0	
5.	Polyurethane—flexible	58·2 (3·6)	11	33·5	
6.	Polyurethane—bonded	77·2 (4·8)	39	101	
7.	Polyurethane—bonded	153·2 (9·6)	50	a	
8.	Polyurethane—flexible, fire retarded	16·4 (1·0)	NI		yes
9.	Polyurethane—flexible, fire retarded	41·4 (2·6)	NI		yes
10.	Polyurethane—rigid	39·0 (2·0)	0·3	6·7	
11.	Polyurethane—rigid	68·19 (4·0)	2·0	14·3	
12.	Polyurethane—rigid, fire retarded	38·6 (2·4)	NI		yes
13.	Isocyanurate—rigid, fire retarded	34·8 (2·2)	NI		yes
14.	Polystyrene—rigid	20·1 (1·3)	NI		no[c]
15.	Polystyrene—rigid	35·5 (2·2)	NI		no[c]
16.	Polyethylene—rigid	36·7 (2·3)	NI		no[c]
17.	Polyvinyl chloride—flexible	117·0 (7·3)	38	157[b]	
18.	Latex—flexible	100·0 (6·2)	0·5	a	

Pilot flame gas: propane; flow: 0·27 ml/s; external radiation: none; orientation: pilot flame tube normal to the sample; distance between sample and pilot flame tube exit: 0·64 cm; distance between ionization probe and pilot flame tube: 35·6 cm.
NI: No ignition.
[a] Only downward burning.
[b] Tentative value.
[c] Only melting.
Reprinted with permission from *Flammability of Cellular Plastics, Part 2*, by C. J. Hilado (Ed.), 1981. Copyright Technomic Publishing Co.

materials commonly used in construction:

	Self ignition temperature	(°C)
PU	approximately	500
Red deal		375
Asphalt		275

However, it is a fact that the burning rate of a low density material such as PU foam is more sensitive to the rate of temperature rise at its surface than is the corresponding factor for a material such as wood.

Cornish *et al.* [4.52] compared the lethal concentration values (LC_{50}) of the released gas during combustion for a variety of polymers under static chamber (rapid combustion) and dynamic chamber (slow pyrolysis) conditions. Table 4.19 shows one set of such comparisons. It is of interest not only in showing the effect of combustion conditions but also in indicating that on this evidence polyurethanes are no more toxic than many other materials.

Table 4.19.

Comparative mortality data of combustion products of polymers [4.51]

$LC_{50}(g)$	Static chamber sample		In order of decreasing toxicity	Dynamic chamber sample	$LC_{50}(g)$
9	Red oak	(most toxic)	1	Wool	0·4
10	Cotton		2	Polypropylene	0·9
21	ABS(FR)		3	Polypropylene (FR)	1·2
23	SAN		4	Urethane foam (FR)	1·3
25	Polypropylene (FR)		5	Polyvinyl chloride	1·4
28	Polypropylene		6	Urethane foam	1·7
31	Polystyrene		7	SAN	2·0
33	ABS		8	ABS	2·2
37	Nylon 66		9	ABS (FR)	2·3
37	Nylon 66 (FR)		10	Nylon 66	2·7
47	Urethane foam (FR)		11	Cotton	2·7
50	Urethane foam		12	Nylon 66 (FR)	3·2
50	Polyvinyl chloride		13	Red oak	3·6
60	Wool	(least toxic)	14	Polystyrene	6·0

Table 4.20.
Comparisons of the relative toxicities of the pyrolysis products of synthetic
and natural building materials in tests with equal volume [4.51]

Materials	Lowest temperature (°C) indicating mortality
Rigid urethane foam	600 and > 600
Rigid isocyanurate foams	500, 600 and > 600
Semi-rigid polyurethane foams	500, 600 and > 600
Foams based on UP resins combined with expanded glass beads	> 600
ABS	400
Polycarbonate	600
Spruce-wood	350
Cork	300

Reprinted with permission from *Developments in Polyurethane—1*, by J.
M. Buist (Ed.), 1976. Copyright Applied Science Publishers Ltd.

Hurd [4.51] shows that many other scientists have obtained comparative critical temperatures for a number of materials used in
building construction (Table 4.20). As may be seen, in terms of
toxicity of pyrolysis products, PU foams compared favourably with
many conventional materials they replace.

Polyisocyanurate (PIR) foams are manufactured by similar processes to PU foams. Their physical properties and the bonds to facings
are often marginally inferior to PU boards, but PIR foams exhibit
superior fire performance [4.39].

Loikkanen [4.53] shows that some data (Table 4.21) can be utilized
when choosing the allowable temperature rise for PU insulation, which
in turn determines the fire resistance time of the protective lining.

Phenol-formaldehyde (PF)
Phenol-formaldehyde (PF) foam has:

—good chemical and thermal resistance;
—high resistance to water transmission and water uptake;
—good dimensional stability;
—high strength to weight ratio;
—less flammability than most polymer foams.

But because of its high open-cell content it has relatively low

Table 4.21.
Results of tests according to method ASTM D 3014 for
different qualities of polyurethane [4.53]

Quality No.	Residual mean (%)	Mean deviation (%)	Burning time (s)	Flame height (mm)
1	54·2	3·4	22·3	250 +
2	69·2	1·7	18·3	250 +
3	83·2	1·7	12·5	250 +
4	85·0	1·1	11·0	250 +
5	67·4	5·9	12·3	250 +
6	73·4	1·2	11·0	250 +
7	62·4	1·9	15·8	250 +
8	39·9	3·9	18·5	250 +
9	60·5	2·5	15·5	250 +
10	17·8	0·4	35·8	250 +
11	18·2	1·4	39·0	250 +
12	18·1	0·2	36·8	250 +

Reprinted with permission from *VTT Symposium 57,
Structural Fire Safety Regulations in Finland and in the
Soviet Union,* 1985. Copyright Technical Research Centre
of Finland.

thermal resistance. Thermal insulation efficiency can be improved by
the application of a skin of hot bitumen or other suitable material.

Phenol-formaldehyde foam achieves the highest classification as the
result of fire tests and produces an optical smoke obscuration of less
than 5%, compared with 50–90% for most commercial grades of PS or
PU foam.

PF foam technology has progressed in recent years so that slabstock
and laminated boards can now be produced using continuous processes. PF composites cannot match the versatility of foam/facing
combinations developed by PUR and PIR manufacturers, but they
offer the construction industry a useful combination of physical and
fire properties [4.54].

PF foams have low thermal conductivity and generate low levels of
smoke in most fire tests. While these characteristics are attractive to
users, manufacturers of PF boards have to overcome two serious
problems in the fire behaviour of some PF formulations. Early PF
foams were very brittle and easily damaged during installation; PF

foams are still friable under ambient conditions but some PF foams exhibit severe spalling (i.e. explosive splintering) during fire tests [4.39]. Other PF foams with a low degree of friability have been reported elsewhere [4.55].

A second fire phenomenon associated with some PF foams is punking, i.e. a slow combustion or smouldering initiated by the localized application of heat. Propagation of the combustion front continues virtually undetectable without further external application of heat until the foam is charred throughout or splits open due to the high exotherm [4.39].

Urea-formaldehyde (UF)

The preparation of urea-formaldehyde resins consists in the following:

an aqueous solution of formaldehyde, known as formalin, is brought to pH 8 by the addition of sodium hydroxide and then urea is added to give a urea : formaldehyde ratio of about 1 : 2 molar. The resulting solution is boiled for about 15 min, acidified (to pH 4) with formic acid and then boiled for a further 5–20 min until the required degree of reaction is attained. The product is neutralized and then evaporated until the required solids content is achieved.

The reactions which occur in the process are:
(a) *Formation of methylol ureas.*

$$
\underset{\text{urea}}{\underset{\overset{|}{\text{NH}_2}}{\overset{\text{NH}_2}{\overset{|}{\text{C}{=}\text{O}}}}} \overset{\text{CH}_2\text{O}}{\rightleftharpoons} \underset{\text{monomethylol urea}}{\underset{\overset{|}{\text{NH}_2}}{\overset{\text{NH}-\text{CH}_2\text{OH}}{\overset{|}{\text{C}{=}\text{O}}}}} \rightleftharpoons \underset{\text{dimethylol urea}}{\underset{\overset{|}{\text{NH}-\text{CH}_2\text{OH}}}{\overset{\text{NH}-\text{CH}_2\text{OH}}{\overset{|}{\text{C}{=}\text{O}}}}} \overset{\text{CH}_2\text{O}}{\rightleftharpoons} \underset{\text{trimethylol urea}}{\underset{\overset{|}{\text{NH}-\text{CH}_2\text{OH}}}{\overset{\text{N(CH}_2\text{OH)}_2}{\overset{|}{\text{C}{=}\text{O}}}}}
$$

(b) *Condensation of methylol ureas.* In the second stage of the polymer preparation outlined above, the reaction is continued under acidic conditions with the formation of the following linear product:

$$\text{HOCH}_2-[-\text{NH}-\text{CO}-\text{NH}-\text{CH}_2-]_n-\text{NH}-\text{CO}-\text{NH}-\text{CH}_2\text{OH}$$

After cross-linking with an excess of formaldehyde or with a source of this aldehyde such as hexamethylenetetramine the following network is obtained:

$$
\begin{array}{c}
\vdots \qquad\qquad\qquad \vdots \\
| \qquad\qquad\qquad | \\
CH_2 \qquad\qquad\quad CH_2 \\
| \qquad\qquad\qquad | \\
\cdots-N-CO-N-CH_2-N-CO-N-CH_2-\cdots \\
| \qquad\qquad\qquad | \\
CH_2 \qquad\qquad\quad CH_2 \\
| \qquad\qquad\qquad | \\
\cdots-N-CO-N-CH_2-N-CO-N-CH_2-\cdots \\
| \qquad\qquad\qquad | \\
CH_2 \qquad\qquad\quad CH_2 \\
| \qquad\qquad\qquad | \\
\cdots-N-CO-N-CH_2-N-CO-N-CH_2-\cdots \\
| \qquad\qquad\quad | \qquad\quad | \\
\vdots \qquad\qquad \vdots \qquad\quad \vdots
\end{array}
$$

The cross-linked polymer is infusible and insoluble. The basic chemistry of the UF polymer is the same as that for other applications, except that foams must cure at room temperature within the first few minutes, as compared to adhesives or mouldings which cure at high temperatures (150°C +).

The preparation of the UF foam implies, in addition to the principal polymer, the use of additives such as: surfactants, catalysts, water, foaming agents, etc.

The surfactant plays an important role in increasing the hydrophobicity of the foam with a view to resist better the hydrolysis processes. Generally these additives are aromatic sodium salts of sulphonic acids.

In cold settings the pH must be lower than in hot setting adhesives. It is usually desirable to have a pH of about 1–2 which may be adjusted with an acidic catalyst such as phosphoric acid. Great care must also be taken to keep the use of acid to a minimum because any excess will hydrolyse the finished foam. The hydrolysis is the most important cause of formaldehyde odour, from older, fully cured foam. Excess acid can cause an initially normal looking foam to, thereafter, crumble into powder.

The water hardness influences foam shrinkage and also the effectiveness of the surfactant. Before preparing the final ready-to-foam and catalyst mixtures, the water hardness must be adjusted with a suitable softener.

In bulk insulating foams, air is used as the foaming gas, yielding 99% open cell structures. If Freon or other foaming agents are used, the cell structure can change, yielding predominantly closed cells.

In the application of UF foam, its compatibility with adjoining materials must be considered. UF foam can be easily peeled from PVC, PE, PS, polyesters and polyisobutylene and behaves as an inert material from the chemical point of view. It sticks to butyl rubber foils, but peels off asphalt papers, used in roofing or foundation water-proofing. Aluminium is discoloured but is not corroded by UF foam, however copper noticeably tarnishes, zinc-plated steel corrodes slightly, and steel is considerably rusted when brought into prolonged contact with UF foam [4.56].

Other additives have been included to modify and tailor-make urea-formaldehyde foam structures to yield specific characteristics, such as: very low density ($11 \cdot 2 \, \text{g/dm}^3$, $0 \cdot 7 \, \text{pcf}$), flexibility, etc. More flexible structures have been made with urea-formaldehyde condensates with resorcinol. Urea-formaldehyde foams modified with alpha, omega-bisepoxies are reported to be less friable and more impact resistant at low densities. Very fine textured low density foams were produced by forming co-condensates with hexamine and melamine.

Urea-phenol condensate foams are excellent binders for concrete and wood [4.7].

Because factory made foam slabs are easily damaged during transportation and occupy excessive and uneconomical shipping volume, UF foams are produced on site from portable foaming equipment. Two separate systems are used for the work:

(a) a resin (prepolymer) stock solution, and
(b) a hardener-surfactant.

Both solutions are usually formulated so that they can be mixed in a $1:1$ ratio. The standard procedure consists in pumping the solutions in accurately measured proportions from open storage drums, or expelling them from pressure vessels (Fig. 4.27).

The key tool for producing a good foam is the gun, an example of which is shown in Fig. 4.28. Hardener and prepolymer are mixed in the barrel packed with glass beads where the components are mixed in turbulent flow. The fully expanded foam emerges from the gun with a consistency of shaving cream.

A good foam sets within the first minute, when it can then be sliced if need be. It hardens fully within 1 day. The residual water vaporizes from the cured foam. If the foam dries too quickly, it tends to warp, shrink or crack. Home insulating foam in closed spaces, on occasion, retains moisture for several years, especially in cold climates [4.56].

Urea-formaldehyde is the lightest of the insulating foams [4.57]; its density is $10-14 \, \text{kg/m}^3$ ($0 \cdot 6 - 0 \cdot 9 \, \text{lb/ft}^3$) [4.56]. It has virtually no

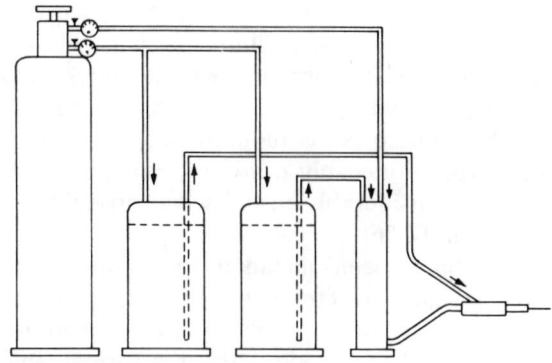

Fig. 4.27. Resin and foam mixing set-up for in-field foam manufacture [4.56]
(USP 2,860,856, Nov. 18, 1958, W. Bauer, Frankenthal, GFR).

mechanical strength even in compression, is white in colour and does
not support combustion; it is very friable. Typical products have a
closed cell content of about 80%. UF foams have a very low thermal
conductivity with a K value of 0·022–0·029 W/m K (0·15–0·20 Btu in
$ft^{-2} h^{-1} °F^{-1}$), thus comparing very favourably with other insulating
materials. Foams with a density of about 12 kg/m^3 (0·75 lb/ft^3) have
the lowest conductivity [4.58]. Astonishingly, the thermal conductivity
changes less than 5% over a density range of 10–30 kg/m^3. If Freon is
used as the blowing agent, the conductivity can be further reduced.
Shrinkage is in the order of 2·98–4%, depending on the formulation
[4.56].

UF foam deteriorates continuously at a moderate to rapid rate so it
has a short life compared with other building materials. The rate of
deterioration depends on the conditions to which it is exposed. In

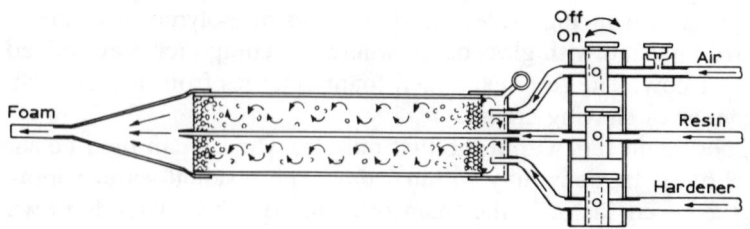

Fig. 4.28. UF foam gun. Aerolite Model [4.56]. (Reprinted with permission
from *Chem. Engng News* **60**(13) (March 29, 1982). Copyright American
Chemical Society.)

residential applications, the factors most likely to accelerate degradation are high temperatures and high humidities. Temperatures such as those encountered in a roof space on a summer day cause rapid deterioration. High humidity resulting from the flow of warm humid air from the living spaces into insulated cavities also causes accelerated degradation. UF foams have a high water absorption and a high water vapour permeability.

Deterioration of the foam leads to breakage of the cell walls and shrinkage of the insulation, decreasing the foam's ability to resist heat and air flow. It has a tendency to undergo considerable shrinkage upon cure and on drying. Formaldehyde gas is released and may be carried into the living space by air infiltration and, at a slower rate, can diffuse through the wall materials. When subjected to elevated levels of formaldehyde for extended periods, occupants can react to the gas and may develop health problems.

Where the cavity containing the foam has not dried properly after installation, fungus may grow on the foam or on other materials in the cavity. Fungus spores can be carried into the living space by air infiltration. Some types of spores are known to cause allergic reactions and sensitization in humans. Fungal growth may also cause deterioration of organic materials in the wall [4.59].

Measurements taken in homes cited in complaints to the US agency CPSC (Consumer Product Safety Commission) found formaldehyde levels averaging 0·12 ppm, whereas measurements from homes without the foam averaged 0·03 ppm formaldehyde. The fact that it takes only a little formaldehyde to cause health problems is a basic assumption of the CPSC.

There is one undesirable consequence of the CPSC's action— economic hardships. The effects of the CPSC investigation have been ruinous to the UF foam industry (Fig. 4.29). From a peak level in 1977 of more than 30 UF foam manufacturers, and 2000 installers, the industry is down to three or four manufacturers and fewer than 200 installers in the USA [4.60].

Exposure to formaldehyde can cause eye, nose and throat irritation, coughing and asthma-like symptoms, headaches, dizziness, nausea, vomiting and nose bleeds. The severity of the reaction depends on the formaldehyde concentration, the duration of exposure and the sensitivity of the individual. Although there may be no reaction to a single exposure, repeated or long-term exposure may increase an individual's sensitivity to the gas. When sensitized, a person may suffer serious

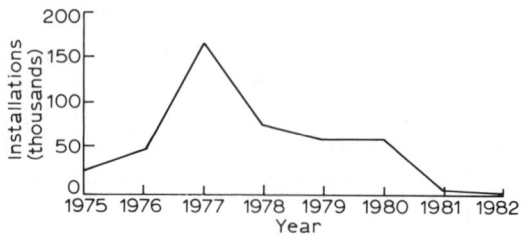

Fig. 4.29. UF foam insulation use has fallen almost as fast as it rose [4.60]. (Reprinted with permission from *Chem. Engng News* (Jan. 3, 1983) 9–16. Copyright American Chemical Society.)

reactions at very low concentrations. Asthmatics may suffer more severe attacks as a result of continued or repeated exposure. High levels of formaldehyde have been found to cause nasal cancer in laboratory animals, indicating that humans should minimize their exposure to the gas. Skin contact with UF foam may cause irritation and sensitization to formaldehyde and other substances.

The symptoms caused by repeated exposure to certain types of fungus spores can be similar to those caused by exposure to formaldehyde; the most common are asthmatic in nature: intermittent breathing difficulties, wheezing, coughing, and a sense of constriction.

Persons suffering from these symptoms should consult a medical doctor to verify the cause [4.59].

Most people are unaffected at concentrations below 0·25 ppm, but maybe 20% of the population will be. There is also a lot of evidence that even at 0·1 ppm a significant number of persons might react adversely to formaldehyde [4.60].

Formaldehyde has a pungent odour at high concentrations but a less definite odour at low concentrations. Most people can detect the presence of the gas at concentrations of 0·1 ppm; few can detect it at levels below 0·05 ppm.

Some foams contain agents which mask the formaldehyde odour but do not remove the gas from the air. Others contain chemical scavengers which react with the formaldehyde and reduce the emission of the gas. As the insulation deteriorates, the effectiveness of the scavengers can decrease. The emission of formaldehyde will increase and the odour may become detectable.

Fungus growth on the foam or on surfaces of the wall cavities may emit a musty odour which is not easily identifiable among other household odours. The ability to detect odour from fungal spores varies from person to person [4.59].

The following factors are considered responsible for formaldehyde high odour:

—incorrect ratio of components (excess formaldehyde);
—excess catalyst;
—excess foaming agent;
—high humidity;
—too high density;
—applications against recommended practice [4.56].

The humidity and temperature also have similar effects (formaldehyde emission) as in the case of UF-bonded boards [4.61].

The OSHA (Occupational Safety & Health Administration) has two options for admissible amounts of formaldehyde:

(a) Exposure limits:
   —current   3·0 ppm
   —proposed
     • carcinogen   1·0 ppm
     • irritant   1·5 ppm
(b) Action level [4.62]:*
     • carcinogen   0·50 ppm
     • irritant   0·75 ppm

Tests conducted on human volunteers have established a low-response curve for the odour threshold (Figs. 4.30, 4.31). Of 64 subjects tested, not one noticed less than 11 ppb, however each one did notice 3 ppm of formaldehyde gas. It is generally believed that most people become desensitized by high formaldehyde levels.

Tests done on 16 healthy volunteers over a period of 4 h have established that the time response curve is dependent on the gas concentration (Fig. 4.32). Formaldehyde is almost always in equilibrium with several of its derivatives and precursors. UF resins contain two formaldehyde sources:

(a) excess formaldehyde dissolved, trapped or absorbed in UF-resins or UF-bonded products; and

* Reprinted with permission from *Chem. Engng News,* Dec. 9, 1985, **63**(49), 4. Copyright 1985, American Chemical Society.

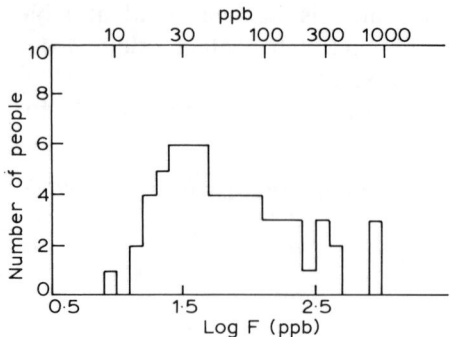

Fig. 4.30. Recognition of formaldehyde by human subjects [4.56]. (Rehn, 1977).

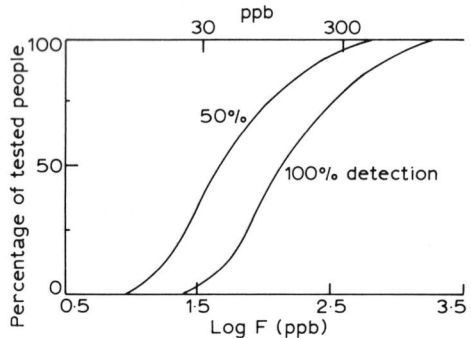

Fig. 4.31. Dose–response curve for formaldehyde odour [4.56] (Rehn, 1977).

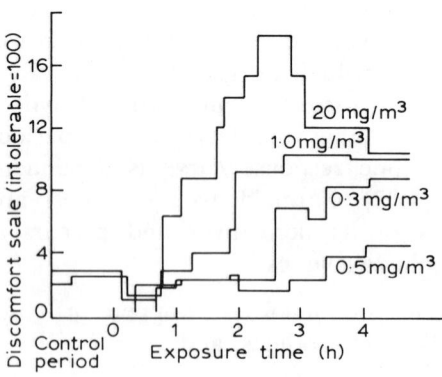

Fig. 4.32. Discomfort response as a function of exposure time for four levels of formaldehyde [4.56] (Andersen, 1979).

(b) hydrolysis products, formed by ageing of UF, especially un-evenly cured resin, such as uneven thin films [4.56].

Although the cancer risk of formaldehyde remains controversial [4.63], a ban on the use of UF foam insulation was announced by National Health and Welfare Canada and the Department of Consumer and Corporate Affairs in December of 1980 [4.59]. Official notice of the ban appeared in the US Register in 1982 [4.60].

The fire behaviour of all cellular polymers, which will be discussed later on, is very much dependent on their degree of exposure to air; while vitiation of the combustion air will affect the type of decomposition reaction, it is the rate of burning, significantly dependent on the degree of oxygen depletion, that will primarily determine the extent of risk to life that they introduce. A balance must then be achieved between the economic design of a protective facing for a given cellular polymer and the time for which that protection must be maintained to ensure safety of life by preventing the emission of heat, smoke and combustion products [4.64].

Epoxy

The technology of epoxy-foam application is similar to that of PU foams. The most important difference consists in the chemistry of the curing phase. Because epoxies react with the curing agent without the evolution of volatiles, they need the addition of a foaming agent. A typical formulation might consist of:

—epoxy prepolymer,
—curing agent (hardener),
—blowing agent (Freon),
—wetting agent (surface-active agent).

The curing agents are highly reactive primary and secondary alphatic polyamines, such as diethylenetriamine or its higher homologues. The high reactivity is necessary for an exotherm to develop and decompose the blowing agent into its gaseous components. The wetting agent is used to give a fine, uniform dispersion of the gas bubbles. An aromatic hydrocarbon (toluene) is beneficial in controlling the heat of reaction and the viscosity of the polymeric system [4.65].

Suitable surface active agents for regulating the pore structure are, in particular, non-ionic products such as lauric acid, palmitic acid, steric acid or oleic acid with polyhydric alcohols, such as sorbitol, and their addition products formed by reaction of ethylene oxide at the free hydroxyl groups [4.55].

To obtain a rigid epoxy foam [4.66] Corbett used a siloxane copolymer as a blowing agent.

Epoxy foams are dispersed either with conventional foam-in-place equipment or with portable spray equipment. The mechanical properties of the poured-in-place and spray-applied low density epoxy foam are given in Table 4.22.

Syntactic epoxy foams, made with different types of microspheres such as poly(vinylidene chloride)-based, phenolic or glass, are well known. The volume of gas per unit weight of spheres determines their bulk efficiency. Thus the less dense the spheres, the lower the weight required to achieve a specific composite density. The poly(vinylidene chloride) microspheres are the most effective on a density basis. Epoxy foams are used primarily for insulation purposes where chemical resistance and adhesion are of prime importance. Syntactic epoxy foams can also be used for this purpose and as a core for laminated sandwich panels. It has been reported that by this technique it is possible to construct an aluminium skin panel that has the same modulus as a conventionally designed structure but is lighter by as much as 50% [4.65].

Epoxy foams have a very good chemical stability, moisture resistance and thermal insulating properties, but because of the high cost their use in building construction is still limited. Special grades of epoxy foams are available that extend their service temperature range to 120°C.

The thermal properties of interest, in addition to thermal expansion and conductivity, are short and long term heat stability and fire resistance. Normally all cast epoxide systems will burn, but can be made self-extinguishing by incorporating halogenated polymers or curing agents [4.68].

Silicone

Silicone foams differ radically from those previously discussed and from other plastic foams based on organic polymers. Generally they are made of poly(organosiloxanes). The main Si—O link in the backbone of these polymers has a number of interesting characteristics which are relevant to the properties of the poly(organosiloxanes).

These polymers are produced by polycondensation of chlorosilanes

Table 4.22.
Typical properties of low density epoxy foams [4.67]

Property	Low formula A	Pour formula B	Spray formula
Density, pcf	2.5	1.55	2.1
Thermal conductivity, Btu/(h)(ft²)(°F/in) (Wm/m²°C)			
Uncut block:			
initial	0.113(16.29)	0.110(15.8)	0.112(16.15)
aged	0.113(16.29)	0.110(15.8)	0.112(16.15)
Cut block:			
initial	0.113(16.29)	0.110(15.8)	0.112(16.15)
aged	0.12–0.16(17.3–23)	0.13–0.16(18.7–23)	0.14–0.16(20–23)
Thermal stability for 48 h (°F)	150	150	200
Volume change (% by vol.)	3	3	+1
Weight change (% by wt.)	−0.3	−0.2	−0.3 to −0.5
Initial compressive strength, psi (kPa)	25(172)	15(103)	15(103)
Aged compressive strength, psi (kPa)	45(310)	—	45(310)
Cell properties:			
closed cells (%)	90	92	90
cell walls (% by vol.)	2	2	3
open cells (% by vol.)	8	6	7
Water absorption (% wt. gain)	0.1	—	0.3
Foam time (min)	3	3	2
Maximum rise completed (min)	3.5	3	2.5
Dry time (min)	3.5	3	2.5
Cure time (min)	20	20	30
Service temp. limit (°F)	175	175	210
Heat distortion	Nil	Nil	Nil
Burning rating	Self-extinguishing	Non-burning	Special

Reprinted with permission from *Plastics Foams*, Vol. 1, by C. J. Benning, 1969. Copyright John Wiley.

with water, which then condense to give the following microstructure:

$$
\underset{\underset{R_1}{|}}{\overset{\overset{R}{|}}{Cl-Si-Cl}} + H_2O \longrightarrow \underset{\underset{R_1}{|}}{\overset{\overset{R}{|}}{HO-Si-OH}} \longrightarrow \cdots-(\underset{\underset{R_1}{|}}{\overset{\overset{R}{|}}{-Si-O-}})-\cdots
$$

Silicone polymer may be prepared by many routes of which four appear to be of commercial value. They are:

—Grignard process,
—direct process,
—olefin addition method,
—sodium condensation method [4.58].

The silicone resins useful in cellular composites are polyorganic siloxanes having the general structure:

$$
\cdots\left[\underset{\underset{R_1}{|}}{\overset{\overset{R}{|}}{-Si}}-O-\underset{\underset{O-H}{|}}{\overset{\overset{R}{|}}{Si}}-O-\underset{\underset{R}{|}}{\overset{\overset{R}{|}}{Si}}-O-\underset{\underset{O-H}{|}}{\overset{\overset{R}{|}}{Si}}-O\right]-\cdots
$$

$$
\cdots\left[\underset{\underset{OH}{|}}{\overset{\overset{R}{|}}{-Si}}-O-\underset{\underset{R}{|}}{\overset{\overset{OH}{|}}{Si}}-O-\underset{\underset{R}{|}}{\overset{\overset{R}{|}}{Si}}-O-\underset{\underset{OH}{|}}{\overset{\overset{OH}{|}}{Si}}-O\right]-\cdots
$$

wherein R or R_1 is an organic group, such as methyl, phenyl, trifluoromethyl, or higher alkyl radicals.

Polymerization takes place by condensation reaction. Water and hydrogen are eliminated during the reaction and are liberated by thermal and catalytic action. Silicone foams may be either rigid or flexible, depending on the structure of the base polymer, and are available in both powder or liquid form. These systems normally contain resin, blowing agents, solvents, or fillers. The powder systems are the oldest and require heating to initiate the foaming and curing reactions. The newer systems require only the addition of catalyst and mixing to initiate the foaming reaction. The phenyl siloxanes are preferred for higher temperature resistance; and many modifications, such as end-blocking and polysiloxane chain, may be employed to modify the polymer structure and the properties of the foam.

Thermosetting silicone foams are available in densities from about

48 to 240 kg/m^3 (3–5 pcf). They can be foamed or sprayed in place and moulded. Board stock is also available [4.67].

The flexible silicone foams are produced by:

—room temperature foaming, or
—foaming with heat.

Generally, the silicones provide materials with a unique combination of both low and high temperature stability.

Fillers and cross-linking agents may improve their physical strength and toughness. Stabilization improves high-temperature performance [4.69].

FLAMMABILITY ASPECTS OF FOAMS

The production growth of both flexible and rigid foams in the last 15 years has been very spectacular due to a combination of factors such as light weight, excellent physical properties and relatively low costs. The wide acceptance of foams in many diversified industries has led to the necessity of developing foams of low combustibility, often combined with low smoke evolution in the case of fire, particularly in industries such as building and construction, transportation, etc. [4.35, 4.37, 4.39, 4.40, 4.64, 4.70–4.79].

A relatively large amount of unfavourable publicity has appeared in the media on the combustion, smoke and toxicity aspects of foams [4.80].

The position achieved by cellular polymers on the market of insulating materials for buildings is mainly due to their unique combination of low density (e.g. less than 50 kg/m^3, 3·13 lb/ft^3) and low thermal conductivity (e.g. less than 0·025 W/m K). Manufacturers of foam insulations have made determined efforts to demonstrate that their products do not present an unacceptable fire hazard when installed in buildings [4.39].

Given sufficient oxygen and heat, all organic polymers burn. All organic polymers evolve toxic products of combustion when burned, if only carbon monoxide. Absolute fire safety of organic polymeric materials does not exist. Yet millions of metric tons of synthetic (plastics) and natural (wood and wool) polymers are used annually in the US without presenting an unmanageable fire safety problem. It is, however, argued that some applications of synthetic polymers may have augmented fire hazards in certain applications.

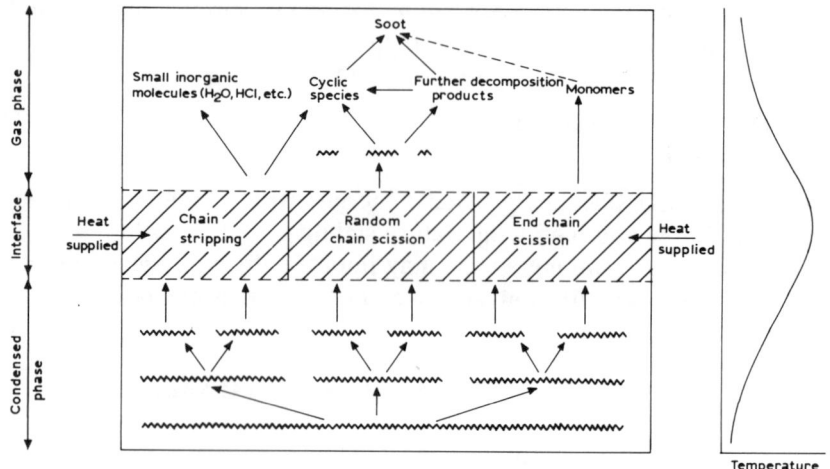

Fig. 4.33. Changes occurring as a result of the supply of heat to the surface of an organic polymer in an inert atmosphere [4.73]. (Reprinted with permission from R. A. Dine-Hart, D. B. I. Parker and W. W. Wright, *Br. Polym. J.*, **3** (1971) 226.)

Most synthetic organic polymers (plastics) burn in a manner different from that of the most familiar of natural polymers, the cellulosics, wood, paper and cotton. Some synthetics burn faster, some slower; some give off more smoke, some less, some under different conditions; a few evolve more toxic gases, some less; and some melt and flow when subjected to heat while others char over extensively. However, the general magnitude of combustibility is of the same order for both synthetic and natural organic polymers. Both burn yet both can be used safely without undue risk [4.72].

The combustion of organic polymers is invariably a complex process which involves a number of interrelated stages. Chemical reaction takes place in three interdependent regions, namely within the condensed phase of the polymer itself, at the interface between the condensed phase and the gas phase. Figure 4.33 shows diagrammatically the changes which may occur when heat is supplied to the surface of an organic polymer in an inert atmosphere and brings about some thermal decomposition of the polymer. When oxygen or other gaseous oxidant is present, the resulting combustion of the polymer is considerably more complicated; even larger temperature and concentration gradients are set up and the balance between the various processes involved results in a net energy flux through the system.

Basically two types of behaviour can be observed during the combustion of organic polymers. With thermoplastics, such as polyolefins, polystyrene and acrylics, thermal decomposition of the polymer leads to the formation of relatively large amounts of combustible volatile products, which subsequently mix with the surrounding air or other oxidant and then burn in the gas phase above the polymer, giving rise to so-called flaming combustion. On the other hand, with some thermosetting polymers (e.g. phenolic resins and polyethers) and with cellulosic materials, the initial step in the combustion process is generally the splitting off of water or other non-combustible species to leave a loose carbonaceous matrix; and this carbonaceous material then reacts exothermically with gaseous oxygen or other oxidant giving rise to non-flaming or smouldering combustion. In other words, the burning process takes place largely in the condensed phase and involves direct chemical interaction of the gaseous oxidant with the surface of the polymer.

Flammability is conveniently expressed in terms of the limiting oxygen index (LOI), which is the minimum percentage of oxygen in the surrounding atmosphere which will support flaming combustion of a substance.

Cullis [4.73], studying the flammability of organic polymers, has found that despite the fact that the monomers are often highly flammable, little or no correlation was found between the amounts of monomer formed during the polymer combustion and its flammability (Table 4.23).

Approaches toward reducing the flammability of polymer systems can be grouped into several categories:

(1) Dilution of the polymer with non-flammable components, such as mineral fillers.
(2) Incorporation of materials which decompose when heated, to give non-flammable gases such as nitrogen or carbon dioxide.
(3) Addition of flame retardants which catalyse char rather than flammable product formation.
(4) Tailoring polymer structures which favour char formation.
(5) Incorporation of additives able to stop the free radical chain reactions which occur during combustion.
(6) Formulation of products which decompose thermally with a net endothermic reaction [4.40].

The most common route to improve the non-flammable behaviour of polymer foams is the addition of flame retardants. A number of

Table 4.23.
Monomer-forming tendency and flammability of some thermoplastic polymers [4.73]

Polymer	Monomer in breakdown products (%)	Limiting oxygen index
Polyformaldehyde	100	15·0
Polytetrafluoroethylene	95	95·0
Poly(methyl methacrylate)	91–98	17·3
Polystyrene	42–45	17·8
Polychlorofluoroethylene	28	95·0
Polyisoprene	5	18·5
Poly(4-methylpent-1-ene)	2	18·0
Polybutadiene	1	18·3
Polypropylene	0·17	17·4
Poly(vinyl chloride)	0·07	47·0
Polyethylene	0·03	17·4

Reprinted with permission from C. F. Cullis, Thermal Stability
and Flammability of Organic Polymers, published in *Br. Polym.
J.* **16** (Dec. 1984) 253–7.

basic types of simple flame retardant additives are in commercial use
today such as:

—phosphorus containing retardants;
—halogen containing additives;
—mixture of halogen compounds with antimony oxide;
—nitrogen and boron compounds;
—alkali metal salts;
—hydrates of metal oxides.

Commercial flame retardant materials are finely tuned systems
including not only resin and flame retardants, but also fillers,
antioxidants, processing additives, pigments, stabilizers and plasticizers [4.72].

In Table 4.24, the average halogen requirements for various
polymers is presented as well as antimony oxide–halogen requirements
to obtain a minimal flame retardant product. Large amounts of such
additives are sometimes necessary.

In tests to determine the effectiveness of flame retardants and smoke
suppressants, a zinc/magnesium complex used as a partial replacement

Table 4.24.

Average percentage requirements of additives to render polymers of minimal fire retardance using halogen or halogen plus antimony oxide [4.72]

Polymers	Limiting oxygen index	Cl (%)	Br (%)	Sb_2O_3 (%) + Cl (%)	Sb_2O_3 (%) + Br (%)
Polyurethanes (foam)	16·5	18–20	12–14	4 + 4	2·5 + 2·5
Acrylates (PMMA)	17·3	20	16	—	7 + 5
Polyolefins (PE and PP)	17·4	40	20	5 + 8	3 + 5
Polystyrene	17·8–18·2	10–15	4–5	7 + 7–8	7 + 7–8
Cellulose (cotton, paper, wood)	18·6	>24	—	12–15 + 9–12	—
ABS	18·8–20·2	23	3	5 + 7	—
Polyacrylonitriles	19·6	10–15	10–12	2 + 8	2 + 6
Epoxies	19·8	26–30	13–15	—	3 + 5
Polyesters	20·6	25	12–15	2 + 16–18	2 + 8–9
Phenolics	21·7	16	—	—	—
Nylon (66)	24·3–28·7	3·5–7	—	10 + 6	—

Reprinted with permission from *Int. J. Polym. Mat.*, **7** (1979) 127.

Table 4.25.

A comparison of the flame-retarding and smoke-suppressing effectiveness of various transitional metal oxides with a zinc/magnesium complex in a semirigid PVC compound[a] [4.81]

Additive	Loading (per hundred)	Limiting oxygen index[b]	Maximum specific optical density[c]	Smoke generation (%)[d]
Antimony trioxide	2·5	32·9	537	−12
Antimony trioxide	5·0	35·0	612	—
Zinc/magnesium complex	2·5	35·2	370	−39
Cadmium oxide	2·5	32·7	561	−8
Niobium oxide	2·5	32·7	594	−3
Cupric oxide	2·5	33·2	577	−6
Cobalt oxide	2·5	33·3	551	−10
Ferric oxide	2·5	33·8	409	−33
Stannic oxide	2·5	34·2	537	−12
Chromic oxide	2·5	32·2	650	+6
Bismuth oxide	2·5	33·9	436	−29
Tungstic oxide	2·5	33·3	560	−8
Molybdenum oxide	2·5	33·2	491	−20
Praseodymium oxide	2·5	32·7	638	+4
Neodymium oxide	2·5	31·9	612	0
Nickel oxide	2·5	32·8	597	−2

[a] Formulations were tested in triplicate in the NBS flaming mode of operation. Sample thickness: 1 mm.
[b] Minimum concentration of oxygen, expressed as vol.%, in a mixture of oxygen and nitrogen that will support flaming combustion of specimen.
[c] A measure of smoke generation used to compare PVC formulations according to ASTM E-622, corrected for soot deposition on optical windows and normalized per 7 g of initial sample weight.
[d] Of the formulation containing 5 phr antimony trioxide.
Reprinted with permission from *Plast. Engng*, **42**(1) (1986) 47. Copyright Society of Plastics Engineers.

for antimony trioxide achieved the best balance of LOI and smoke formation. A smoke reduction of almost 40% occurred without any loss in the LOI (Table 4.25). The addition of 2·5% of antimony trioxide to the base formulation used in Table 4.24, i.e.

PVC	100%
Dioctylphthalate	30%
Tri-basic lead sulphate	5%

Table 4.26.
Effect of dioctyl phthalate on LOI
[4.82]

DOP(%)	Limiting oxygen index
0	48
30	31
40	26
60	23

Reprinted with permission from *Cell. Polym.* **4** (1985) 347–65. Copyright Elsevier Applied Science Ltd. (These are data originally published by C. Mathews & G. S. Plemper, *Brit. Polymer J.*, **15** (1983) 95–103.)

Calcium stearate	1%
Antimony trioxide	2·5%
Smoke suppressant	2·5%

making a total of 5% antimony trioxide, raised the LOI by 2·1 percentage points, but smoke generation increased by about 12% [4.81].

PVC itself contains, in theory, 56·7% of chlorine. It is no surprise therefore, that rigid PVC is self-extinguishing with an LOI of 52, and a nil ASTM burning rate. The addition of plasticizers which are flammable, can cause flexible PVC to be no longer self-extinguishing and can lower the LOI significantly. The effect of phthalate level on the LOI is shown in Table 4.26.

Besides the LOI, some other foam characteristics need to be known, such as:

—smoke generation;
—toxic gases;
—flaming drips.

Although polymers tend to have a higher ignition temperature than wood and other cellulosics, some are easily ignited with a small flame and burn vigorously.

The burning of some polymers is characterized by the rapid

Table 4.27.
Factors affecting the fire behaviour of building elements insulated with rigid
foam boards [4.39]

Component	Chemical factors	Physical factors
Foam	Polymer structure	Thickness
	Fire retardants	Density
	Reinforcing fibres	Thermal conductivity
		Friability
Facings	Inorganic, organic or metallic	Thickness
	Water content	Density
	Reinforcing fibres	Thermal conductivity
	Surface coatings	Vapour permeability
	Adhesives	Melting point
		Tensile strength
Joints	Sealant composition	Vapour permeability
	Intumescence	Melting point
		Bond strength
		Tensile strength

generation of large amounts of very dense, sooty, black smoke. Additives used to inhibit the flammability may increase smoke production. Smoke generation from a given plastic foam may vary widely depending on the nature of the polymer, the additives used, whether fire exposure was flaming or smouldering, and what ventilation was present.

Fire will generate lethal products of combustion, principally carbon monoxide. Depending on the polymer and the particular fire conditions, toxic gases may also be evolved.

Thermoplastic foams tend to melt and flow when heated. In a fire situation, this characteristic may cause melting away from the flame front and inhibit further burning, or it may produce flaming and tar-like dripping which is difficult to extinguish and which may start secondary fires.

Briggs [4.39] summarizes (Table 4.27) some general factors affecting the fire behaviour of building elements insulated with rigid foam boards.

The comparative fire performances of four types of rigid foams such as:

—expanded polystyrene (EPS),
—polyurethane (PU),

Table 4.28.

Typical results of small-scale fire tests on rigid foams with density of 35 kg/m³ [4.39]

Test	Units	EPS	PU	PIR	PF
Ignitability					
1.1 Oxygen index (ISO 4589)	%	28	24	30	46
1.2 Setchkin furnace (ASTM D1929)					
Flash ignition temperature	°C	345	285	415	490
Self-ignition temperature	°C	490	500	510	450
Flame propagation					
2.1 Horizontal burning (BS 4735)	Extent burnt (mm)	37	125	10	14
2.2 Butler chimney (ASTM D3014)	Weight retained (%)	66	20	86	95
Smoke					
3.1 XP2 Box (BS 5111, Part 1)	Max. obscuration (%)	80	85	30	3
3.2 NBS Box (BS 6401)					
Smoke density in non-flaming mode	D_m (corrected)[a]	25	120	70	8
Smoke density in flaming mode		300	330	130	10

[a] D_m, maximum specific optical density.

—polyisocyanurate (PIR),
—phenol-formaldehyde (PF),

are summarized in Table 4.28.

These data must only be used for general comparisons since significant variations can occur due to polymer microstructure and composition changes within individual classes of foams.

Model building codes require that foamed plastic used as interior wall insulation be covered with a thermal barrier, such as 13 mm thick ordinary gypsum board, or other method which reduces the risk of ignition and the subsequent flash-fire propensity. The use of foamed plastics in the cavities of hollow masonry walls, such as perimeter insulation around the foundation of a building, as insulation below concrete slabs on ground, and for roof insulation under certain conditions, is generally without thermal barrier protection. All such plastics in the thickness and density used must have a smoke-developed rating no greater than 75 under ASTM E84 tunnel test (NFPA 255) conditions.

In other than non-combustible or fire-resistive building types, foamed plastic-insulated, steel or aluminium-sheathed building panels are permitted by the model building codes, provided the foam core has a flame spread classification of 25 or less and the space is protected by automatic sprinklers.

When used as interior wall and ceiling finish, plastic materials other than foamed plastic generally are not subject to any special requirements. As for any other materials, plastics are subject to limitations on surface flame spread and often on the smoke generated, as measured by standard test procedures, such as the Steiner Tunnel Test or radiant panel test. These or special limitations may be applied to plastics used as diffusers in lighting fixtures where it is often acceptable to have a plastic which deforms and drops out of the fixture at an elevated temperature still well below its ignition point. Special limitations may also be set by building codes for use of plastic glazing instead of glass in exterior walls of buildings, where the normal requirements for conformance to limitations for interior finish may be waived.

Plastic laminates for countertops, kitchen cabinets, table tops, etc., are not usually included in the definition of interior finish, as regulated by building codes. Even when it is not regulated by local code, the flammability level should be limited to that encountered when natural products are used in these applications. Plastic laminates are commonly available with low surface flame spread classifications, in the range of 25–75.

REFERENCES

[4.1]   Frisch, K. C., *Int. J. Polym. Mater.*, **7** (1979) 113–25.
[4.2]   Richardson, M. D. W., *Plast. Rubber Int.*, **5**(3) (1980) 115.
[4.3]   Pascoe, M. W., In *Polymer Engineering Composites*, ed. M. D. W. Richardson. Applied Science Publishers, London, 1977, pp. 437–9.
[4.4]   Puterman, M., Narkis, M., & Kenig, S., *J. Cell. Plast.*, July/Aug. (1980) 223–8.
[4.5]   Narkis, M., Puterman, M., & Kenig, S., *J. Cell. Plast.*, Nov./Dec. (1980) 326–30.
[4.6]   Hilyard, N. C., & Young, J., In *Mechanics of Cellular Plastics*, ed. N. C. Hilyard. Macmillan, New York, 1982, pp. 1–26.
[4.7]   Benning, C. J., *Plastic Foams: The Physics and Chemistry of Product Performance and Process Technology, Vol. II*. Wiley-Interscience, New York, 1969.
[4.8]   Titow, W. V., *PVC Technology*, 4th edn. Elsevier Applied Science, London and New York, 1984.
[4.9]   Lasman, H. R., & Scullin, J. P., In: *Encyclopaedia of PVC, Vol. 2*, ed. L. J. Nass. Marcel Dekker, New York and Basel, 1976, pp. 802–6.

[4.10] Allcock, H. P., & Lampe, F. W., *Contemporary Polymer Chemistry*. Prentice Hall Inc., Englewood Cliffs, N.J., 1981, p. 526.

[4.11] Rohsenow, W. M., & Hartnett, P., *Handbook of Heat Transfer*. McGraw-Hill, New York, 1973, pp. 282–5.

[4.12] Rodrigues, F., *Principles of Polymer Systems*, 2nd edn. Hemisphere Publishing Co., Washington, 1982, pp. 362–74.

[4.13] Villamizer, C. A., & Han, C. D., *Polym. Engng Sci.*, **18**(9) (1978) 699–710.

[4.14] Holloway, L., *Glass Reinforced Plastics in Construction*. Surrey University Press, 1978, pp. 89–141.

[4.15] Price, H. L., & Nelson, J. B., *J. Comp. Mater.*, **10** (1976) 314–18.

[4.16] Scala, E., *Composite Materials for Combined Functions*. Hayden Book Co. Inc., Rochelle Park, N.J., 1973, pp. 190–9.

[4.17] Gluck, D. G., Hagan, J. R., & Hipchen, D. E., *Advances in Urethane Science and Technology*, Vol. 7, ed. K. C. Frisch & D. Klempner. Technomic Publishing Co. Inc., Lancaster, PA, 1980.

[4.18] Feldman, D., *Polymer Technology*. Technica, Bucharest, 1974.

[4.19] Kennedy, R. N., & Harsha, P., In *Plastics in Building*, ed. J. Skeist. Reinhold, New York, 1966, pp. 169–87.

[4.20] Blaga, A., *Canadian Building Digest 167*. National Research Council—Canada, 1974.

[4.21] Frisch, K. C., & Saunders, J. H., *Plastic Foams*. Marcel Dekker, New York, 1972.

[4.22] Renz, R., & Ehrenstein, G. W., In *Cellular Plastics*, ed. J. M. Buist. Applied Science Publishers Ltd, London, 1982, pp. 5–13.

[4.23] Meinecke, E. A., & Clark, R. C., *Mechanical Properties of Polymeric Foams*. Technomic Publishing Co. Inc., Westport, Conn., 1973.

[4.24] Han, C. D., Kim, Y. W., & Malhotra, K. D., *J. Appl. Polym. Sci.*, **20** (1976) 1583–95.

[4.25] Dominick, G., *J. Elast. Plast.*, **11** (1979) 133–9.

[4.26] Blaga, A., *Canadian Building Digest 168*. National Research Council—Canada, 1974.

[4.27] Rogers, T. H., & Hecker, K. H., In *Rubber Technology*, 2nd edn, ed. M. Morton. Van Nostrand Reinhold Co., New York, 1973, p. 459.

[4.28] Dransfield, A., In *Developments in Polyurethane—1*, ed. J. M. Buist. Applied Science Publishers, London, 1978, p. 149.

[4.29] Ball, G. W., Hurd, R., & Walker, M. G., *J. Cell. Plast.*, **6**(21) (1970) 66.

[4.30] Schmidt, W., *Appl. Phys.*, **11**(4) (1968) 19.

[4.31] Doherty, D. J., Hurd, R., & Lester, G. R., *Chem. Ind.*, **30** (1960) 1340.

[4.32] Buist, J. M., Doherty, D. J., & Hurd, R., Progress in Refrigeration Science and Technology. In *Proceedings of the International Congress of Refrigeration*. Pergamon Press, Oxford, 1965, p. 271.

[4.33] Brandreth, D. A., & Ingersoll, H. G., *Proceedings of the SPI 25th Annual Urethane Division Technical Conference*, Scottsdale, Arizona (Oct. 1979).

[4.34] Gaboriaud, F., & Vantelon, J. P., *J. Polym. Sci. Chem. Ed.*, **19** (1981) 139–50.

[4.35] Zabski, L., Walczyk, W., & Weleda, D., *J. Appl. Poly. Sci.*, **25** (1980) 2659–80.

[4.36] Robertson, E., Waszeciak, P., Sherman, M., & Conroy, A., *J. Cell. Plast.*, Sept./Oct. (1980) 279–92.

[4.37] Neywick, C. V., Yoerger, R. E., & Peterson, R. F., *Proceedings of the SPI 25th Annual Urethane Division Technical Conference*, Scottsdale, Arizona (Oct. 1979).

[4.38] Grassie, N., Zulfigar, M., & Guy, M. J., *J. Polym. Sci., Polym. Chem. Ed.*, **18** (1980) 265–74.

[4.39] Briggs, P. J., *Cell. Polym.*, **4** (1985) 261–77.

[4.40] Jenkins, A. D., & Kennedy, J. F. (Eds), *Macromolecular Chemistry, Vol. 2.* The Royal Society of Chemistry, London, 1982, pp. 73–5.

[4.41] Shorter, G. W., & McGuire, J. H., *Canadian Building Digest 178.* National Research Council—Canada, 1976.

[4.42] Rogers, F. E., Ohlemiller, T. J., Kurtz, A., & Summerfield, M., In *Flammability of Cellular Plastics, Part 2*, ed. C. J. Hilado. Technomic Publishing Co. Inc., Westport, Conn., 1981, pp. 1–9.

[4.43] Annamalai, K., & Sibulkin, M., ibid., pp. 27–40.

[4.44] Holmes, C. A., ibid., pp. 41–62.

[4.45] Ohlemiller, T. J., & Rogers, F. E., ibid., pp. 63–83.

[4.46] Bonsignore, P. V., & Levendusky, T. L., ibid., pp. 84–103.

[4.47] Tatem, P. A., Powell, F. L., & Williams, F. W., ibid., pp. 125–36.

[4.48] Herrington, R. M., ibid., pp. 137–54.

[4.49] Deleon, A., & Snider, S. C., In *Advances in Urethane Science and Technology, Vol. 9*, ed. K. C. Frisch & D. Klempner. University of Detroit, 1984, p. 169.

[4.50] Pearce, E. M., In *Contemporary Topics in Polymer Science, Vol. 5.* Plenum Press, New York and London, 1984, p. 401.

[4.51] Hurd, R., In *Developments in Polyurethane—1*, ed. J. M. Buist. Applied Science Publishers Ltd, London, 1978, pp. 175–98.

[4.52] Cornish, H. H., *et al., International Symposium on the Toxicity and Physiology of Combustion Products*, University of Utah, Mar. 1976.

[4.53] Loikkanen, P., VTT Symposium 57, Structural Fire Safety Regulations in Finland and in the S.U. Technical Research Center of Finland, ESPOO, 1985, pp. 84–95.

[4.54] Anon., *Modern Plast. Int.*, Oct. (1981), 48–50.

[4.55] Meltzer, Y. L., *Foamed Plastics. Recent Developments.* Noyes Data Corp., Park Ridge, N.J., 1976.

[4.56] Meyer, B., *Urea Formaldehyde Resins.* Addison-Wesley, London, Amsterdam, 1979.

[4.57] Patton, W. J., *Materials in Industry*, 3rd edn. Prentice Hall, Englewood Cliffs, N.J., 1986, pp. 297–8.

[4.58] Brydson, J. A., *Plastic Materials*, 4th edn. Butterworths Scientific, London, Boston, 1982, pp. 611–12, 730–65.

[4.59] Chown, G. A., Bowen, R. P., & Shirtliffe, C. J., *Urea Formaldehyde Foam Insulation*, Department of Building Research, No. 19. National Research Council—Canada, 1981.

[4.60] Hanson, D. J., *Chem. Engng News*, March 29 (1982) 34–37.

[4.61] Myers, G. E., *For. Prod. J.*, **35**(9) (1985) 20–31.
[4.62] Anon., *Chem. Engng News*, Dec. 9 (1985) 4.
[4.63] Anon., *Chem. Engng News*, Apr. 30 (1984) 17–20.
[4.64] Rogowski, B. F. W., *Cell. Polym.*, **4** (1985) 325–38.
[4.65] Breslau, A. J., In *Epoxy Resins, Chemistry and Technology*, ed. C. A. May & Y. Tanaka. Marcel Dekker Inc., New York, 1973, pp. 563–74.
[4.66] Ranney, M. W., *Epoxy Resins and Products. Recent Advances*. Noyes Data Corp., Park Ridge, N.J., 1977, p. 214.
[4.67] Benning, C. J., *Plastic Foams, Vol. 1*. Wiley-Interscience, New York, 1969.
[4.68] Potter, W. G., *Epoxide Resins*. Iliffe Books, London, 1970, pp. 98–100.
[4.69] Critchley, J. P., Knight, G. J., & Wright, W. W., *Heat-Resistant Polymers*. Plenum Press, New York and London, 1983.
[4.70] Hilado, C. J. (Ed.), *Flammability of Cellular Plastics, Part 2*. Technomic Publishing Co. Inc., 1981.
[4.71] Schaffer, E. L. (Ed.), *Behavior of Polymeric Materials in Fire*, ASTM Technical Publ. 816, 1983.
[4.72] Nelson, G. L., *Int. J. Polym. Mater.*, **7** (1979) 127–45.
[4.73] Cullis, C. F., *Br. Polym. J.*, **16** (1984) 253–7.
[4.74] Paul, K. T., *Polym. Engng Sci.*, **22**(13) (1982) 793–7.
[4.75] Rogowski, B. F. W., Building Research Establishment, CP.39/96, Borehamwood, 1974.
[4.76] Kourtides, D. A., *Poly. Plast. Technol. Engng*, **11**(2) (1978) 159–98.
[4.77] Woolley, W. D., *J. Macrom. Sci.-Chem.*, **A11**(8) (1977) 1509–17.
[4.78] Stark, G. W. V., Field, P., & Pitt, A., Fire Research Note No. 1017, Fire Research Station, Borehamwood, 1974.
[4.79] Prado, G., Jagoda, J., & Lahaye, J., *Fire Research*, **1** (1977/78) 229–35.
[4.80] Rawis, R. L., *Chem. Engng News*, Jan. 3 (1983) 9–16.
[4.81] Fletcher, F. J., & Dogherty, A., *Plast. Engng*, **42** (1986) 47–50.
[4.82] Hitch, M. J., & Rolph, D. C., *Cell. Polym.*, **4** (1985) 347–65.

Chapter 5

Adhesives and Sealants

INTRODUCTION

Vegetable mucilage and animal glues have been used as adhesives since the beginning of history. They supply a good bond under dry conditions but fail if the joints become damp. In addition they bond only certain materials such as wood and paper. The introduction of casein and blood glues was a marked improvement, but these protein materials are subject to fungus attack and are not water-resistant.

The advent of polymeric adhesives and sealants was a tremendous improvement. It is now possible to bond materials with adhesives so tenaciously that the bond is stronger than the cemented materials.

The molecular weights of these new adhesives and sealants are very high, and in the case of rigid products, the whole adhesive fibre uniting the surfaces may be regarded as connected by primary valency bonds into a single large irregular molecule.

Most adhesives are organic in nature, and the traditional glues and gums derived from natural sources are long chain molecules, soluble in water, and are either proteins or polysaccharides.

The advent of synthetic polymers has led to the wider use of adhesives, as the newer materials, when properly used, can produce bonds of much greater strength than has been possible in the past.

The notion 'adhesion' refers to the attraction between substances whereby they are brought into contact; it is necessary to do work in order to separate them.

The science of adhesion is concerned to understand the nature of this attraction between substances, to measure its magnitude and to study the effect of outside influences on it.

The total adhesive force holding two materials together is the sum of two factors, namely, specific adhesion and mechanical adhesion. The first is chemical; it is the molecular attraction between two materials. The actual bond may be a chemical union, such as the surface linkage in rubber-to-metal bonding, or it may be simply an electrical attraction between electrons of the two substances.

Mechanical adhesion is the bonding force provided by interlocking action. Specific adhesion can be considered as an active force holding the materials together. It is effective under tensile, shear, and peel type loadings, whereas mechanical adhesion is passive and not very effective until acted upon by an outside force [5.1].

Comprehensive theories need to operate at various levels and one general explanation for the behaviour of macromolecular compounds as adhesives has been formulated; it is necessary to consider why polymers rather than other materials confer the desired properties and then why industrially important adhesives are not single macromolecular compounds but mixtures of basic polymers with other products.

To a certain extent the formulation of adhesives and sealants is an art in which various and conflicting characteristics are balanced by intuitive judgement, but there are general principles involved which can be related to a descriptive if not a quantitative theoretical background and which are not usually discussed in literature dealing with practical aspects of adhesives and sealant formulation.

Adhesives and sealants are used in almost every phase of the building industry. The use of adhesives in construction began with the finishing trades.

Flooring materials, wallpaper and roofing cements were the first volume applications.

Paralleling the growth of certain wall constructions has been the increase of a whole new family of higher-performance speciality caulking or sealing compounds, i.e. sealants. The term 'sealant' was first used to differentiate these new polymer based compounds from the older oil-based caulks. Modern usage, however, has extended the meaning of the word, so that the term 'sealant' is now used to include all types of weatherproofing joint materials currently in use.

Polymer sealants are designed to prevent the passage of moisture, air and heat through all the joints and seams in the structure [5.2].

As already stated, the class of substances from which adhesives and sealants are drawn, are those of high molecular weight, which, if they crystallize, do not, except in very special circumstances, give single, discrete crystals but form crystals imbedded in and still continuous as molecular chains with other, less ordered structures.

Modern sealants are composed of pigmented or unpigmented synthetic elastomers, which in the uncured state, constitute pourable or easily extendable putty-like mastics. The consistency of sealants in the non-cured state can be adjusted according to the requirements of application, e.g. self-levelling for horizontal joints and non-sagging for vertical ones. In the cured state, after a vulcanization process, these sealants become transformed into elastomers. When cured, they bind the structural elements and are able to contract or expand with the motion of the elements [5.3].

ADHESIVES

In more recent years there has been a rapid development of adhesive bonding as an economic and effective method for the fabrication of various components and assemblies [5.4].

According to the 1980 Book of Standards [5.5], an adhesive is a substance capable of holding materials together by surface attachment. The holding together of two surfaces by interfacial forces which may consist of valence forces or interlocking action or both, is known as adhesion.

Kinloch [5.6] defines the adhesive as a material which when applied to substrate surfaces can join them together and resist separation.

The use of adhesives offers advantages in comparison with conventional techniques such as brazing, welding, riveting, bolting, etc. Some of the advantages are:

—the ability to join efficiently thin sheets, or dissimilar materials;
—an increase in design flexibility;
—an improved stress distribution in the joint which leads to an increase in fatigue resistance of the bonded component;
—a convenient and cost effective technique.

There are a number of scientific disciplines that have much to contribute to adhesive bonding technology. On consideration of polymers in structural adhesive joint applications, several researchers [5.7] describe the advantages of the multidisciplinary approach. It is suggested that surface physical chemistry, polymer science and mechanics are integrally involved. A pattern that includes adhesion as well as

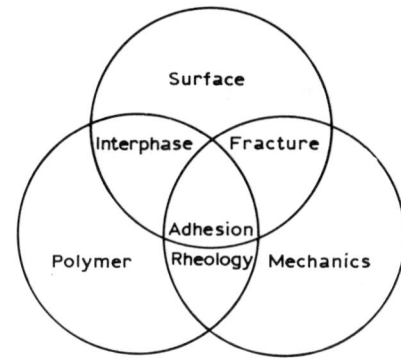

reinforcing related areas seems to emerge when considering the interrelation of these disciplines. The pattern is illustrated in Fig. 5.1.

Although the interfacial properties can be critical to joint strength, it is not yet possible to predict these properties quantitatively. We do know that they depend on the surface characteristics of the adhesive and the adherend prior to bonding and on the surface phenomena that occur when the two are brought together [5.8]. Modifications of polymer surfaces for adhesion oriented applications have necessitated a careful analysis of the surface region morphology (surface physics) and chemical properties of the surface layer (surface chemistry) [5.9].

The adhesion scientist often needs to consider aspects of surface chemistry, polymer chemistry, stress analysis, polymer physics and fracture phenomena to interpret fully his data [5.6].

The interaction of solid surfaces with gases or liquids leads to physical adsorption or chemisorption of molecules or atoms on the solid surface. The character of this adsorption depends on the surface energy of the solids and the chemical nature of the adsorbents. Physical adsorption is brought about by dispersion forces, while chemisorption is due to the exchange of electrons between the solid and the adsorbed molecule, leading to the formation of a chemical bond, ionic or covalent. Because of this the chemisorbed layer is usually a single molecule thick while, in physical adsorption, successive molecular layers result. These molecules adjacent to the solid surface are subject to much greater attraction forces than the subsequent layer of molecules. Usually the layers close to the solid surface are in a more orderly arrangement, which gradually disappears with the increasing distance of the subsequent layers from the solid surface.

Solid surfaces can be classified as low-energy and high-energy surfaces. High-energy solid surfaces, as exemplified by most metals, various metallic oxides, diamond, quartz, glasses, and similar have surface energies ranging from 0·5 to 5 J/m², the values being higher the greater the hardness and the higher the melting point. Low-energy solid surfaces, which are characteristic of organic polymers, resins, waxes and most organic compounds, have specific surface energies of less than 0·1 J/m². Uncontaminated high-energy solid surfaces are wetted by all pure liquids (excluding liquid metals) because they have surface energies greater than 0·1 J/m². Low-energy solid surfaces are not wetted completely by a wide variety of pure liquids. Every liquid having a low specific surface energy always spreads freely on a specularly clean, high-energy surface at ordinary temperature unless the film adsorbed by the solid converts it to a low-energy surface having a surface tension lower than that of the liquid. Liquids which cause the formation of an adsorbed and oriented layer on the solid surface resulting in a low-energy surface (even lower than that of the spreading liquid) are called autophobic.

As a measure of the wetting ability of a solid by a liquid, Young introduced the concept of the contact angle and offered the following relation (Fig. 5.2)

$$\gamma_s = \gamma_{sL} + \gamma_L \cos \theta \tag{5.1}$$

and

$$\cos \theta = \frac{\gamma_s - \gamma_{sL}}{\gamma_L} \tag{5.2}$$

The relation between the contact angle and wetting abilities is illustrated by Figs. 5.3(a) and 5.3(b).

The work of adhesion for a solid–liquid interface can be estimated in a similar way as for liquid–liquid adhesion.

$$W_a = \gamma_L + \gamma_s - \gamma_{Ls} \tag{5.3}$$

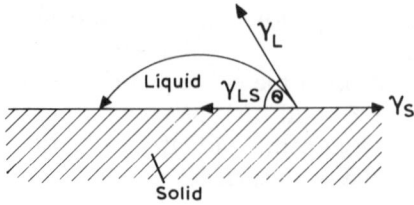

Fig. 5.2. Derivation of Young's equation. Subscripts L, S and LS stand for liquid–air, solid–air and liquid–solid interfaces [5.10]. (Reprinted with permission from *The Nature and Properties of Engineering Materials*, 2nd edn, by Z. D. Jastrzebski, 1977. Copyright John Wiley.)

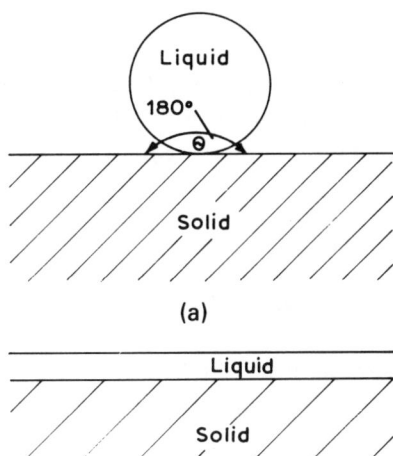

where subscripts L, s and Ls stand for liquid–air, solid–air and liquid–solid interfaces. Combining with Young's equation (5.1), we get

$$W_a = \gamma_L + \gamma_L(\cos \theta)$$

$$W_a = \gamma_L(1 + \cos \theta) \tag{5.4}$$

Equation (5.3) indicates that the work of adhesion is enhanced by a high value of γ_s and γ_L and a low value of γ_{sL}. To keep the value of γ_s high, the surfaces of solid adherents should be thoroughly clean. Increasing γ_s can be secured by a change in surface chemistry using chemical treatment or by surface roughening. A low value of γ_{sL} is usually evident when the liquid has a chemical affinity toward the solid substrate (adherent) [5.10]. This means that the polymeric adhesive needs to be able to spread over the solid substrate surface, and needs to displace air and any other contaminants that may be present on the surface. According to Kinloch [5.6] an adhesive which conforms ideally to these conditions must:

—when liquid, exhibit a zero or near zero contact angle;
—at sometime during the bonding operation have a viscosity that should be relatively low, i.e. no more than a few centipoises;
—be brought together with the substrate at a rate and in a manner that should assist the displacement of any trapped air.

Thermodynamically, wetting will occur whenever the free energy change (ΔF) for producing the liquid–solid interface is negative compared with the free energy changes for loss of the solid–air and liquid–air interfaces. That is, when

$$\Delta F_{\text{wetting}} = \Delta F_{\text{Ls}} - (\Delta F_{\text{L-air}} + \Delta F_{\text{s-air}}) \tag{5.5}$$

is *negative,* wetting will occur [5.11].

The mechanism of adhesion has been the subject of scientific inquiry and debate for many decades. Kinloch [5.12] mentions that the mechanisms of adhesion are still not fully understood and many theories are to be found in the technical literature. The four main mechanisms of adhesion which have been proposed are:

—mechanical interlocking;
—diffusion theory;
—electronic theory;
—adsorption theory.

On the other hand, Schneberger [5.11] shows that,

—chemical reaction;
—interdiffusion;
—dipole–dipole attractions;
—Van der Waal's forces; and
—mechanical interlocking, have all been advanced as the explanation for adhesion.

None of these theories fit every situation, and frequently several of them appear to play a role in bonding. It has also been suggested that acid–base interactions at the adhesive–adherend interface could play a very important role in adhesion. Evidently, the practice of adhesion science reveals that it is not a simple phenomenon comprehensible with a single model. The physical reality tends to suggest that several models which have been postulated operate at the same time.

There exist between scientists many different points of view. Although in the past many scientists tended to think almost exclusively in terms of an adsorption phenomenon, scientists from the USSR think exclusively in terms of diffusion or electrostatic theory, and generally a far more electric attitude is common now.

Chemical bonds, once regarded as unnecessary to explain joint strength or adhesive action, are now seen to be quite common and for ionic bonds to be closely linked with the electrostatic theory. Diffusion as the mechanism of auto-adhesion of elastomeric polymers and adsorption admitted to explain the wetting of surfaces by liquids, are essential preconditions for adhesive action. Adsorption, or the sense of orientated molecules on fixed sites is applied for many cases of polymeric adhesives on high energy solids.

The physical properties of a synthetic adhesive are greatly influenced by the way atoms are arranged within the macromolecule. Useful polymers have molecular weights that range from 10^3 to 10^6. In these giant molecules, atoms are usually arranged in a linear or cross-linked fashion.

Primary chemical bonds along polymer chains are entirely satisfied. The only forces between linear macromolecules are secondary bond forces of attraction, which are weak relative to primary bond forces. The high molecular weight of macromolecular compounds allows these forces to build up enough to impart excellent strength, dimensional stability, and other mechanical properties to the substances.

Cross-linked polymers result when the linear structures, in effect, branch out and combine in three dimensions. The cross-linked structures are the result of strong chemical bonds between the linear segments of the macromolecules.

In practice there are two ways to apply macromolecular adhesives or sealants. The adhesive may be applied as a non-polymerized liquid (monomer, oligomer) that reacts after wetting the surface to form a polymer. This approach is typical of epoxies, phenolics, polyesters and alpha-cyanoacrylates. The other technique is to apply the fully polymerized product in a liquid or rubbery state. This requires that the macromolecular compound be applied above the melting range, above its glass transition temperature or in solution. Bond strength develops as the adhesive solidifies [5.13].

Adhesive bonding has many advantages to offer the building industry. No other method of attachment is satisfactory for so many applications. It would be absurd to consider nailing a ceramic wall tile into position or to use plywood panelling which had its wood plies stapled together. Even sandpaper depends on an adhesive to hold the grit to its paper backing. When all the applications of adhesives are taken into account, adhesive bonding must be considered as the most widely used method of holding various materials together. The importance of the surface polarity and other surface characteristics for polymer adhesion has been considerably discussed in recent years [5.7, 5.9, 5.11, 5.14–5.19].

A useful generalized theory of adhesion however, can be built upon the basis of electrical attractions. These electrical attractions, resulting from uneven surfaces, which are not normally considered to be 'electrical', participate easily in attractive interactions if adhesives can be found which will wet them. The reason that polyethylene and poly(tetrafluoroethylene), i.e. Teflon, are difficult to bond is simply that available adhesives are thermodynamically more stable if their molecules attract one another than if they interact with low energy surfaces. The key to this dilemma is to modify the surfaces of difficult-to-bond materials so that they become capable of strong electrical attractions. A number of proprietary processes are routinely used industrially for this purpose.

Surfaces

Solid surfaces have a texture which is characterized by waviness, roughness, and indentations such as scratches and pits. Waviness and roughness, shown in Fig. 5.4, represent coarsely and finely spaced surface irregularities. They have little effect on adhesive spreading but may affect the bond strength if the peaks are sharp-edged and act as stress concentration points. Such stress risers should be less harmful with adhesives pliable enough to absorb stress energies. This behaviour may explain the effect of surface roughness on bond strength (Fig. 5.5).

Surface indentations may interfere with wetting if they retain trapped air or the concentrated solvents from the adhesive. Such air or solvent pockets may result in expansion pressure stresses when the bond is cured or used at high temperatures [5.11].

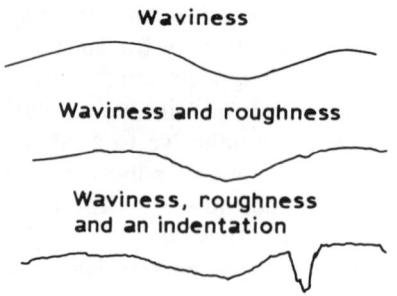

Waviness

Waviness and roughness

Waviness, roughness and an indentation

Fig. 5.4. Solid surfaces are characterized by waviness, roughness and indentations [5.11]. (Reprinted with permission from *Adhesives Age*, **23**(1) (1980) 42. Copyright Communication Channels Inc.)

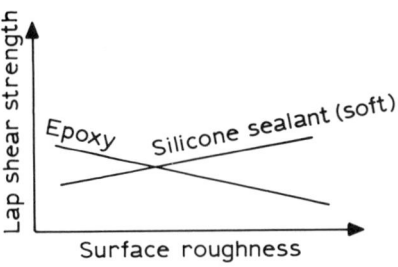

Fig. 5.5. The effect of surface roughness on the lap shear bond strength of a hard (epoxy) and a soft commercial adhesive [5.11]. (Reprinted with permission from *Adhesives Age,* **23**(1) (1980) 42. Copyright Communication Channels Inc.)

Adhesives are applied in a fluid state so as to facilitate the interfacial contact and also its penetration in the pores and crevices of the substrate. Shrinkage during polymerization or solvent release and differences in expansion between adhesive and adherend can lead to serious stresses at the bond interface. The interfacial shear stresses which arise because of expansion differences sometimes make particular bonds risky or impossible.

Mittal [5.17] shows that the maximum values of adhesive strengths are attained when the surface free energies of the adherends and adhesives are approximately equal. The author states that if the nature of the adhesive is the same, the interfacial free energy is the most important surface-chemical parameter in the determination of adhesive strengths. If different adhesives are used, then the situation is somewhat different, i.e. for the same values of γ_{sL} or W_a, different adhesive strengths are possible depending on the adhesives used.

Lurie [5.20] has measured the velocities of deformation just great enough to cause adhesive, rather than cohesive failures between a number of metals and a tacky adhesive. His results are plotted in Fig. 5.6 which suggests a correlation between the mechanical performance of joints and the thermodynamic work of adhesion. His contact angle values are based upon the use of a liquid less viscous than, but chemically similar to, the adhesive.

Levine *et al.* [5.21] studied the adhesion between films (e.g. PVC, PVDC, PVF, etc.) and adhesive composed of an epoxy prepolymer (Epon 828) cured with diethylenetriamine. The adhesive exhibited a surface energy of 50 erg/cm². The tensile tests were carried out according to ASTM 1205. The design of the experiment was such that failure in all cases occurred between the plastic film and the adhesive. They found that as the value of the critical surface tension of wetting,

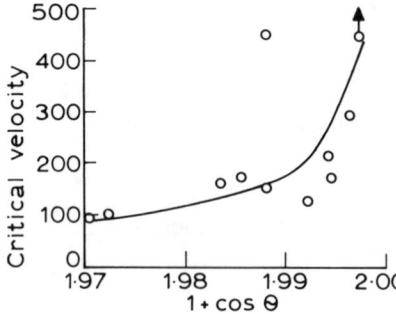

Fig. 5.6. Plot of mechanical performance of adhesive joints and $(1 + \cos \theta)$ [5.20]. (R. M. Lurie, MIT PhD thesis, 1955.)

γ_c, decreases, the tensile strength also decreases linearly as shown in Fig. 5.7.

The calculation of the values of the interfacial free energy between the liquid adhesive and the various films led to a linear relation between the tensile strength and the reciprocal of the free energy [5.17].

Gent [5.22] mentioned that coupling agents, or adhesion promoters, are widely used to make adhesive joints between polymers and metal or glass that are able to withstand severe conditions of high temperature and high humidity. They have a dual functionality so that they are able to interlink the two adherends by reacting with the surface atoms of both materials. As already mentioned in the case of glass reinforced composites, silanes are considered excellent coupling agents. One part

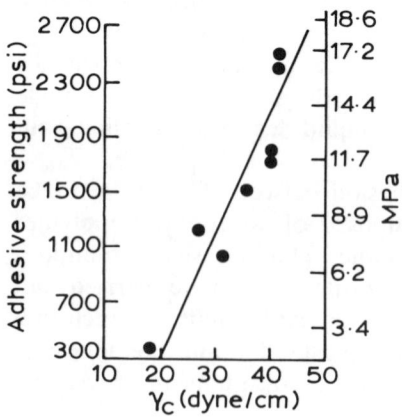

Fig. 5.7. Plot of adhesive tensile strength and γ_c of polymers using epoxy adhesive of surface free energy 50 erg/cm² [5.21]. (Reprinted with permission from *J. Polym. Sci.*, **B-2**(915) (1964). Copyright John Wiley.)

of the coupling agent molecule is able to react with —OH groups on the surface of glass, and another part is chosen to react with a specified polymer.

Ahagon & Gent [5.23] studied the adhesion of polybutadiene (PB) elastomer to a silane-treated glass surface and varied the amount of interfacial bonding in a systematic way. Vinyltriethoxysilanes are capable of forming covalent bonds with PB during free-radical cross-linking, while ethyltriethoxysilane is unreactive. The amount of interfacial interlinking between the elastomer and the glass surface was varied from no chemical bonding (only Van der Waals attraction forces when ethyltriethoxysilane was used alone on the glass surface) to increasing amounts of interfacial bonding when increasing proportions of vinyltriethoxysilane were used (Fig. 5.8). A previous study by Gent & Hsu [5.24] had shown that strong chemical covalent bonds are formed between silanes and glass.

As the technology of adhesive bonding has progressed, the development of chemical agents for treating adherend surfaces (i.e. coupling agents, primers and adhesion promoters) has also progressed. Their function is to protect the adherend from moisture and other contamination prior to bonding, increase the adherend wettability, protect the bond from moisture and/or improve the interfacial bond strength. These adhesion promoters are usually applied to the adherend in such a way as to form a thin (<1000 Å) film into which the resin can penetrate and presumably interact. The fact that the adhesive polymer

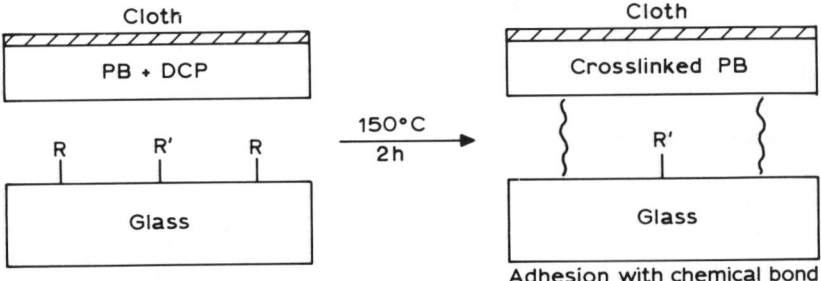

Fig. 5.8. Bonding polybutadiene (PB) to glass treated with a mixture of vinylsilane (R) and ethylsilane (R'). Elastomer layer is cross-linked and interlinked *in situ* by free radical producing agent, dicumyl peroxide (DCP) [5.22]. (Reprinted with permission from *Adhesives Age,* **25**(2) (1982) 27. Copyright Communication Channels Inc.)

and one of the adhesion promoters, i.e. the primer or promoter coating, can interdiffuse, implies that there exists an interfacial layer whose properties are considerably different from either of the components it is bonding. Bascom & Patrick [5.8] consider that this is probably the key to the mechanism of adherend surface treating agents. Different types of surface treatments are discussed by Liang & Dreyfuss [5.25], Eckstein & Berger [5.26] and Leary & Campbell [5.27]. Theoretical aspects of adhesion are reported in many papers, books and conference presentations [5.6, 5.12, 5.28–5.35].

Adhesive Classifications

There exists a considerable number of different types of adhesives which are currently in use and there is no adequate single system of classification for all products. The adhesives industry has generally employed classifications based on end-uses, such as metal-to-metal adhesives, paper and packaging adhesives, general purpose adhesives, and so on.

Other criteria for adhesives classification may be: the chemical composition, the method of application, the physical form, suitability for particular service requirements or environments, etc.

The scheme adopted here (Table 5.1) is based on the origin and chemical composition of the main component of the most known adhesive formulations. It must be pointed out however that many synthetic polymers used as adhesives may at times be used in other adhesive formulations. This applies to sealants as well.

Some of the synthetic polymers marked in Table 5.1, such as polystyrene, polyamides, saturated polyesters and others (ethylene copolymers, polypropylene, etc.) are usually in the form of hot-melt adhesives.

Properties of some Adhesives

Thermoplastics

Poly(vinyl acetate) (*PVAc*). The properties of poly(vinyl acetate) do not deteriorate at low temperatures or in sunlight and are unaffected by bacteria, fungi or insects; but this polymeric adhesive will not sustain prolonged stress, particularly at elevated temperatures. It swells slightly and becomes opaque on long immersion in water, but recovers on drying; more recently synthesized appropriate copolymers show greater water resistance [5.37].

Table 5.1.
Adhesive classification [5.4, 5.36]

Occurrence	Origin	Adhesive
(1) Natural	(a) Animal	Albumen, casein, animal glue, shellac
	(b) Vegetable	Resin: gum arabic, tragacanth, colophony, Canada balsam, natural rubber
		Oils: linseed oil
		Waxes: carnauba wax
		Proteins: soya bean
		Polysaccharides: starch, dextrine
	(c) Mineral	Waxes: paraffin[a]
		Resins: copal, amber
		Other materials: silicates, phosphates, sulphur, magnesia, litharge, bitumen
(2) Synthetic	(a) Thermoplastic	Vinyl polymers and copolymers: poly(vinyl acetate), poly(vinyl alcohol), poly(vinyl acetal)s, polystyrene
		Acrylic polymers: polyacrylates, polymethacrylates, polyacrylamides, poly(cyanoacrylate)s polyamides and saturated polyesters polyurethanes
		Cellulose derivatives:[b] Cellulose acetate, cellulose nitrate, cellulose acetate–butyrate, methyl cellulose, hydroxy ethyl cellulose, carboxy ethyl cellulose, and others
	(b) Thermosetting	Phenolic resins: phenol–formaldehydes, resorcinol formaldehydes
		Amino plastics: urea–formaldehydes, melamine formaldehydes
		Polyepoxides and derivatives: polyepoxides, epoxy–polyamide, epoxy–bitumen, epoxy–polysulphide, etc.
		Unsaturated polyesters
		Polyaromatics: Polybenzimidazole, polyimide, polybenzothiazole

[a] May also be classified as an organic compound.
[b] Products derived from chemical modification of natural polysaccharides (cellulose).
Reprinted with permission from *Adhesives Handbook*, 2nd edn, by J. Schields, 1976. Copyright Sira Ltd, South Hill, Chislehurst, Kent, UK.

Some researchers [5.38] underline the variation among poly(vinyl acetate) adhesives in their resistance to moisture and to weathering; however, all proved vulnerable to some degree to continuous stressing on the glued joints.

Developments in the formulation of poly(vinyl acetate) adhesives have greatly reduced the sensitivity to water and involve some degree of cross-linking of the polymer chains.

Due to its solubility, this polymer may be used dissolved in various solvents as well. It is soluble in acetic acid, benzyl alcohol, 95% ethyl alcohol, methanol, acetaldehyde, furfural, aniline, dichloroethylene, trichloroethylene, *n*-butyl acetate, ethylacetate, isopropyl acetate, dioxane, benzene, toluene, acetone, etc. [5.39].

This macromolecular compound has excellent adhesion properties, due to the nature of its functional groups, and adheres to a host of different substrates such as: metal, porcelain, wood, paper and various textiles. The adhesive is quick-setting, a property which shortens press and assembly times in manufacturing processes. The latex, which one obtains through emulsion polymerization, is free of odour, is non-flammable, mould-resistant, has a uniform viscosity, and a good shelf-life. It may be used with or without a plasticizer; the presence of a plasticizer increases the adhesion properties.

For some applications, such as for making sealable coatings, for grease-proofing paper, or for the manufacture of laminating plastics, poly(vinyl acetate) resins are sold as a solution in the organic solvents.

Poly(vinyl alcohol). For commercial purposes, poly(vinyl alcohol) is obtained from poly(vinyl acetate) which is readily hydrolysed by treating an alcoholic solution with aqueous acid or alkali. The process occurs through the following mechanism:

$$\cdots -CH_2-CH-CH_2-CH- \cdots \longrightarrow \cdots -CH_2-CH-CH_2-CH- \cdot$$
$$\qquad\qquad OCOCH_3 \quad OCOCH_3 \qquad\qquad\qquad OH \qquad OH$$

The time of reaction and the catalyst concentration have a great influence on the degree of alcoholysis. This characteristic has a decisive effect on the properties of poly(vinyl alcohol). The replacement of acetate groups by hydroxylic ones which are more polar, increases the polarity of the resulting polymer.

The most common commercial types of poly(vinyl alcohol) are the so-called partially hydrolysed grades in which 87–89% of the acetate groups have been replaced, and the completely hydrolysed grades in which 99–100% of the acetate groups have been replaced.

The completely hydrolysed types of poly(vinyl alcohol) are soluble in hot water and have good film-forming characteristics as well as good adhesive properties. These types are available in the high, medium and low viscosity grades depending on the molecular weight of the polymer. Film-forming and adhesive properties of these grades are roughly proportional to viscosity, but the penetration of water solutions in different materials decreases with increasing molecular weight. Poly(vinyl alcohol) containing a small percentage of acetyl groups is cold water soluble; a high percentage results in water insolubility. A variation in saponification number (or degree of saponification) or acetyl content, alters the thermoplasticity, the heat sealing and other important properties.

To convert poly(vinyl alcohol) adhesives in water-resistant products it is necessary to cross-link its linear chains with formaldehyde, boric acid, some salts and other unsolubilizing agents.

Poly(vinyl acetal)s. This group of vinyl polymers is derived from poly(vinyl alcohol) through its condensation with aldehydes, under the influence of heat and an acid catalyst. A molecule of the aldehyde condenses with two hydroxyl groups on alternate carbon atoms to produce a six-membered heterocyclic ring with an acetal structure:

$$\cdots -CH-CH_2-CH-CH_2- \cdots$$
$$O-CH-O$$
$$R$$

Because the reaction conditions are such that not all hydroxyl groups participate in the acetalization process, the resulting microstructure of poly(vinyl acetal)s is generally more complicated:

$$\cdots -CH_2-CH-CH_2-CH-CH_2-CH-CH_2-CH- \cdots$$
$$OH \quad\quad O-CH-O \quad\quad OCOCH_3$$
$$R$$

This indicates that poly(vinyl acetal)s contain not only acetal groups but also hydroxyl and acetate groups as well. By varying the proportion of any of these groups, different types of material and adhesives can be obtained.

Poly(vinyl formal)s and poly(vinyl butyral)s are by far the most important commercial products [5.36, 5.39, 5.40]. Poly(vinyl formal) is used especially for electrical insulation whereas poly(vinyl butyral) is a very important and relatively new adhesive product which is used in the fabrication of safety-glass laminate. It adheres well to glass due to the presence of hydroxyl groups (18–19% —OH) which render it strongly polar, and it also possesses excellent transparency characteristics.

Poly(vinyl chloride) (*PVC*). PVC adhesive is not affected by acids, alkalis or aqueous solutions; even strong oxidizing agents such as chromic and nitric acids have little effect on PVC.

Vinyl chloride is copolymerized with a number of monomers to produce materials which are extensively used as adhesives. Probably the most important are the vinyl acetate–vinyl chloride copolymers; these are used in packaging and generally for fixing PVC to various substrates such as paper, wood, leather, or other plastic materials. A terpolymer in which a small proportion of acrylic acid is included is used for adhesion to metals, ceramics and glass [5.1].

Acrylics. These polymers are prepared from the esters of acrylic acid and methyl acrylic acid, and have the following microstructure:

$$\cdots -CH_2-CH-CH_2-CH- \cdots$$
$$\underset{\displaystyle COOR}{|} \qquad \underset{\displaystyle COOR}{|}$$

$$R = CH_3-; \; CH_3-CH_2-; \; etc.$$

$$\overset{\displaystyle CH_3}{\underset{\displaystyle |}{}} \qquad \overset{\displaystyle CH_3}{\underset{\displaystyle |}{}}$$
$$\cdots -CH_2-C-CH_2-C- \cdots$$
$$\underset{\displaystyle COOR}{|} \qquad \underset{\displaystyle COOR}{|}$$

The acrylic and methacrylic monomers react in free-radical type polymerizations to form polymers which vary with the ester group [5.39]. Acrylic monomers as a class, do not find as wide a use as the methacrylate derivatives.

In the methacrylate series, the most widely used member is the methyl esters. The higher esters give polymers which are generally softer and have a lower softening temperature.

The mass polymerization of this monomer has been carefully studied since large volumes of this resin are consumed to manufacture glass-like plates of remarkable clarity [5.40]. Indeed, the high clarity and excellent light transmission characteristics of poly(methyl methacrylate) coupled with its resistance to sunlight and its low density, make this resin ideally suited to the production of various types of windows in aircraft manufacture.

A further outstanding property of poly(methyl methacrylate) is its good outdoor weathering, in which respect this material is markedly superior to most other thermoplastics.

These resins are soluble in ketone and ester solvents as well as aromatic hydrocarbon–alcohol mixtures.

Copolymerization in the field of acrylic monomers, due to the presence of esteric functional groups, leads to products with good adhesive properties; among the most important are cyanoacrylates and anaerobic adhesives.

Poly(cyano methylacrylate) is capable of providing strong bonds between metals, glass, ceramics and certain plastics.

Alkyl α-cyanoacrylates polymerize rapidly through an ionic mechanism and solidify at room temperature.

The polymer is resistant to many common solvents, including oils, but dissolves in dimethylformamide, and is also attacked by dilute acids, alkalis, hot water or steam. Bonds are serviceable at temperatures as low as $-17°C$ and up to $80°C$ and above. Bonds formed with glass, rubber and wood are generally stronger than the material being bonded [5.41]. It is also important to note that poly(cyanomethacrylate)s are relatively expensive adhesives, however, their use precludes the need for expensive heating and pressure equipment in a fabrication process.

Anaerobic adhesives are obtained with aliphatic diols (e.g. tetramethylene glycol dimethacrylate). These monomers have the property, when in contact with air, of remaining freely fluid in the presence of a free-radical catalyst, but of polymerizing at room temperature when penetrating between closely-fitting surfaces of metal, glass, ceramics and certain polymers. The particular features of the reaction enable it to be used for locking thread grooves, the resulting bond being stronger than can be achieved by locknuts [5.37].

Some vinyl monomers such as vinyl chloride, vinyl acetate and vinylidene chloride have been used in acrylic copolymers as the minor constituent; their presence permits the formulation of different adhesives.

Acrylamide polymerizes to yield poly(acrylamide), a white odourless powder, soluble in water; its solutions are stable over a wide pH range and may be used as an adhesive, thickening agent, or dispersant.

Polyamides. Substituted nylons are soluble in some alcohols, alcoholic mixtures and alcohol–water mixtures and have been used to provide solution adhesives with good rust resistance. Water- and solvent-resistance is poor. Nylon 6, 66 and 610, as well as N-methylmethoxy polyamide, are suitable polymers for solution adhesives.

Soluble nylons may be compounded with many thermosetting polymers to improve their properties to yield such outstanding characteristics as their resilience and peel strength [5.41].

Among the most important copolyamides used as adhesives, the following systems are to be noted:

  (a) branched diamines, cyclohexane dicarboxylic acid, linear aliphatic dicarboxylic acids;
  (b) nylon 6, 12, 66, through transamidation activated with caprolactam, sebacic acid or AH salt;
  (c) copolycondensation in acetic acid medium of caprolactam with ω-amino-undecanoic acid [5.42–5.44].

Strong and versatile protective adhesives and coatings are produced by mixing solutions of some polyamides with epoxy resins. The coatings are tough, hard, flexible and strongly adhering to a wide variety of surfaces such as wood, concrete, steel, glass, and many different kinds of plastics.

Viscous non-fibrous polyamides, which harden with liquid epoxy resins, can be used to provide inks and adhesives (including heat-sealing compositions).

Polycarbonates. Related to polyesters are the polycarbonates which are sometimes used in hot melt adhesives but are mostly used in the joining of materials which themselves are polycarbonates. The most common and simplest adhesive used is that based on bisphenol A:

$$\left(-O-C_6H_4-\overset{\overset{\displaystyle CH_3}{|}}{\underset{\underset{\displaystyle CH_3}{|}}{C}}-C_6H_4-O-\overset{}{\underset{\underset{\displaystyle O}{\|}}{C}}- \right)_n$$

which is joined by solvent bonding; the best bonds are obtained with chloroform and ethane 1,2-dichloride [5.1].

Polyurethanes. The wide ranging technology of the polyurethanes depends upon two major differences in the approach to their formulation. Either a prepolymer can be formed and the ratio of the starting products chosen to give either an isocyanate or a hydroxyl terminated chain which is later reacted in some fashion in a curing reaction, or, some other form of prepolymer is made and terminated with a group containing an active hydrogen to react with a di-isocyanate or one of higher functionality.

Suitable prepolymers are: polyesters, polyethers, polylactones, hydroxy terminated polydienes and a range of naturally occurring ester oils or their derivatives [5.45, 5.46].

The choice exists between using a one-part or a two-part formulation. One-part formulations are based on polyurethanes of very high molecular weight which are still soluble in petroleum–ester mixtures. Two-part adhesives are normally used where high cohesive strength and improved heat resistance are required and these comprise with the prepolymer terminated in some active group and as the second part either a multifunctional isocyanate or an isocyanate terminated polymer of rather lower molecular weight [5.1, 5.47].

Polyurethane adhesives are most widely used for bonding elastomers, fabrics and thermoplastics; they continue to represent an excellent choice for bonding metal to plastics or rubbery adherents. But the moduli of polyurethane adhesives, although high enough, are too low to provide the tensile shear strengths and levels of creep resistance which are required in most metal-to-metal bonding. Polyurethanes have also the drawback of poor chemical and thermal stability, losing bond strength at temperatures above 80–100°C. Some authors have pointed out that the best of the polyurethane adhesives are superior in performance to virtually any other adhesive type at cryogenic temperatures, maintaining their shear strength and tough-

ness at temperatures far below those which cause serious embrittlement of adhesives designed for use at ambient temperatures and above
[5.48].

The properties of these polymers can be varied almost without limit
by varying the ratio of polyester or polyether to di- or tri-isocyanate or
by changing the character of the polyhydric compound. Addition of
water, followed by further isocyanate reaction, provides polyurethane-
polyureas with modified properties [5.49].

Polyurethane two-part adhesives are prepared by mixing the components and applying them to the surface of the adherents. After
assembly the joint is permitted to cure [5.50]. The pot life of the
mixture is short. Excellent adhesion is obtained to wood, paper,
aluminium, glass and many plastics. Satisfactory surface coatings have
also been applied to metal, concrete, plastic and wood.

The adhesive properties of this group of polymers may be attributed
to the polar nature of the macromolecular compounds used in its
formulation. Furthermore, the isocyanates present in the polyurethane
compositions may react with any active hydrogen present in the
adherent or with thin films of water often present on the surfaces of
materials such as ceramics, glass and metals. These types of reactions
result in enhanced keying of the adhesive and permit the attainment of
high bond strengths [5.36].

Cellulose derivatives. Cellulose is the principal component of nearly
all forms of plant life, and as such, is one of the most widely
distributed and available chemical raw materials.

The microstructure of this polymer corresponds to a chain of
β-D-glucose anhydride units linked together by primary valences
through oxygen bridges at the 1–4 positions:

Although cellulose has a high stability to organic solvents and water,
its ether and ester type derivatives being soluble products, may form
various kinds of adhesives. In this way, acetone solutions of cellulose

acetate are used as adhesives, while cellulose acetate-butyrate is used for lacquers and hot melt applications.

Cellulose nitrate is a general adhesive, a pigment binder, waterproof finish and lacquer; it deteriorates when heated and is highly flammable. All forms of cellulose acetate are unaffected by bacteria, fungi or insects.

Among cellulose ethers, both alkali and water soluble types, carboxymethyl cellulose (sodium salt) is used as an adhesive (high viscosity non-staining) for wallpaper, binder for pigments, and it is also used to improve the strength of unfired ceramics. Hydroxyethyl cellulose is used as an adhesive and a paper sizing additive.

Alkali-soluble ethers, as plasticized compositions, can be calendered onto fabric, and can be applied from a solution or a hot melt adhesive (containing mineral oil) for protective lacquers [5.37].

A new carbohydrate polymer has been introduced that has properties useful in the compounding of adhesives [5.51]. It is an excellent building block for remoistening type adhesives. This polymer is compatible with all the usual bases for aqueous adhesives, starches, dextrins, natural gums and poly(vinyl alcohol).

Thermosetting Plastics

Phenolics. The raw materials used in the manufacture of phenol-formaldehyde polymers are phenol, phenol derivatives, such as resorcinol or paratertiarybutyl phenol, cresols, formaldehyde and hexamethylene tetramine [5.36].

Phenolic resins can be a complex formulation of products that include plasticizers, extenders, stabilizers, curing agents, hardeners, etc. The large volume products such as phenolic glues for plywood, are often made as needed, *in situ* at the manufacturing site. Unlike laminating resins, the adhesive must not saturate the wood veneer or chips, but must remain in the glue line of the surface of the ply or chips. In general, the adhesives are made with high viscosity so that they will remain within the glue line. Water present in the adhesive formulation does not need to be removed before curing since it can diffuse through the wood.

Phenolic adhesives based on resorcinol may be cured at room temperature by adding extra formaldehyde in the form of paraformaldehyde. This kind of adhesive may be prepared by first reacting resorcinol with aqueous formaldehyde without catalyst; the reaction is then completed by heating the mixture in the presence of a catalyst [5.52].

The physical properties of resorcinol resins are in the same range as the phenolics, but they have yet to replace them in industry due to their higher cost. They are also considerably darker in colour, which precludes their use for other than materials shaded black or the very dark red and browns.

Certain resorcinol–formaldehyde resin adhesives have been developed which will cure at room temperature with the addition of a suitable hardening agent. This type of adhesive has an indefinitely long shelf-life because it does not have the tendency to harden unless the hardener is added. However, once the hardener is added, the reaction then progresses quite rapidly such that within 4–5 h at 24–26°C the mixed adhesive is permanently set. Different curing temperatures can be obtained by adjusting the proportions of resin and hardener.

These adhesives were originally developed for the bonding of wood members at room temperature where higher temperatures could not be obtained. They have since been used in the bag moulding industry and in the manufacture of laminated wood beams and other types of wooden articles where higher temperatures had been employed. Good results have also been obtained in bonding other materials such as plastics, glass and ceramic wares. In the field of metal bonding and particularly the bonding of metal to other materials, these adhesives produce excellent bonds if the metal surface has been adequately prepared with a special adhesive primer.

Aminoplastics. Commonly called 'amino resins', the two main groups of polycondensation polymers are:

—urea–formaldehyde;
—melamine–formaldehyde.

The development of these resins occurred in conjunction with the growth in production of the inexpensive base polymers urea and formaldehyde.

About 40% of urea–formaldehyde polymer production are used for plywood bonding and other adhesive uses. The other applications are in moulding, textile treating and coating, paper treating and coating, and protective coatings, to name a few.

The urea–formaldehyde resins are used to produce both the so-called 'cold-setting' adhesives and the various grades of powders and water solutions requiring heat for final bonding. The excellent storage life and simple application methods have made these adhesives extremely important. The cold-setting type has a catalyst incorporated in the powder, which does not become active until water is added. The inclusion of catalyst makes it possible to use this adhesive at a temperature of 24°C, but it should not be used for lower temperatures, as the final cure is too slow.

Because of the similarities in the reaction of urea and melamine with formaldehyde, the processes used for the production of urea resins are readily adaptable to the production of melamine resins. Generally, the reaction of melamine with formaldehyde is a little more difficult to handle; it is regulated by controlling conditions of pH and temperature.

The requirements for a resin adhesive call for a soluble type, preferably soluble in water, which has adequate viscosity to permit spreading without penetration.

Although superior properties are obtained through the use of melamine resins, those used in the largest volume are urea resins and urea–melamine combinations.

Melamine and phenolic adhesives have excellent water resistance. Urea–formaldehyde adhesives are lower in cost, but somewhat more water sensitive. Often the best all-round balance between cost and performance is achieved with a blend of urea–formaldehyde and melamine–formaldehyde resins.

Compositions containing all three components, phenol, urea and melamine have been proposed.

It is not uncommon for urea–formaldehyde adhesives to contain a modifier such as benzyl alcohol or furfural cross-linked and better able to relieve stresses imposed by shrinkage due to cure and loss of water.

Urea–formaldehyde adhesives may also be mixed with poly(vinyl acetate) emulsion; such a formulation will combine the short clamping time of the vinylic adhesive with the heat and water resistance of the cross-linked urea–formaldehyde adhesive [5.52].

Epoxies. The amounts of epoxy polymers manufactured and consumed are insignificant in comparison with the quantities of polyethylene, polystyrene and other polymers manufactured today. Even when polymers such as polyesters and polyurethanes are considered, the amounts of epoxy resins produced are comparatively small.

However, in terms of variety and breadth of application, polyepoxides are surely superior to all other polymers.

Polyepoxides can be used as adhesives, sealants, casting resins, dipping compounds, moulding powders, paints and varnishes, powder coatings, including the matrix in reinforced composites [5.53].

Most commercial epoxy resin intermediates are derived from epichlorohydrin and bisphenol A. The basic process involves the reaction of epichlorohydrin and bisphenol A at 65°C in the presence of a caustic soda solution. Epoxy prepolymers can be cross-linked either catalytically or with reactive curing agents.

Among the first group of these reactive curing agents are Lewis acids and bases. The most important curing agents are: carboxylic acids, acid anhydrides, primary and secondary amines, dicyandiamide, amido–amines, substituted imidoazoles, ketimines, mercaptan terminated compounds, isocyanates, etc.

The physical properties of a cured epoxy resin depend upon many factors, including type and amount of curing agent, cure history, type of filler, type of reinforcement, and other modifiers. The relationship between the cure and the resultant properties has been the object of a considerable amount of research work in the last decade [5.54–5.58].

Polyepoxides have proven to be very versatile adhesive materials because of a combination of properties such as:

—excellent adhesion to a wide variety of different surfaces;
—chemical stability due to ether linkages in the polymer chains;
—outstanding electrical properties;
—curing without the evolution of noxious by-products;
—curing at low or moderate temperatures;
—low shrinkage, optical clarity;
—good abrasion resistance.

Some of the properties which have made epoxy resins so suitable for surface coatings have also made these resins applicable for use as adhesives. Since the formulations cure with little shrinkage, they bond well to metals, wood, glass and ceramics.

The chemical stability of the adhesive line prevents breaking of the bond, except under the most severe acidic conditions. Flexibility and toughness are also important when considering the properties of the epoxy adhesives.

A significant increase in the use of epoxy adhesives in the civil engineering industry has taken place over the last few years. This has been chiefly due to two factors:

(1) The increasing realization that the bond strengths of epoxy adhesives are considerably greater than the cohesive strength of concrete, which is the most common load-bearing construction material used in civil engineering. The tensile strength of good quality concrete varies between 1·75 and 5 MN/m^2 (253·7–725 psi) according to the mix used, whereas the tensile strength of epoxy adhesives exceeds 5 MN/m^2 (725 psi) and is often much greater, in the range of 56 MN/m^2 (8120 psi).

(2) The ability to make significant savings in time by using an adhesive. The rate of development of mechanical strength of epoxy adhesives is therefore important in this context, and it can vary widely depending upon the epoxy system used, the ambient temperature and other factors. However, it is certainly faster than the rate at which concrete develops comparable proportions of its final strength.

The uses of epoxy adhesives and grouts can be classified broadly into:

(a) Remedial work such as strengthening and repair of existing structure (concrete crack repair, bonding concrete to concrete, bonding reinforcements).

(b) New work, where the use of adhesive was envisaged at the design stage [5.53].

Various modifiers, bituminous or synthetic polymers, have been used to improve the properties of polyepoxides. Furthermore, different thermoplastics and thermosetting polymers including elastomers have also been incorporated into polyepoxides to modify their properties [5.59]. A polyamide soluble in an ethanol–water mixture is used in epoxy–nylon film adhesives to obtain high peel strength as well as good heat resistance. A thermoplastic polyurethane-modified epoxy resin has been developed which is reported to give better peel strength at cryogenic temperatures than that obtained with epoxypolyamide.

Poly(vinyl acetal)s show good compatibility with epoxy resins and improve the peel strength of adhesives. Among the thermosetting resins, phenolics have long been blended to obtain heat resistant adhesives. Epoxy–nitrile rubber blends yield high-peel-strength adhesives [5.52].

With carboxyl-terminated butadiene–acrylonitrile copolymer, the thermal characteristics of an epoxy have also been modified [5.60].

Unsaturated polyesters. Unsaturated polyesters form the basis of glass reinforced plastics used for a multitude of construction purposes and are also marketed as adhesives.

Cross-linked polyesters have good heat stability, showing little weight loss up to about 200°C. They are also resistant to a wide range of organic solvents but they are attacked by chlorinated hydrocarbons, esters and ketones. The ester groups provide sites for hydrolytic attack and strong alkalis cause appreciable degradation [5.52].

Polyaromatics. In recent years a new group of carbon polymers, having inherently rigid chains, has been developed. Their synthesis involves the introduction of highly stable, rigid, aromatic or heterocyclic ring systems into the polymer chains [5.41]. A listing of the polymer types being investigated or in limited commercial use is given in Tables 5.2 and 5.3 [5.36].

Table 5.2.
Heat resistant adhesives—all carbon rings [5.36]

Polymer	Properties
Polyphenyls	Decomposes at 530°C; infusible, insoluble polymer
Poly-*m*-phenoxylene	Decomposes close to 500°C; heat cures above 150°C to elastomer
Polyphenylene sulphide	Melts at 270–315°C; cross-linked polymer stable to 450°C in air
Polybenzyls; polyphenethyls	Fusible, soluble and stable at 400°C; low molecular weight
Poly-*p*-xylene	Melts above 520°C; insoluble; capable of forming films
Polyterephthalamides	Melting points up to 455°C
Polysulphanyldibenzamides	Melting points up to 380°C
Polyhydrazides	Dehydrate from 200 to over 400°C to form polyoxadiazoles
Polyoxamides	Some melting points above 400°C; give clear, tough, flexible films
Phenolphthalein polymers	Melting points from 300 to over 400°C; formable into fibre and film
Hydroquinone polyesters	Soluble polymers with melting points of 385 to over 400°C
Poly(hydroxybenzoic acids)	Films melt at 380–450°C; stable to oxidation but not to hydrolysis

Table 5.3.
Heat resistant adhesives with nitrogen, sulphur, or oxygen in rings [5.36]

Polymer	*Properties*
Polyimides	Commercial film, coatings and resin stable up to 600°C, continuous use up to 300°C
Polybenzothiazoles	Cured polymer soluble in concentrated sulphuric acid. Stable in air at 600°C
Polybenzimidazoles	Developmental laminating resin fibre and film, stable at 300°C in air
Polyquinoxalines	Stable in air at 500°C; tough, flexible resins make film, plastic
Polyphenylenetriazoles	Thermally stable to 400–500°C; make film, fibre coating
Polydithiazoles	Decompose at 525°C; soluble in concentrated sulphuric acid
Polyoxadiazoles	Decompose at 450–500°C; can be made into fibre or film
Polyamides	Stable to oxidation up to 500°C
Poly(vinyl isocyanates)	Soluble polymer that decomposes at 385°C; prepolymer melts above 406°C

The leading candidates for adhesives are those that can be formed initially as a soluble prepolymer which can be subsequently cured by thermally induced condensations of reactive groups on the prepolymer chains. These types include:

—polybenzimidazoles;
—polyimides;
—polyoxadiazoles.

Polybenzimidazoles are used especially for bonding metals and more particularly for the bonding of honeycomb sandwich type metal assemblies currently used in the aircraft industry. Copolymers of polybenzimidazoles have been applied in solution for heat-resistant fibrous laminations and adhesives.

Properties of polybenzimidazole and polyimide adhesives are compared in Table 5.4 with those of epoxy–phenolic adhesives.

Heat resistant adhesives are still expensive and must be cured at a much higher temperature (processing temperature above 316°C) than standard systems, but have properties comparable to 'classic' high temperature adhesives, including good low temperature properties.

Table 5.4.
Heat resistant adhesives—General properties [5.36]

Property	Polyimide	Polybenzimidazole	Epoxy–phenolic
Joint cure	2 h @ 300°C 140 N/cm² (203 psi)	1 h @ 316°C 140 N/cm² (203 psi)	1 h @ 177°C 35 N/cm² (51 psi)
Glueline thickness (mm)	3–6	3–5	8–10
Tensile shear strength at 25°C, N/cm² (psi)	1 540 (2 233)	2 800 (4 060)	2 800 (4 060)
Tensile shear strength after 7 days in 71°C water, N/cm² (psi)	910 (1 319)	1 330 (1 928)	1 820 (2 639)
Tensile shear strength at 41°C, N/cm² (psi)	2 540 (3 654)	2 800 (4 060)	2 800 (4 060)
Peel strength at 25°C (climbing dram, metal–metal), N/cm² (psi)	35 (50)	86 (124)	>77 >(111·7)

Structural Adhesive Bonding

There is no generally accepted definition of a structural adhesive. Kinloch [5.12] considers structural adhesives as adhesives based upon monomer compositions which polymerize to provide high modulus, high strength adhesives between relatively rigid adherends, so that a load-bearing joint is constructed. In the construction industry, not only metals, but wood, concrete and other building materials are also frequently bonded using adhesives.

This group of adhesives is often based on phenolic, epoxy or acrylic prepolymers (oligomers) which polymerize or cure to provide a highly cross-linked adhesive, e.g. thermosetting polymers. Their main advantages are:

(a) low viscosity to flow over an adherend surface without the need to employ solvents;
(b) being polar polymers assists in removing atmospheric contamination which is present on metallic surfaces and increases the degree of intrinsic adhesion.

The building industry is now applying 'structural bonding', a method of structural fabrication originating in the aircraft industry. Although not universally successful, factory application has worked, for example, in the case of timber beams for use as primary structural members. It has also been used in the manufacture of prefabricated panels where brick slips, or other decorative facings, are bonded to epoxide resin 'concrete' backings. On-site application of the technique has generally been unsuccessful since the substrate preparation is critical and the length of time that the adhesive requires to attain its full strength is dependent on the temperature. At low temperatures it may never reach its maximum strength [5.61].

A rational basis for structural engineering must be based on the ability to predict the loads and hence the stresses which are likely to be encountered in service.

Adams & Wake [5.31] show (Fig. 5.9) some typical classifications of joints which are commonly found in current engineering practice. Any joint occurring in practice is designed to carry a given set of loads; in most cases the substrates are loaded in tension. The subsequent loads on the adhesive are then a function of the geometry of the joint. The single-lap joint (Fig. 5.9(a)) is one of the most commonly occurring joints and is the configuration most often used for testing adhesives. However, the stress state is more complex. The loads in the single lap

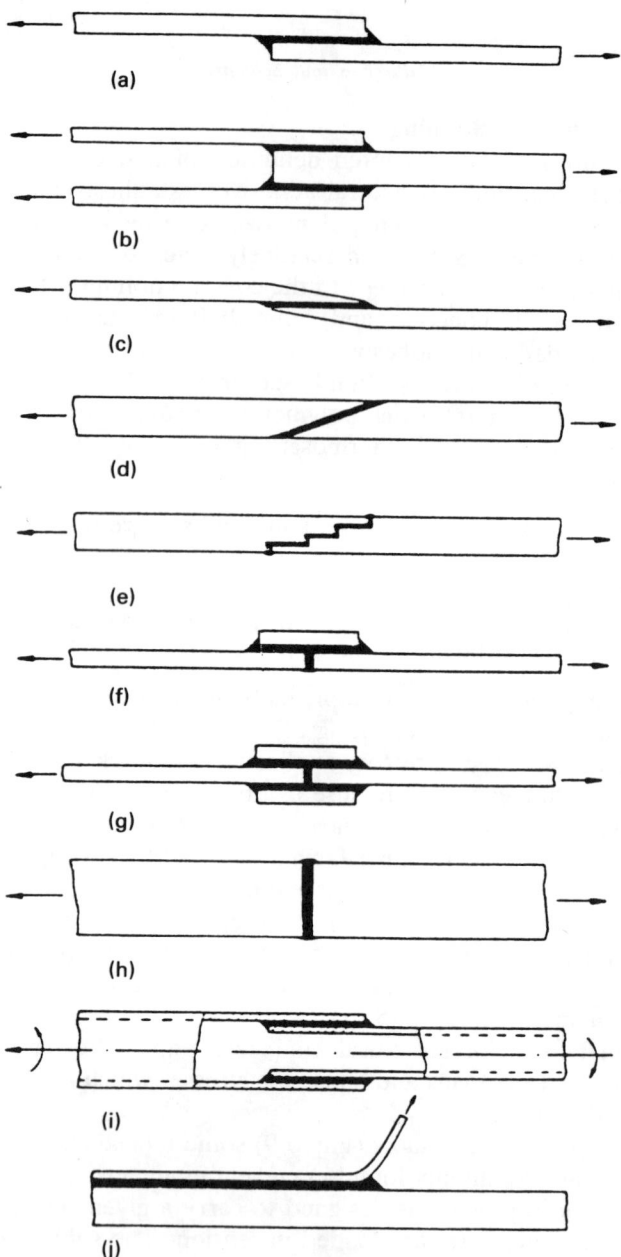

Fig. 5.9. Some common engineering adhesive joints [5.31]. (a) Single lap; (b) double lap; (c) scarf; (d) bevel; (e) step; (f) butt strap; (g) double butt strap; (h) butt; (i) tubular lap; (j) peel. (Reprinted with permission from *Structural Adhesive Joints in Engineering* by R. D. Adams & W. C. Wake, 1984. Copyright Elsevier Applied Science.)

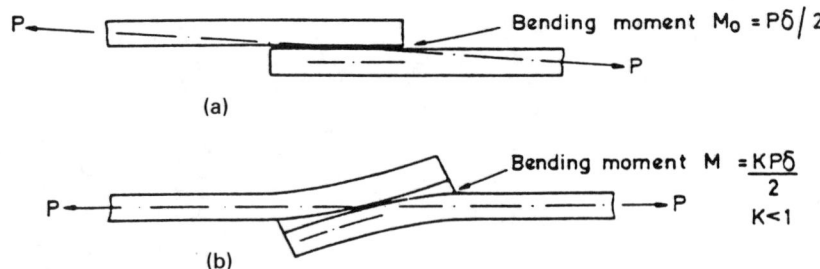

Fig. 5.10. Illustrating a way of representing the Goland and Reissner bending moment factor geometrically [5.31]. (Reprinted with permission from *Structural Adhesive Joints in Engineering* by R. D. Adams & W. C. Wake, 1984. Copyright Elsevier Applied Science.)

are not colinear. A bending moment must therefore exist and the joint will rotate (Fig. 5.10). Clearly the adhesive layer will no longer be solely in shear, but will have tearing stresses at the end of the joint. Furthermore, the adherents are no longer in single tension, but are bent. It is quite possible that both the adhesive and the adherent may have become plastic, particularly in the highly-stressed region.

It has been said that by using a double-lap joint the bending effect can be eliminated. While it is true that there is no gross rotation of the joint, it is no more than a back-to-back arrangement of two single lap joints, each of which has its own system non-colinear in the region of the joint. The other configurations of joints shown in Fig. 5.9 are largely designed to improve load-transfer so as to minimize stress concentration and peel, the last one being the most important enemy of the joint designer. Whereas the adhesive in the single- or double-lap joint is certainly at low loads, largely in shear, that in Fig. 5.9(j) is experiencing principally transverse, tearing loads which we call peel. Indeed, the configuration shown in Fig. 5.9(j) is that used as the basis of a number of peel tests.

Some physical properties of a modern rubber-toughened structural epoxy adhesive cured with dicyandiamide (DICY) are shown in Table 5.5. There are some noteworthy features:

(1) All the joints failed by cohesive fracture through the adhesive, except for the carbon fibre reinforced plastic (CFRP)-steel

Table 5.5.

Representative physical properties of a rubber-toughened epoxy-based structural adhesive measured in bulk and in various adhesive joints [5.31]

Test[a]	Adherends	Property	
Bulk adhesive			
Flexure	—	Modulus	2·8 GPa (406×10^3 psi)
	—	Failure stress	74·5 MPa ($10·8 \times 10^3$ psi)
	—	Failure strain	2·7%
		Glass transition temp.	120°C
Adhesive joints[b c]			
Torsional shear	Aluminium alloy	Failure stress	61 MPa ($8·84 \times 10^3$ psi)
Axially-loaded butt joints	Steel	Failure stress	58 MPa ($8·41 \times 10^3$ psi)
Single-lap joint, in tension	Aluminium alloy	Failure load	9 kN (2 016 lbf)
		Failure stress	28 MPa ($4·06 \times 10^3$ psi)
	Steel	Failure load	12·3 kN (2 755 lbf)
		Failure stress	38 MPa ($5·51 \times 10^3$ psi)
Double-lap joint, in tension	CFRP–steel	Failure load	24 kN (5 376 lbf)
		Failure stress	6·9 MPa (1 000 psi)
	CFRP–steel; tapered adherends	Failure load	80 kN ($17·9 \times 10^3$ lbf)
		Failure stress	19·7 MPa ($2·85 \times 10^3$ psi)
Peel tests			
90° peel	Aluminium alloy	Peel strength	5 kN/m (342 lb/ft)
	Steel	Peel strength	0·6 kN/m (41 lb/ft)
135° peel	Aluminium alloy	Peel strength	4 kN/m (273 lb/ft)
Precracked, TDCB[d]	Steel	Fracture energy, $G_I{}^c$	0·9 kJ/m² (0·078 Btu/ft²)
Precracked, compact shear	Steel	Fracture energy, $G_{II}{}^c$	2·2 kJ/m² (0·191 Btu/ft²)

[a] Tests conducted at 23°C and a moderate rate.
[b] See Fig. 5.11.
[c] All joints by cohesive fracture through the adhesive, except for the CFRP–steel double-lap joints which failed in the CFRP adherend.
[d] TDCB: tapered-double-cantilever-beam specimen.

Reprinted with permission from *Structural Adhesive Joints in Engineering* by R. D. Adams & W. C. Wake, 1984. Copyright Elsevier Applied Science.

double-lap joints which failed by interlaminar fracture in the CFRP adherend. Thus, in these imaged joints, interfacial failure between the adherend and the adhesive was never recorded. An interfacial locus of joint failure is usually only found after environmental attack.

(2) It may be seen from the data of Table 5.5 that very high joint strengths may be attained, but the actual values are greatly dependent upon the detailed joint geometry and the substrate material. This arises, of course, from the stress distributions in loaded adhesive joints generally being non-uniform [5.31].

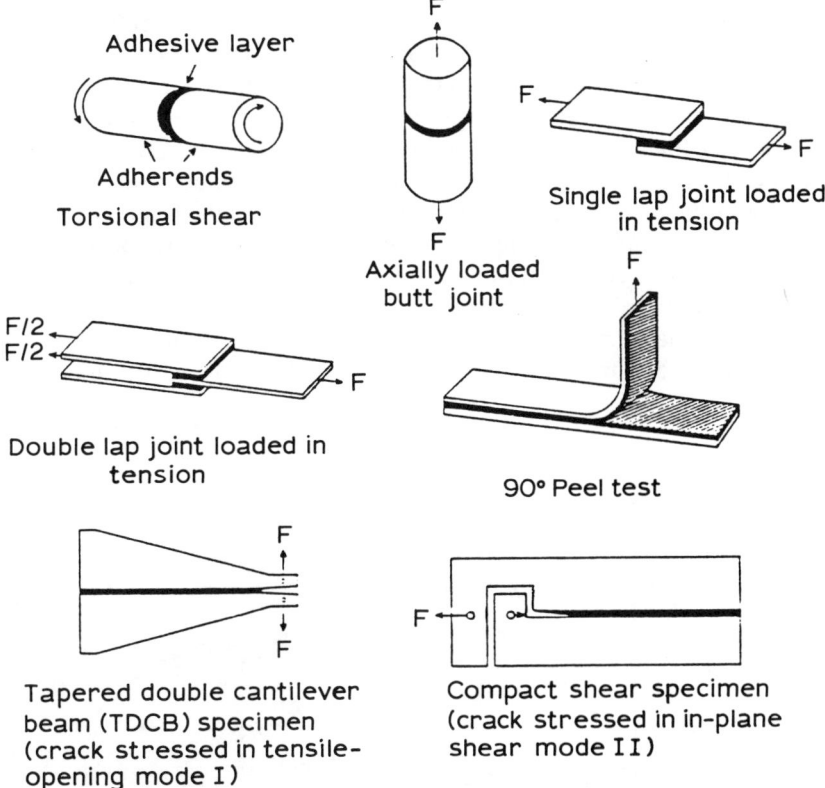

Fig. 5.11. Sketches of joint geometries employed to obtain the data shown in Table 5.5 [5.31]. (Reprinted with permission from *Structural Adhesive Joints in Engineering* by R. D. Adams & W. C. Wake, 1984. Copyright Elsevier Applied Science.)

The peel mechanics and fracture criteria on peeling and the peel rate are discussed in the literature [5.35, 5.62–5.64].

It is known that the mechanical properties of uncross-linked and unfilled polymers change drastically at or around their T_g. Modern structural adhesives are made tougher to improve their low temperature performance and cross-linked to extend their usefulness in the upper temperature ranges [5.15]. As shown in Fig. 5.12, below their T_g they are generally brittle; much above their T_g they are thermoplastic, and under stress show creep and flow.

Pocius *et al.* [5.65] undertook a 10-year study with a view to making qualitative correlations of the durability of metal-to-metal adhesive bonds as a function of adhesive type (phenolic, epoxy, polyurethane), temperature of cure, type of adherend, type of surface preparation, and the amount of load placed on the specimen. The authors concluded that the high temperature curing adhesives have a better durability than the room temperature cures. This is likely to be due to a higher cross-linked density capable of being achieved with the high temperature cures and the additional effect of toughening which would resist stress cracking. The phenolic adhesives seem to be especially good with respect to durability. Phenolics are known to bond very strongly to aluminium oxide.

The primary problem with epoxy adhesives appears to be the swellability of the resin in moisture. This is likely to be due to the

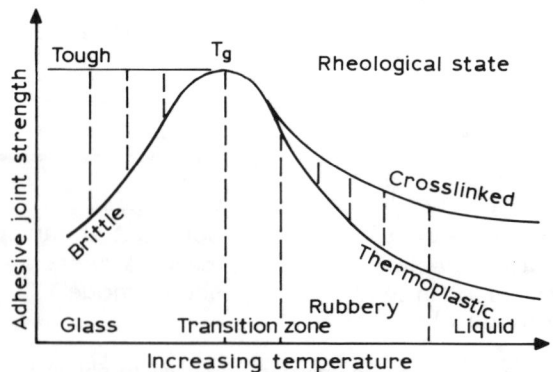

Fig. 5.12. Schematic diagram of bond strength and rheological properties [5.15]. (Reprinted with permission from *Adhesives Age,* **16**(10) (1979) 34. Copyright Communication Channels Inc.)

preponderance of hydrophilic groups in the cured epoxy adhesive. Polyurethanes have the additional problem that metal ion catalysed hydrolysis of the urethane bond can take place.

Drake & Siebert [5.66] have presented a review of the morphology, fracture toughness, viscoelastic effects, durability and fracture mechanisms for formulating epoxy structural adhesives.

The adhesive fracture of elastomer-modified epoxy polymers and commercial structural adhesives has been determined in combined shear and cleavage loading and found to be lower than the corresponding opening-mode (pure cleavage) fracture energy by as much as a factor of ten [5.67].

Immersion in water has been shown to considerably reduce both the strength and durability of structural joints consisting of different substrates. The mechanism for the environmental failure of these joints is the displacement of adhesive on the substrate surface by water. The kinetics of this mechanism indicate that the rate of diffusion of water to the interface is an important factor [5.68].

SEALANTS

Introduction

Modern sealants are composed of pigmented or unpigmented synthetic elastomeric polymers, which, in the non-cured state, constitute pourable or easily extrudable putty-like mastics. The consistency of sealants in the non-cured state can be adjusted according to the requirements of application, e.g. self-levelling for horizontal joints and non-sagging for vertical joints. In the cured state, these sealants become transformed into elastomeric materials.

Sealant is the name now given to the new materials that have, in general, replaced putties and caulks. They contribute significantly to keeping out rain, air and dust, and even improve the thermal performance of a wall. Sealants are developing in conjunction with the construction industry's increasing use of large panels. Movements of these large units caused by changes in temperature and humidity necessitate new types of putties, as the old ones are not able to cope with the larger joint movements, especially after long periods of service. The new polymeric materials, known as sealants, have to provide material continuity between building elements while the joint may change in dimension by as much as 50% as it opens in winter and closes in summer.

The term sealant refers specifically to a liquid, paste, or dough-like solid which, when suitably applied to neighbouring components of a structure, forms a resilient seal. The sealant may or may not act as an adhesive, but it is sufficiently resilient to accommodate relative movement of the structural members caused by expansion, contraction, vibration, and other differential stresses.

The characteristics that define a sealant are:

—it should be easily deformable during application and in service;
—to absorb cyclic movement without permanent distortion it should possess recovery properties;
—it should not flow from the joint;
—it should adhere to a wide range of surfaces;
—its properties should not vary greatly across the service temperature range;
—it should be durable [5.69].

Sealants are marketed in three main forms:

—putty-like mastics;
—non-cured tapes;
—cured gaskets.

The sealants in mastic form are employed in one- and two-component formulations. The main advantages of one-component sealants are that they require no mixing before application, and have an extended pot life. The consumption of one-component sealants is growing.

Curing of one-component sealants depends upon the moisture content of the atmosphere; during dry periods, the curing process may be slowed down. When applied, the one-component sealants form a skin in a few hours and cure through to a rubbery elastomeric material within several days.

From the perspective of the rubber chemistry, there is a natural inclination to classify sealants in terms of their chemical composition or physical properties. Hence from this point of view one can distinguish the following sealant groups:

—silicones;
—polyurethanes;

—polysulphides;
—polymercaptans;
—chlorosulphonated polyethylenes;
—polyacrylics;
—polychloroprenes;
—butyl rubbers;
—halogenated butyl rubbers;
—polyisobutylenes;
—polybutenes; and
—drying and non-drying oil based caulks.

The most significant classification is based on the balance of properties that the architect or design engineer would consider, separately and in combination, when specifying the requirements for a particular joint. These factors are:

—joint movement;
—durability;
—cost.

As already mentioned, building sealants are a group of macro-molecular compounds whose function is to prevent the passage of moisture, air, dust, gases and heat through building joints. Since 70% of sealant use is for exterior applications, sealants prevent the exchange of outdoor climatic conditions with those found indoors. Thus these polymers are susceptible to weather-induced degradation (Fig. 5.13).

Sealants may be in the form of viscous liquids, mastics, tapes or preformed gaskets. As viscous pourable fluids they are used in horizontal joints where they fill the voids between the substrates and thus provide better adhesion than would be possible with gun-applied sealant. Mastics, which are applied by means of a gun, trowel or knife, contain thixotropic agents to control the flow of the sealant and prevent sagging and flowing out of vertical or near vertical joints. Tapes, cured or uncured, are used as bedding compounds or in glazing work. Preformed cured gaskets are extruded or foamed, and are available in a multitude of shapes and forms. They are installed by forcing the gasket into the joint opening; a tight seal is dependent on the pressure of the gasket against the joint walls. They are generally fabricated from high recovery elastomers.

Exterior sealants can act in either one-stage or two-stage weath-

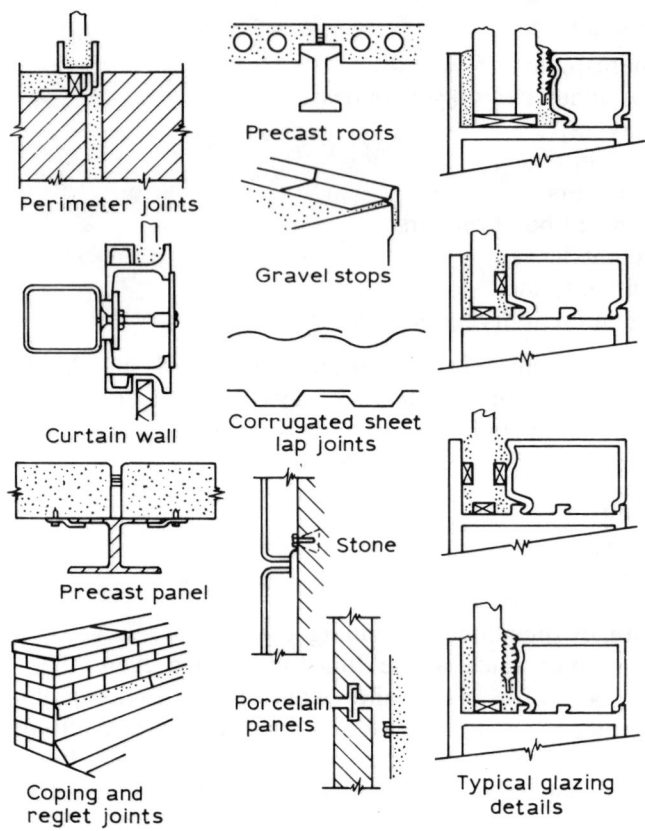

Fig. 5.13. Typical sealant applications [5.2]. (Courtesy of Tremco Manufacturing Co.)

erproofing joint configurations. The first one uses the seal at the exterior side of the joint, as both a rain and an air seal. Two-stage weatherproofing uses the seal at the interior face of the joint, where it acts as an air seal only (Fig. 5.14). A non-airtight rain seal is used in an exterior location of the joint. This procedure protects the polymeric sealant from weathering factors, extreme cold, and wet/dry cycles.

The majority of joints fall into the following categories: working, non-working, butt and lap joints. Working joints change size or shape with the relative movements of the substrates (control joints, expansion joints, etc.). Non-working joints have movement that is mini-

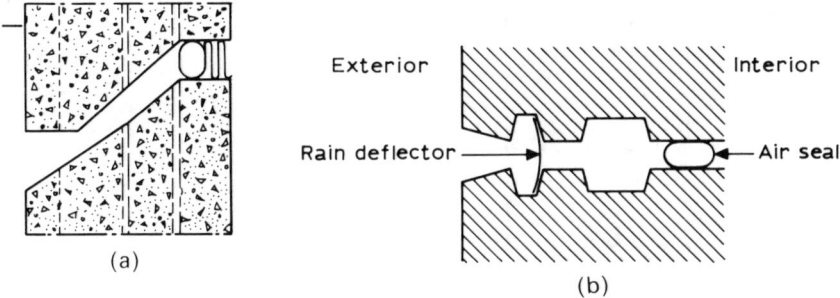

Fig. 5.14. (a) Horizontal joint-offset panel edge with rain seal. (b) Vertical joint with rain seal and air seal [5.2]. (Joint designs by T. Isaksen, Norwegian Building Research Institute.)

mized or eliminated. Butt joints subject the sealant to alternative tensile and compressive stresses. Lap joints subject the sealant primarily to shear stresses (Fig. 5.15) [5.2].

The shape and dimension of the sealant cross-section is of primary importance in determining the movement induced stresses and strains on the sealant and on the substrate. The optimal geometry of a seal is twice as wide as it is deep, but a ratio of one is more commonly used

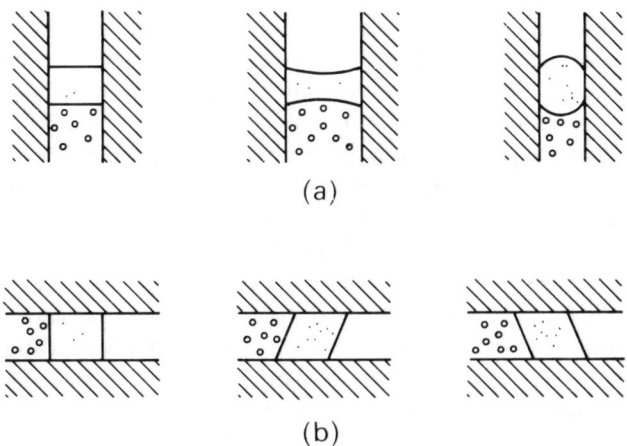

Fig. 5.15. Deformation of sealant beads in (a) butt joints, and (b) lap joints [5.2]. (Design by T. Gjelsvik, Norwegian Building Research Institute.)

Polymeric Building Materials

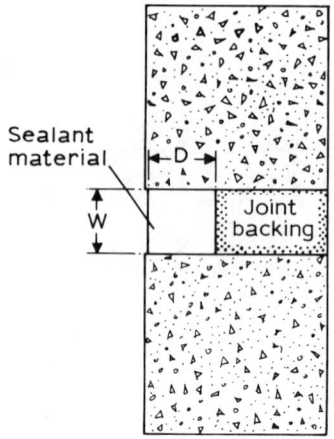

Sealant material

W

Fig. 5.16. Joint with shape factor of one $(W/D = 1)$ [5.2]. (Reprinted with permission from *Construction Sealants and Adhesives* by J. P. Cook, 1970. Copyright John Wiley.)

in practice to reduce the frequency of failure (Fig. 5.16). Elastic sealants must be bonded to only two opposite sides of the joint, allowing the bottom surface to deform freely.

The height and mass of the structure, wind loads, moisture absorption, amount of shade, compass orientation, ambient temperature, and colour of substrate dictate the movement that building joints must undergo. The higher the intensity of movement, the more elastic the joint must be.

The properties and performance of a sealant are based on its components. Rubbery sealants that have high recovery also have lower tear resistance. Deformable sealants have instantaneous elasticity under the short-term loads, but will creep or flow under long-term loading. These sealants include polysulphides, polymercaptans, butyls, solvent based acrylics, and latex caulks. As noted earlier, preformed gaskets are high recovery elastomers that are precompressed into the joint, hence the seal is dependent on the force exerted on the joint by the gasket. As the joint opens, due to substrate contraction, the gasket may fall out once compression is reduced to zero. Adhesives may be used to allow gasket operation under tensile loading. Tapes can be high recovery (cured) or low recovery (uncured), and the stresses under load vary according to the type of tape being used.

Failures of the sealant depend on the type of sealant used, the installation, and service conditions. Adhesive failure, which is the loss of bond between the sealant and its substrate is most common in mastic type sealants. Mastic sealants are also subject to cohesive failure, which is denoted by failure within the body of the sealant, and spalling failure, where the overall strength of the sealant exceeds the cohesive strength of the substrate. Spalling failure occurs in instances where a stiff, high recovery polymeric sealant has been applied to a low strength substrate. Deformable sealants can fail with a change in sealant shape due to flow. Failure can also occur if there is a large joint movement before the seal is fully cured.

Preformed gaskets fail due to a loss of interface pressure between the seal and its substrate. The seal must be designed so that pressure continues to be exerted when the joint is at its maximum expansion. When gaskets are made with a high quality elastomer, failure rarely occurs within the polymer, however, the gasket may crack due to weathering (ageing), a trend generally attributable to poor rubber compounding. Foamed gaskets, usually closed cell urethane or neoprene (polychloroprene), are less rigid than extrusions. They exhibit little or no lateral spread when compressed and do not extrude from the joint. Failure is often due to flow and stress relaxation which occur after long periods of time.

Sealants can fail due to ageing and weather exposure. This type of failure is often characterized by discoloration and crazing and/or stiffening of the sealant surface and is a result of the individual or combined effects of solvent evaporation, ozone attack, migration of plasticizers, ultra-violet radiation, water immersion, etc.

To prevent failure and promote certain performance characteristics, additives such as adhesion promoters, fillers, pigments, plasticizers, thixotropic agents, etc., can be introduced into the polymer. To promote better performance, external accessories used include: primers, release agents and backup materials. Elastomeric sealants usually consist of a base polymer, filler (to control consistency and lower the cost), plasticizer (to modify the hardness and modulus), and a curing agent.

Fillers can improve the physical properties of a sealant. Typical fillers are: carbon black, calcium carbonate, talc, clays and ground silica. For non-sagging sealants, thixotropic agents such as bentonite, asbestos fibre, or colloidal silica are used. Fillers may be classified as either reinforcing agents or consistency regulators, of which the former includes carbon blacks and ground silica and the latter, clays, talc and calcium carbonate.

Plasticizers lower the modulus of elasticity and increase ultimate elongation. Certain plasticizers can also improve the low temperature flexibility. Plasticizers also aid in releasing trapped air in the sealant rendering it more workable. Also, they permit higher filler loadings to be achieved for the same mass of base polymer.

Sealants cure by a number of different chemical systems including: catalytic action, moisture absorption, or solvent evaporation. Two-component sealants are cured by catalytic action whereas one-component sealants are often cured by moisture absorption or solvent release. Solvents increase the gunnability of one-component sealants, however their levels are kept low to avoid excessive shrinkage. The principal solvents used are: toluene, xylene, naphthas, petroleum derivatives and water.

Pigments are additives used to match the colour of the sealant to the substrate, but they can also change a sealant's properties. Carbon black and titanium oxide are the most commonly used pigments although carbon black also serves as a reinforcing agent.

Primers provide better adhesion of the sealant to the substrate. The more elastic the sealant, the greater the need for a primer. The primer provides a better wetting of the surface, and the sealant adheres to a thin film of primer. The primer must be compatible with both the sealant and the substrate.

Backup materials are used to control the depth of the joint, and must be unaffected by any solvent that is contained in the sealant. Typical backup materials include: neoprene, polyurethane, poly-ethylene, etc.

Release agents are placed on top of the backup material to prevent the adhesion of the sealant to the backup. Lubricant-adhesive additives are often used as release agents in conjunction with preformed gaskets. The lubricant helps ease the gasket into the joint, while the adhesive helps maintain the gasket in its proper position.

Properties of some Sealants

Silicones

The silicone polymers exhibit low interchain attractions as well as having a highly flexible backbone. These characteristics contribute to the ability of silicone products (rubber, sealant, etc.) to remain elastic down to very low temperatures, due to their high molar volume, low solubility parameter, low surface tension and high compressibility.

The high thermal stability of polydimethylsiloxane rubber is a reflection of the good stability of the bonds involved (which comprise —Si—O—, —Si—C—, —Si—H—) and lack of weak bonds, rather than any special outstanding feature [5.70].

The silicone products were first developed for high temperature insulation and for encapsulation of electrical parts, but due to their relative inertness, new applications emerged in other fields (e.g. biomedical). Today silicone products range from liquids to reinforced plastics. In the case of liquids, it is necessary to use highly purified dialkylchlorosilanes and monofractional trialkylchlorosilanes as chain stoppers to keep the viscosity of the liquids constant. These fluids have the advantage of oxidative stability, low freezing point, and a small temperature coefficient of viscosity.

Silicone elastomers are produced from linear polymers based on polydimethylsiloxane by partial cross-linking. The cross-linking mechanism used includes free radical initiated cross-linking, incorporation of trifunctional monomers, and incorporation of unsaturated groups for further reaction Usually, 10% of the methyl groups in polydimethylsiloxane are replaced by vinyl groups, and these types of copolymers not only allow cross-linking with peroxide but may also be used as substrates for grafting. It has been established that 3% silicone rubber is equivalent to 8% polybutadiene in toughness, probably due to its low T_g ($-130°C$).

Bouncing putty is based on a polydimethylsiloxane polymer modified with boric acid, additives, fillers, and plasticizer, to give a product that on shock behaves like an elastic material but flows as a viscous liquid on slow application of pressure [5.71].

Silicone sealants are high quality, high recovery elastomers which are stable in both the cured and uncured states. They are easy to work with and provide a very neat joint. Their outstanding property is high recovery, and after 1 year of service the sealant may show as much as 98% recovery. The drawback of high recovery is the inherent low tear resistance, which is about half the value of the tear resistance of polysulphides or polymercaptans [5.72]. Once the deformed sealant is cut or punctured, the existing internal stresses will extend a tear rapidly, leading to sealant failure. Silicones require no solvent for workability, thus curing shrinkage is negligible due to almost 100% solids.

Silicone elastomers consist of a polyorganosiloxane (usually polydimethylsiloxane, due to its low cost and wide range of attainable properties), additives which provide strength and other desirable properties, and a cross-linking system. Sealants are classified as room temperature vulcanizing (RTV), because they cure at room temperature and do not require heat to cure, as is the case for heat-cured rubbers [5.73]. The polymer chain used in heat-cured rubber has a DP of 7000, while the DP of RTV elastomers is 1000. The shorter chain length is necessary because lower viscosity is required for RTV applications, but the tensile strength and elongation are lower than for heat-cured rubber. Varying the organic groups attached to the silicon atoms can expand the capabilities of silicone elastomers. Substituting methyl with phenyl groups, for example, provides greater low temperature flexibility, which is required for sealants used in the aerospace industry. Fillers can be used to enhance desirable properties, reduce cost, add colour, or to increase resistance to degradation.

One-component silicone sealants are condensation curing, which means that they cure by reacting with moisture present in the atmosphere. A trifunctional organosilane is the cross-linking system, while a metal soap acts as a catalyst for the condensation reaction. The sealant is stable until it is exposed to the atmosphere, and when exposed it undergoes hydrolysis with atmospheric moisture. The cross-linked network between polysiloxane chains is formed in a reaction which is catalysed by the metal soap. This reaction liberates an acid such as acetic, which gives off a slight odour. If the sealant is allowed to develop its entire cross-linked network through complete cure, all of its inherent characteristics and properties can be achieved.

Curing is accelerated by hot humid weather, and retarded by cold weather. If cured in cold weather, the sealant remains more flexible at low temperatures (although it may harden and become more brittle than organic elastomers). The range of colour tints is virtually unlimited and colour stability is good, thus substrates are not stained. Silicone sealants exhibit excellent resistance to weathering, moisture, ultra-violet radiation, and are not affected by ozone. After 8000 h of accelerated weathering, there is no discoloration, appreciable hardening, loss of flexibility, or staining of substrates. This makes them prime candidates for exterior one-stage sealing. The tensile adhesive strength is low in relation to the tensile cohesive strength. This necessitates the use of a primer for working joints, which is one of the primary applications of silicone sealants. Installation requires a very clean joint, thus silicone is a more labour intensive sealant than others. Cook [5.2] has enumerated the advantages and disadvantages of silicone sealants as shown in Table 5.6.

Table 5.6.

Advantages and disadvantages of silicone sealants [5.2]

Advantages	Disadvantages
1. One-component systems	1. Too expensive
2. Available in the widest range of colours	2. Require critical surface preparation and priming (adhesion highly dependent on both)
3. Colour stability	3. Poor adhesion to concrete
4. High temperature resistance	4. Inferior elongation relative to polyurethanes and polysulphides
5. Excellent handling characteristics at ambient temperatures of $-35°F$ ($-37°C$) to $+140°F$ ($+60°C$)	5. Inferior tensile strength relative to polyurethanes
6. Good UV light and ozone resistance	6. Poor tear resistance
7. Non-sagging on vertical walls	7. Joint width limitations
8. Good adhesion to metal and glass	8. Shelf-life too short
9. Exhibit no shrinkage	9. Require contact with atmospheric moisture to cure
10. Become tack-free in a short period of time (i.e. fast curing)	10. Cohesion too great for their adhesion
11. Excellent flexibility; permanently flexible	11. Pick up dirt
12. Almost 110% recovery from elongation or compression (very low compression set)	12. Unpleasant odour (release of acetic acid)
13. Good heat and chemical resistance	
14. Excellent durability or long life (approx. 30 years)	
15. Longer pot life (when mixed) than two-part polysulphides	
16. Good resistance to digging and gouging	
17. Non-staining to most materials	

Reprinted with permission from *Construction Sealants and Adhesives* by J. P. Cook, 1970. Copyright John Wiley.

The mechanical properties of silicone sealants as a function of temperature and rate of movement have been studied by Karpati [5.74–5.76]. The same author has exposed different silicone sealants on strain-cycling racks simulating movements in external building joints. Following the use of differential scanning calorimetry (DSC), it was found that a quick outdoor exposure screening test could identify unsatisfactory materials within two months [5.77, 5.78].

Silicone sealants were found to be resistant in a 1-year weather-o-meter test, are more resistant to oxidation than any other sealants, have hydrolysis occur at high temperatures (200°C), but have poor resistance to dilute alkalis [5.79].

In a more recent paper, Fisher [5.80] underlines the excellent ultra-violet, ozone and heat resistance of silicone sealants.

Room temperature curing silicone elastomers, according to Green-lee [5.81], will adhere well to unprimed, unanodized aluminium and steel if they contain alkoxysilyl groups and if the substrate has been brought into contact with certain anions either before or during cure.

Polyurethanes

The polyurethane sealants constitute a part of the low modulus elastomers. They are prepared from isocyanate terminated pre-polymers and polyols (polyesters or polyethers) which are blended at room temperature.

One-component sealants are based on isocyanate terminated pre-polymers with an isocyanate equivalent of 1000–2000, which are made from isocyanate and active hydrogen materials with an excess of isocyanates. Their curing relies on the atmospheric moisture and proceeds with the formation of urea linkages:

$$2O=C=N-R-N=C=O + H_2O \longrightarrow$$

$$O=C=N-R-\underset{\underset{H}{|}}{N}-\underset{\underset{O}{\|}}{C}-\underset{\underset{H}{|}}{N}-R-N=C=O + CO_2$$

(urea unit linkage)

Tertiary amine or a metallic catalyst may enhance the curing of one-component polyurethane systems.

In the case of two-component PU sealants, the curing proceeds by formation of urethane linkages:

$$\ldots NCO + HO \ldots \longrightarrow \ldots N\!-\!\!\underset{\underset{H}{|}}{C}\!\!\overset{\overset{O}{\|}}{-}\!O \ldots$$

(urethane linkage)

Polyurethanes are typical high recovery sealants. No by-products separate in their production or curing; even with moisture-cured one-component sealants, the only by-product is a small amount of CO_2.

Cured PU sealants are composed of high MW fragments with a low density of cross-linkage which results from the interaction of polyfunctional polymer building segments or from the reactions of the isocyanates and the urea units. The cure kinetics of some PU have been studied by Carlson *et al.* [5.82].

The low modulus of PU sealants is important especially for their use under severe conditions such as continuous cycling between expansion and contraction.

Polyurethanes are good polymers for sealant production because of some outstanding properties such as:

—abrasion resistance;
—oil resistance;
—stability to biodegradation;
—excellent elasticity and, consequently, a high recovery;
—good flexibility at temperatures as low as $-54°C$ ($-65°F$);
—no loss of elasticity with ageing [5.3].

Polyurethanes suffer from adhesion failures upon immersion in water and tend to yellow on exposure to ultra-violet radiation.

A good PU sealant would have the following general formulation:

Polyurethane	34–45%
Fillers	30–40%
Colorants	2–3%
Thixotropic agents	1–2%
Adhesion additives	1–3%
Plasticizers	15–25%
Solvent	0–4%

The two-component sealant offers greater versatility in compounding and colour matching and for a number of years was always the better sealant. One-component PU sealants are usually more expensive, since they are more difficult to produce, involve more reaction steps, and may involve additional catalysts. The packaging of one-component PU requires moisture-free sealant, and additional steps in dehydrating the compound and components through all stages in its manufacture [5.83]. Schematic representation of one- and two-component PU sealants are presented in Figs. 5.17 and 5.18. Toluene-2,4-diisocyanate (TDI) is the isocyanate component.

Doyle [5.85] discusses in detail the advantages and the drawbacks of these two groups of PU sealants and Panek & Cook [5.83] have compiled a list of the same as shown in Table 5.7.

In recent years polyblending has proven to be an interesting method for modifying and improving some of the properties of PU [5.86, 5.87].

Successful blends of polyurethanes with different polymers have been known for a few years. Feldman [5.88, 5.89] carried out a

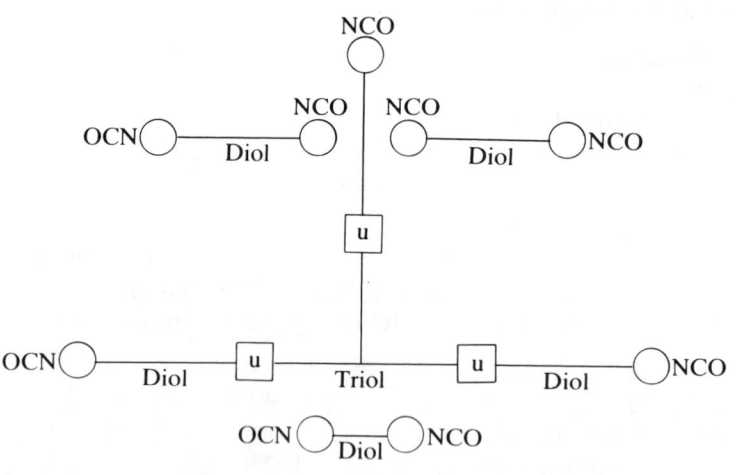

$\boxed{u}$ = urethane linkage, $-NH-\overset{\overset{\displaystyle O}{\|}}{C}-O-$ $\bigcirc$ NCO = Toluene-2,4-diisocyanate

Fig. 5.17. Schematic structure of one-component urethane sealants. Composition: 6 moles diols, 1 mole triol, 12 moles TDI [5.84]. (Reprinted with permission from *J. Appl. Polym. Sci.*, **9** (1965) 2965. Copyright John Wiley.)

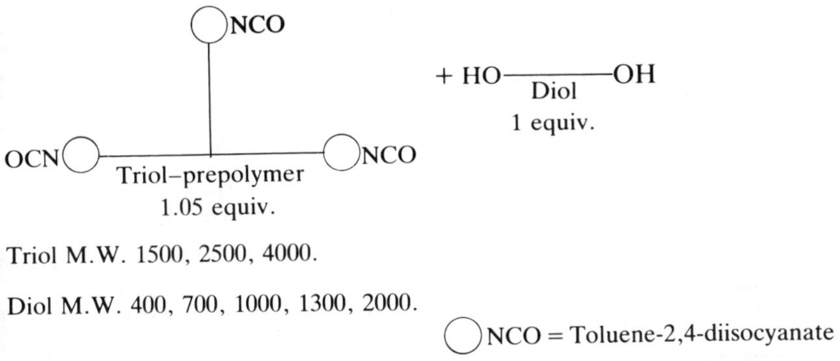
NCO
+ HO—Diol—OH
1 equiv.
OCN — Triol–prepolymer — NCO
1.05 equiv.
Triol M.W. 1500, 2500, 4000.
Diol M.W. 400, 700, 1000, 1300, 2000.
NCO = Toluene-2,4-diisocyanate

Fig. 5.18. Schematic structure of two-component urethane sealants [5.84]. (Reprinted with permission from *J. Appl. Polym. Sci.*, **9** (1965) 2965–83. Copyright John Wiley.)

weathering study on a PU elastomeric adhesive and its polyblends with the following vinyl polymers: PVC, poly(vinyl alcohol) (PVA), poly(vinyl acetate) (PVAc), and vinyl acetate–vinyl chloride copolymer. In the case of artificial weathering the best results were obtained with the PU–PVA polyblend; this may be due to some supplementary interactions between the hydroxylic groups of PVA and various PU functional groups. The decrease in the mechanical properties of the specimen kept in artificial weathering conditions is explained by the decrease in MW as a result of a cryolytic process.

In another paper [5.90], PU-polyorganosiloxane blends and mixtures with some fibre reinforcing agents or with plasticizers were studied. It was established that the elastic properties of PU-silicone polyblends increase and the tensile stress decreases with the amount of siliconic component.

The polyblends made of PU and an acrylic terpolymer show a higher strain than that of PU, and a lower tensile strength. At a 2/1 ratio of PU to the acrylic terpolymer the resistance of this polyblend to thermal cycling exposure is better than that of PU [5.91].

Table 5.8 contains the DSC (Differential Scanning Calorimetry) results obtained by Blaga & Feldman [5.92] in the case of PU polyblends with other polymers, plasticizers (DBP—dibutylphthalate) or fibres. SEM observations and DSC study of these mixtures led the authors to some of the following conclusions.

Table 5.7.

Advantages and disadvantages of urethane sealants [5.83]

Advantages	Disadvantages
1. Can be used in joints up to 15 cm	1. Light colours can discolour
2. ±25% movement capability	2. Poor water immersion resistance
3. Excellent recovery	3. Not recommended for wet joints
4. Excellent UV resistance	4. May require more priming
5. Excellent ozone resistance	5. Limited package stability for one-component
6. Fast cure of multicomponent	6. One-component requires more cure time
7. Long work life for multicomponents	7. Multicomponent requires mixing
8. Negligible shrinkage	8. One-component not recommended for traffic
9. Excellent tear resistance	areas
10. Excellent chemical resistance	9. Not recommended for stopless glazing
11. Excellent durability (20–30 years)	
12. Can meet ASTM C-920 for all systems	
13. Much better than polysulphides	

Reprinted with permission from *Construction Sealants and Adhesives* by J. R. Panek & J. P. Cook, 2nd edn, 1984. Copyright John Wiley.

SEM observations indicate that addition of thermoplastic polymers and reinforcement (glass or cotton fibre) to polyether diol prior to reacting it with MDI enhances the foamability of the formulation, resulting in products with a larger number of cells than is possible with the unmodified reactants. In contrast, addition of DBP plasticizer considerably reduces foamability, as evidenced by a smaller number of cells of comparable size in the blends than in the product from the unaltered formulation. Like the matrix of the unmodified PU foam, the matrix of the cellular blends produced by the addition of DBP plasticizer at weight ratios of PU:DBP down to 6·6:1 was homogeneous. Thus the two components are miscible in blends of these proportions. From the DSC curves reported, soft phase miscibility or interaction cannot be deduced. The usual criterion for compatibility is a single T_g varying linearly with composition between those of the pure phases. This has not been observed either because of T_g overlap or, possibly, because of 'cold crystallization' of the PU phase at low temperatures, the resulting exotherm offsetting the base line shift (due to T_g). The DSC results presented do indicate some loss in the crystal structure organization due to plasticization.

A homogeneous matrix was also observed by SEM in binary blends of PU and PVA, PVAc, or VAc–VC copolymer with a weight ratio of 40:1, and in PU containing glass fibre (20:1) or cotton fibre (40:1)

Table 5.8.
Thermal transitions of polymer blends and blend components [5.92]

Sample	Weight ratio	Thermal transitions (°C)	Sample	Weight ratio	Thermal transitions (°C)	Sample	Weight ratio	Thermal transitions (°C)
PU[a]	—	12; 60	PU-PVC	40:1	9; 60	PU-GF[d]	—	12; 60
PU-DBP	20:1	3; 60	PU-PVC	20:1[b]	8; 60	PU-CF[d]	—	12; 60
PU-DBP	10:1	7; 56	PU-PVC	20:1	15; 60	PVA	—	11; 48
PU-DBP	6-6:1	1; 20	PU-PVC	10:1	7; 61	PVAc	—	11; 50
PU-PVA	40:1	12; 60	PU-PVC	5:1	8; 60	VAc-VC copolymer	—	9; 30; 73
PU-PVA	20:1	18; 60	PU-PSO	2:1	-42; 9; 60			
PU-PVAc	40:1	8; 32	PU-PSO	2:1[c]	-42; 9; 57	PVC	—	9; 86
PU-(VAc-VC)	40:1	9; 60	PU-PSO	1:1	-42; 9; 57	PSO	—	-42; 2 (sh)
			PU-PSO	1:2	-42; 8; 56			

[a] The T_{gs} (glass transition temperature of the soft segment) at -45°C was not affected by plasticization, blending or reinforcing; the transition at 12°C is due to crystalline melting of soft segments and that at 60°C is associated with breakup of short-range ordered hard segment.
[b] Slow mixing was used during preparation.
[c] This sample contained a small amount of carbon black.
[d] GF and CF designate glass fibre and cotton fibre, respectively.
PSO = polysiloxane.

Reprinted with permission from *J. Appl. Polym. Sci.*, **28** (1983) 1033–44. Copyright John Wiley.

reinforcement. The DSC results could not be used to determine the miscibility of the polymer components. Blends produced by the addition of PVA at a weight ratio of 20:1, PVC in ratios from 20:1 down to 5:1, or polysiloxane polymer were heterogeneous, as indicated by SEM observations. With the exception of the PU–PSO mixtures, the thermal behaviour of these blends did not permit any conclusions regarding miscibility of the components.

Polysulphides

Polysulphides are based on alkyl polysulphide polymers which can be cured by various means to give products with rubbery characteristics; their microstructure is:

$$(-R-S_x-)_n$$

n being in the range of 2–4.

In this case the polycondensation process consists of:

$$n(Cl-R-Cl) + n(Na_2S_x) \rightarrow (-R-S_x-)_n + 2nNaCl$$

The base polysulphide polymers are usually prepared by a suspension process.

Polysulphides have elastomeric properties and are distinguished by outstanding resistance to oils and solvents. The resistance improves as the sulphur content of the polymer increases. Products of high sulphur content are also less permeable to vapours and gases. Polysulphide elastomers are also noted for oxygen and ozone resistance. The polymers are attacked by strong oxidizing acids and strong alkalis [5.93].

Liquid polysulphides for sealants are generally made of dichloro-ethyleneformal and sodium polysulphide:

$$n(ClC_2H_4OCH_2OC_2H_4Cl) + n(NaS_{2.25}) \rightarrow$$

$$HS(C_2H_4OCH_2OC_2H_4SS_{2.25})_n - C_2H_4OCH_2OC_2H_4SH + nNaCl$$

Dichloroethyleneformal is slowly added to a solution of sodium polysulphide plus emulsifying agents, and a latex is thus formed. After the latex is washed clean, then the splitting salts consisting of sodium sulphhydrate and sodium sulphite are added, and the mixture is then acidified, coagulated, and finally washed clean. The amount of sodium sulphhydrate determines the MW of the liquid polymer.

The cure mechanism is a simple oxidation of the mercaptan groups:

$$HS-R-SH \xrightarrow[\frac{1}{2}O_2]{} R-S-S-R + H_2O$$

There are various catalysts which can be used to initiate the curing process. With a metallic peroxide (metal O_2), the peroxide is reduced to a simple oxide. PbO_2 is most commonly used for fast curing sealants, and gives good polymeric properties for use as building sealants. Manganese dioxide (MnO_2) gives better heat resistance and stability to ultra-violet radiation, and is used for aircraft sealants and for insulating glass.

Generally fillers and plasticizers are needed to improve the physical properties. First, the fillers improve the physical properties, but then plasticizers are needed to lower the modulus and hardness. Both these additives, fillers and plasticizers, greatly reduce the cost of the sealant. A basic sealant formulation consists of the following [5.83]:

Polysulphide	40% wt.
Fillers	35% wt.
Plasticizers	20% wt.
Others	5% wt.

Some examples of fillers used for some polysulphide sealants are presented in Table 5.9. Plasticizers improve the workability of the polysulphide sealant while lowering its modulus. Control of modulus is extremely important in an elastomeric sealant to ensure that the adhesive strength of the compound does not exceed the structural strength of the substrate to which it is applied.

The amount or type of plasticizer must be compatible in the cured sealant and its volatility should be adequate for its intended use [5.94]. Dibutyl phthalate (DBP), chlorinated paraffins and coal tar fractions of high aromatic content are used as plasticizers for polysulphide sealants.

When compared with other additives, these sealants may cure very fast or very slow depending on the ingredients and curing system used. It is essential to regulate the curing of commercial usage, which is achieved by using retarders or accelerators.

The cure rate of liquid polymers is affected strongly by moisture and pH; increasing the moisture content of the compound generally speeds the cure time. The curing is slow in acidic pH and fast in alkaline media.

Table 5.9.
Mineral and black fillers for LP liquid polysulphide compounds [5.94]

Filler	Composition	Spec. gr.	Particle size (μm)	pH	Free moisture (%)
Pelletex (SRF No. 3)	Pelleted semi-reinforcing furnace	1·80	0·08	8·5	1·0
Sterling MT	Medium thermal	1·80	0·46	9·5	0·5
Calcene TM	Pp'td calcium carbonate (coated)	2·55	0·07	9·0–9·7	0·4
Multiflex MM	Pp'td calcium carbonate	2·65	0·06	9·0	0·5
Witcarb RC	Pp'td calcium carbonate (coated)	2·55	0·06	7·3–7·5	2·0
York White	Dry ground limestone	2·71	5–10	9·4–9·7	—
CalWhite	Wet ground calcium carbonate	2·71	5	9·0–9·5	—
OMYA BSH	Surface-treated chalk (1% stearic acid)	2·70	1–3	7·0	0·1
Icecap K	Anhydrous calcine clay	2·63	1·0	5·0–6·0	0·5
Cabosil	Fumed silica—99% SiO_2	2·2	0·015	3·5–4·2	1·0
HiSil 233	Pp'td silica—87% SiO_2	2·0	0·02	6·5–7·3	6·0
Permolith 40M (Lithopone)	29% zinc sulphide 71% barium sulphate	4·3	99·8% through 325 mesh	8·0–9·0	—
Blanc Fixe	Pp'td barium sulphate	4·4	0·18	8·8–9·5	0·3
Titanox 2020	Titanium dioxide	4·1	0·3	7·0–8·0	0·7
Superlith XXXHD	Pure zinc sulphide	4·1	—	6·8–7·8	—

Reprinted with permission from Shulman, M. A. & Yazujian, A. D. In *Plastic Mortars, Sealants and Caulking Compounds*, ed. R. B. Seymours. ACS Symposium Series 113; American Chemical Society, Washington, D.C., 1979, p. 129. Copyright 1979, American Chemical Society.

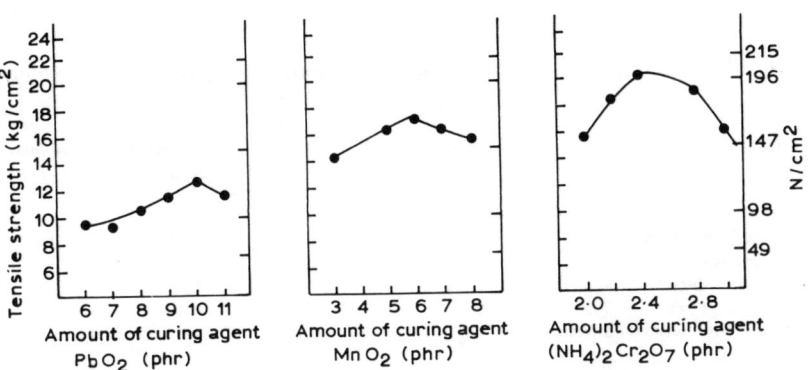

Fig. 5.19. Optimum mix ratios of curing agents for polysulphide sealants [5.95]. (Reprinted with permission from *J. Appl. Polym. Sci.,* **30** (1985) 3569–78. Copyright John Wiley.)

Ramaswamy & Sasidharan Achary [5.95] studied the influence of some curing agents such as PbO_2, MnO_2 and $(NH_4)_2Cr_2O_7$ on the properties of a polysulphide sealant. Based on the experimental results obtained, it has been shown that the curing agent has an influence on the mechanical (Figs. 5.19, 5.20 and Table 5.10), adhesive and thermal (Tables 5.11, 5.12) properties of the sealant. These changes are due to the variations in cross-linking density of the cured sealant produced by the curing systems. Fuel resistance of the polysulphide sealant is unaffected by the curing agents (Table 5.13).

Stearic acid, lead, zinc and aluminium stearates and anhydrides of

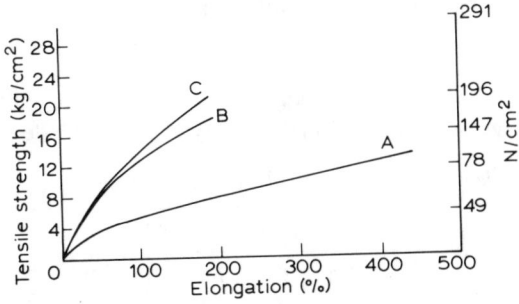

Fig. 5.20. Stress–strain curve for polysulphide sealant [5.95]. Cured with: A, PbO_2; B, MnO_2; C$(NH_4)_2Cr_2O_7$. (Reprinted with permission from *J. Appl. Polym. Sci.,* **30** (1985) 3569. Copyright John Wiley.)

Table 5.10.
Effect of curing agents on the mechanical properties of polysulphide sealants
[5.95]

Polysulphide sealant	Tensile strength kg/cm^2 (kPa)	Elongation (%)	Stress at 100% elongation	Hardness
(i) Cured with PbO$_2$	12·6 (1 234)	440	5·4	50
(ii) Cured with MnO$_2$	17·3 (1 695)	178	12·6	58
(iii) Cured with (NH$_4$)$_2$Cr$_2$O$_7$	20·6 (2 018)	182	14	64

Reprinted with permission from *J. Appl. Polym. Sci.*, **30** (1985) 3573. Copyright John Wiley.

Table 5.11.
Effect of curing agents on the adhesion of polysulphide sealants [5.95]

Polysulphide sealant	Peel strength kg/cm (N/cm)	Shear strength^a kg/cm^2 (kPa)
(i) Cured with PbO$_2$	10·4^c (101)	11·6^c (1 136)
(ii) Cured with MnO$_2$	5^c (49)	15·2^c (1 489)
(iii) Cured with (NH$_4$)$_2$Cr$_2$O$_7$	8·14^c (79)	20·4^c (2 000)

$^{a\,c}$ indicates cohesive failure in sealant.
Reprinted with permission from *J. Appl. Polym. Sci.*, **30** (1985) 3574. Copyright John Wiley.

propionic, benzoic and succinic acid are used as retarders. A number of amino compounds are used as accelerators for the curing process of polysulphide rubbers and sealants. Using primers, the peel strength (adhesion) might be improved (Table 5.14) [5.96].

Table 5.12.
Effect of curing agents on the thermal stability of polysulphide sealants [5.95]

Polysulphide sealant	Temp. of various % weight loss (°C)						IDT (°C)	T_{max} (°C)
	10%	20%	30%	40%	50%	60%		
(i) Cured with PbO$_2$	215	250	265	275	285	292	140	290
(ii) Cured with MnO$_2$	270	280	292	295	300	305	160	300
(iii) Cured with (NH$_4$)$_2$Cr$_2$O$_7$	275	290	300	305	307	320	200	305

Reprinted with permission from *J. Appl. Polym. Sci.*, **30** (1985) 3577. Copyright John Wiley.

Table 5.13.

Effect of curing agents on the fuel[a] resistance of the polysulphide sealant [5.95]

Polysulphide sealant	Swelling in fuel (%)	Fuel contamination (%)
(i) Cured with PbO_2	1·49	0·044
(ii) Cured with MnO_2	1·27	0·051
(iii) Cured with $(NH_4)_2Cr_2O_7$	1·25	0·045

[a] ATFK-50.
Reprinted with permission from *J. Appl. Polym. Sci.*, **30** (1985) 3577.
Copyright John Wiley.

An alkyl phosphatotitanate, a mercapto functional silane, an oxirane functional silane, and polyfunctional isocyanate type primers were evaluated by Usmani *et al.* [5.97]. While there was no evidence of toxicity shown by the tests, it was advisable, as with most chemicals, to use caution and to avoid repeated or prolonged contact with the skin.

The polysulphides are noted for the fact that the polymers have a distinct and disagreeable odour. However, the fully cured compounds in use are virtually odour free. Various masking agents have been used in an attempt to eliminate the odour but none has been markedly successful.

The properties of the polysulphides vary so widely with cross-link density and filler loading that it is very difficult to state a definitive temperature operating range. The practical temperature range to be considered in most sealant work, ranges from −40°C to 60°C. All of the liquid macromolecular compounds used in formulating sealants have a T_g outside this range, so that the measurable change in hardness becomes the major factor for consideration. This group of sealants is stable above 60°C. Long term exposure has shown that polysulphide sealants have excellent resistance to ozone, ageing, sunlight and weathering. Under accelerated heat ageing, these materials show only a modest loss in properties, due probably to the oxygen saturated backbone.

Damusis [5.3] considers that the most informative tests that can be made of polysulphide sealants are the creep and stress relaxation tests. The creep test is a recording of elongation to time at a constant load. The curve which is shown in Fig. 5.21 is a typical creep curve for a

Table 5.14.
Use of primers for non-adhesive polysulphide sealants[a] [5.96]

Substrate	Aluminium		Steel		Glass		Concrete	
	A[b]	B[c]	A	B	A	B	A	B
Primer								
Polyclad 932	—	—	2 630 (178)	Poor —	2 280 (155)	Poor —	1 750 (119)	1 750 (119)
PR 1099	2 630 (178)	Poor —	2 100 (142)	Poor —	2 800 (190)	Poor —	2 100 (142)	1 930 (131)
PR 1422	2 630 (178)	1 750 (178)	2 630 (178)	2 280 (155)	3 150 (214)	2 800 (190)	Poor —	Poor —
Ty-Ply S	1 750 (119)	Poor —	1 580 (107)	1 930 (131)	2 100 (142)	Poor —	1 750 (119)	2 100 (142)
Chemlock-220	2 450 (166)	2 100 (142)	1 580 (107)	2 800 (190)	3 850 (261)	3 680 (250)	Poor —	Poor —
N-15	2 450 (166)	—	2 800 (190)	—	3 330 (226)	—	—	—

[a] Formulation: LP-2, 100; SRF black, 30; stearic acid, 1; C-5 paste, 15.

[b] A = after 7 days at room temperature; peel strength, N/m (lb/ft); ASTM D903-49T.

[c] B = after 7 days at room temperature plus 7 days immersion in water; peel strength, N/m (lb/ft).

Reprinted with permission from *Rubb. Chem. Technol.*, **54** (1981) 197.

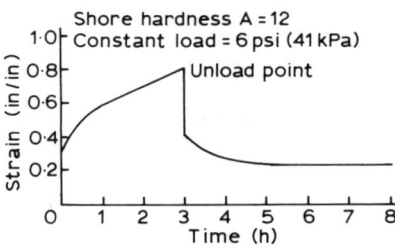

Fig. 5.21. Creep curve for a polysulphide sealant [5.3]. (Reprinted with
permission from *Sealants* by A. Damusis (Ed.), 1967. Copyright Reinhold
Publishing Co.)

filled and plasticized polysulphide system. The mechanical behaviour
of the polysulphide sealants is partially elastic and partially viscous.
Stress relaxation may be thought of as the converse of creep. While
creep represents elongation versus time at constant load, stress
relaxation is stress versus time at constant elongation. The stress
relaxation curve is a plot of time versus the decay in stress. Such a
curve has been found by many investigators to be a negative
exponential function of the form:

$$f = f_0 \exp(-t/T) \qquad (5.6)$$

where
  f = stress at any time
  f_0 = initial stress
  t = time
  T = stress relaxation time, which is defined as the time necessary for a
    specimen to decay to $1/e$ or 36·8% of its initial value.

For polysulphide sealants, the stress relaxation time varies with the
cross-link density of the polymer. Some of the lower-modulus sealants
may have relaxation times as short as 20 min, while those of the harder
materials may be 8 h or longer.

The fields of polysulphide polymer applications are shown in Fig.
5.22 [5.98].

The polysulphides were the first polymeric-type sealants to be used
in the construction industry. The weatherproofing of buildings is
perhaps the application that demands the greatest versatility of a
sealant.

Under extreme stresses in tension, compression, transverse or
longitudinal shear, the polysulphide sealants have a very good
performance record in many notable structures all over the world.
Usmani [5.99] has investigated the effect of the MW and the degree of
cross-linking on a series of properties of polysulphide sealants.

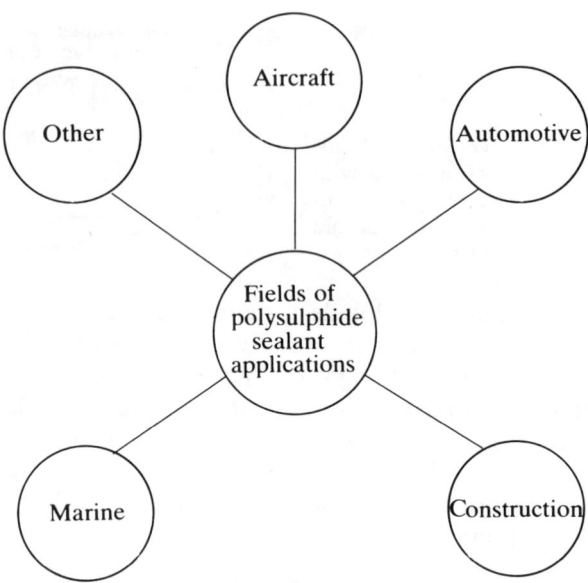

Fig. 5.22. Depiction of major fields of polysulphide sealant applications [5.98].
Reprinted with permission from *Polym. Plast. Technol. Engng*, **19**(2) (1982)
165–99. Copyright Marcel Dekker.)

Although there have been many successful applications of polysulphide sealants in highways, many of the early test applications showed up very poorly. Much of this can be traced to poor formulation with regard to conditions of movement, poor application and poor inspection [5.3].

Chlorosulphonated Polyethylene

For the synthesis of chlorosulphonated PE the low density PE, with an $MW_n = 20\,000$, is treated with chlorine in the presence of SO_2. Generally, the synthesis is carried out in solution in hot CCl_4. Both chlorine and sulphonyl chloride groups are introduced into PE, the degree of substitution and the ratio of the types of groups depending on the reaction conditions.

Chlorinated polyethylenes containing a small proportion of chlorosulphonyl ($-SO_2Cl$) groups provide useful synthetic elastomers. The chlorine content is generally 25–35% and the sulphur content 0·9–

1·7%, thus out of every 100 monomer units, some 25 to 42 are chlorinated.

Such a polymer may be represented as follows:

$$\left(-((CH_2)_6-\underset{\underset{Cl}{|}}{CH})_{12}-\underset{\underset{\underset{Cl}{\|}}{O=S=O}}{CH}-\right)_n$$

Introduction of chlorosulphonyl groups supplies reaction sites for cross-linking, which is usually realized with oxides such as MgO or PbO in the presence of a little water. The cross-linked product has good resistance to chemical attack, especially attack by ozone, oxygen and other oxidizing agents.

Chlorosulphonated PE is used as a rubber and sealant and also where resistance to weathering, corrosion or abrasion is required, e.g. flooring, linings for chemical plants, roofing, rollers, cables, etc.

Chlorosulphonated PE is sometimes used in unvulcanized form, generally with high amounts of fillers and plasticizers, for flooring.

Typical sealants (the trademark of related polymers produced by Dupont is Hypalon) would contain:

Chlorosulphonated PE	25%
Plasticizer	25%
Filler	25%
Solvent	10–15%
Curing agents, stabilizers	10–15% [5.83].

Table 5.15 presents the main advantages and disadvantages of a chlorosulphonated polyethylene sealant.

Polyacrylics

Polyacrylic adhesives and sealants are formulated from functional acrylic monomers, which achieve excellent bonding upon polymerization. Characteristic examples are the cyanoacrylates and ethylene glycol dimethacrylates. Both are one-component systems, and polymerization occurs upon exposure to the atmosphere. Cyanoacrylate based polymers and copolymers are marketed as contact adhesives and they have found numerous applications [5.71]. These are highly polar liquids based on alkyl–cyanoacrylates, $(CH_2=C(CN)COOCH_3)$; the

Table 5.15.
Advantages and disadvantages of Hypalon sealants [5.83]

Advantages	Disadvantages
1. Impervious to water	1. Slow cure of 1–4 months
2. Remain flexible	2. Poor packages stability
3. Fair recovery	3. Higher cost
4. Good chemical resistance	4. High shrinkage
5. Excellent UV resistance	5. Not for interior use
6. Excellent ozone resistance	6. Not for traffic joints
7. Can meet TT-S-00230C standard for ±12·5% movement capability	7. Not for sidewalks
8. Can meet ASTM C-920 standard for ±12·5% movement capability	8. Limited to ±12·5% movement
9. Can meet proposed specification for solvent based caulks	9. Tough gunnability
10. One-component	

Reprinted with permission from *Construction Sealants and Adhesives,* 2nd edn. by J. R. Panek & J. P. Cook, 1984. Copyright John Wiley.

formulation of these products includes:

—a plasticizer;
—a thickener;
—a stabilizer (to inhibit the polymerization of the monomer).

These monomers polymerize and rapidly set solid at room temperature simply on pressing into a thin film between metals, ceramics, rubbers or certain plastics. The bonds are strong, particularly to polar substrates and after ageing. The polymerization is accelerated by alkalis; the formed polymer resists many common solvents, including oils, but dissolves in dimethyl formamide and is slowly attacked by dilute acids, alkalis, hot water or steam [5.37].

In order to achieve the desirable properties such as elasticity, recovery, adhesion, hardness and ultra-violet stability, it is an advantage to use copolymers instead of homopolymers. A good typical acrylic sealant would have the following composition [5.83]:

Acrylic copolymer	35–40%
Filler	40–45%
Solvent	10–15%
Plasticizer	1–5%
Thixotropic agent	2–3%
Catalyst	1%

Solvent based acrylic sealants fall into a semi-elastomeric class of sealants, since they do exhibit thermoplastic properties. The polymer is usually supplied as a 85% solution in xylene. The lower the weight loss the better the sealant. It must be apparent that the values of 10% loss in weight must result in approximately 20% volume shrinkage, which can affect the geometry and hence the performance of the sealant.

Solvent based acrylic sealants are used effectively in a wide variety of joints. They are recommended for perimeter caulking (windows, doors, panels), control joints, precast concrete joints, panel-to-panel joints, bedding of mullions, panels and frames, etc. In glazing they are used for filling the complete channels, cap beads over tape, heel beads under glass and back bedding [5.100] (Fig. 5.23).

Because of their slow setting (a partial set is obtained in 2–5 months, but a final cure may require up to 1–2 years) and cold flow, the acrylic sealants are not recommended for use in traffic bearing joints found in plazas, highways, bridge decks, etc., or in underwater applications.

Complete cure is accompanied by a large increase in hardness. A good acrylic sealant should not exceed a Shore 'A' hardness of 55 after long time exposure or after heat ageing.

It is generally agreed that acrylic sealants can tolerate on the average a total of 15% movement, or ±7·5%. Being thermoplastics, they can take less movement when they are cold, and more movement when they are warmer. This behaviour recommends this class of sealants for southern climatic areas. They possess excellent weatherability, with good colour stability and adhesion.

The mechanical properties given in Table 5.16 compare the tensile and elastomeric properties of the chemically-cured, solvent-based acrylic sealants with those of a room temperature, gunnable, plasticized, conventional air-drying, thermoplastic, solvent-based acrylic sealant [5.101].

In terms of elastomeric properties, the recent development of chemically-cured acrylic sealants presents a significant improvement over the air-drying thermoplastic acrylic sealants. This behaviour permits their use in joints of higher movement (at least 25% total movement in compression and extension).

On exposure to low temperatures the acrylic sealants become quite brittle and crack at a temperature of −18°C (0°F) [5.102]. The flexibility of some modified acrylics is discussed by Hauser & Loft [5.103].

Some advantages and disadvantages of solvent-based acrylics are

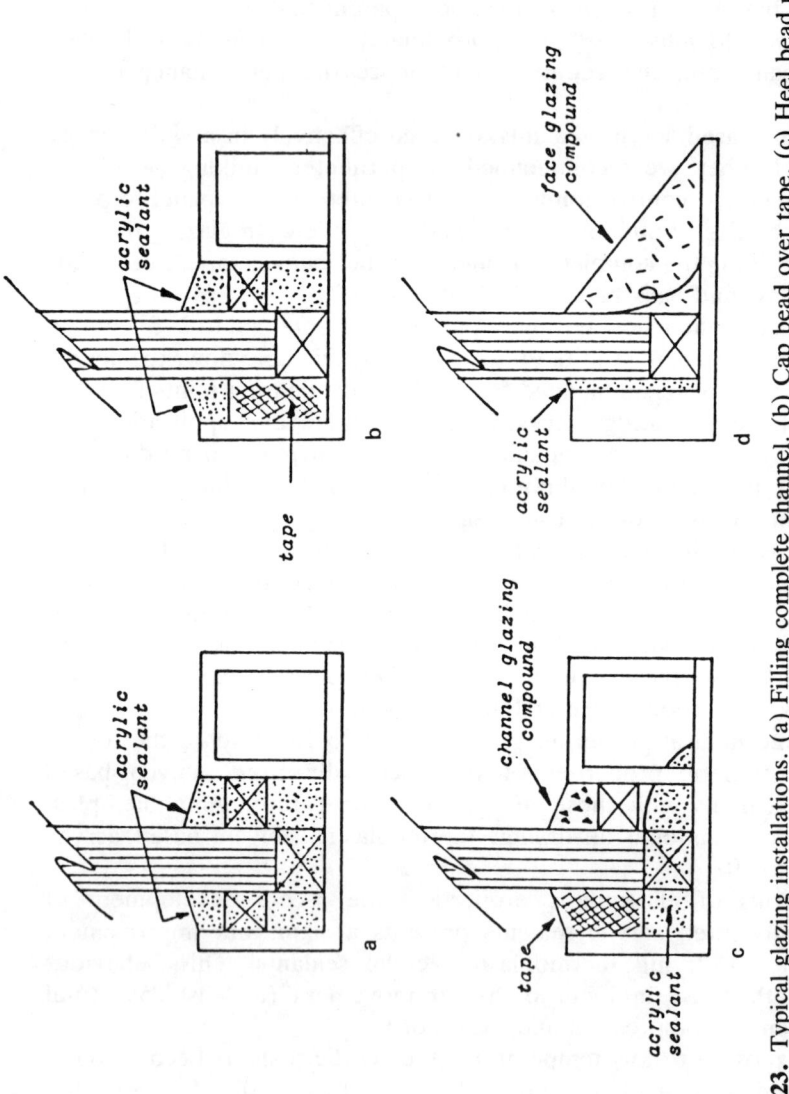

Fig. 5.23. Typical glazing installations. (a) Filling complete channel. (b) Cap bead over tape. (c) Heel bead under tape and glass. (d) Back bedding [5.100]. (Reprinted with permission from *Building Seals and Sealants* by J. Panek (Ed.), 1976. Copyright ASTM.)

Table 5.16.
Mechanical and stability properties[a] of chemically curing versus air-drying acrylic solvent-based sealants [5.101]

Property	Chemically curing acrylic resin based sealants		Air drying, conventional acrylic sealant formulation[b] (unplasticized)
	Unplasticized formulation	Plasticized formulation	
Mechanical properties[c]			
Tensile strength, psi (kPa)	87·2 (601)	48·3 (332)	19·6 (135)
Elastic recovery[d] (%)	66·3	67·2	25·0
Elongation @ max. stress (%)	221·0	233·0	57·0
Elongation @ break (%)	253·0	250·0	195·0
Caulk stability			
Consistency,[e] seconds initial	6·0	5·0	6·5
after 1 week @ 50°C	3·0	4·0	—
after 4 weeks @ 50°C	5·0	5·0	9·0
Resistance to cracking[f]			
after 1 week at 70°C	Excellent	Excellent	Cracks
after 1 week at 90°C	Excellent	Excellent	Cracks

[a] Unless otherwise noted, Rohm and Haas Research Laboratories sealant test procedures as described in detail in *Resin Review*, **XVI**(3) (1966), were employed to obtain these data.
[b] Room-temperature gunnable, solvent-based sealant.
[c] After 1 week @ 90°C.
[d] At 25% extension.
[e] Extrusion rate at 23°C from a standard, 6-oz Semco cartridge with 3/8-in orifice.
[f] Aluminium or wooden channels.

Reprinted with permission from *Building Seals and Sealants* by J. Panek (Ed.), 1976, STP 606. Copyright American Society for Testing and Materials.

Table 5.17.
Advantages and disadvantages of solvent-based acrylics [5.83]

Advantages	*Disadvantages*
1. Require no priming	1. Poor low-temperature elasticity
2. Require minimum surface preparation	2. Cannot be used in joints over 20 mm
3. Excellent adhesion	(3/4 in) wide
4. Good colour stability	3. Poor recovery
5. Cure tough	4. Slow skinning and cure
6. One-component system	5. Remain tacky for several days and pick up
7. Excellent UV resistance	dirt
8. Self-healing	6. Limited flexibility
9. Good durability of over 20 years	7. Strong, offensive odour until skin forms
10. Good chemical resistance	8. Poor water resistance
11. Non-staining	9. Not recommended for water immersion
12. Some sealants can take ±12·5% movement	10. High cold flow
	11. Not recommended for traffic joints
	12. Some systems only have ±7·5% movement
	capability

Reprinted with permission from *Construction Sealants and Adhesives*, 2nd edn, by J. R. Panek & J. P. Cook, 1984. Copyright John Wiley.

investigated by Faud [5.104], and a list has been compiled by Panek & Cook [5.83], as shown in Table 5.17.

Adhesive sealant compositions with improved properties and enhanced outdoor resistance were prepared by Blaga *et al.* [5.105] by blending an acrylic terpolymer (AT) with PVC. The morphology of these blends was studied by SEM, energy-dispersive X-ray analysis and DSC. The blends are heterogeneous. They consist of a continuous phase, which is either pure or mixed AT, and a particulate phase having the morphology of the added component. The particulate phase of AT–PVC contains mixed AT. The AT–PVC blends have improved mechanical properties (e.g. ultimate tensile strength, adhesive strength, etc.). The improvement in mechanical properties is strong, probably because the added PVC has strong specific interaction capabilities with AT. Whereas the unblended AT has very low outdoor durability, the AT–PVC blends display enhanced resistance to weathering, as evidenced by substantially higher ultimate tensile strength of weathered specimens than those of the control, unweathered specimens.

SEM and energy dispersive X-ray analysis (EDXA) observations indicated that the addition of POS to AT in concentrations of 25–50%, inclusive, results in homogeneous blends with improved mechanical properties [5.106]. Whereas the mechanical properties of unblended AT deteriorate markedly in outdoor exposure, those of AT–POS blends remain unchanged or improve substantially during weathering.

The glass fibres in AT–GF mixtures are well distributed, as indicated by SEM observations. Failure during fracture of blend specimens is cohesive in nature, thus indicating good bonding between the matrix (AT) and the glass fibre reinforcement. Reinforcing of AT with glass fibre leads to blends with considerably improved tensile properties and enhanced durability.

Blending the same AT with some other polar polymers such as poly(vinyl alcohol), poly(vinyl acetate) or vinyl acetate–vinyl chloride copolymer resulted in sealant compositions with improved properties and enhanced outdoor weathering resistance. The improvement in the ultimate tensile strength is greatest in AT–vinyl acetate–vinyl chloride copolymer blends, because the partners have greater specific interaction capabilities. For example, in the vinyl chloride monomer unit the α-hydrogen atom (H-bond donor) can interact with the ester carbonyl (H-bond acceptor) of the acrylic terpolymer to form a hydrogen bond, and is the key factor in achieving miscibility with many polymers containing acceptor groups. Furthermore, the possibility of charge-transfer and dipole–dipole interactions involving the chlorine atom of the vinyl chloride repeat unit and the ester oxygen of AT is also possible, as has been suggested for blends of PVC and polymers containing an ester group. Smaller improvements in the mechanical properties of AT polyblends with PVA and PVAc may be the result of weaker specific interactions between the molecules of the components [5.107].

Polychloroprenes

Polychloroprene (2-chloro-1,3-butadiene) is commercially produced in an emulsion polymerization process. A typical formulation might be [5.93]:

Monomer	100 pbw
Water	150 pbw
Rosin	4 pbw (stabilizer)
Sodium hydroxide	0·8 pbw (stabilizer)
Methylene bis(napthalene-sulphuric acid) sodium salt	0·7 pbw (emulsifier)
Sulphur	0·6 pbw (modifier)
Potassium persulphate	0·2–1·0 pbw (initiator)

Polymerization is carried out at 40°C and is allowed to continue until a 90% conversion is reached. After the regulation of the MW with tetraethylthiuramdisulphide, the coagulation takes place by acidification (acetic acid) and freezing.

There are today about 15–20 different types of chloroprene homo- and copolymer latices available. The polymers are partially cross-linked, insoluble in aromatic solvents and have a very high Mooney viscosity. These types of latices are produced by making a sulphur-modified polymer by incorporation of sulphur in the polymerization formulation, leading to polysulphide linkages [5.108].

In the chloroprene macromolecular chain, the individual monomeric repeating units can take up to four basic configurations as shown in Fig. 5.24 [5.109].

There are two types of polychloroprene sealants: one-component and two-component. The one-component cure very slowly to a Shore A hardness of 35–40. Their shelf-life can be prolonged by storage at low temperatures. The cured sealants become tough, have good water immersion properties, have a movement capability in the order of ±12·5%, and have good chemical stability.

Details of a typical two-component sealant are presented in Table 5.18. This is a solvent-free, non-shrinking, slump resistant sealant. It sets fast but cures slowly.

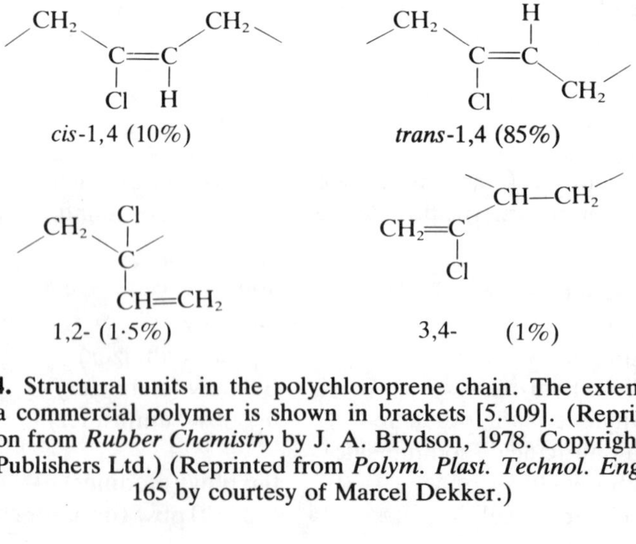

cis-1,4 (10%)   trans-1,4 (85%)

1,2- (1·5%)   3,4-   (1%)

Fig. 5.24. Structural units in the polychloroprene chain. The extent of each type in a commercial polymer is shown in brackets [5.109]. (Reprinted with permission from *Rubber Chemistry* by J. A. Brydson, 1978. Copyright Applied Science Publishers Ltd.) (Reprinted from *Polym. Plast. Technol. Engng*, **19**(2) 165 by courtesy of Marcel Dekker.)

Table 5.18.
Polychloroprene two-component sealants [5.3]

Designation of formulation	1	2
	(parts by weight)	
First component		
Polychloroprene–neoprene FB	100	100
Antioxidant—'Neozone A'	2	2
Thermal black	100	100
Hard clay	30	30
Dioctyl sebacate–plasticizer	50	50
Zinc oxide	10	10
Second component		
'Accelerator 833', a butyraldehyde–butylamine condensation product	7–10	—
Tetraethylenepentamine	—	5–8
Pot life (h)	6–4	25–1
Physical properties		
Tensile strength (psi)	920	560
Elongation (%)	400	320
Shore A hardness	38	44

Polychloroprene sealants are made with a solvent content of approximately 40%, which results in high shrinkage, and have a limited adhesion to some surfaces. Their main desirable property is that they are one of the few sealants compatible with asphaltic concrete and bitumen [5.83].

As can be seen from Table 5.18, the formulation of a polychloroprene sealant contains, as well as the macromolecular compound, plasticizer, filler and metallic oxides which regulate the cure rate and serve as acceptors for the trace amounts of hydrogen chloride released during processing, curing or ageing, and also serve as antioxidants and accelerators.

In Table 5.19 the advantages and disadvantages of polychloroprene sealants are listed.

Polyisobutene and Butyl Rubber

Polyisobutene or polyisobutylene, as it is sometimes referred to, is considered the precursor of butyl rubber and is produced by low temperature cationic polymerization of isobutylene. Its microstructure

Table 5.19.

Advantages and disadvantages of polychloroprene sealants [5.83]

Advantages	Disadvantages
1. Compatible with bitumens	1. High shrinkage, up to 40% by volume
2. Compatible with asphalts	2. Only dark colour
3. Compatible with neoprene gaskets	3. Very slow cure
4. Low cost	4. Not recommended for dynamic joint
5. Good water resistance	movement
6. One-component	5. No specifications
7. May have ±12·5% movement capability	6. Stains stone
8. Good adhesion to bitumens	7. Stains wood
9. Good adhesion to asphalts	
10. Good adhesion to metals	

Reprinted with permission from *Construction Sealants and Adhesives*, 2nd edn, 1984, by J. R. Panek & J. P. Cook. Copyright John Wiley.

may be represented as:

$$\ldots CH_2-\underset{\underset{CH_3}{|}}{\overset{\overset{CH_3}{|}}{C}}\left(CH_2-\underset{\underset{CH_3}{|}}{\overset{\overset{CH_3}{|}}{C}}\right)_n-CH_2-\underset{\underset{CH_3}{|}}{\overset{\overset{CH_3}{|}}{C}}\cdots$$

Polyisobutene is not vulcanizable by normal means and cannot be converted to a practical elastomer because its degree of saturation is 100%.

Polyisobutenes of low MW are viscous liquids, but the more important polymers $(10 \times 10^4 – 40 \times 10^4 \, MW)$ are highly extensible, tough rubberlike materials. These high MW products are used in unvulcanized adhesive rubber compositions, in caulking and sealing compositions, to plasticize and tackify polyethylene (for melt coating and heat sealing applications), as an additive to asphalts and waxes, and in vulcanizable rubber mixes.

The lack of unsaturation of polyisobutene has been overcome by copolymerization of the isobutene with a small amount of a diene monomer to yield a polymer vulcanizable by conventional sulphur-based systems. For many reasons isoprene is preferred. The isoprene units, randomly distributed, provide the unsaturation that permits the copolymer, known as butyl rubber, to be vulcanized; the small number of double bonds remaining in the vulcanizate render it more resistant to chemical attack.

Polyisobutene·and butyl rubber are obtained by cationic polymerization. Commercial rubbers are sometimes described as having a low, medium or high degree of unsaturation, i.e. containing approximately 0·6–1·0, 1·0–1·8 or 2–2·5 mol% isoprene respectively. The microstructure of butyl rubber may be represented:

$$\cdots-\left(-CH_2-\underset{\underset{CH_3}{|}}{\overset{\overset{CH_3}{|}}{C}}-\right)_n-CH_2-\underset{}{\overset{\overset{CH_3}{|}}{C}}=CH-CH_2-\cdots$$

Butyl rubber sealants can be obtained using butyl crude rubber, which can be dissolved in an appropriate organic solvent to give a solvent-based butyl sealant. Non-shrinking caulking compounds can be obtained by using low MW polyisobutene oils to soften the butyl rubber.

As can be seen in Table 5.20, the formulation of such a sealant is complex because of the numerous additives necessary for its compounding.

Table 5.20.
Butyl-rubber-based caulking compound [5.83]

Ingredient	% by weight
Polyisobutylene	2·5
Butyl rubber, 50% solution in mineral spirits	20·5
Talc powder	30
Calcium carbonate filler	20
TiO_2 pigment	3
Adhesion resin	4
Polybutene	10
Plasticizer	2
Thixotropic agents	2
Drying catalysts	0·05
Mineral spirits	6
Solids content 84% by weight	

Reprinted with permission from *Construction Sealants and Adhesives*, 2nd edn, 1984, by J. R. Panek & J. P. Cook. Copyright John Wiley.

In the above formulation butyl rubber is used for structure but polyisobutene and an adhesive resin are used for tack. The trace driers are used to give a tack-free surface and slowly cure. The total butyl composition is approximately 22%, but the total solvent is 16%, which could result in approximately 30% volume shrinkage. Most butyl caulking compounds cure by solvent release. Some compositions are based on crude rubber solvated by mineral spirits. Some more solvent may be used to incorporate more filler into the compound [5.83].

Cured butyl rubbers have exceptional resistance to ageing by oxidation, ozone and heat, also to the action of corrosive chemicals in general, though many organic liquids (hydrocarbons, chlorinated hydrocarbons) have a swelling action similar to that of other hydrocarbon products (rubbers, sealants). Cured butyl has a relatively low permeability to gases and serviceability in the range of $-50°C$ to $\pm125°C$ (or even higher for short periods) [5.37].

Besides adhesives and sealants, butyl rubber is used for weather seals and gaskets in buildings.

Butyls require some joint cleaning, but extensive surface preparation is not required. Laitance and loose concrete should be removed from the faces of the concrete joints, but priming is unnecessary. Metal surfaces may be wiped with an oil-free cloth wet with solvent to remove contaminants. Wood surfaces and glass should be wiped clean. Butyl caulks work best in relatively non-moving joints. They are used on the mating surface of heating and air-conditioning ducts and for sealing openings where pipes and ducts pass through roofs and partitions. Butyl caulks can be used below grade on masonry joints. They may be used in lap joints of sheet siding, metal roofs, glazing, wooden doors and windows, and protective sidings. Butyls adhere to aluminium, steel and vinyl siding. Some advantages and disadvantages of butyl caulks are presented in Table 5.21 [5.83].

Hot melt adhesives and sealants are the prevalent 100% active products in that they can be automated, used with minimal waste, and provide virtually immediate bonds [5.36, 5.110, 5.111]. Some of the synthetic polymers such as polystyrene, polyamides, saturated polyesters, ethylene copolymers, polypropylene, and other polymers are used usually as hot melt adhesives or sealants [5.112]. As can be seen from Table 5.22, polyisobutylene and hot melt butyl have lower moisture vapour transmission (MVT).

Halogenated Butyl

The low density of unsaturation of butyl elastomer, while having

Table 5.21.
Advantages and disadvantages of butyl caulks [5.83]

Advantages	Disadvantages
1. Reasonable cost	1. Very slow cure
2. Availability	2. High shrinkage
3. Good flexibility	3. High compression set
4. Good adhesion to most substrates	4. Limited to joints with ±7·5% joint
5. One-component	movement
6. Little surface preparation	5. Not recommended for expansion joints
7. Good water resistance	
8. Good colour stability	
9. Four colours available	
10. Only material for capping neoprene gaskets	

Reprinted with permission from *Construction Sealants and Adhesives* by J. R. Panek & J. P. Cook, 2nd edn, 1984. Copyright John Wiley.

Table 5.22.
MVT values of various sealants [5.83]

Polymer system	MVT $(g/m^2/24\,h)$
Silicone	16–24
Polysulphide	6–16
Hot-metal butyl	1–4
Urethane	6–16
Polyisobutylene	0·1–0·2

Reprinted with permission from *Construction Sealants and Adhesives* by J. R. Panek & J. P. Cook, 2nd edn, 1984. Copyright John Wiley.

certain advantages, has as a consequence a low cure rate when using the conventional vulcanization systems used in common diene elastomers. Butyl copolymers containing a small amount of combined chlorine or bromine vulcanize more quickly than normal butyl rubber because the halogen atoms provide additional sites for cross-linking. For this reason they can be blended with more unsaturated rubbers, e.g. natural, styrene–butadiene copolymer, polychloroprene. Chlorinated butyl rubber contains 1·1–1·3% combined chlorine and 1–2 mol% of unsaturation.

One of the drawbacks of butyl rubber is its poor adhesive properties to metals and other rubbers. This is largely due to the lack of polar groups in the microstructure of this copolymer. Halogenation remedies this deficiency. Bromination appears to involve substitution rather than addition.

Drying Oil-based Caulks

During the first half of the 20th century, oil-based caulking compounds were employed almost exclusively for sealing joints in buildings against the infiltration of air, dust and water. They were adequate for sealing the structures made of massive stone and brick with walls several feet thick and relatively small lights of glass framed in wood.

These caulking compounds are presently supplied in gun grade or knife grade formats. Gun grade type is a viscous semi-liquid, suitable for application by air or hand-operated caulking guns, or dispensed from disposable cartridges with the aid of a drop-in cartridge gun. Knife grade caulking compounds are more viscous and somewhat similar to mortar in consistency. The use of this type of caulk has diminished to a much greater extent than the gun grade caulks.

The oils used in sealant formulations are essentially triglyceride esters of long chain fatty acids with an MW in the range of about 700–900.

$$
\begin{array}{lll}
CH_2OH & HOOC\!-\!R_1 & CH_2\!-\!OOCR_1 \\
CHOH & +\ HOOC\!-\!R_2 & \longrightarrow\ CH\!-\!OOCR_2 + 3H_2O \\
CH_2OH & HOOC\!-\!R_3 & CH_2\!-\!OOCR_3 \\
\text{Glycerine} & \text{Fatty acids} & \text{Triglyceride}
\end{array}
$$

R_1, R_2, R_3 are straight hydrocarbon chains. Fatty acids vary in the length, degree of unsaturation and number of double bonds:

Saturated (C_nH_{2n})	$-CH_2-CH_2-CH_2-CH_2-CH_2-CH_2-$
Mono-unsaturated (C_nH_{2n-2})	$CH_2CH\!=\!CH-CH_2-CH_2-CH_2-$
Di-unsaturated (C_nH_{2n-4})	$-CH_2CH\!=\!CH-CH_2-CH\!=\!CH-$
	(Unconjugated)
Di-unsaturated (C_nH_{2n-4})	$-CH_2CH\!=\!CH-CH\!=\!CH-CH_2-CH_2-$
	(conjugated)
Tri-unsaturated (C_nH_{2n-6})	$-CH\!=\!CH-CH\!=\!CH-CH_2-CH\!=\!CH-$
	(mixed)

Table 5.23.
Typical saturated and unsaturated fatty acids [5.3]

Name	Double bonds	Empirical formula
Lauric	0	$C_{11}H_{23}COOH$
Palmitic	0	$C_{15}H_{31}COOH$
Steraric	0	$C_{17}H_{35}COOH$
Oleic	1	$C_{17}H_{33}COOH$
Linoleic	2	$C_{17}H_{31}COOH$ (unconjugated)
Linolenic	3	$C_{17}H_{29}COOH$ (mixed)
Elaeostearic	3	$C_{17}H_{29}COOH$ (conjugated)
Ricinoleic	1	$C_6H_{13}CH(OH)C_{10}H_{18}COOH$

One key to the performance of oil-based sealants is the unsaturated double bonds. These are the sites for oxidation and polymerization reactions necessary to cross-link fatty acid molecules and to transform the fluid or semi-plastic material in a solid product [5.3]. Some typical fatty acids are listed in Table 5.23.

The amount and the type of fatty acids contained in different oils varies from one oil to another. Fatty acid composition of a few animal and vegetable oils is presented in Table 5.24.

The formulation of oil and resin based caulk parallels the formulation of other caulks and sealants, in that there has to be a proper ratio of filler to binder, and the sealant or caulk has to set or cure within a desirable period of time to give a product according to standards [5.83].

A good quality gun-grade oil-based caulk might have the general formulation presented in Table 5.25.

Oil-based caulking compounds can be applied at all normal working temperatures. Below 5°C (40°F), due to the viscosity, the application becomes difficult. The surfaces should be clean, dry, free of frost, and free from contaminating substances such as oil, grease, wax, tar, asphalt, rust, etc. [5.113].

Table 5.26 presents the advantages and disadvantages of oil-based caulks.

Table 5.24.

Fatty acid composition of some important oils [5.3] (average percentage)

	Saturated				Unsaturated						
	Lauric	Palmitic	Stearic	Other saturates	Oleic	Linoleic	Linolenic	Elaeostearic	Ricinoleic	C_{20-24} acids	Other unsaturates
Castor			1·5	0·7	7·7	3·1			87·0		
Coconut	46·7	9·7	2·2	32·2	6·6	2·6					
Cottonseed		21·6	2·4	2·4	28·6	45·0					18·0
Herring		8·0		7·0	9·0	13·0				45·0	
Linseed		6·3	2·5	0·7	19·1	21·8	49·6				
Oiticica		6·5	5·2		5·6			4·5			78·2[a]
Palm		41·7	4·1	1·4	45·2	7·6					
Peanut		7·9	3·8	6·8	57·0	24·5					
Rape		1·0	1·0	4·0	22·3	15·0	1·8				54·9[b]
Safflower		3·5	2·5	0·8	19·7	70·1	3·4				
Sardine		7·4	2·7	10·1	13·1	7·4				46·4	12·9
Soybean		9·7	2·4	1·2	29·9	50·8	6·0				
Tung		4·0	0·9		6·7	0·7		87·7[c]			

[a] Licanic acid.
[b] Erucic acid.

Reprinted with permission from *Sealants* by A. Damusis (Ed.), 1967. Copyright Reinhold Publishing Co.

Table 5.25.
General formulation for an oil-based caulk [5.83]

Ingredients	Percentage by weight
Boiled linseed oil	25–30
Ground calcium carbonate	45–50
Polybutene	5–10
Colorants	3–5
Thixotropic agents	2–4
Paint dryers	0·1–0·3
Mineral spirits	10–12

Reprinted with permission from *Construction Sealants and Adhesives*, 2nd edn, by J. R. Panek & J. P. Cook, 1984. Copyright John Wiley.

Table 5.26.
Advantages and disadvantages of oil-based caulks [5.83]

Advantages	Disadvantages
1. Can remain plastic	1. No recovery
2. Tool easily	2. Little flexibility
3. Apply easily	3. Can harden or crack with poor quality
4. One-component	4. Movement limited to ±2% to ±5%
5. No primers needed	5. Not recommended for moving joints
6. Lowest-cost caulk	6. Slow cure rate
7. Good colour stability	7. Can stain substances
8. Can last over 10 years	8. Few good-quality caulks
9. Good package stability	9. Shrinkage can reach 20%
10. Some low-shrinkage caulks	
11. Good for low movement	
12. Fast skinning	

Reprinted with permission from *Construction Sealants and Adhesives*, 2nd edn, by J. R. Panek & J. P. Cook, 1984. Copyright John Wiley.

REFERENCES

[5.1] Wake, W. C., *Adhesion and the Formulation of Adhesives*. Applied Science Publishers Ltd, London, 1976.

[5.2] Cook, J. P., *Construction Sealants and Adhesives*. Wiley-Interscience, New York, 1970.

[5.3] Damusis, A. (Ed.), *Sealants*. Reinhold, New York, 1967.

[5.4]   Shields, J., *Adhesives Handbook,* 2nd edn. Newnes–Butterworths, London, 1976.

[5.5]   DeLollis, N. J., *Adhesives, Adherends, Adhesion.* R. E. Krieger Publishing Co., Huntington, New York, 1980.

[5.6]   Kinloch, A. J., *J. Mater. Sci.,* **15** (1980) 2141–66.

[5.7]   Eddy, S. R., Lucarelli, M. A., Helminiar, T. E., Jones, W. B. & Picklesimer, L. G., *Adhesives Age,* **23**(2) (1980) 18–24.

[5.8]   Bascom, W. D. & Patrick, R. L., *Adhesives Age,* **17**(10) (1974) 25–32.

[5.9]   Shonhorm, H., Frisch, H. L. & Gaines, G. L. Jr., *Polym. Engng Sci.,* **17**(7) (1977) 440–9.

[5.10] Jastrzebski, Z. D., *The Nature and Properties of Engineering Materials,* 2nd edn. John Wiley, New York, 1977, pp. 161–3.

[5.11] Schneberger, G. L., *Adhesives Age,* **23**(1) (1980) 42–6.

[5.12] Kinloch, A. J., In *Durability of Structural Adhesives,* ed. A. J. Kinloch. Applied Science Publishers, London and New York, 1983, pp. 1–42.

[5.13] Schneberger, G. L., *Adhesives Age,* **17**(4) (1974) 17–23.

[5.14] Baszkin, A. & Ter-Minassian-Saraga, *Polymer,* **19** (Sept., 1978) 1083.

[5.15] Mahoney, C. L., *Adhesives Age,* **16**(10) (1979) 34–40.

[5.16] Mittal, K. L., *Polym. Engng Sci.,* **17**(7) (1977) 467–73.

[5.17] Mittal, K. L., In *Adhesion Science and Technology, Vol. 9A,* ed. L. H. Lee. Plenum Publishing Co., New York, 1975, pp. 129–68.

[5.18] Huntsberger, J. R., *Adhesives Age,* **15**(12) (1978) 23–4.

[5.19] Lee, L. H., In *Adhesive Chemistry,* ed. L. H. Lee. Plenum Press, New York and London, 1984, pp. 5–62.

[5.20] Lurie, R. M., Adhesion of a High Molecular Weight Polymer. MIT, Ph.D. Thesis, 1955 (quoted by K. L. Mittal in Ref. 5.17).

[5.21] Levine, M., Ilkka, G. & Weiss, P., *J. Polym. Sci.,* **B-2** (1964) 915 (quoted by K. L. Mittal in Ref. 5.17).

[5.22] Gent, A. N., *Adhesives Age,* **25**(2) (1982) 27–31.

[5.23] Ahagon, A. & Gent, A. N., *J. Polym. Sci., Polym. Physics ed.,* **13** (1985) 1285–300.

[5.24] Gent, A. N. & Hsu, E. C., *Macromolecules,* **7** (1974) 933–6.

[5.25] Liang, F. & Dreyfuss, P., In *Adhesive Chemistry, Developments and Trends,* ed. L. H. Lee. Plenum Press, New York and London, 1984, pp. 121–38.

[5.26] Eckstein, Y. & Berger, E. J., ibid., pp. 139–64.

[5.27] Leary, H. J. Jr & Campbell, D. S., ibid., pp. 517–24.

[5.28] Kinloch, A. J., *J. Mater. Sci.,* **17** (1982) 617–51.

[5.29] Lee, L. H. (Ed.), *Adhesion Science and Technology, Parts A & B.* Plenum Press, New York and London, 1975.

[5.30] Lee, L. H. (Ed.), *Adhesion and Adsorption of Polymers.* Plenum Press, New York and London, 1980.

[5.31] Adams, R. D. & Wake, W. C., *Structural Adhesive Joints in Engineering.* Elsevier Applied Science, London and New York, 1984.

[5.32] Gent, A. N., *Rubb. Chem. Technol.,* **55** (1982) 525–35.

[5.33] Bodnar, M. J. (Ed.), *Durability of Adhesive Bonded Structures.* John Wiley, New York, 1977.

[5.34] Runge, M. L. & Dreyfuss, P., *J. Polym. Sci., Polym. Chem. ed*, **17** (1979) 1067–72.

[5.35] Gent, A. N. & Hamed, G. R., *J. Appl. Polym. Sci.*, **21** (1977) 2817–31.

[5.36] Feldman, D., Polymer adhesives Part I. *Canad. Plast.*, **37**(9) (1979) 36–42; Polymer adhesives Part II. *Canad. Plast.*, **37**(10) (1979) 38–9; Polymer adhesives Part III. *Canad. Plast.*, **38**(1) (1980) 40–1; Polymer adhesives Part IV. *Canad. Plast.*, **38**(2) (1980) 44–5; Polymer adhesives Part V. *Canad. Plast.*, **38**(3) (1980) 59–69; Polymer adhesives Part VI. *Canad. Plast.*, **38**(4) (1980) 57–9.

[5.37] Roff, W. J. & Scott, J. R., *Handbook of Common Polymers*. Butterworths, London, 1971, p. 70.

[5.38] Beech, J. C., *J. Inst. Wood Sci.*, **7**(6) (1977) 7–17.

[5.39] Leonard, E. C., *Vinyl and Diene Monomers, Part I*. John Wiley, New York, 1970, p. 187.

[5.40] Feldman, D., *Macromolecular Compounds Technology*. Technica, Bucharest (in Romanian), 1974.

[5.41] Boening, H. V., *Structure and Properties of Polymers*. Georg Thieme Publishers, Stuttgart, 1973, pp. 22, 27.

[5.42] Feldman, D. & Marculescu, B., *Industria Usoara*, **12** (1977) 509.

[5.43] Mark, H. F. (Ed.), *Encyclopedia of Polymer Science and Technology*, Vol. 11. Wiley–Interscience, New York, 1969, pp. 61–128.

[5.44] Billmeyer, F. W. Jr, *Textbook of Polymer Science*, 3rd edn, Wiley–Interscience, New York, 1984.

[5.45] Lyman, D. J., Heller, H. & Barlow, M., *Die Makromolekulare Chemie*, **84** (1965) 64.

[5.46] Myuller, B. E., Apukhtina, N. P. & Klebanski, A. L., *Rubb. Chem. Technol.*, **38**(2) (1965) 452.

[5.47] Broyer, E., Makosko, C. W., Critchfield, F. E. & Lawler, L. F., *Polym. Engng Sci.*, **18**(5) (1978) 382.

[5.48] Patrick, R. L. (Ed.), *Treatise on Adhesion and Adhesives, Vol. 3*. Marcel Dekker, New York, 1973.

[5.49] Morton, M. (Ed.), *Rubber Technology*. Van Nostrand Reinhold Co., New York, 1973.

[5.50] Anon., *Mod. Plast.* (June, 1971) 44.

[5.51] Jwinski, D. J. & Janke, R. A., *Adhesives Age* (Sept., 1971) 42.

[5.52] Updergaff, J. H. & Suen, T. J., In *Polymerization Processes*, ed. C. E. Schildknecht & J. Skiest. John Wiley, New York, 1977, pp. 497–534.

[5.53] Potter, N. G., *Use of Epoxy Resins*. Chemical Publishing Co., New York, 1976.

[5.54] Morgan, R. J. & O'Neal, J. E., *J. Mater. Sci.*, **12** (1977) 1966–80.

[5.55] Morgan, R. J. & O'Neal, J. E., *J. Macromolec. Sci., Physics B*, **15**(1) (1978) 139–69.

[5.56] Morgan, R. J. & O'Neal, J. E., *Polym. Plast. Technol. Engng*, **10**(1) (1978) 49–116.

[5.57] Dearlove, T. J., *J. Appl. Polym. Sci.*, **22** (1978) 2509–21.

[5.58] Dearlove, T. J. & Ulicny, J. C., *J. Appl. Polym. Sci.*, **22** (1978) 2523–32.

[5.59] Yorkgitis, E. M., Eiss, N. S. Jr, Tran, C., Wilkes, G. L. & McGrath, J. E., In *Epoxy Resins and Composites,* ed. I. K. Dusek. Springer Verlag, Berlin, 1985, p. 80.

[5.60] Walker, J. M., Richardson, W. E. & Smith, C. H., *Mod. Plast.* (May, 1976) 62.

[5.61] Anon., Adhesives—Products in Practice, CI/SFB, *Architects Journal,* May, 1979, pp. 923–36.

[5.62] Gent, A. N. & Hamed, G. R., *Polym. Engng Sci.,* **17**(7) (1977) 462–5.

[5.63] Igarashi, T., In *Adhesion and Adsorption of Polymers,* ed. L. H. Lee. Plenum Press, New York, 1980, pp. 421–37.

[5.64] Tsuji, T., Masuoka, M. & Nakao, K., ibid., pp. 439–53.

[5.65] Pocius, A. V., Wangsness, D. A., Almer, C. J. & McKown, A. G., In *Adhesive Chemistry, Developments and Trends,* ed. L. H. Lee. Plenum Press, New York, 1984, pp. 617–42.

[5.66] Drake, R. S. & Siebert, A. R., ibid., pp. 643–54.

[5.67] Bascom, W. D., Jones, R. L. & Timmons, C. D., In *Adhesion Science and Technology,* ed. L. H. Lee. Plenum Press, New York, 1975, pp. 501–11.

[5.68] Kinloch, A. J., Dukes, W. A. & Gledhill, R. A., ibid., pp. 597–614.

[5.69] Jackson, B. S., In *Industrial Adhesives and Sealants,* ed. B. S. Jackson. Hutchinson Benham, London, 1976, p. 53.

[5.70] Brydson, J. A., *Rubber Chemistry.* Applied Science Publishers Ltd, London, 1978, pp. 396–406.

[5.71] Ulrich, H., *Introduction to Industrial Polymers.* Hanser Publishers, München, 1982.

[5.72] Beznaczuk, L. M., Master's thesis, Concordia University, Montreal, 1985.

[5.73] Klowsowski, J. M. & Gant, G. A. L., In *Plastic Mortars, Sealants and Caulking Compounds,* ed. R. B. Seymour. ACS Symposium Series 113, Washington, 1979, p. 133.

[5.74] Karpati, K. K., *J. Paint Technol.,* **44**(565) (1972) 55–62.

[5.75] Karpati, K. K., *J. Paint Technol.,* **44**(569) (1972) 58–66.

[5.76] Karpati, K. K., *J. Paint Technol.,* **44**(571) (1972) 75–85.

[5.77] Karpati, K., *Adhesives Age* (Nov., 1980) 41–7.

[5.78] Karpati, K., *J. Coat. Technol.,* **56**(710) (1984) 29–32.

[5.79] Karpati, K., *J. Paint Technol.,* **40**(523) (1968) 337–47.

[5.80] Fisher, T., *Progress. Architect.,* **2** (1985) 105–10.

[5.81] Greenlee, T. W., In *Adhesion Science and Technology,* ed. L. H. Lee. Plenum Press, New York, 1975, pp. 339–54.

[5.82] Carlson, G. M., Neag, C. M., Kuo, C. & Provder, T., In *Advances in Methane Science and Technology, Vol. 9,* ed. K. C. Frisch & D. Klempner. Technomic Publishing Co., Lancaster, PA, 1984, pp. 47–64.

[5.83] Panek, J. R. & Cook, J. P., *Construction Sealants and Adhesives,* 2nd edn. John Wiley, New York, 1984.

[5.84] Damusis, A., Ashe, W. and Frisch, K. C., *J. Appl. Polym. Sci.,* **9** (1965) 2965–83.

[5.85] Doyle, E. N., *The Development and Use of Polyurethane Products*. McGraw-Hill, New York, 1971, pp. 125–85.
[5.86] Wang, C. B. & Cooper, S. L., *J. Appl. Polym. Sci.*, **26** (1981) 2989–3006.
[5.87] Hourston, D. J. & Zia, Y., *J. Appl. Polym. Sci.*, **28** (1983) 2139–49.
[5.88] Feldman, D., *Polym. Engng Sci.* **21**(1) (1981) 53–6.
[5.89] Feldman, D., *J. Appl. Polym. Sci.*, **26** (1981) 3493–501.
[5.90] Feldman, D., *J. Appl. Polym. Sci.*, **27** (1982) 1933–44.
[5.91] Feldman, D., *Polymer*, **24** (1983) 359–64.
[5.92] Blaga, A. & Feldman, D., *J. Appl. Polym. Sci.*, **28** (1983) 1033–44.
[5.93] Saunders, K. J., *Organic Polymer Chemistry*. Chapman and Hall, London, 1973.
[5.94] Shulman, M. A. & Yazujian, A. D., In *Plastic Mortars, Sealants and Caulking Compounds*, ed. R. B. Seymour. ACS Symposium Series, 113, 1979, p. 129.
[5.95] Ramaswamy, R. & Sasidharan Achary, P., *J. Appl. Polym. Sci.*, **30** (1985) 3569–78.
[5.96] Ghatge, N. D., Vernekar, S. P. & Lonikar, S. V., *Rubb. Chem. Technol.*, **54** (1981) 197.
[5.97] Usmani, A. M., Chartoff, R. P., Warner, W. M., Butler, J. M., Salyer, T. O. & Miller, D. M., *Rubb. Chem. Technol.*, **54** (1981) 1081.
[5.98] Usmani, A. M., *Polym. Plast. Technol. Engng*, **19**(2) (1982) 165–99.
[5.99] Usmani, A. M., *Polym. News*, **10** (1985) 231–6.
[5.100] Waters, W. J., In *Building Seals and Sealants*, ed. J. Panek. ASTM STP 606, Philadelphia, 1976, pp. 54–61.
[5.101] Young, H. C., ibid., pp. 62–77.
[5.102] Mraz, J. S., ibid., p. 78.
[5.103] Hauser, M. & Loft, T., *Adhesives Age* (Dec., 1980) 21.
[5.104] Faud, R. H., In *Sealants*, ed. A. Damusis. Reinhold Publishing Co., New York, 1967, pp. 226–34.
[5.105] Blaga, A., Feldman, D. & Banu, D., *J. Appl. Polym. Sci.*, **29** (1984) 3421–30.
[5.106] Feldman, D., Blaga, A. & Banu, D., *Polymer*, **25** (1984) 1603–6.
[5.107] Feldman, D., Blaga, A. & Banu, D., *J. Macromolec. Sci. Chem.*, **A22**(10) (1985) 1385–412.
[5.108] Feast, A. A. J., In *Polymer Latices and their Applications*, ed. K. O. Calvert. Applied Science Publishers Ltd, London, 1984, pp. 21–47.
[5.109] Brydson, J. A., *Rubber Chemistry*. Applied Science Pubishers Ltd, London, 1978.
[5.110] Baker, T. E. *et al.*, *Mater. Synerg.*, **10** (1978) 845.
[5.111] Grover, M. M., *TEC Annual Review*, **2** (1974) 393.
[5.112] Wake, W. C., *Adhesion and the Formulation of Adhesives*, 2nd edn. Applied Science Publishers Ltd, London, 1982, p. 267.
[5.113] Becker, J. W., In *Building Seals and Sealants*, ed. J. Panek. ASTM STP 606, Philadelphia, 1976, p. 3.

Chapter 6

Polymers in Solar Energy Conservation

Solar energy is an ideal source of energy not only for space heating, but also for power generation. This energy is practically inexhaustible and convertible, and its use does not result in pollution as with oil, coal or other forms of energy sources.

Polymers have a share in the future development of solar energy conservation systems. They offer not only potentially lower costs, easier processing, light weight but also a greater design flexibility. Polymers are used in most of the solar systems and equipment such as covers, thin-film honeycombs and housing for flat-plate collectors, reflecting surfaces and optical lenses for concentrating collectors, reflector shells, structural and support members, insulation, piping, moisture barriers, adhesives, sealants, etc. [6.1].

The widespread use of polymers is evident from the information displayed in Table 6.1, where it can be seen that applications of polymers are listed for each major solar technology. For the solar applications mentioned, cost, performance and durability must be optimized.

COVERS FOR SOLAR COLLECTORS

A non-concentrating solar heat collector consists of a transparent cover, an absorber, tubes or ducts for the heat transport fluid and an insulated case which limits the thermal losses to the environment (Fig. 6.1). The cover is selected for the highest transmittance of the air mass-to-solar spectrum of radiation and the lowest transmittance of the infra-red re-radiated from the absorber. The cover should also

Table 6.1.
Different polymer applications [6.3]

Solar energy systems	Mirror glazings	Flat-plate glazings	Encapsulation	Seals—adhesives	Structural members	Heat transfer—energy storage	Paints—coatings	Piping	Thermal and electrical insulation	Moisture barriers	Fresnel lenses	Membranes
Solar thermal conversion	×	×	×	×	×		×	×	×	×	×	
Photovoltaics	×	×	×	×	×	×	×		×		×	
Solar heating and cooling of buildings	×	×	×	×	×		×	×	×	×	×	
Wind				×	×	×	×		×	×		
Ocean thermal				×	×		×	×	×			
Biological/chemical	×		×	×	×	×	×		×		×	×

Reprinted with permission from *Polymers in Solar Energy Utilization*, ACS Symposium Series 220. Copyright, 1983, American Chemical Society.

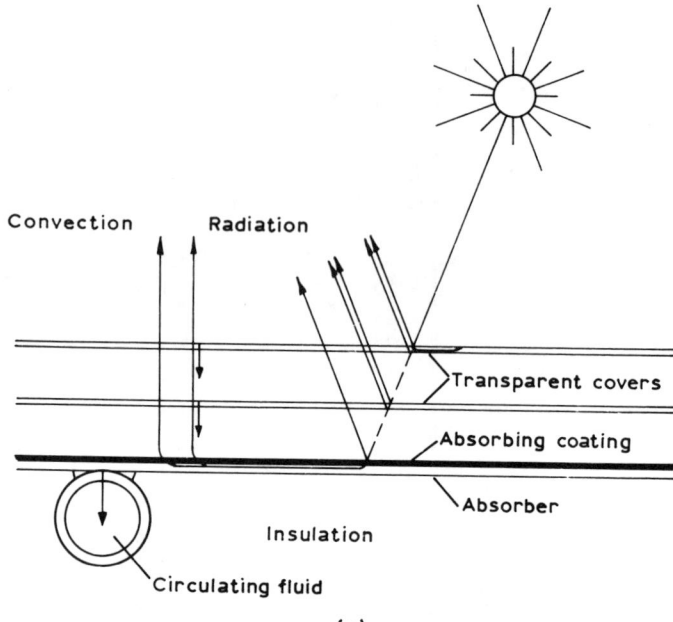

Convection   Radiation

Transparent covers

Absorbing coating

Absorber

Insulation

Circulating fluid

(a)

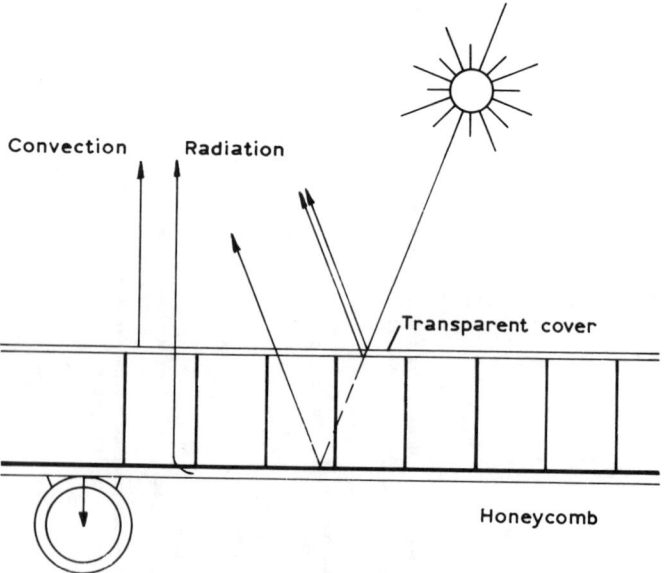

Convection   Radiation

Transparent cover

Honeycomb

(b)

minimize the heat loss by conduction and convection and for this reason may be composed of several separate sheets and may include a honeycomb structure with cells parallel to radiation direction.

The collector cover serves several important functions; it permits the passage of the solar radiation, it prevents loss of heat and it protects the absorber from mechanical and weathering damage. The transmittance of a cover is reduced by two factors: the reflectance, which increases with a rising difference in refractive index from that of air and which depends on the angle of incidence, and the absorptance, which is increased by certain impurities and by the scattering at internal surfaces. The cover can reduce the heat loss by having a low conductivity and a low emittance of infra-red radiation from the absorber. Multiple covers decrease the heat loss but also decrease the amount of light entering; the heat balance between these effects usually lies at two sheets for a selective absorber in operation during a Canadian winter. Honeycomb lattices with channels parallel to the sunlight have proven effective in decreasing both the convection within the collector and the radiation loss. Finally, to protect the absorber the cover must be able to resist both miscellaneous impact and uniform weather loads even at fairly high temperatures arising from the heating up of the absorber. The cover has to be resistant to degradation by solar radiation, temperature and weather conditions. Figure 6.2 presents one of the possible locations of the solar collector and some other jobs for which polymers can be applied.

Optical Characteristics

Glass, which has been utilized for centuries because of its ability to transmit visible light (Table 6.2) and withstand the weather, is an obvious glazing for collectors. The absorptance of glass increases from 3 to 6% as impurities such as manganese or iron increase (indicated by a greenish tinge when viewed edge-on).

Polymers are being used as glazing for solar collectors, although they have not generally displaced glass from its traditional uses for

Fig. 6.1. Cross-sections of two flat collectors showing the principal components: (a) two covers, absorber plate with fluid duct and insulation; (b) a honeycomb replaces one cover. The main losses restricting efficiency are reflection and absorption of the solar radiation and thermal losses due to infra-red radiation, convection and conduction [6.4]. (Reprinted with permission from *J. Mater. Energy Syst.*, **2** (1980) 65. Copyright Metallurgical Society.)

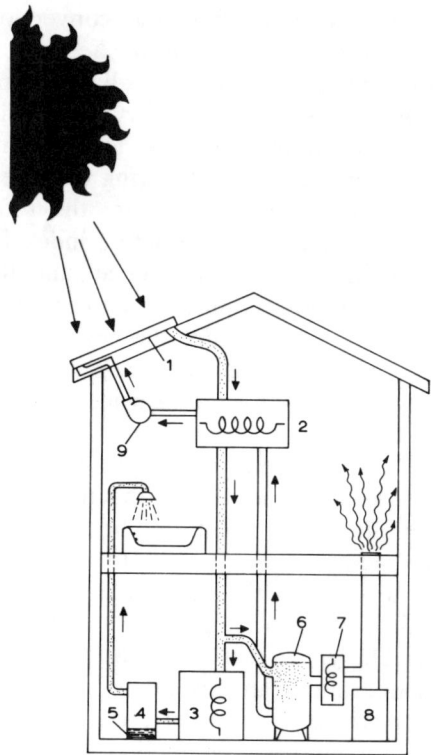

Fig. 6.2. Components of one type of solar house where plastics can be applied to the job. 1. Solar collectors. 2. Heat exchanger (working fluid/water). 3. Heat exchanger (water/potable water). 4. Water tank. 5. Auxiliary water heater. 6. Heat storage tank. 7. Heat exchanger (water/air). 8. Auxiliary air heater. 9. Working fluid pumps [6.5]. (Reprinted with permission from *Mod. Plast.,* (1976) 52. Copyright McGraw-Hill.)

window glazing. Among the reasons for this displacement, aside from the optical properties above, are the lightness, impact resistance and low cost compared to glass.

Polymeric materials have a low absorptivity for solar radiation and having no absorption bands are intrinsically colourless (Table 6.2, Fig. 6.3); the absorptance can be decreased compared to glass by using thinner sheets or films. Furthermore, plastics have refractive indices less than that of ordinary glass, which reduce their reflectance and thus contribute to a transmittance greater than that of glass. When a

Glazing materials; optical properties [6.4]

Cover material identification	Trade names	Formula	Light transmittance %	Cost ($/m²)	Longwave transmission %ᵃ	Daily shortwave transmission	Typeᶜ	0°	45°	67°
								Angle of incidence		
FEP Fluorinated ethylene–propylene copolymer (FEP)	Suntek Teflon		95, 96 (1 mm film)	3-8, 5-3, [0.54]ᵉ	13		Tot.	93	90	65
							Dir.	85	78	48
Poly(vinyl fluoride) (PVF) 0·07 mm	Tedlar	$[CH_2-CHF]_m$		1-83, [0.5]ᵉ	43	91·0				
Poly(methyl methacrylate) (PMMA)	Plexiglass	$[-CH_2-C(CH_3)(COOCH_3)-]_m$	91 after 10 years	[0.16]ᵉ			91.0			
Polycarbonate (PCO) 1·6 mm	Tufflak (Lexan)	$[-O-C_6H_4-C(CH_3)_2-C_6H_4-O-C(=O)-]_m$	82–89 depending on thickness		6	84·4 Dir.	Tot.	86	83	72
Poly(ethylene terephthalate) (PETP) 0·12 mm	Mylar	$[-O-CH_2-O-C(=O)-C_6H_4-C(=O)-]_m$		1-94, [2.2]ᵉ	32	86·5	Tot.	88	86	72
							Dir.	87	84	72
Fibre glass reinforced polyester (FRP) 1 mm	Sun Lite premium		80 after 20 years	4.52	6	83·1	Tot.	81	80	57
							Dir.	65	60	34
Corrugated fibre glass 1 mmᵈ				5·27	7–8	78·1–	Tot.	81	75	43
				5·92		79·2	Dir.	52	45	21
Glass, double-strength 3 mm				5·38 [0·03–0·18]ᵉ	3	87·8	Tot.	89	85	74
							Dir.	88	85	70

ᵃ Longwave > 2·8 m.
ᵇ Shortwave < 2·8 m.
ᶜ Type of solar energy flux transmission measurement (January).
ᵈ Coated with PVF.
ᵉ $/m² (25 μm thickness) or 0·1 $/ft² (1 mil thickness), *Mod. Plast.* (Jan. 1980) 126.
Tot. = total solar energy.
Dir. = direct beam component of the total solar energy.
Reprinted with permission from *J. Mater. Energy Syst.*, **2** (1980) 65–82. Copyright, 1980, Metallurgical Society.

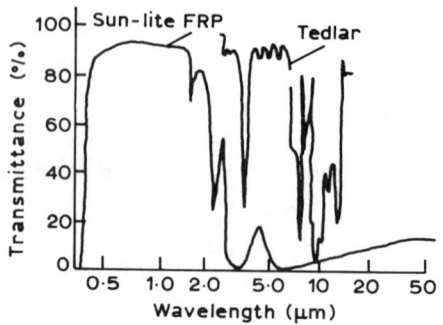

Fig. 6.3. The transmittance curve for Sun-Lite, a glass fibre reinforced polyester, and for Tedlar [6.4]. (Reprinted with permission from *J. Mater. Energy Syst.*, **2** (1980) 65. Copyright Metallurgical Society.)

polymer film is stretched in fabrication it usually becomes crystalline with a refractive index higher than that of the amorphous state because its density is higher. Crystalline polymers, especially when oriented due to internal structure and the shape of the crystallites, exhibit birefringence which has no appreciable effect on the total transmittance of the polymer for collector purposes. Polymers as a group have a much greater variety in composition and structure than glass, and consequently have a wider range of optical, mechanical and endurance properties than glass (Tables 6.2, 6.3). Some films have been specially compounded for solar service, such as that of Suntek, or Vistron's Filon (PMMA, PETP, PVF, glass) and offer a unique combination of properties.

In comparison with glass, polymers suffer in general from having low absorptance for thermal radiation even though they usually have some infra-red absorption bands related to certain functional groups. The polymers which have low long wave (>2.8 μm) transmittance are polymethylmethacrylate (PMMA), polycarbonate (PC), polyvinyl-fluoride (PVF), Teflon (PTFE), and polyesters, which contain strong polar groups such as oxygen or halogen atoms. However, except for the first two, infra-red transmittance is some five times greater than that of glass. The simpler plastics such as polyethylene (PE) and polypropylene (PP) have even greater long wave transmittance, almost equal to that for solar radiation. The long wave transmittance of polyesters can be reduced substantially by adding a suitable filler, usually glass fibres, but these also reduce the solar transmittance by about 10%, mainly because of interface scattering. The transmission of infra-red can be reduced through its reflection by heat mirror coatings such as SnO_2 or In_2O_3 to as little as 0·25 or 0·08 respectively. Such coatings have an effect equivalent to that of an absorber coating which has low emittance of infra-red, and hence need not be used with selective absorber coatings. The heat mirrors unfortunately reduce the solar radiation transmission by as much as 20% for SnO_2 and 5% for In_2O_3 [6.4].

Honeycomb Covers

Transparent honeycombs have been introduced in order to prevent convection and radiation losses while reducing transmittance less than the additional cover sheet that they replace. Light rays not parallel to the channel axis reach the absorber partially by multiple transmissions and partially by multiple reflections; the factor of transmissions decreases from 0·99 at 0° to 0·84 at 70° incidence. Honeycombs have been successfully constructed from glass, PVF, PTFE and polycarbonate. On the positive side, a honeycomb made with infra-red absorbing materials acts as a selective structure. If the aspect ratio is 4 to 1 or more, multiple absorptions and re-emissions reduce the overall infra-red emission by 80–90%. On the negative side, it conducts heat through its walls in proportion to their thickness. For a single window and a non-selective absorber surface a honeycomb improves efficiency by 50%, and sometimes at less expense than a selective surface. Instead of a honeycomb a foamed translucent plastic has been used between two windows to reduce thermal losses by 40% while reducing transmittance by 5%.

Loadbearing Capability

The glazing material has to be able to resist wind, snow and gravitational loads over the range of temperatures achieved by the collector. This can be in the range of −40–120°C in normal operation and as high as 200°C when the fluid is not flowing in the absorber. With glass, there is no difficulty in supporting normal weather imposed loads in addition to its own weight over spans of 1 m at temperatures well in excess of the above. The polymer materials have adequate strength at 25°C but soften at a variety of temperatures to be discussed in Table 6.3. The mechanical properties of polymers are also dependent on the processing history, on additives such as fillers, reinforcements, plasticizers, etc., and on strain rate, so that published data

Polymeric Building Materials

Table 6.3.

Glazing materials; mechanical, thermal and other properties [6.4]

	Mechanical properties					Thermal coefficients and other properties							
Cover material	Weight per unit area kg/m^2 (lb/ft^2)	Tensile strength MPa ($psi \times 10^3$)	Impact strength (J)	Elongation (%)	Tensile modulus $kPA \times 10^5$ ($psi \times 10^5$)	Conductivity ($J/m\,s\,°C$)	Expansion ($°C^{-1} \times 10^{-5}$)	Flammability	Chemical resistance	Surface weathering	Max. oper. temp. (°C)	Heat shrink capability and temp. (°C)	Outdoor longevity (years)
FEP Suntek (0·025 mm)	0·098 (0·02)	20-100 (2·9-14·5)	—	100-190	20 (2·9)	0·25 0·028	10·8				148 204		30
PVF (0·12 mm)	0·14 (0·028)						4·8	Nil	High	Undamaged in 20 years	208	148	N/A
PMMA (3·2 mm)	3·68 (0·75)	55-75 (7·9-10·8)	11	207	24-35 (3·5-5·1)	0·19	7·3	Combustible	Alkali, weak acids, oils	Nil	82-93	0·002-0·008 %, at 154-261	10-20

PCO 1·2 (0·24)	55-65 (7·9-9·4)	22	60-100	24 (3·5)	0·19	6·84	Combustible, self extinguishing	Weak acids, aromatic solvents	Resistant	121-132	0·005-0·007 %, at 261-315	5-7
PETP film	3·4-7 (0·49-1·0) / 75-79 (10·8-11·4)	7	50-100	0·6 (0·008) / 0·066-0·079 / 0·009-0·011	0·06-0·08	0·72-2·23	<3·8 mm/m ASTM D-635	Most airborne chemicals	20 years expectancy	93		20 min maint.
Sunwall two layers 1 mm, 12 mm air gap			—				No spec.					
PETP + PMMA Flexigard (composite 0·0175 mm) 0·25 (0·05)						5-9			Highly resistant	121	30%	7-10
FRP	75-79 (10·8-11·4)	14			0·13	2·52	<3·8 mm/m	Acids, alkalis	Good colour stability	93	No shrink	20
Glass (3·2 mm) 7·8 (1·59)	40-90 (5·8-13·05)	0	0	700 (101·5)	0·75	0·85	Nil	Highly resistant	Nil	600	No shrink	100

must be scrutinized for applicability to the precise material and conditions of service. In particular, the strength and creep resistance of cast sheet may be considerably lower than those of extruded, stretched and crystallized film.

The mechanical properties of a polymer, such as modulus, yield strength and tensile strength at fracture are usually found to degenerate with rising temperature, or diminishing rate of strain. Changes in the stress–strain performance generally arise from mechanisms related to the flexibility of chain segments. These are enhanced by the addition of plasticizers and by increasing temperature and are reduced by increasing strain rate, crystallization and orientation. In other words, the strength properties depend largely upon the difference between the test temperature and the glass transition temperature, T_g; raising the former or lowering the latter have similar effects. In the temperature region -40–$70°C$, vinyl polymers (PVC) and polyolefins display a fairly linear strength dependence on temperature; whereas Teflon and linear aromatic polyesters exhibit little change.

Related to the strength at high temperatures (50–200°C), yet distinct because of the much lower strain rate, is creep at a low stress (e.g. 450 kPa, 65 psi). Polyethylene and PVC soften and sag above 60°C and polypropylene, polyester, PMMA and PVF are only slightly better (above 90°C). The superior plastics in this respect are the polycarbonates and Teflon which have maximum operating temperatures of 130 and 200°C respectively. The glass reinforced plastics do not deflect as much as unreinforced polyester and may be used at somewhat higher temperatures.

In combination with the mechanical properties, the coefficient of thermal expansion can have an important influence. Glass has a modulus of 70 GPa, similar to non-ferrous metals, and has a linear expansion of $10 \times 10^{-6}°C^{-1}$, about half that of aluminium. The worry here, then, is whether gaps will be left between the glazing and the frame on heating or whether either the glazing or the frame will damage the other on cooling. In a sound design, the changes are taken up with the help of proper gasket shape and material. In the thermoplastics the coefficient of expansion is between 90 and 190 $\times$ $10^{-6}°C^{-1}$ so that sheets tend to buckle, becoming wavy, and films to sag as a result of repeated rises from night to operating temperatures of 100–120°C. Under stagnation conditions the film can sag sufficiently to come into contact with the absorber, possibly adhering to it; this can be prevented by an automatic emergency venting system or avoided by using only the more creep resistant materials (possibly glass) as the inner cover. To overcome the sagging, polymer film has been stretched while hot over aluminium frames with intermediate supports. This has been partially successful, but on cooling has occasionally led to permanent distortion of the frames. Creep during cycles of overheating will gradually loosen the film. A loose or sagging film (or sheet to a lesser extent) is capable of being vibrated by the wind, which ultimately can lead to fatigue failure and tearing of the film from the frame. The use of reinforcing glass fibres and corrugated sheet can eliminate this problem, with some loss in cover transmittance.

Impact resistance is desirable to withstand occasional hail storms or random missiles thrown by children. The brittle failure of glass in response to such stressing is well known and the use of thick glazing is an effort to reduce the problem. The use of tempered glass with surface residual compression stresses, or wire-mesh reinforced glass, is likely to increase the life considerably. As discussed below, thermoplastics generally have a good impact resistance but the thermosetting plastics are as fragile as glass and hence are not feasible for use in collectors even though they often have superior thermal resistance.

The impact strengths of amorphous polymers are similar if they have the same average molecular mass and are at the same temperature differences from the glass transition; this fact derives from the principle of corresponding viscoelastic states. For most applications it is desirable for the material to exhibit a high modulus and high impact strength. Amorphous polymers have this desirable combination of properties only at about 70° above the glass temperature. Above this temperature range the modulus becomes too low, and below T_g the material becomes too brittle. The impact strengths of thermosetting polymers are rather low and vary little with temperature over a wide range, e.g. between -80 and 200°C. For thermoplastic material, the impact strength is usually high (Table 6.3) but very temperature dependent, markedly decreasing as the temperature approaches T_g. As a general rule crystalline polymers exhibit high impact strength, which decreases as the degree of crystallinity is increased.

Durability

Weathering effects have an important bearing on the choice of a material. However, it is only for materials which have been commercially available for many years that there is significant weathering information and this has often been accumulated from cases of mis-application rather than from systematic investigations. Poor design and fabrication techniques probably contributed to some of the faults and weaknesses observed.

Many studies have established that the prime factors causing breakdown are the ultra-violet portion and heating effects of solar radiation. The outdoor degradation of polymers is mainly the result of photo-oxidation which is activated by solar UV (0·295–0·400 μm). Although this region of the solar spectrum contains less than 5% of the solar energy, there is sufficient photon energy only in this wavelength range to break bonds in the polymers. One group of polymers are completely transparent and hence stable to UV. These include PMMA, PTFE and polymethylsiloxane. The last has a low tensile strength so needs to be reinforced or used as a coating. The simple polymers such as PE or PVC are subject to rapid dissociation due to UV. However, these polymers can be protected by a coating, or outer window which screens out the UV. The first group can be rendered opaque to UV and the second group stabilized by the addition of stable molecular absorbers of UV such as hydroxybenzophenones or phenyl esters of salicylic acid. High temperatures speed up the process of degradation.

The resistance to natural weathering is very good for glass, PMMA and fluorocarbon plastics (Tedlar and Teflon) and there is confirming experience over periods of many years. Weathering of polycarbonates results in discoloration and in fine cracks which lead to moisture damage and finally to further cracking. This attack can be slowed by adding a protective layer of PVF or FEP. Polyester film (Mylar) and sheet (Fig. 6.4) are also subject to weathering which in the glass-reinforced polyester results in pop-out of the glass fibres. This can also be solved by application of a coat of fluorocarbon resin. Industrial air pollutants, which occur in high concentrations in localized areas, may result in damage to any of the above materials but is beyond the scope of this discussion.

While heavy weathering damage leads first to the loss of impermeability and later to the failure of the cover, low levels of damage can reduce the transmittance by as much as 10%. It is of interest that a cover of PMMA can have its original transmittance restored after a decade of exposure by polishing away the weathered layer. In addition to environmental alteration to the smoothness and integrity of the window surface, the transmittance can also be diminished by the deposit of particulate pollutants which can, for example, lower the

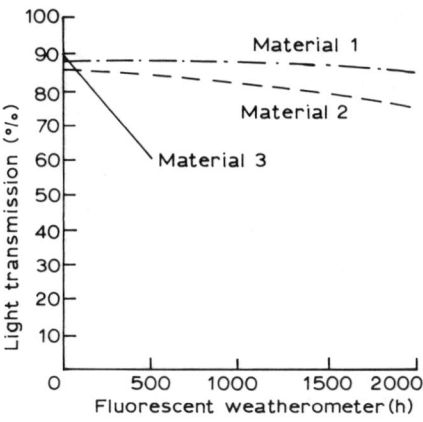

Fig. 6.4. The decrease in light transmission of fibre glass reinforced polyester with increasing accelerated radiation. Sun-Lite Premium FRP (1) is more highly stabilized to UV than Sun-Lite Regular (2) and ordinary FRP (3) [6.4]. (Reprinted with permission from *J. Mater. Energy Syst.*, **2** (1980) 65. Copyright Metallurgical Society.

transmittance of glass by 8% over a 1 month period. Many polymers have electrostatic properties which attract dust and thus suffer a greater decrease in transmittance than glass. Plastics such as Teflon require a surface treatment to render them wettable so that rain will uniformly wash down the surface.

Many of the plastics optically suited for use with flat plate collectors are very soft and subject to scratching and abrasion from dust particles. Their resistance to such damage can be greatly improved by hard coatings not in themselves suitable for films, e.g. polymethylsiloxane for use on PMMA and polycarbonate. Tedlar and Teflon have good resistance to abrasion [6.4–6.8].

Durability characteristics of fluorocarbon, silicone, acrylic, ethylene–acrylic, ethylene–propylene (EPDM) polymers and copolymers were studied inside a solar collector by Mendelsohn *et al.* [6.9]. Although none was found to be entirely satisfactory, the fluorocarbon polymer displayed the best durability and thermal stability overall. The silicones were second best. Wood [6.10] has mentioned that fluorocarbons, such as Teflon, could be used in the form of thin films of 25 μm (1 mil) as inner glazing in solar collectors. Tests reportedly indicate that panels with inner glazing of this type can be 30% more efficient than all-glass panels because one film's lower refractive index allows greater amounts of heat-producing light to pass through, thereby enabling the collector to pick up heat earlier and later in the day.

When an acrylic polymer is substituted for the glass outer glazing, the panel is improved more significantly. A $1 \times 2 \cdot 2$ m (3×7 ft) all-glass panel transmits 77% of the available solar energy and weighs about $22 \cdot 5$ kg (50 lb); a similar glass–Teflon version transmits 85% of the energy and weighs $11 \cdot 25$ kg (25 lb); while a comparable acrylic-fluorocarbon unit transmits 88% and weighs only $5 \cdot 6$ kg ($12 \cdot 5$ lb). Teflon was exposed to the equivalent of 15 years of solar radiation in Florida with no measurable deterioration in any of its performance properties.

Luck & Mendelsohn [6.11] have found that the chemical products evolved through thermal degradation and outgassing usually condense on the glass or polymer glazing surfaces where they significantly reduce the transmittance of solar light and thereby reduce the efficiency of the solar collector.

Figure 6.5 shows the condensable–time diagrams for different polymers that are being used as glazing (cover plates) for thermal solar

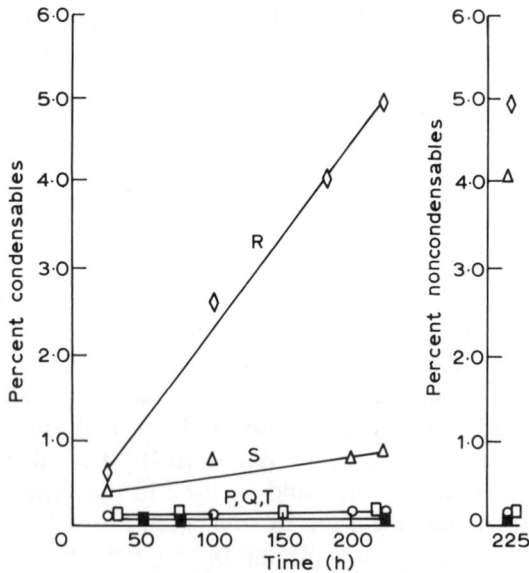

Fig. 6.5. Percentage condensables and non-condensables of glaze polymers as a function of time at 150°C: O, PMMA; □, PC; ◇, CAB; △, GRP; ■, FEP [6.11]. (Reprinted with permission from *Polymers in Solar Energy Utilization.* ACS Symposium Series 220. Copyright, 1983, American Chemical Society.)

collectors. These polymers produce linear curves and only one condensable product was evolved from each polymer. The polycarbonate (Q), the PMMA (P), and the fluorinated ethylene–propylene copolymer (T), essentially did not evolve any condensable products. The glass reinforced polyester (S) gave a linear curve indicating a single type of condensable product identified as an aromatic ester. The cellulose acetate butyrate (R) also produced a linear condensable–time curve which consisted of CAB fragments.

The same authors showed that the evolved products initially condense on the glazing material as liquids, and produce little adverse effects upon the relative light transmittance. However, as these thin condensate films are exposed to heat and oxygen, over long periods of time, they slowly degrade and oxidize, and they are eventually converted into solid products which greatly reduce light transmittance. These effects are presented in Table 6.4. Several of these values were obtained from working collectors while others were obtained from laboratory prepared samples.

The outgassing effects of some other polymers used as sealants, absorber plates and thermal insulation were also studied [6.11].

Brauman *et al.* [6.12] have studied the reactivity with mirror materials of eight polymers used as protective coatings for solar mirrors, such as:

—ethylene–vinylacetate copolymer (EVA),
—ethylene–propylene copolymer (EP),
—polyisobutylene (PIB),
—poly(methylmethacrylate) (PMMA),

or in film form:

—poly(vinylidene fluoride) ($PVDF_f$)
—biaxially oriented $PVDF_f$($PVDF_f$BIAX),
—acrylate–methacrylate copolymer ($Korad_f$),
—poly(ethylene terephthalate) (PET_f).

Using loss in reflectance as an indication of mirror failure and change in tensile properties as an indication of polymer failure, the authors have ranked the polymer/mirror assemblies studied for overall performance. In Table 6.5 the assemblies are ranked in terms of poor, intermediate and good performance. Polymethyl methacrylate (3M)/Al/adhesive is the most durable of the polymer/mirror systems studied. The results indicate that both types of failure (physical and

Table 6.4.
Effect of glaze deposits on relative light transmission [6.11]

Deposit	Quantity g/m^2 (oz/ft^2)	Relative light transmittance % reduction
Water leaching of glass		
Salts	0·151(0·004)	20
Salts	0·237(0·007)	34
Silicone rubber fragments		
Liquid	0·4 (0·012)	1
Powder	0·129(0·004)	18
Butyl rubber fragments		
Liquid	0·2 (0·006)	1
Solid varnish	0·08(0·002)	4
Acrylic fragments		
Solid film	0·02(0·0006)	1
Stearic acid		
Liquid	0·4 (0·012)	1
Solid	0·4 (0·012)	5
Processing oils		
Liquid	0·4 (0·012)	1
Solid varnish	0·2 (0·006)	10
EPDM fragments		
Liquid	0·1 (0·0032)	1
Solid varnish	0·1 (0·0032)	4
PVC fragments		
Liquid	0·05 (0·0016)	1
Varnish	0·1 (0·0032)	3

chemical) can often be reduced by proper selection of the polymer/mirror system and the additives in the polymer.

Failure occurs as delamination or degradation and yellowing at the polymer–mirror interface. Physical failure was evident as delamination of polymer and mirror during weathering of PMMA/Ag/glass. These mirrors remained shiny and clear throughout the exposure period, but they showed some loss in reflectance after 9 days of exposure, when some samples also began to delaminate (Fig. 6.6).

Table 6.5.
Ranking of the weathering performance of polymer/mirror assemblies [6.12]

Category	Assembly[a]
Poor	EVA/Ag/glass
	EP/Ag/glass
	EP_c/Ag/glass
	PIB/Ag/glass
	PIB_c/Ag/glass
	$Korad_f$/Ag/Antique
Intermediate	PMMA/Ag/glass
	$PVDF_f$/Ag/sandwich
	$PVDF_f$/Ag/Glidden
	$PVDF_f(BIAX)$/Ag/Glidden
	PET_f/Al
Good	$PMMA_f$ (3M)/Al/adhesive/glass

[a] c = polymer compounded with 3 wt% carbon black; f = polymer in film form.
Reprinted with permission from Brauman, S. K., MacBlane, D. B. & Mayo, F. R. In *Polymers in Solar Energy Utilization*, ed. C. G. Gebelin, D. J. Williams, & D. Deanin. ACS Symposium Series 220, American Chemical Society, Washington, D.C., 1983, p. 128. Copyright, 1983, American Chemical Society.

Reaction at the polymer/mirror interface, observed only for silver mirrors coated with EP, EVA and PIB, was indicated by varying degrees of loss in reflectance and degradation of the polymer. The most noticeable change was yellowing of the mirrors (indicated by the arrow in Fig. 6.7) that preceded loss of reflectance.

Based on their findings the authors concluded that combinations containing $PMMA_f$ and possibly $PVDF_f$ and PET_f were the most durable polymer coatings.

The superiority of acrylic polymers used for concentrating collector mirrors or lenses was also underlined by other authors [6.5].

At the Brookhaven National Laboratory several solar collectors have been built and tested [6.13]. The major contribution of this design towards high performance and low cost evolves from the use of high performance polymer films in the window and absorber/heat exchanger portions of the solar panel. Present efforts in the development of this technology have identified polymer films which can meet the requirements of performance and cost for solar energy collectors

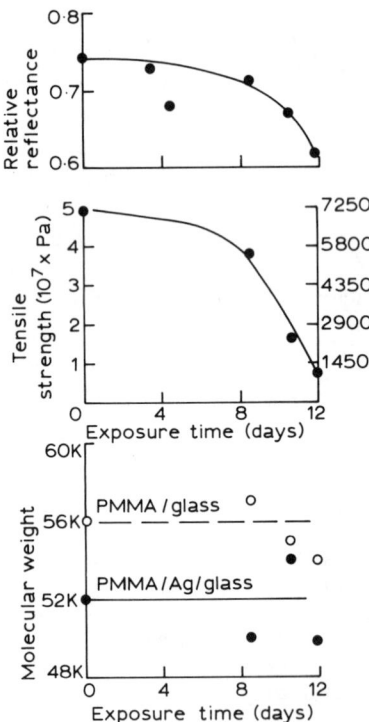

Fig. 6.6. Weathering of polymethyl methacrylate systems. ●, PMMA/Ag/glass; ○, PMMA/glass [6.12]. (Reprinted with permission from *Polymers in Solar Energy Utilization.* ACS Symposium Series 220. Copyright, 1983, American Chemical Society.)

for residential and commercial buildings. Figure 6.8 presents an example of a low cost installation.

Berry & Dursch [6.14] studied the optical and mechanical characteristics of several commercially available polymer films. Test results to date indicate that the fluoropolymers exhibit the best weathering characteristics. Little or no optical or mechanical degradation was observed after the equivalent of 16 years. The metallized polyesters and polycarbonates, although stabilized, were severely degraded in mechanical properties in both real time and accelerated exposures.

Wischmann [6.15] reported about the ageing of different protective coatings for back surfaces of mirrors such as polyurethane, polysulphide, poly(vinylidene chloride) and styrene–butadiene block copolymer. The author has found that mirror corrosion was manifested in a variety of forms, as will be discussed later on.

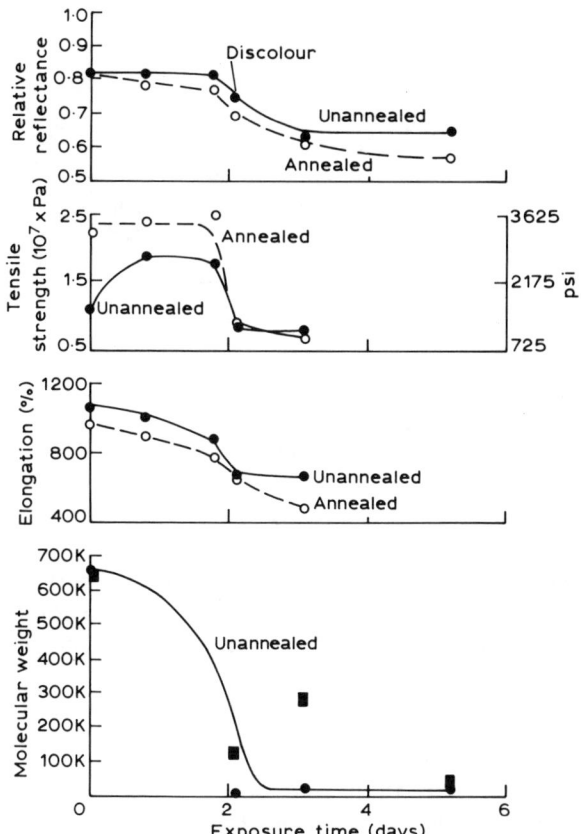

Fig. 6.7. Weathering of EVA/Ag/glass systems. ●, EVA/Ag/glass unannealed; ○, EVA/Ag/glass annealed; ■ EVA/glass unannealed [6.12]. (Reprinted with permission from *Polymers in Solar Energy Utilization*. ACS Symposium Series 220. Copyright, 1983, American Chemical Society.)

Different aspects of polymer films for mirror glazings are discussed in technical literature [6.6, 6.8, 6.16–6.31].

ENCAPSULATION

The encapsulation of photovoltaic cells is well known and has been thoroughly investigated but much less has been written on the use of polymers as encapsulants for phase change materials (PCM).

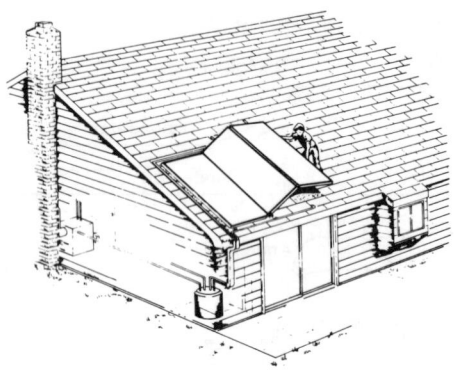

Fig. 6.8. Example of low cost installation [6.13]. (Reprinted with permission from *Polymers in Solar Energy Utilization.* ACS Symposium Series 220. Copyright, 1983, American Chemical Society.)

Encapsulants are necessary for electrical insulation of the photovoltaic circuit; they also provide mechanical protection for the solar cell wafers and corrosion protection for the metallic contacts and circuit system.

Cuddihy *et al.* [6.32] have studied four polymers used as pottant (i.e. material which is the actual encapsulation medium in a module) such as:

—ethylene–vinylacetate copolymer (EVA),
—ethylene–methylacrylate copolymer (EMA),
—poly-*n*-butyl acrylate (P-*n*BA),
—aliphatic polyether–urethane (PU).

The module assembly and the position of the pottant macromolecular compound are shown in Fig. 6.9.

The optical, mechanical, electrical and chemical requirements for such materials were also discussed by Lewis [6.33], who also mentioned that block or graft thermoplastic elastomers with relatively low MW amorphous segments of a weather resistant saturated backbone have the potential of being superior polymers for potting solar cells. The cross-link forming crystalline segments make relatively soft, low MW, rubbery polymers handle well. They exhibit high cohesive strength or toughness and low surface tack when the crystalline domains are solidified (Fig. 6.10).

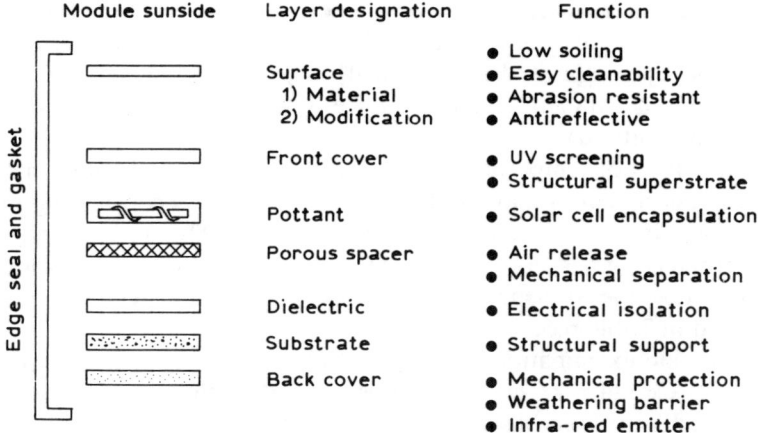

Fig. 6.9. Construction elements of photovoltaic encapsulation systems [6.32]. (Reprinted with permission from *Polymers in Solar Energy Utilization.* ACS Symposium Series 220. Copyright, 1983, American Chemical Society.)

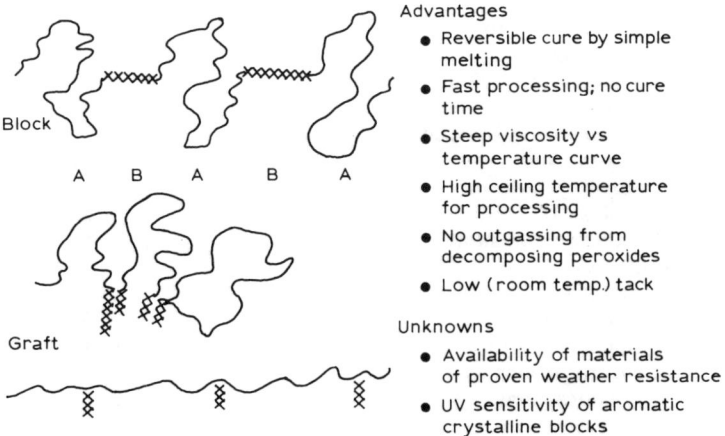

Fig. 6.10. Structures of thermoplastic elastomers [6.33]. (Reprinted with permission from *Polymers in Solar Energy Utilization.* ACS Symposium Series 220. Copyright, 1983, American Chemical Society.)

The ageing behaviour of several encapsulant candidates for photo-voltaic module designs with a plastic front surface were studied in the field and by accelerated ageing [6.34]. The results suggest that of the two pottants studied, EVA is more resistant to degradation than poly(vinyl butyral).

The effects of temperature and ionizing particle irradiation on some polyorganosiloxanes and epoxy polymers used as encapsulating materials have been studied by Beatty *et al.* [6.35]. The design, cost, and performance goals for encapsulation are discussed by Sarbolouki [6.36]. For the transparent polymeric pottants the costs are summarized in Table 6.6.

From an examination of all existing and conceived flat module designs, Cuddihy *et al.* [6.37] observed that these designs can be separated into three basic classes as illustrated in Fig. 6.11. These are designated as substrate bonded, superstrate bonded and laminated. The designations refer to the design method by which the solar cells are fixed or contained within the encapsulated module.

Table 6.6.
Transparent polymeric pottants [6.36]

Candidate materials	Cost ($/kg)	Cost[a] ($/m²) of module)
Weatherable with UV protection		
Ethylene/propylene rubber (EPDM)	1·1	0·71
Ethylene/vinyl acetate copolymer	1·1	0·71
Polyvinyl chloride plastisol	1·3	1·2
Weatherable without protection		
Acrylics	3·3	3·2
Silicone gels	8·2	6·5
Silicone elastomers	19·8	16·1
Fluorocarbons and halocarbons	14·8–44	>16

[a] Assuming 0·76 mm (0·030 in) of pottant. To obtain cost in 0·01$/ft² of module, multiply by 10.
Reprinted with permission from *Mod. Plast.* (Jan. 1980) 122.
Copyright McGraw-Hill.

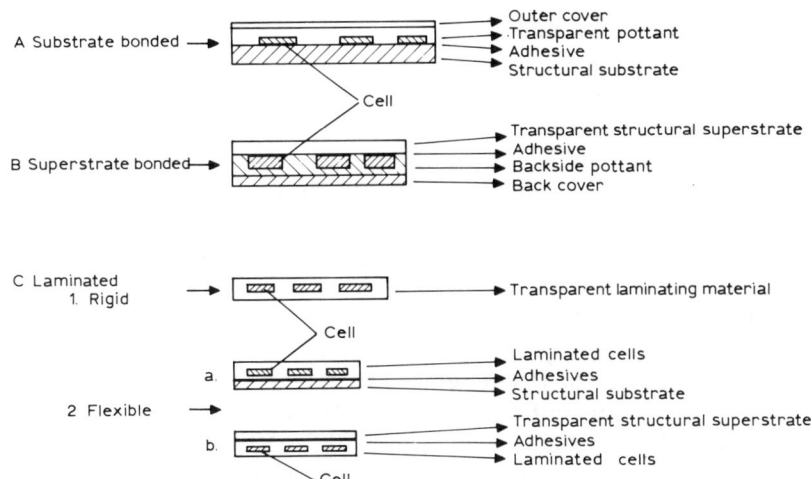

Fig. 6.11. Flat module design classification [6.37]. (Reprinted with permission from *Solar Energy*, **22** (1979) 389–96. Copyright Pergamon Press.)

From these three design classes, the authors identified the following six basic construction elements:

(1) outer covers;
(2) transparent pottants (also the laminating material);
(3) adhesives;
(4) structural substrates;
(5) structural transparent substrates;
(6) backside pottants and/or covers for the substrate bonded design.

A material survey revealed the following transparent elastomers could be considered as viable pottant candidates:

—ethylene–vinylacetate copolymer (EVA), $1.10/kg;
—poly(vinyl chloride) plastisol (PVC), $1.32/kg;
—ethylene–propylene rubbers (EPR), $ 1.10/kg.

As a result of an economic analysis on low cost encapsulation materials for terrestrial solar modules Cuddihy *et al.* [6.37] indicated that to meet a low cost goal, the following material technologies must be developed or advanced:

—UV screening outer covers;
—elastomeric acrylics;
—transparent UV stabilizers for UV sensitive transparent pottants;
—cost effective utilization of silicone and fluorocarbon polymers.

Feldman *et al.* [6.38] have recently published a study aimed at developing cost effective polymeric thermal storage modules easy to retrofit in buildings. The goal is to reduce the consumption of fuels and electricity for heating and cooling buildings by utilizing solar and internal heat gains (from lights, people, etc.) and night air cooling with low energy fans rather than refrigeration machines.

The modules, tiles of about 12 mm (0·5 in) depth, are applied to the ceilings or upper walls of rooms over various areas depending on the heating or cooling requirements of the particular room. To obtain a large thermal storage capacity, about half the weight of each tile is composed of materials which melt near room temperature with high latent heats of melting (50–60% of ice). The heat storage capacity is then equivalent to the specific heat stored in about 100 mm (4 in) of concrete cycled through 10°C, but the fluctuation in room temperature required for storage in the latent heat tiles is much smaller and can be kept within the comfort range (about 5°C), provided heat transfer to the tiles is sufficiently rapid at small temperature differences.

Sufficient heat transfer is ensured in part by the large surface area-to-volume ratio of the tiles, but also by the polymer matrix that makes up most of the rest of the tile material. In a monolithic block of PCM subjected to a small temperature difference, the melting or freezing plane advances very slowly because the large latent heat ensures nearly isothermal conditions near the melting or freezing plane. In addition, for the freezing part of the cycle, heat transfer to air is impeded because the material between the freezing plane and the air is solid, so convective heat transfer to the outside of a monolithic PCM block cannot occur. The polymer matrix serves as a thermal short circuit between liquid at the back of the tile and the air at the front, increasing the effective area for freezing and the average temperature gradient across the convoluted freezing surface. It also serves, of course, to provide structural strength when the PCM is liquid. Good heat transfer at night when the PCM is freezing is important to ensure maximum utilization by day of waste or solar heat in cold climates or maximum cooling capacity in warm climates. The more PCM that freezes at night the more that can melt the next day.

The study reported on some mechanical and flammability tests for the tiles, but primarily on some evidence of interactions between various polar polymers used for the tile matrix, and PCM (mixtures of saturated fatty acids). These interactions increase the surface tension between the fatty acids and the polymer matrix so that, when liquid, the fatty acids do not seep out of the tiles even with large pores. This allows a tile with no outside surface covering to impede heat transfer between the air and the PCM. Thus, the polymer serves as an internal packaging material rather than an external container, as well as a heat transfer aid.

Several candidate PCMs have been studied recently for use as thermal storage in buildings or with solar heaters. Most of the work has been done on salt hydrates such as Glauber's salt and calcium chloride hexahydrate. These are attractive because the hydrates themselves are very cheap and they have high latent heats of fusion. However, the salt hydrates are corrosive to most building materials. Additives must be used with them to prevent supercooling and segregation of water from unhydrated salt after incongruent melting. The additives can reduce the latent heat per unit volume of a packaged storage module by 25%. Containers for salt hydrates cannot be metal, because of corrosion problems, but must be impermeable to water vapour to prevent escape of the water of crystallization by diffusion. One package for Glauber's salt consisted of an aluminium foil pouch to prevent water loss lined with polyethylene film to protect the aluminium, placed in a polymer concrete panel for mechanical strength [6.28]. Since the concrete was heated well above room temperature by direct adsorption of sunlight, no heat transfer problem occurred, but that would not be the case for the small temperature swings to be utilized in some of the applications for these tiles.

Fatty acids were originally chosen as PCMs for these tiles because none of the above problems are encountered with their use. They are not corrosive to any common building material, hardly supercool at all, and melt congruently. Water vapour has virtually no effect on them. The fatty acids cost more than salt hydrates on a bulk basis, but are cheaper to package, so the final module costs are comparable. For passive thermal storages intended to operate with small temperature gradients, the extra heat transfer for a package with no outer container is an additional advantage.

This is one more example of the important role that polymers can play in the development of energy conservation and solar utilization systems. They offer potentially lower costs, easier processing, lighter weight and greater design flexibility than conventional materials such as glass, metals, wood and concrete [6.2, 6.39]. The market for polymers in these industries has grown rapidly [6.10]. Polymers have been used as external containers for salt hydrate thermal storage because of their resistance to corrosion and to the diffusion of water vapour. Some examples are: polyester film pouches, polyethylene tubes containing calcium chloride hexahydrate and high density polyurethane trays containing Glauber's salt [6.17, 6.40, 6.47].

The polymers, used either in the powdered state or as granules (polyethylene), were the following:

—poly(vinyl chloride) (PVC), Geon resin, B. F. Goodrich;
—poly(vinyl alcohol) (PVA), Mowiol 66-100, Hoechst Can.;
—poly(vinyl acetate) (PVAc), Mowilith DS, Hoechst Can.;
—vinyl acetate–vinyl chloride copolymer (VAc–VC), Hostaflex 131, Hoechst Can.;
—polyethylene, high density (HDPE), Union Carbide.

The phase change materials used were:

—lauric acid (LA), Anachemia Chemicals Ltd;
—stearic acid (SA), American Chemicals Ltd;
—pressed stearic acid (PSA), Emery 400, Emery Industries Ltd.

Table 6.7 shows the properties of Emery 400. This product is 60%

Table 6.7.
Pressed stearic acid–emery E400: properties [6.38]

Melting temperature (°C)	58
Heat of fusion (kJ/kg)	162
Heat of storage density (MJ/m^3)	125
Density (kg/m^3) (solid)	949
Density (kg/m^3) (liquid)	823
Specific heat (kJ/kg°C) (solid)	1·8
Specific heat (kJ/kg°C) (liquid)	2·4

stearic acid, 30% palmitic, with 10% of various fatty acids, depending on the batch.

Bleached cellulosic fibres or glass fibres were used as reinforcing agents and paraffin chloride (Cereclor 70 powder, Canadian Industries Ltd) and antimony trioxide (Anachemia Chemicals Ltd) were used as fire retardants. Using these materials, mixed in a blender, $200 \times 200 \times 20$ mm tiles were moulded with different ratios between the matrix and the PCM. Some of these formulations are presented in Table 6.8.

Generally, these modules keep their shape and dimension up to 37–43°C, depending on the composition, without losing any fatty acid. Those made of HDPE showed no deformation up to 51°C. The samples were kept under cold conditions for 1 month in a humidity chamber at 3°C and 69% RH. No deformations and only small losses in weight, due to spalling, as in the case of sample No. 9 (Table 6.8), 0·47% and sample No. 11, 2·59% (the maximum measured), were observed.

Cellulose fibres and glass fibres were used as reinforcing agents to improve the mechanical properties of some of the tiles. Tests for compressive strength and shrinkage under load were done according to ASTM C-109-80. Some of the results are presented in Table 6.9. Better mechanical properties were usually obtained with a larger amount of long glass fibres.

Flammability tests, with a Bunsen burner in a fume hood, according to ASTM D-635, show that some of the fire retardants usually used with polymers give safe tiles. The best results were obtained with formulations such as:

PVC	1 part
Fatty acid mixture (LA + SA 3:1)	1·4 parts
Cellulose fibres	0·14 parts
Antimony trioxide	0·1 parts
Paraffin chloride	0·3 parts

The IR spectrum of SA–PVC systems shows a slight decrease of the frequency of the C=O group from 1700 to 1690 cm^{-1} which might be interpreted as a decrease of carboxyl association in SA in favour of a new association in SA–PVC. The 1430 cm^{-1} band in SA, due to a coupling of the deformation vibration of OH with the C=O group vibration and the valency vibration at 1295 cm^{-1}, also shows shifts towards lower frequencies when PVC is added, namely to 1420 and

Table 6.8.
Examples of formulations used for tiles [6.38]

Tile No.	Polymer	Phase change material	Polymer/PCM w/w ratio	Reinforcement
1	PVA	SA + LA 3:1 mixture	1/1·25	—
2	VAc–VC	SA + LA 1:1 mixture	1/1·66	—
3	PVC	SA + LA 3:1 mixture	1/1	—
4	PVA + PVC 1:1 polyblend	SA + LA 3:1 mixture	1·25/1	—
5	PVA + VAc–VC 8:10 polyblend	SA + LA 3:1 mixture	1/1·25	—
6	PVA	SA + LA 1:3 mixture	1/1·2	—
7	VAc–VC	SA + LA 1:3 mixture	1/2	—
8	PVAc	SA + LA 1:3 mixture	1/1·8	5% cellulose fibres
9	PVAc	SA	1/1	5% glass fibres
10	PVC	SA	1/1	10% glass fibres
11	PVC	SA	1/1	—
12	HDPE	Emery E400	1/1	—

Reprinted with permission from *Polym. Engng Sci.*, **25**(7) (1985) 406. Copyright Society of Plastics Engineers.

Table 6.9.
Compressive strength and shrinkage under load according to ASTM C-109-80 [6.38] (50 mm cube specimens were used)

Polymer	PCM	Polymer/PCM w/w ratio	Reinforcement[a]	Ultimate compressive strength kN (psi)	Shrinkage under load (%)
PVC	SA	1:1·25	2·5% short glass fibres	2·8(0·40)	1·2
PVC	SA	1:1·25	5% short glass fibres	4·4(0·63)	1·4
PVC	SA	1:1·25	2·5% long glass fibres	4·6(0·66)	1·4
PVC	SA	1:1·25	5% long glass fibres	8·5(1·23)	1·8

[a] The short glass fibres have lengths in the range 15–25 mm, and the long ones between 85 and 110 mm.
Reprinted with permission from *Polym. Engng Sci.*, **25**(7) (1985) 406. Copyright Society of Plastics Engineers.

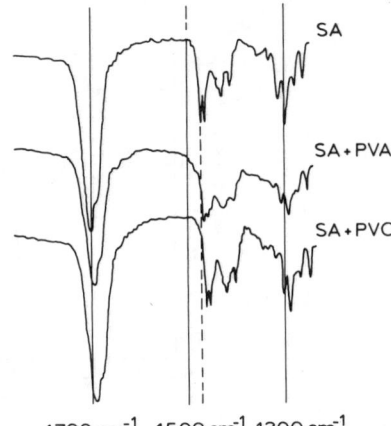

Fig. 6.12. Shifts in the main IR spectral peaks of the C=O groups in SA and SA–PVC and SA–PVA systems [6.38]. (Reprinted with permission from *Polym. Engng Sci.,* **25**(7) (1985) 406. Copyright Society of Plastics Engineers.)

$1290 \ cm^{-1}$. This also reflects a decrease in the association of carboxylic groups (Fig. 6.12, Table 6.10).

Similar observations can be made for the SA–PVA system (Fig. 6.12, Table 6.10).

Indirect evidence of polymer–fatty acid interaction may be indicated by shifts in the glass transition temperature, T_g, in the DSC results [6.42]. The DSC curves for PVC and the 1:1 PVC–SA system are presented in Fig. 6.13. Figure 6.14 shows mixtures of PVC with increasing amounts of SA. The shifts of the T_g of PVC could indicate

Table 6.10.

The main peaks of the C=O group in stearic acid and in binary SA–PVC 1:1 and SA–PVA 1:1 systems [6.38]

Vibration type	SA	SA–PVC 1:1	SA–PVA 1:1
C=O dimer	1 700	1 690	1 695
Deformation vibration of OH with the C=O group	1 430	1 420	1 425
Valency vibration	1 295	1 290	1 293
—OH deformation	940	930	935

Reprinted with permission from *Polym. Engng Sci.,* **25**(7) (1985) 406. Copyright Society of Plastics Engineers.

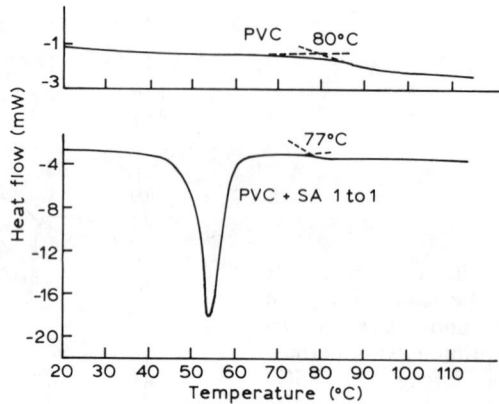

Fig. 6.13. The DSC diagrams for PVC and 1:1 PVC–stearic acid systems [6.38]. (Reprinted with permission from *Polym. Engng Sci.*, **25** (7) (1985) 406. Copyright Society of Plastics Engineers.)

polymer–fatty acid interaction [6.43]. These interactions may be either of the dipole–dipole type or the hydrogen bonding type. Hydrogen bonding is responsible for more miscible polymer-additive systems then any of the other types of interactions, and thus is more likely to be the cause of these effects.

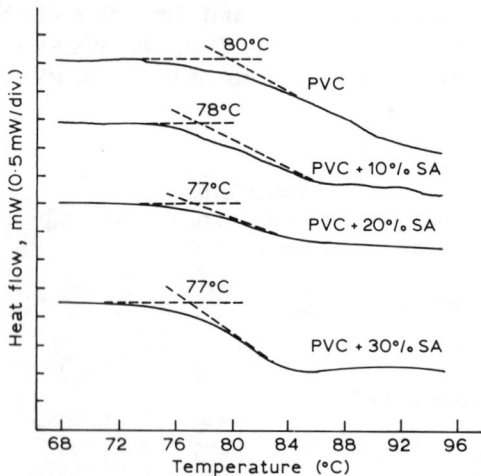

Fig. 6.14. The DSC diagrams for PVC–stearic acid mixtures [6.38]. (Reprinted with permission from *Polym. Engng Sci.*, **25**(7) (1985) 406. Copyright Society of Plastics Engineers.)

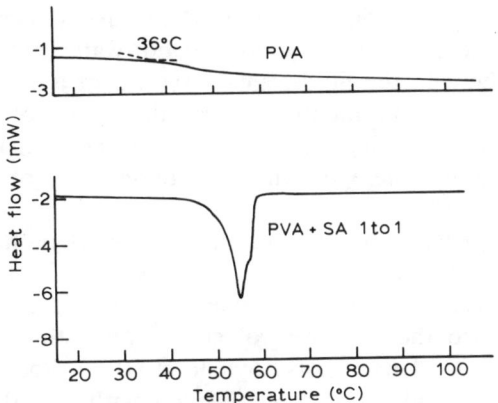

Fig. 6.15. The DSC diagrams for PVA and the 1:1 PVA–stearic acid system [6.38]. (Reprinted with permission from *Polym. Engng Sci.*, **25**(7) (1985) 406. Copyright Society of Plastics Engineers.)

In addition to the decrease of the T_g in the PVC–SA systems, a slight decrease of the melting point of SA from 56 to 54°C and modification of its enthalpy of fusion from 160 to 148 J/g (PVA) and 146 J/g (PVC) was observed. Similar effects were observed in the cases of PVC mixtures with other fatty acids. The initial T_g of PVA (48°C in pure PVA) cannot be observed in the mixture curve because it is masked by the melting endotherm of the SA (Fig. 6.15). These modifications of the PVA:SA system may be explained by interactions (hydrogen bonding) between these components. Such interactions could also lead to a possible tendency of the system towards crystallization at lower temperatures.

Further investigations are in progress to clarify the nature of the interactions in these systems and of the improvement of the thermal behaviour of polymer–PCM modules.

PIPES

Properties such as low thermal conductivity, light weight, corrosion resistance and ease of installation make some types of polymer pipe very attractive for use in solar energy equipment [6.1]. It is clear that glass has decreasing prospects for retaining its status as a material for solar still covers [6.44]. Some authors [6.5] have shown that conventional chlorinated PVC or polybutylene hot-water pipe can be used almost everywhere in the fluid-transfer system of a solar house. Two exceptions have to be mentioned: in the collector itself, where temperatures are too high for continuous performance, and, in heat-exchange units where the thermal conductivity of plastics is too low.

Polyethylene, PVC and more recently polybutylene have been used for pipes in ground coupled heat pump systems. Metz [6.45] has reported that the use of rigid pipe materials such as PVC has diminished due to the labour involved in connecting joints and the risks of leaks. Low density polyethylene (LDPE) pipe has been used widely in heating-only systems in Sweden with an ethylene glycol–water solution, because it is unsuitable for higher temperatures. Medium density polyethylene (MDPE) pipe has been used in experiments in the US at temperatures between -10 and $+50°C$ using water and ethylene glycol–water solutions. High density polyethylene (HDPE) and polybutylene have better pressure resistance and durability at elevated temperatures.

Process piping accounts for a small but rapidly growing fraction of the total. The low installation cost of thermoplastic pipe systems is an advantage, but it is not decisive. In most cases reliable performance and durability are so important that the cost of the system is negligible compared with the consequences of failure. In the selection process of polymeric pipes we have to take into account the following:

—Engineers are often justified in preferring to select a well established material with a known, although possibly limited durability, than one which promises increased durability at lower costs but is untried.
—Thermoplastics suffer from comparatively low strength and rigidity, especially at elevated temperatures. At 100°C few thermoplastics can be considered for selection and those remaining have design stresses as low as $1·0–3·4$ MN m^{-2} (145–500 psi) [6.46].

COATINGS, ADHESIVES, SEALANTS

Protective coatings are being developed to improve the abrasion resistance of metallized plastic mirrors and they offer the following advantages over the currently available materials:

—low cost,
—light weight,
—improved optical performance.

Between the existing coatings are [6.47]:

—high silica coatings; require high temperature cure (>100°C);
—abrasion resistant coating applied to acrylic sheet material;
—thermoset silicone coating for copper and aluminium.

Wischmann [6.15], using an artificial weathering chamber, has investigated the use of some polymers as protective coatings for back surfaces of mirrors and solar heliostat edge seals.

For the former application, polyurethane, polysulphide, poly(vinylidene chloride) and styrene–butadiene block copolymer were used. The coatings were evaluated on:

(a) mirrors which contained only the silver and standard sacrificial copper layer; and
(b) mirrors having the familiar protective grey paint.

The sulphur of the polysulphide almost immediately attacked the silver. The poly(vinylidene chloride) did not manifest satisfactory adhesion towards the substrate and began to peel after 7 days. The polyurethane offered some protection, but with time, it also began to peel, and thereafter mirror corrosion ensued. The styrene–butadiene copolymer afforded the greatest protection, exceeding all other coated samples in durability; this result was expected since this copolymer is a hydrophobic hydrocarbon. The comparison shows that the hydrocarbon coatings behave best when they are used as moisture barriers.

In the latter application, three different silicone and a butyl rubber sealant were tested as edge seals. Test specimens were prepared by bonding a Gardner mirror to an appropriate substrate with an amine cured polyurethane and then sealed around the mirror edge with a bead of sealant. The different substrates used were:

—cellular glass,
—polystyrene foam,
—polystyrene foam with an added butyl rubber pad,
—paper honeycomb sealed with epoxy–fibre glass or melamine,
—sine-wave fibre glass sealed with epoxy–fibre glass or melamine.

The findings in this ageing study resulted in the following conclusions which should have a bearing on heliostat designs:

(1) Edge seal choices seemed of minor importance compared to the overall module design and bonding integrity. Although they age well, silicone edge seals appear to bond poorly and/or permeate moisture easily.

(2) To minimize moisture diffusion pathways either a better barrier is needed, that is, an improved design, or to ensure the integrity of the various bond interfaces, i.e. edge seals to substrates, silver to glass, etc.

(3) Mirror corrosion was manifested by a variety of forms such as pitting, spotting, delaminations and discolorations with no particular pattern. Efforts were not made to elucidate corrosion mechanisms, yet it is believed the first step in the degradation process is delamination or creation of a void area at some critical interface where harmful reactants, i.e. water, collect, thereby precipitating degradation.

(4) Compatibility of various materials in these heliostat designs is an important consideration which, if properly addressed, can add longevity to the system. For example, a product which outgasses a relatively benign by-product like alcohol (as opposed to acetic acid) would be a logical choice as an edge sealant. In general, butyl rubber sealants were superior to silicones, provided a UV stabilized product is used.

An efficient solar collector is one which has a high absorptance in the visible and ultra-violet range of the spectrum, and a low emittance in the infra-red range. Between the two groups of coatings used for solar collectors, the so-called 'selective' ones have a high absorptance in the wavelength range of 1–3 μm and low emittance in the range of 3–15 μm. 'Non-selective' coatings have both high absorptance and high emittance in those ranges. Paul & Gumbs [6.17] studied whether the dark, thermally stable polymers produced on heating (PAN or polybutadiene) could be used in the preparation of solar energy-absorbing coatings. It was suggested that the colour-producing mechanism observed during 1,2-polybutadiene or PAN pyrolysis in the range of 200–300°C involves the formation of ladder polymers containing conjugated double bonds (Fig. 6.16). The objective was to find whether thin films of these ladder polymers coated on metal surfaces could be heated to give the optical characteristics necessary for solar energy absorbers.

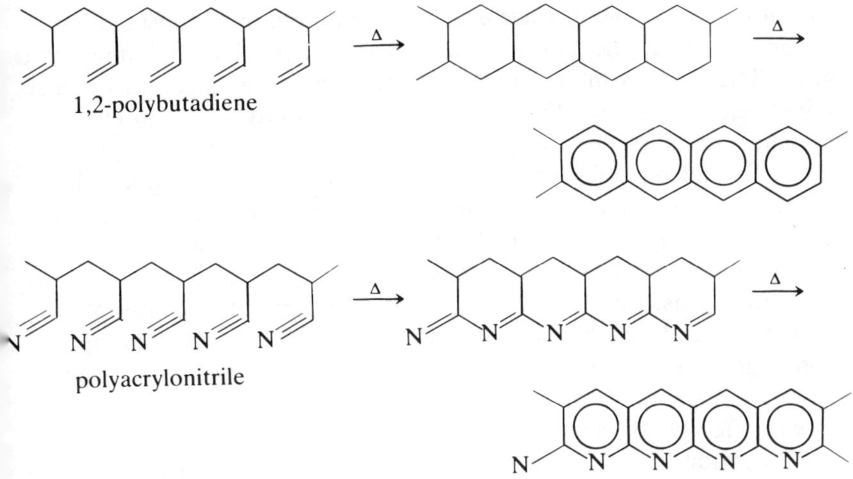

1,2-polybutadiene

polyacrylonitrile

Fig. 6.16. Formation of ladder polymers [6.17]. (Reprinted with permission from *J. Appl. Polym. Sci.*, **21** (1977) 959. Copyright, 1977, John Wiley.)

The results obtained indicate that it is feasible to prepare solar energy-absorbing coatings by heating thin films of these polymers on metallic surfaces.

Pohlman [6.48] underlines the importance of the interfacial adhesion of materials with respect to their durability. Good interfacial adhesion between materials, i.e. metal/glass, metal/polymer and metal/metal, is of primary importance for long term reliability of solar concentrator systems. Reflective metals on glass or polymers must maintain good adhesion to have a cost-effective lifetime. Protective organic coatings over metallized surfaces must adhere to them in order to provide protection from moisture, atmospheric pollutants, etc. Structural adhesives used to join the structural back-up panel to the reflective surface must adhere in order to maintain the structural integrity and dimensional stability of the subsystem. Sealants must adhere intimately to panel edges (glass, plastic, metal, wood, etc.) to stop moisture or airborne pollutants from inward migration and eventual attack of the metallized reflective surface.

POLYMERS FOR SOLAR PONDS

Flynn *et al.* [6.49] reported a few years ago that the solar pond is one of the most promising energy technologies for large scale collection and storage of solar energy. The purchase and installation of a liner is one of the major capital costs incurred in building solar ponds. The authors underline that more research is needed to realize an inexpensive, reliable liner with low permeability and high chemical resistance to chlorides at temperatures up to the boiling point of water. The advent of such a liner would bring the solar pond firmly into the market place as a viable alternative to the present, non-renewable energy source methods for generating electricity.

Many plastics and elastomers have been used as flexible membrane liners for ponds, pits, lagoons, canals and reservoirs. Thermoplastics are readily and reliably seamed, but are usually protected against photodegradation. Some vulcanized elastomers have good weatherability but are not readily seamed after vulcanization. Thermoplastic rubbers offer good weathering in addition to easy seaming. Woodley [6.50] mentioned the following polymers as those most used as liners for solar ponds:

(a) thermoplastics: PE, PVC;
(b) vulcanized rubbers: butyl rubber, ethylene–propylene–diene comonomer terpolymer (EPDM);
(c) thermoplastic rubbers: chlorinated polyethylene, chlorosulphonated polyethylene.

The requirements for impermeable liners include:

(1) Chemical resistance to salt brine; ranging from saturated solutions diluting down to essentially fresh water.
(2) Heat resistance; serviceable from freezing or below, to over 100°C.
(3) Weatherability; resistant to UV and ozone and capable of continuous outdoor exposure.
(4) Durability; cost effective performance for 20 years or more.
(5) Reliability; pin-hole free construction, resistance to mechanical damage, delamination and blistering.
(6) Repairability; easy to seam with seams stronger than parent sheet. Homogeneous seams that do not require special equipment or skilled operators for field repairs.

Woodley emphasizes the importance of the selection of the lining polymer because leaks cause loss of brine, loss of heat, loss of insulation, and are potential environmental pollutants. Following his study, the combination of easy seaming and repairs, outstanding weathering, and performance in high temperature, high concentration brine solutions makes chlorosulphonated polyethylene thermoplastic elastomer a prime candidate for lining salt-gradient solar ponds.

REFERENCES

[6.1] Blaga, A., *Solar Energy*, **21**(4) (1978) 331–38.
[6.2] Caroll, W. F., & Schissel, P., Polymers in Solar Technologies: A R & D Strategy, SERI/TR-334-601. Solar Energy Research Institute, Golden, Co., 1980.
[6.3] Caroll, W. F., & Schissel, P., In *Polymers in Solar Energy Utilization*, ed. C. G. Gebelein, D. J. Williams & R. D. Deanin. ACS Symposium Series 220, 1983, p. 3.
[6.4] McQueen, H. J., Shapiro, M. M., & Feldman, D., *J. Mater. Energy Syst.*, **2** (1980) 65–82.
[6.5] Martino, R., *Mod. Plast.*, **53** (1976) 52–5.
[6.6] White, J. S., *Solar Energy Catalog*. Solar Vision Inc., USA, 1977, p. 46.
[6.7] White, J. S., *Polym. News*, **3** (1977) 239.
[6.8] McQueen, H. J., SESCI—National Conference on Solar Energy, Montreal, 1981.
[6.9] Mendelsohn, M. A., Navish, F. W., Jr, Luck, R. M., & Yedman, F. A., In *Polymers in Solar Energy Utilization*, ed. C. G. Gebelein, D. J. Williams & R. D. Deanin. ACS Symposium Series 220, 1983, 39–80.
[6.10] Wood, A. S., *Mod. Plast.*, **53** (1976) 42.
[6.11] Luck, R. M., & Mendelsohn, M. A., In *Polymers in Solar Energy Utilization*, ed. C. G. Gebelein, D. J. Williams & R. D. Deanin. ACS Symposium Series 220, 1983, pp. 81–98.
[6.12] Brauman, S. K., MacBlane, D. B., & Mayo, F. R., ibid., pp. 125–42.
[6.13] Wilhelm, W. G., ibid., pp. 27–38.
[6.14] Berry, M. J., & Dursch, H. W., ibid., pp. 99–114.
[6.15] Wischmann, K. B., ibid., pp. 115–24.
[6.16] Temple, P., & Kohler, J., *Solar Age* (April 1979) 25.
[6.17] Paul, D. F., & Gumbs, R. M., *J. Appl. Polym. Sci.*, **21** (1977) 959.
[6.18] Hall, A., *Plastics World* (May 1984) 52.
[6.19] Fayet, P., Paillous, A., & Sable, C., *Proceedings of the International Congress on Sun in the Service of Mankind*, UNESCO & Centre Nationale d'Etudes Spatiales, Paris, July 1973, p. 283.
[6.20] Godbey, L. C., Bond, T. E., & Zornig, H. F., *Proceedings of the Annual Meeting of American Society of Agricultural Engineers*, 1977, No. 77–4013.
[6.21] Garg, H. P., *Solar Energy*, **15** (1974) 299.
[6.22] Selkowitz, S., *Proceedings of the 2nd Passive Solar Conference, Vol. 2*, International Solar Energy Society, American Section, Philadelphia, PA., 1978, p. 329.

[6.23] Broder, J. D., & Mazaris, G. A., *Proceedings of the 10th Photovoltaic Conference,* Palo Alto, CA, Institute of Electrical & Electronics Engineers. IEEE, Piscataway, N.J., 1973, p. 272.
[6.24] Treble, F. C., *Proceedings of the 7th Photovoltaic Conference,* Pasadena, CA, Institute of Electrical & Electronics Engineers. IEEE, Piscataway, N.J., 1969, p. 226.
[6.25] Davis, A., Deane, G. H. W., Gordon, D., Howell, G. V., & Ledbury, K. J., *J. Appl. Polym. Sci.,* **20** (1976) 1165.
[6.26] Davis, A., & Gordon, D., *J. Appl. Polym. Sci.,* **18** (1974) 1173.
[6.27] Lindrose, A. M., & Guess, T. R., *Proceedings of the National SAMPE Symposium Exhile (NSSED 2),* Society for Advancement of Materials and Process Engineering. Azusa, CA. Vol. 23, 1978, p. 386.
[6.28] White, J., *Proceedings of the 32nd Annual Technical Conference,* Reinforced Plastics, S.P.J. Inc., Section 1F, 1977, p. 1.
[6.29] Burkhardt, W. C., *Proceedings of SPE Cleveland Section Regional Technical Conference,* Technical Papers, Ohio, Society of Plastics Engineers, Greenwich, CT. 1974, p. 81.
[6.30] Ratzel, W. J., *Kunststoffe,* **68**(10) (1978) 611–15.
[6.31] Bollerwedel, H. W., *Kunststoffe,* **68**(10) (1978) 660–2.
[6.32] Cuddihy, E. F., Coulbert, C. D., Willis, P., Baum, B., Garcia, A., & Minning, C., In *Polymers in Solar Energy Utilization,* ed. C. G. Gebelein, D. J. Williams & R. D. Deanin. ACS Symposium Series 220, 1983, pp. 353–66.
[6.33] Lewis, K. J., ibid., pp. 367–86.
[6.34] Lewis, K. J., & Megerle, C. A., ibid., pp. 387–407.
[6.35] Beatty, M. E., Woerner, C. V., & Crouch, R. K., *J. Mater.* **5**(4) (1970) 972–84.
[6.36] Sarbolouki, M. N., *Mod. Plast.* (Jan. 1980) 122–6.
[6.37] Cuddihy, E. F., Baum, B., & Willis, P., In *Solar Energy,* **22** (1979) 389–96.
[6.38] Feldman, D., Shapiro, M. M., & Fazio, P., *Polym. Engng Sci.,* **25**(7) (May, 1985) 406–11.
[6.39] Johnson, T. E., MIT Solar Building No. 5; The Third Year Performance. *Passive Solar Journal,* **1**(3) (1982) 175.
[6.40] Levy, M., & Vofsi, D., *Am. Chem. Soc., Div. Polym. Chem.,* Polymer Preprints, **23** (1982) 197.
[6.41] Edesess, M., & Flynn, R. P., *Am. Chem. Soc., Div. Polym. Chem.,* Polymer Preprints, **23** (1982), p. 199.
[6.42] Ceccorulli, G., Pizzoli, M., Scandoza, M., & Pezzin, G., *J. Macromol. Sci., Phys.,* **B20** (1981) 519.
[6.43] King, L. F., & Noel, F., *Polym. Engng Sci.,* **12** (1972) 112.
[6.44] Hay, H. R., *Solar Energy,* **14** (1973) 393–404.
[6.45] Metz, P. D., In *Polymers in Solar Energy Utilization,* ed. C. G. Gebelein, D. J. Williams & R. D. Deanin. ACS Symposium Series 220, 1983, pp. 211–16.
[6.46] Butt, L. T., & Wright, D. C., *Use of Polymers in Chemical Plant Construction.* Applied Science Publishers Ltd, London, 1980.

[6.47] Petit, R. B., & Roth, E. P., In *Solar Materials Science,* ed. L. E. Murr. Academic Press, New York, 1980, p. 191.

[6.48] Pohlman, S. L., ibid., p. 337.

[6.49] Flynn, R. P., Short, T. H., & Edesess, M., In *Polymers in Solar Energy Utilization,* ed. C. G. Gebelein, D. J. Williams & R. D. Deanin. ACS Symposium Series 220, 1983, pp. 187–94.

[6.50] Woodley, R. M., ibid., pp. 195–210.

Chapter 7

Polymer Applications in Roofing and Flooring

ROOFING

The roof of a building is a complex system with the function of sealing the building for a long time against a series of factors like light, wind, rain, snow load, temperature variations, hail storms, abrasion, etc.

The basic parts of a roof are the deck, the thermal barrier (insulation) and the impervious roofing membrane which seals the roof complex structure.

Most probably, this part of the envelope was the first structural element made by man to protect himself from the implacable forces of nature.

The built-up roof membrane is made of:

—bitumen (asphalt) or other more modern materials,
—the roofing felts (reinforcement),
—the aggregates for protection of bitumen against ultra-violet light and oxidation.

Figure 7.1 presents the roof system components.

Bitumen materials have been used since 3500 B.C. With their characteristics of being at the same time waterproofer, preservative and binder, they have been utilized by the ancients for the construction of their houses and roads.

The molten asphalts used for waterproofing are a mixture of mineral fillers and bitumens. Such materials are found in nature in the form of sands impregnated by residues of crude oil which has lost its volatile components over the centuries and formed asphaltic rocks.

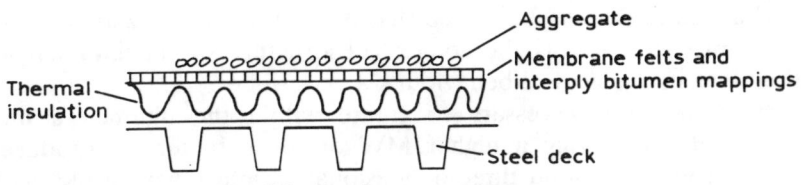

Fig. 7.1. Roof system components.

With time, hot asphalts for roofing were replaced by multilayer, tar-based waterproofers, then came distilled bitumen followed by oxidized bitumen interlaid with roofing felt and later alternating with mineral based sheet.

Modification of Bitumen with Polymers

Because of its limited characteristics, bitumen was the object of many studies aimed at improving its physical and mechanical properties. Several approaches were tried, such as chemical treatment and blending with natural or synthetic rubbers or latices.

Van Beem & Brasser [7.1] underlined that the criteria for the proper selection of such polymers are the following:

(a) The polymer must improve the susceptibility to temperature loading rate, resistance to permanent deformations and resistance to fracture over the widest range of temperatures, loading rates or times, and strains.

(b) The compatibility of polymer and bitumen must be such as to exclude any tendency to separate either while maintained at high temperatures or in normal service.

(c) The polymer must have only a slight influence on the workability (viscosity) of molten bitumen so that the rubber–bitumen blend can be applied like a conventional bitumen.

Polymers such as:

—polyisobutylenes,
—polybutadienes,
—polyisoprenes,
—natural rubber,
—styrene–butadiene copolymer,

(having a solubility parameter between 8 and 9 and an MW below 10^6) may dissolve at an elevated temperature in bitumen and form a stable microscopic dispersion at a much lower temperature.

Hameau & Druon [7.2] noted that the degree of peptization, similar to a micelle, controls the sol or gel structure of the bitumen and therefore its rheological behaviour.

It is therefore necessary to create within the bitumen a more structural gel, having a higher MW or, even better, to produce a bitumen with a second three-dimensional structure having the elastic characteristics of an elastomer. In this respect, consideration must be given not only to the phenomena of interfacial tension, polarity, aromaticity and configuration, but also to modification of the visco-elastic and thixotropic properties of bitumen.

If a significant decrease in bitumen brittleness at low temperatures is required, the T_g of bitumen must be made lower.

It follows that some polymer groups, such as:

—polymers with a high degree of cross-linking,
—polymers highly branched,
—high polar polymers, and
—polymers with much steric hindrance,

are unsuitable.

Therefore polymers with low cross-linking or with moderate inter-molecular (Van der Waals) forces are preferred to obtain suitable elastic properties; also polymers having macromolecular chains sufficiently long to permit rotation and increased flexibility should be used [7.3].

So far, in practice there are various polymer–bitumen mixtures such as: PE–bitumen [7.4] or elastomer–bitumen [7.5] including, styrene–butadiene–styrene block copolymer, [SBS]–bitumen [7.6] or butadiene–styrene copolymer [SBR]–bitumen [7.7].

Polymers like PE, natural rubber and polybutadiene, characterized by low intermolecular forces, must be added in sufficiently high proportions to improve substantially the mechanical strength of bitumen.

To obtain important elastic deformations, only macromolecular compounds with an atactic configuration or a limited amount of microstructural regularity should be used.

All these recommendations must be related to the characteristics of the bitumen and the condition of application, which should be modified as little as possible [7.3].

The use of waterproof roofing membranes based on the thermoplastic elastomer–bitumen mixtures is increasing, especially in northern climate countries.

The classic thermoplastic elastomers are the ABA type triblock copolymers, in which the A blocks consist of PS while the B block is polybutadiene or polyisopropene (SBS or SiS). They may be represented schematically as follows:

SSSSSSSSBBBBBBBBBSSSSSSSS

SSSSSSiiiiiiiiSSSSS

Because of the incompatibility of these blocks, a phase separation occurs whereby the polystyrene, being the minor component, forms a dispersed phase within a polydiene matrix. This two-phase material shows a remarkably uniform morphology, because of the near-monodisperse molecular weights of the blocks, a result of the anionic polymerization system used in the synthesis of these polymers [7.8].

Such triblock thermoplastic elastomers contain about 25% styrene. Block copolymers of this sort possess an interesting combination of properties. They are clear and tough and exhibit a level of flexibility somewhat higher than that of polypropylene.

At temperatures above the glass transition point of the 'hard' component, this system is an extrudable and mouldable plastic because its rubbery part does not interfere with its plasticity. On cooling to lower temperatures it exhibits the properties of rubber because the elastomeric blocks are still capable of rapid conformational changes. This remains the case until, at very low temperatures, the glass transition point of B is reached and the system turns into a hard and eventually brittle mass.

In the case of a styrene–butadiene–styrene block copolymer, the T_g of polystyrene is around 90°C and the T_g of polybutadiene is about −60°C. As a consequence, above 90°C we have a plastic; between 90°C and −60°C we have a rubber; below that we have glass. It can be readily understood that in the rubbery range between the two transition points the 'frozen in' polystyrene segments act as cross-links and prevent the polybutadiene chains from making irreversible displacements that would lead to a permanent set. We have, as a result of the two-component character of the system, a cross-linking principle that is not represented by highly localized single bonds but by inelastic domains of supermolecular dimensions. This is a distinct advantage because it reduces the danger of stress accumulation. On the other hand, this cross-linking system weakens as we approach the glass transition temperature of the plastic and disappears entirely above it.

The most efficient use of this bicomponent principle is the

combination of hard segments with a very high T_g, or even better, a high T_m (crystalline melting point), and soft segments having a very low T_B (brittle point) [7.9].

SBS block copolymers are two-phase systems. Hence, since PS and polybutadiene are incompatible polymers, the two types of segments (blocks) are separated into two distinct phases linked chemically to form a three-dimensional network (Fig. 7.2).

Suitable organic solvents or heating above the PS transition point (T_g) may destroy such a kind of network. It may be restored by releasing solvents or by cooling below the T_g of PS.

It is supposed that when the SBS is dispersed in hot bitumen, the cohesion of PS fragments diminishes and may disappear, more or less completely. Thus, at high temperatures the thermoplastic elastomer allows the SBS–bitumen mixture to behave like a viscous liquid, and at the same time, because of the relatively low MW of the macromolecular chains, it increases the viscosity only to slightly greater than

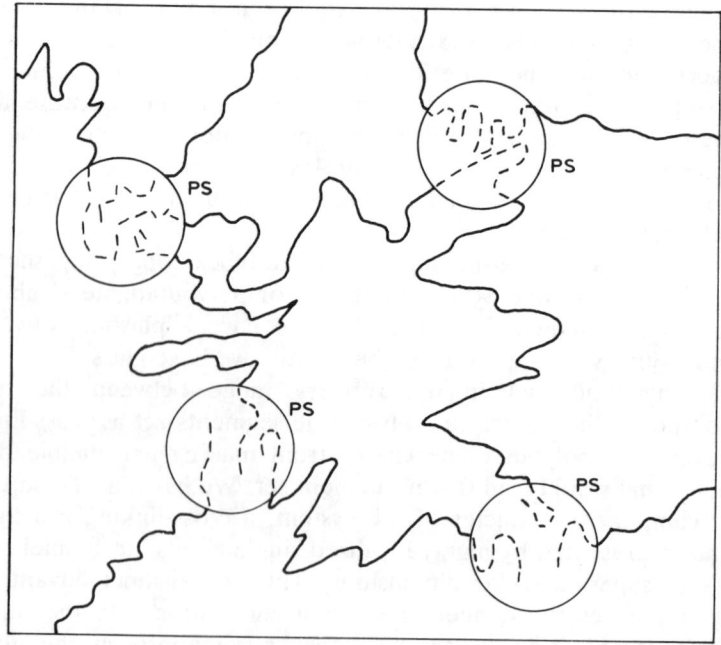

Fig. 7.2. Schematic representation of SBS triblock copolymer [7.3].

that of plain bitumen at the same temperature. When cooled, PS fragments associate and, below their T_g they act as cross-links. Only if the SBS has been thoroughly mixed with the bitumen and the elastomeric segments remain well dispersed at ambient temperatures, will the structure be continuous and rubbery and form entirely within the bitumen [7.3].

The mandatory conditions for obtaining improvements in bitumen properties are:

—a good compatibility of SBS and bitumen, and
—the use of a bitumen with a certain composition (asphaltenes and maltenes).

Van Beem & Brasser [7.1] showed that to dissolve SBS in bitumen is a complex problem because of the complexity of the composition of bitumen. Modern theories on solvent–polymer systems are not readily applicable to the polymer–bitumen systems.

It is evident that addition of the optimum quantity of a thermoplastic rubber substantially improves the physical-mechanical properties of bitumen; the softening points are considerably elevated, the low temperature resistance markedly improved and the penetration values are lower, as we may see from the data of Table 7.1.

Bitumens with too high or too low aromaticity are unsuitable since the first kind gives rise to a single phase compound in which the PS fragments appear dissolved even at room temperature, while bitumens

Table 7.1.

Influence of SBS (Europrene AG) on the properties of an 80/100 bitumen [7.3]

Bitumen 80/100	100	100	100	100	100	100
SBS (Europrene AG)	—	5^a	$7 \cdot 5^a$	10^a	$12 \cdot 5^a$	15^b
Penetration @ 25°C, 0·1 mm	92	58	53	48	46	45
Softening point IP (°C)	45·6	94	104	112	115	117
Fraass breaking point (°C)	−10	−15	−18	−20	−26	−31
Penetration index	−0·6	6·5	7·3	7·6	7·7	7·8
Viscosity @ 150°C (cSt)	160	—	—	3 437	—	—
Viscosity @ 175°C (cSt)	61	—	—	1 705	—	—
Viscosity @ 200°C (cSt)	34	—	—	920	—	—

[a] Mixed 1 h @ 200°C.
[b] Mixed 1 h @ 210°C.

with low content of aromatic hydrocarbons yield mixtures unstable during storage and with a coarse dispersion.

Bitumens with a high content of heavy asphaltenes (which because of their high MW are insoluble in *n*-heptene) are to be avoided [7.3].

SBR–Bitumen Mixtures

Kraus & Rollman [7.7] have studied the manufacture of SBR modified bitumen; the possible applications as waterproofing and roofing materials are considered. Properties investigated were phase separation, dynamic mechanical characteristics, creep, morphology, low temperature resistance, thermo-oxidative stability, and weathering resistance.

SBR is the most widely used rubber in the entire world; it is a random copolymer of styrene and butadiene made by free radical copolymerization in emulsion.

Ethylene–Vinyl Acetate Copolymer (EVA)–Bitumen Mixtures

Adler & Goebel [7.10] show that due to the compatibility between EVA and bitumen, roof sheeting with improved properties such as: tear strength, water permeability, ozone resistance and thermal stability can be attained.

EVA is an ethylene–vinyl acetate random copolymer. The presence of vinyl acetate units in the polyethylene chains reduces crystallinity, and at 20–30% concentration produces rubbery copolymers; products with more than 75% vinyl acetate units are however hard and rigid.

Depending on the composition and molecular weight of EVA copolymers, they may have a low softening point and in this case are used as wax additives or they may be rubbery products resembling low density PE in appearance.

Both filled and unfilled EVA copolymers have good low temperature flexibility and toughness and the absence of leachable plasticizer provides a clear advantage over plasticized PVC in some applications.

Synthetic Roofing Membranes

The commercial roofing market in the USA in 1982 consisted of:

65·7% build up roofing
23·3% single ply roofing
5% metal roofing and
6% other types.

A forecast suggested that by 1987, single ply roofing would amount to 49% of the market, and buildup roofing to 40% [7.11].

A lot of polymers proved suitable as high quality roofing membrane, such as:

—PVC reinforced or without reinforcement [7.12–7.14];
—ethylene–propylene–diene monomer terpolymer (EPDM) [7.15–7.19];
—chlorosulphonated polyethylene (Hypalon) [7.15–7.23];
—acrylics [7.24];
—polyester and glass fibre reinforced polyester [7.25–7.27];
—reinforced polyurethane [7.24];
—reinforced polytetrafluoroethylene (PTFE) [7.28];
—butyl rubber;
—polychloroprene (neoprene) [7.17].

For transparent roofing the following are recommended:

—polycarbonate [7.30];
—fluoroplastics including PTFE coated glass fibres [7.31];
—poly(methyl methacrylate), Plexiglass [7.32–7.33].

Elder [7.34] and others [7.35] discuss the use of PVC single ply roofing membranes with particular reference to perforation resistance, ease of installation, weathering, building code requirements, wind blow-off and processing methods. Details of long term and accelerated weathering tests are given.

Some advantages and disadvantages of polymeric membranes are pointed out by Griffin [7.36].

Advantages
Ultra-lightweight (for unballasted, adhered membranes).
Adaptability to irregular roof surfaces.
Isotropic (or nearly isotropic) physical behaviour (as opposed to the anisotropic weakness of organic and asbestos felt membranes in the transverse direction).
Superior architectural quality (notably colour).
Generally superior heat reflectivity (for cooling-energy conservation).
Better elongation (up to 800% at 70°F, 21°C).
Superior performance at sub-zero temperatures (flexibility down to −75°F (−59°C) for some materials).

Easier application (plus greater potential for shop prefabrication of large sheet membranes, flashing, hood, and so forth).

Fewer weather restrictions on application (permissible at sub-freezing temperature for some loose-laid sheets).

Less hazard from moisture entrapment during installation (for loose-laid, permeable synthetic sheets).

Easier repair of punctures, splits, tears.

Easier flashing application at corners and irregular surfaces, where stiffer built-up materials are difficult to form.

Possibly greater reliance on factory-manufacture material quality.

Disadvantages

Greater importance of good workmanship.

More limited range of suitable substrates (especially for fluid-applied systems).

Less puncture resistance (and consequently greater vulnerability to traffic damage).

Lack of performance and design criteria comparable to those available for built-up systems.

Poor reporting of technical data.

As we can see from Table 7.2 the synthetic polymer membranes offer a broad range of roofing systems.

EPDM is a terpolymer made of ethylene, propylene and a diene monomer which is usually: 5-ethylidene-2-norbornene (ENB), 1,4-hexadiene and dicyclopentadiene (DCPD).

Due to this third monomer EPDM is unsaturated but the unsaturation is located in a pendant side-group and not in the polymer backbone as in most elastomers. This unsaturation permits the use of sulphur as vulcanizing agent. The amount of unsaturated monomer is no more than 1·2%.

EPDM has a low polarity; after vulcanizing, it gives rubbery properties of elastic recovery to a relatively inert polymer backbone. This inertness makes EPDM outstanding for weathering but makes it equally difficult to bond because traditional bonding agents do not wet or spread on the surface nor are they likely to chemically bond to the EPDM due to the lack of chemical sites for attachment. Westley [7.37] gives a detailed state of the art of bonding of EPDM single ply roofing. His conclusions and recommendations are:

—EPDM roofing is the most widely used single ply in the business

Table 7.2.
Single ply synthetic roof membranes [7.36]

Type of membrane	Membrane material	Acceptable substrates	Thickness mil[a]	Thickness mm
Liquid applied[b]	Acrylic	Concrete, plywood, spray-in-place urethane foam, remedial roofing	20	0·5
	Butyl	Concrete, plywood, insulation board, remedial roofing	15–30	0·4–0·8
	Chlorosulphonated polyethylene	Spray-in-place foam, weatherproofing coating for elastomeric membranes	20–45	0·5–1·1
	Neoprene/chlorosulphonated PE	Concrete, plywood, spray-in-place urethane foam	20	0·5
	Polyvinyl chloride (PVC) and vinyl	Concrete, plywood, spray-in-place urethane foam	15–30	0·4–0·8
	Rubberized asphalt	Concrete	150–180	3·8–4·6
	Silicone	Spray-in-place urethane foam	20	0·5
	Urethane[c]	Concrete, plywood, spray-in-place urethane foam, remedial roofing	20–60	0·5–1·5
Preformed sheets[d]	Chlorosulphonated PE/asbestos backed	Plywood decks on industrialized or modular construction	35	0·9
	EPDM (ethylene propylene diene terpolymer)	Concrete, plywood, insulation board, remedial roofing	45	1·1
	Neoprene	Concrete, plywood, insulation board	63	1·6
	Polyvinyl chloride (PVC)	Concrete, spray-in-place urethane foam, remedial roofing	48	1·2
Composite preformed sheets	Nylon-reinforced PVC backed with neoprene or butyl	Concrete, plywood, insulation board, remedial roofing	30	0·8
	Non-woven glass reinforced PVC	Concrete, plywood, insulation board	47	1·2

[a] 1 mil = 0·001 in.
[b] Liquid membranes are commonly applied by a number of techniques, including brush, roller, squeegee, trowel, and conventional and airless spraying.
[c] A number of widely varying membrane systems are classified under the title urethane, including asphalt and coal tar modified urethane.
[d] Sheet membranes are applied bonded to or loose-laid on the surface.
Reprinted with permission from Rossiter, W. J. & Mathey, R. G., *Elastomeric Roofing. A Survey.* NBS Tech. Note 972, July 1978.

today. Its future is bright with growth projected to average 25% per
year for the next 3–5 years before levelling out.

—Proper techniques for seam cleaning and preparation must be
employed to maximize adhesion. No adhesive will bond to releasing
agents.

—EPDM membranes can be seam bonded by several commercially
available adhesive systems as described in this chapter. Data
generated by this study show that a consistently high level of
adhesion and comparable environmental resistance are provided by
100% solids preformed adhesive systems.

—Whereas none of the adhesives tested result in rubber tear, the
preformed tape provides an elastomeric bond, which after cure,
approaches that for a monolithic structure; one whose composite
properties match that of the sheet.

—Chemlok TXL, a new preformed adhesive, is just being introduced
to the industry. It can be used as a bonding tape, self-adhering strip
for mechanically attached roofs, or self-adhering flashings. Unlike
other adhesives, its properties are designed to suit the intended use.

—TXL Tapes, Strips and Flashing address the problem of health and
safety on the roof. With no volatile components, health and
flammability hazards are sharply reduced. As the roofing community
becomes more aware of the hazards of flammable solvent carrying
systems, they will seek out systems that will both perform and be
safe to use.

—As with everything, there is a choice. For the supplier who answers
long term questions with short term answers, we predict short term
rewards. Stated otherwise, a 10 year roof needs a 10 year adhesive.
Ten year adhesives cannot be commodity solvent-borne adhesives
whose very vulnerability to failure through cycling undermines the
credibility of the EPDM roof itself.

A very quick picture of the growth of the single ply roofing
membrane market and of EPDM single ply is presented in Table 7.3.

Other references [7.15, 7.18] discuss EPDM properties such as:
mechanical strength, ozone resistance, abrasion resistance, weatherability, thermal stability, etc.

Chlorosulphonated polyethylenes are amorphous vulcanizable elastic polymers marketed under the name Hypalon. The high reactivity of
the sulphonyl chloride cross-linking sites affords a wide choice of
practical curing systems. Recently there have been introduced per-

Table 7.3.
Commercial/industrial roofing market [7.37]

Type	1983 $(ft^2 \times 10^6)$	(%)	1984 $(ft^2 \times 10^6)$	(%)	1988 $(ft^2 \times 10^6)$	(%)	Avg. annual growth[b]
BUR[a]	1 426	62	1 378	58	968	36	—
Single ply	736	32	850	36	1 515	57	18·4%
Other	138	6	142	6	187	7	—
Total	2 300	100	2 370	100	2 670	100	—
Single ply							
EPDM	420	57	506	60	1 062	70	25·0%
Other	316	43	344	40	453	30	—
Total	736	100	850	100	1 515	100	—

[a] BUR = Built Up Roofing.
[b] Growth rates are calculated by the Trend Line method.
Reprinted with permission from: *Proceedings of 126th Meeting of the Rubber Division*, ACS, Denver, 1984.

oxide and metallic oxide/rubber accelerator systems in the presence of acrylic type coagents. The litharge/magnesia/dipentamethylene/thiuram tetrasulphide/nickel dibutyl dithiocarbamate (NiBD) system gives the best overall heat resistance. Conventional fillers, carbon black, softeners, plasticizers and release agents are used. Vulcanization is accomplished by normal methods as well as by moisture, ammonia and UHF radiation [7.38].

Hypalon in roof covering applications has a series of advantages such as: good mechanical properties, light weight, weldability, photostability, ozone abrasion and chemical resistance, and availability in a range of colours [7.22]. White single ply roofing made from Hypalon can cut the energy load associated with a building's roof from 30 to 60% compared with conventional black roofing. According to projections, the result is a significant reduction in energy bills. A computer study based on a conductive heating model, projects the energy savings achieved because of the reflectivity of Hypalon [7.20].

Combinations of Hypalon and neoprene/asbestos sheet may also be used [7.22].

Important aspects of elastomeric and PVC roofing are discussed in other references [7.34, 7.39].

FLOORING

Consideration of costs in use is particularly important in those parts of a building which, for functional reasons, are exceptionally vulnerable to soiling, wear and damage. This is especially true for the floors and the interior walls. The range of flooring materials increases in direct proportion to the number of new polymers daily added to the market.

As well as the common floor surfacing materials (e.g. ceramic tiles, wood block, wood strip and board, granolithic, terrazzo, marble, asphalt, linoleum, cork tiles), new materials are used, such as: PVC tiles, vinyl chloride–vinyl acetate copolymer, vinyl–asbestos tiles, PVC welded sheet, synthetic fibre–epoxy polymers, PP, polyurethane.

PVC Flooring

PVC and vinyl–asbestos tiles were found to be the most economic materials for use in flooring applications, while others (excluding marble) had broadly similar costs, though 20–50% higher than these two materials.

The characteristics of PVC were found to be entirely suitable for flooring applications, and few complaints have been registered [7.40].

The most widely used synthetic flooring material is produced by calendering compounds based on vinyl chloride–vinyl acetate copolymer filled with asbestos fibres. In general, the copolymers for flooring applications contain about 13–15% vinyl acetate.

Characteristic formulations of these types of products are presented in Table 7.4.

Because the plasticizer originally used for PVC tiles has led to straining problems, the use of internal plasticization through the copolymerization of vinyl chloride with vinyl acetate was implemented in the formulation of the tile.

The so-called homogeneous vinyl is processed by laminating precalendered webs produced in a common PVC calendering operation. The laminate product is made of several sheets of varying thickness: the top layer is a clear, relatively stiff formulation; the middle one contains more filler and is usually opaque; and the bottom sheet contains high filler loadings. After laminating, the product is die-cut into tiles and packaged.

PVC flooring materials can occasionally suffer embrittlement, shrinkage and cracking under conditions of heavy use applications (some of these effects have been attributed to the use of solvent-gel

Table 7.4.
Basic vinyl tile formulations [7.41]

	Vinyl asbestos (wt%)	Homogeneous vinyl (PVC) (wt%)
Vinyl chloride–vinyl acetate copolymer (87/13)	16	—
Vinyl chloride–vinyl acetate copolymer (95/5)	—	25
Plasticizer (DOP)	5	10
Epoxy plasticizer	0·5	1
Coumarone–indene resin	5	—
Asbestos shorts	31	—
Ground limestone	39	31
Talc	—	30
Pigment (TiO$_2$)	3	2·5
Stabilizer	0·5	0·5

Reprinted with permission from *Encyclopedia of PVC, Vol.* 1 by L. I. Nass (Ed.), 1976. Copyright Marcel Dekker.

cleaners and strong detergents). However, they are widely used in hospitals and schools, and domestic kitchens and bathrooms, where they offer a wide choice of colours and patterns, ease of cleaning, good cushioning, insulation and reasonable price (the last three features represent advantages over linoleum).

PP Flooring

Heavy-duty, lightweight PP duck boarding is claimed to provide a versatile, easily cleaned work platform, increasing operator comfort and safety. PP flooring is non-corroding and resistant to bacteriological attack. The upper surface is ribbed for non-slip characteristics, with slotted, self-draining squares allowing free passage for contaminants [7.42].

Epoxy Flooring

Epoxy polymer is frequently used as the covering or topping of a sub-floor. However, due to its low level of sound insulation and its lack of pleasing appearance, epoxy flooring is not used for domestic purposes. In consideration of both technical performance and cost, epoxy floors are not likely to succeed in the domestic market, which is moving strongly towards the use of soft flooring and carpets.

But for industrial flooring, epoxy floors are clearly competitive on a performance basis with all the floor types considered. On a cost performance assessment, epoxy floorings remain competitive with ceramics, but are probably not competitive with cheaper finishes such as asphalt mastics and granolithic concrete, where the performance of these systems is acceptable.

There are three types of epoxy flooring systems:

—pourable self-levelling seamless floors,
—trowelled floors, and
—terrazzo floors.

Furthermore, epoxy primers may be used with other overlays, e.g. with polyester or PU overlays. These epoxy flooring systems can be used as new floor coverings over a sub-floor of concrete, wood or steel, and can also be used for remedial work and be applied over existing floors.

The following are some important properties of epoxy polymer floors:

(1) High resistance to a wide range of acids, alkalis, solvents, fuels and other corrosive materials.
(2) They are laid as continuous, jointless screeds or toppings, so that there can be no seepage through joints to the sub-floor, or shaped into channels, gullies, skirtings, etc.
(3) They can be laid quickly with the minimum interruption of work and alteration to existing floor levels.
(4) They can cure rapidly (12–24 h) and down-time for machinery is therefore minimal.
(5) The floors are highly resistant to abrasion and virtually eliminate the problem of dust formation. They also have high mechanical strength and favourable strength-to-weight ratios, especially in comparison with concrete.
(6) They adhere with immense strength to all common flooring substrates when the proper surface preparation is carried out.
(7) They can be produced with a non-slip surface that is effective even when wet, and they are readily cleaned [7.43].

Epoxy polymers can also be used as tile grouts. Some new epoxy flooring products can provide the same type of finish as PU systems and at a competitive price [7.44].

Epoxy flooring has a proven durability of over 25 years [7.45].

The growth of seamless floors has had an exciting and profound effect on both the polyurethane and flooring industries. To the polyurethane industry, it has been an important factor in the recognition of the superb properties of urethane elastomers and the rapid growth of moisture curing urethane resins. It has added a new dimension to the flooring industry as a unique flooring concept which can produce durable, attractive and imaginative decorative effects at the job site, by embedding a variety of different coloured particles or fillers into a liquid polyurethane resin, just poured from the can.

The features of seamless floors are well established. These floors are easily installed, durable, lightweight, flexible, slip and dent resistant, scratch and scuff resistant, stain and dirt resistant, fungus resistant, heel mark resistant, and have superior chemical resistance compared to many floor materials.

The application of seamless flooring is relatively uncomplicated: after carefully preparing the floor surface, a basecoat is applied by roller spray or trowel. This coat is usually pigmented to camouflage voids or chips which may have occurred while preparing the surface. The layup is allowed to cure. When cured, the layup is sanded to remove the sharp edges and smooth the surface, and then vacuumed to remove the sanding dust.

Depending on the application and the amount and type of traffic, two or more urethane glaze coats are applied as the wear surface. For industrial floors where appearance is not critical, sanding and glazing are not necessary.

There are a number of coloured materials which the imaginative mind can use to provide decorative effects for seamless floors. Any material, considered for its possible decorative effects in these floors, should not dissolve, swell or curl in the resin, and should not cause side effects with the resin, such as discoloration, bleeding and retarded cure [7.46] (e.g. flakes of PVAc have been used in combination with PU resin to provide a resistant as well as a decorative flooring material).

REFERENCES

[7.1] Van Beem, E. J., & Brasser, P., *J. Inst. of Petrol.*, **59**(566) (1973) 91–7.
[7.2] Hameau, G., & Druon, M., *Bull. Liason Lab. Pet. Chem.*, **81** (Jan.–Feb., 1976), Ref. 1756, RILEM, 121–9.

[7.3]   Piazza, S., Arcozzi, A., Balestrazzi, F., & Verga, C., *Poliplasti*, **27**(257) (1979) 91–102.
[7.4]   Goetze, H., *Kunst. Bau.*, **17** (1982) 76–7.
[7.5]   Anon., *Rubb. Plast. News*, **12**(12) (1983) 13.
[7.6]   Meynard, J., *Proc. Int. Inst. Synthetic Rubber Producers*, New Orleans, LA (April 19–23, 1982), Part 2.8, 1–6.
[7.7]   Kraus, G., & Rollman, K. W., *Kant n. Gummi Kunst.*, **34**(8) (Aug., 1981) 645–57.
[7.8]   Morton, M., *Rubb. Chem. Technol.*, **56**(5) (Nov.–Dec., 1983) 1096.
[7.9]   Mark, H. F., *Chemtek* (April, 1984) 220.
[7.10] Adler, K., & Goebel, K., *Kunst. Bau.*, **14**(4) (1979) 179–81.
[7.11] Anon., *Plast. Bldg Constr.*, **6**(8) (1983) 3.
[7.12] Anon., *Rubb. Plast. News*, **13**(7) (1983) 70.
[7.13] Schneider, H., *Kunst. Bau.*, **18**(2) (1983) 62–4.
[7.14] Anon., *Plast. Rubb. Wkly*, **986** (1983) 9.
[7.15] Anon., *Elastomerics*, **114**(2) (1982) 26–7.
[7.16] Anon., *Rubb. Wld*, **186**(1) (1982) 19–20.
[7.17] Anon., *Chem. Week*, **133**(17) (1983) 16.
[7.18] Strong, A. G., American Chemical Society, 124th Rubber Division Meeting (Fall), International Rubber Conference and Expo., Houston, Texas, Oct. 25–28, 1983, Paper 82, p. 35.
[7.19] Anon., *Chem. Engng News*, **60**(22) (1982) 39.
[7.20] Anon., *Plast. Bldg Constr.*, **7**(9) (1983) 8.
[7.21] Anon., *Elastomers, Notebook*, No. 137 (1983), 8–9.
[7.22] Sweet, G., *Mater. Plast. Elast.*, No. 10 (Oct., 1982), 566–9.
[7.23] Baseden, G. A., American Chemical Society, 124th Rubber Division Meeting (Fall), International Rubber Conference and Expo., Houston, Texas, Oct. 25–28, 1983, Paper 94, p. 13.
[7.24] Anon., *Polym. News*, **9**(5) (1983) 151–2.
[7.25] Anon., *Plast. Bldg Constr.*, **6**(5) (1983) 3.
[7.26] Anon., *Plast. Rubb. Wkly*, No. 986 (May, 1983) 9.
[7.27] Anon., *Plast. Ind.*, **4**(12) (1979) 24–9.
[7.28] Anon., *Mod. Plast. Int.*, **12**(7) (1982) 26.
[7.29] Anon., *Plast. Int. News (Jap.)*, **29**(11) (1983), 163.
[7.30] Anon., *Mod. Plast. Int.*, **8**(11) (1978) 14–16.
[7.31] Fitz, H., *Techn. Rdsch. (Bern)*, **75**(51/52) (1983) 10.
[7.32] Buck, M., *Kunst. Bau.*, **18**(3) (1983) 105–8.
[7.33] Anon., *Kunst. Bau.*, **18**(2) (1983) 78–85.
[7.34] Elder, B., *Proceedings of the Technical Conference, Vinyl in Building and Construction*, Naperville, IL., Sept. 20–22. The Society of Plastic Engineers, Vinyl Division, 1982, pp. 75–101.
[7.35] Anon., *J. Vinyl Technol.*, **5**(2) (1983) 79–84.
[7.36] Griffin, C. W., *Manual of Built-Up Roof Systems*. McGraw-Hill, New York, 1982, pp. 247–75.
[7.37] Westley, S. A., *Proceedings of the 126th Meeting of the Rubber Division*, ACS, Denver, CO., Oct. 25, 1984.
[7.38] Blow, C. M., & Hepburn, C., *Rubber Technology and Manufacture*, 2nd edn. Butterworth Scientific, 1982, p. 128.

[7.39] Rossiter, W. J., Jr, & Mathey, R. G., *Plastics in Building, Regional Technical Conference,* November 9–10, 1976, SPE, Boston, p. 78.
[7.40] Taylor, J. A., *Br. Plast. Rubb.,* (January, 1985) 13–15.
[7.41] Nass, L. I. (Ed.), *Encyclopedia of PVC, Vol. 1.* Marcel Dekker, New York, 1976, p. 126.
[7.42] Anon., *Manufacturing Chemist* (June, 1984) 83.
[7.43] Potter, W. G., *Uses of Epoxy Resins.* Newnes-Butterworths, London, 1975, pp. 157–65.
[7.44] Anon., *Plast. Rubb. Wkly,* No. 980 (1983) 6.
[7.45] Neffgen, B., *Int. J. Cem. Comp. Lightweight Concr.,* **7**(4) (1985) 253–60.
[7.46] Anon., *Plast. Rubb. News* (May, 1983) 14–23.

Chapter 8

Polymer Degradation

The term polymer degradation is used to denote changes in physical properties caused by processes involving bond scission in the backbone of the macromolecule. In linear polymers, these reactions lead to a reduction in molecular mass, i.e. to a diminution of chain length.

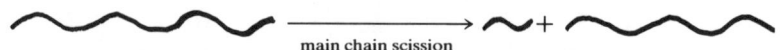

main chain scission

When considering biopolymers, the definition of polymer degradation is extended to include changes of physical properties caused not only by chemical but also by physical processes involving the breakdown of higher ordered structures.

In both cases the degradation involves a deterioration in the functionality of polymeric materials, which in the case of biopolymers is usually called denaturation.

This chapter is essentially devoted to different aspects of polymer degradation. In this connection, it must be pointed out that alterations in physical properties are, of course, not only caused by bond scissions in the polymer backbone, but also very often by simultaneously occurring modifications in pendant groups or side chains. With the exception of intermolecular cross-linking, however, reactions in pendant groups of linear polymers affect the physical properties only to a minor extent relative to reactions in the backbone.

Cross-linking may be considered the opposite of degradation as it leads to an increase in the molecular weight and size.

Since the splitting of the bonds in the backbone or main chain of linear polymers, i.e. main chain degradation, is the chief objective of this chapter, it seems appropriate to point out that the expressions *scission, rupture, breakage* and *lesion* are used synonymously to indicate bond fracture.

Expressions in other languages, equivalent to the term degradation as defined above are: Degradacion (Spanish), Degradation (French), Degradizione (Italian), Abbau (German), Bunkai (Japanese) [8.1].

For practical reasons, however, it is useful to sub-divide this field according to its various modes of initiation. These comprise:

—thermal degradation,
—photodegradation (caused by light),
—radiodegradation (X-rays, high energy sources in general),
—chemical degradation,
—biodegradation (biological degradation),
—mechanical or mechano-chemical degradation.

Thermal degradation refers to the case where the polymer, at an elevated or relatively elevated temperature starts to undergo chemical modifications without the simultaneous involvement of another compound. Often it is rather difficult to distinguish between thermal and thermo-chemical degradation because polymers are only rarely chemically 'pure'. Traces of catalyst additives or impurities present in the material might react with the polymer if the temperature is high enough.

Photodegradation concerns the physical and chemical modifications caused by irradiation of polymers with ultra-violet or visible light. The existence of chromophoric (light absorbing) groups in the macromolecules (or in the additives) is a prerequisite for the initiation of photochemical reactions. Generally, photochemically important chromophores absorb in the UV range (i.e. at wavelengths below 400 nm).

High energy radiation such as electromagnetic radiation (X-rays, γ-rays) or particle radiation is not specific with respect to absorption. The existence of chromophoric groups is not a prerequisite (as in the case of photodegradation) since all parts of the molecule are capable of interacting with the radiation. The extent and character of physical and chemical changes depend on the chemical composition of the irradiated material and on the nature of the radiation.

The effectiveness of radiation in modifying polymers is influenced by the presence of oxygen or additives, type of radiation, degree of crystallinity, and presence of solvent [8.2].

High energy radiation-induced alterations of polymers are important for their utilization in fields of high radiation flux.

Chemical degradation refers, in its strict sense, exclusively to processes which are induced under the influence of chemicals brought into contact with high polymers. Depending on the nature of the chemical agent, this type of degradation is known as oxidation, acidolysis, hydrolysis, alcoholysis, etc.

Biodegradation (biologically initiated degradation) is related to chemical degradation as far as microbial attack is concerned. Microorganisms produce a variety of enzymes capable of reacting with natural and synthetic polymers.

Mechanical degradation generally refers to macroscopic effects brought about under the influence of shear forces (during extrusion, degradation in ball-mill, forced flow of concentrated polymer solutions, etc.).

The strong inter-relationship between the various modes of polymer degradation should be emphasized. Frequently, circumstances prevail that permit the simultaneous occurrence of various modes of degradation. Typical examples are;

(1) environmental processes, which involve the simultaneous action of ultra-violet light, oxygen and harmful atmospheric emissions; or

(2) oxidation deterioration of polymers (thermoplastics, elastomers, etc.); this degradation is based on the simultaneous action of heat and oxygen.

MECHANICAL CORROSION

Mechanical corrosion by particulate matter is very common for polymers used outdoors. There is a large variety of particulate material in the air, primarily solid matter rather than liquid. Such kinds of particles are of different shape, size and chemical composition and come from a variety of sources. They may be tiny metal particles from metallurgical fumes, porous conglomerates of sooty carbon, soil particles, fly ash and fly dust of all types. Their characteristics (size, shape, composition, etc.) determine the surface corrosion of polymers used for outdoor applications.

Particles larger than about 10 μm are emitted as a result of some physical processes and tend to settle out on the surface of different materials rather rapidly. This dustfall plays an important role in the abrasive corrosion of polymer surfaces, however, it also decreases the effect of photochemical processes by the scattering and absorption of sun irradiation.

Unlike dustfall, particles smaller than about 10 μm in size are also able to corrode surfaces, but they do not protect surfaces against irradiation from the sun as is the case with larger particles. Particles having sizes less than 0·1 μm do not have any corrosion effect.

The speed at which the particulate matter contacts the polymeric surfaces is also of great importance. At relatively low velocities, small particles hitting a polymeric surface will abrade it and alter its properties. At some critical velocity (e.g. of a vehicle or blowing air) the target surface is penetrated and the particles are deformed or break up as they pass into the material. Penetration of the polymeric surfaces by such kinds of particulate is usually accompanied by the formation of deep cylindrical cavities in the target plastic.

As a result of the high velocities of impact, the transfer of a tremendous amount of kinetic energy takes place within an extremely small region. This energy is partitioned into heat, light, explosive mass ejection, and shock wave generation. The conditions produced are of sufficient magnitude to lead to thermal degradation and/or pyrolysis of the macromolecular surface.

The magnitude and modes of damage to the polymers by impact of particulate matter is dependent upon a number of parameters such as: mass, size, shape, velocity, physical state and trajectory of the particle. Soft and rubbery polymers have higher spallation threshold and less crater damage than the rigid ones. Greater penetration frequency is also obtained with the impact of small particles on soft plastics or ablating plastics containing a molten surface layer. Fibre reinforced polymers containing many interfaces are better able to withstand impulsive stress of short magnitude, because they distribute energy over a wide area.

Puncture, craters, cracks and other types of mechanical corrosion which are noted in particle impact on polymers may cause serious degradation of the material [8.3].

POLYMER SERVICE LIFE IN POLLUTED ATMOSPHERES

The deterioration of polymers as a result of pollutants takes place in the following stages:

—loss of gloss;
—minute crazing;
—severe crazing and cracking;
—leaching;
—loss of reinforcement from the structure;
—gradual breakup.

The timescale over which these processes occur is dependent on the microstructure of the polymer as well as the type and amount of additives.

First, the loss of surface gloss is a sign of deterioration. It can be lost by abrasion, surface leaching, minute and severe crazing.

Discoloration of polymers is usually the result of complicated photo-chemical processes and the result of the attack of some strong pollutants such as SO_2, NO_2, O_2, O_3 and singlet oxygen [8.3].

Polymer Life Phases

Degradation may take place during every phase of the life of a polymer, i.e. during its synthesis, processing and application.

Depending on the conditions of synthesis (polymerization, polycondensation, etc.), during the process, depolymerization may take place. Depolymerization is the inverse of synthesis through polymerization, namely, a stepwise separation of the monomers from the growing chain end. Polymerization is possible only below the equilibrium temperature of the system, the so-called *ceiling temperature.*

$$polymerization \rightleftarrows depolymerization$$

Depending on the conditions (parameters) of polymerization, 'weak sites' may appear on the polymer backbone which later may cause its degradation. For example, the presence of double bonds or of tertiary chlorines in PVC (resulting from branch formation during polymerization) may reduce PVC stability. The presence of catalyst residues, of some polymerization additives, or some impurities may be very decisive with respect to the polymer life span (e.g. polyethylene). The amount of additives used during VC polymerization increases in the following sequence: Block→ Suspension→ Emulsion; the stability of PVC decreases in the same sequence.

The *processing* of many high polymers occurs at relatively high temperatures (180–300°C) and under mechanical stresses. Such kinds of conditions may produce their degradation and modification of the expected properties. The damage to the product may result in the formation of some defects which will act as weak sites during its subsequent service life, which will ease its deterioration.

In the case of the processing of polyethylene in the presence of some oxygen traces, carbonyl groups (—C═O) may appear on its backbone; these groups, which are ultra-violet light absorbers, will constitute weak sites in photodegradation processes:

$$\cdots \text{—CH}_2\text{—CH}_2\text{—CH}_2\text{—CH}_2\text{—} \cdots \text{O}_2 \longrightarrow$$

$$\cdots \text{—CH}_2\text{—}\underset{\underset{\text{O}}{\|}}{\text{C}}\text{—CH}_2\text{—CH}_2\text{—} \cdots$$

To avoid the production of such defects as potential sources of deterioration it is necessary to introduce special stabilizers during processing.

The best known appearance of polymer degradation is related to its *application*, especially outdoor application. Some kinds of polymer degradation, such as outdoor ageing of claddings, door and window frames, roofs, etc., are well known. Because the scale of polymer uses is very broad, the stability requirements are highly diverse. In most cases as in the case of construction, there is a demand for a long service life. For instance, some transparent plastic foils for greenhouses may be completely destroyed after a few months. Underground cables for telecommunication or current applications are, however, expected to last several decades. In both of these cases, the polymers come in contact with the soil and the same types of bacteria probably attack them; the *biodegradability* of the polymers used for these applications should thus be different [8.4].

There are a number of important trends in materials utilization, such as:

—the use of alternative raw materials;
—lower energy usage in processing and manufacturing;
—the stimulation of productivity increases;
—the use of extreme operating conditions;

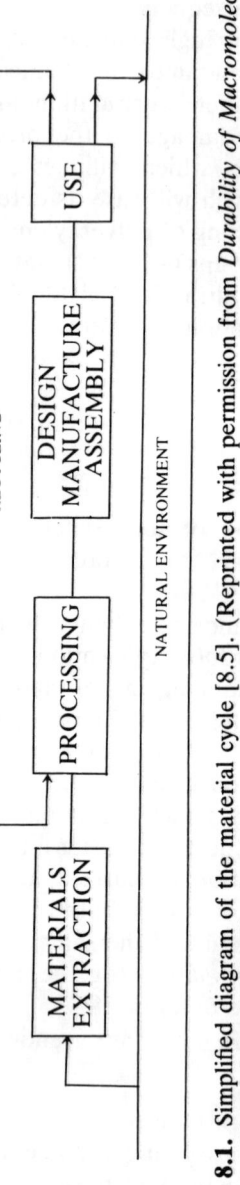

Fig. 8.1. Simplified diagram of the material cycle [8.5]. (Reprinted with permission from *Durability of Macromolecular Materials* by R. K. Eby (Ed.), 1979. ACS Symposium Series 95. Copyright American Chemical Society.)

—operation close to design limits;
—the increasing complexity of engineering materials;
—the concerns with health, safety and the environment.

Each of these trends places increased performance demands on materials. Trends in materials use throughout the cycle place increased performance demands on materials (Fig. 8.1).

The problem of waste disposal is increasing with the use of increasing quantities of polymers (plastics, rubbers, synthetic fibres, caulking materials, etc.). Organized recuperation of remains presently exists especially in the case of the production wastes of polymer processing factories [8.5].

THERMAL DEGRADATION

Changes in temperature dramatically alter the properties of materials. The strength of most materials decreases as the temperature increases (Fig. 8.2). Furthermore, sudden catastrophic changes may occur when heating above critical temperatures. High temperatures may change the structure of engineering materials or cause polymers to melt or char.

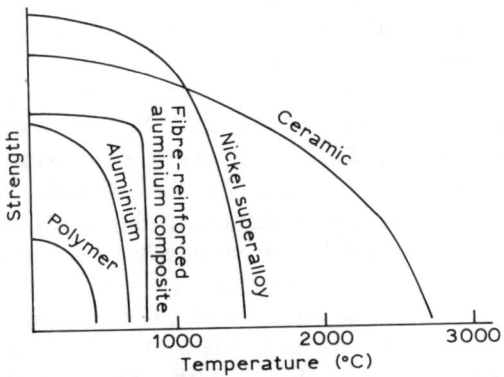

Fig. 8.2. Increasing temperature normally reduces the strength of a material. Polymers are suitable only at low temperatures. Some composites, special alloys and ceramics have excellent properties at high temperatures [8.6]. (Reprinted with permission from *The Science and Engineering of Materials* by D. R. Askeland, 1984. Copyright Brooks/Cole.)

Thermal degradation of polymers can be divided roughly into two general categories, (1) random chain scission and (2) depolymerization.

(1) Random chain scission can be visualized as a reaction sequence approximating the reverse of polymerization. Chain scission occurs at random points along the chain, leaving chain fragments of relatively high molecular weight (oligomers); scission of a high molecular weight chain produces macromolecules of x and $n - x$ degrees of polymerization.

$$[M]_n \rightarrow [M]_x + [M]_{n-x}$$

where x and $n - x$ reflect the number of units (monomer units) in the macromolecules remaining after the scission. For all practical purposes it can be assumed that no monomer is liberated and the weight loss to volatile product is nearly negligible.

(2) The depolymerization process is essentially one in which monomer units are released step by step from the chain ends. This is viewed as the opposite of the propagation step in addition polymerization. It is encountered most prevalently with vinyl polymers and polymers produced from cyclic monomers.

Both types of degradation can occur simultaneously or only one of them may occur, depending on (a) the nature of the polymer, (b) the temperature and some other factors. The relative importance of these two processes can be deduced from MW versus percentage of degradation to monomer curves. Figure 8.3 idealizes three distinct types of behaviour. Degradation via a route exactly the reverse of polymerization corresponds to line A→C. In such a case, the chains unzip in a rapid reaction when activated. Pure random chain scission would be represented by line A→B. As noted previously, the MW of the polymer drops rapidly until a very low average MW is reached. Line A→D corresponds to the rupture of the bonds near chain ends.

Thermal ageing can induce purely physical phenomena (change of morphology, migration of gases, etc.) or can promote chemical phenomena, the most important of which is oxidation [8.8].

Stable polymeric materials have been rationalized by the synthetic polymer technologist on the basis of the ability of the material to retain its mechanical properties. Modulus of elasticity, tensile strength and hardness as short range properties mean that the macromolecular compound should not melt or undergo extensive softening at the temperatures required by a specific application.

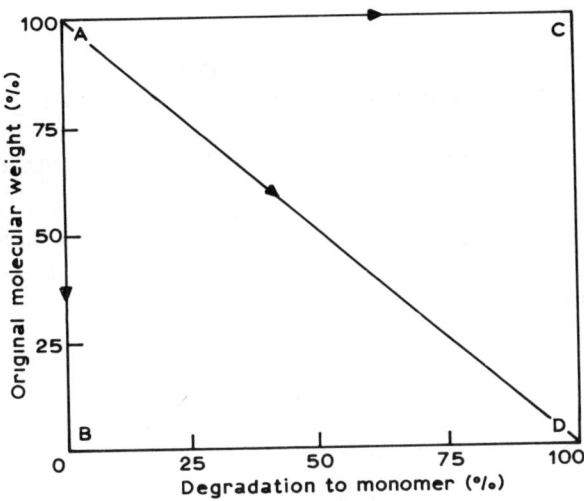

Fig. 8.3. Relative importance of three distinct degradation routes as noted from MW changes versus percentage degradation to monomer. Line AB, random chain scission; line AC, depolymerization; line AD, rupture of weak links at or near chain ends [8.7]. (Reprinted with permission from *Thermal Stability of Polymers* by R. T. Conley (Ed.), 1970. Copyright Marcel Dekker.)

Longer periods of service temperatures require that under external forces the creep of the polymer must be quite slow. If extended service is required, degradation becomes the most serious problem to be contended with. In general, a polymer must have the following characteristics to fall into the stable polymer category:

(a) It must possess a high melting temperature range or at least have a very high softening temperature.
(b) It must be capable of exhibiting high resistance to spontaneous thermolytic decomposition.
(c) It must be relatively unaffected by chemical reagents or radiation.

Obviously no such materials exist and therefore the task is to trace the relative behaviour of existing materials in terms of their structure in order to arrive at some optimum performance level for particular polymer systems as the basis for new engineering materials development. To date, three approaches have been used to produce highly stable polymers:

(1) Crystalline materials have been prepared;
(2) Polymers capable of a high degree of cross-linking have been examined;
(3) Rigid, linear, multistrand polymers have been synthesized [8.7].

Usually crystalline polymers are thermoplastics. The cross-linked ones are thermosetting polymers. In the last group, the polymers are polyaromatics or heterocyclics consisting essentially of sequences of aromatic rings with few, if any, single bonds as part of the backbone chain (Fig. 8.4).

poly-*p*-phenylene

polyimide
(heterocyclic polymer)

The criterion established for defining the thermal behaviour of polymers is a weight loss criterion such as can be obtained from thermogravimetric analysis (TGA).

Figure 8.5 indicates the thermographs of several heterocyclic polymers in static air and exemplifies some of the comparative aspects of their degradation processes.

It is not the aim of this chapter to discuss the mechanisms of thermal degradation of polymers in great detail. Table 8.1 shows the major decomposition products for a number of polymers. It is clear that segmental elimination of monomer units (zipping) is favoured where tertiary carbon atoms are present in the polymer backbone.

Thus poly(methyl methacrylate) and poly(α-methyl styrene) degrade in this manner, as do polymethacrylates in general. If no energetically favourable zipping type of degradation exists and if all side groups are stable, more complex reactions occur. Sometimes several reactions occur simultaneously and exact analysis of the

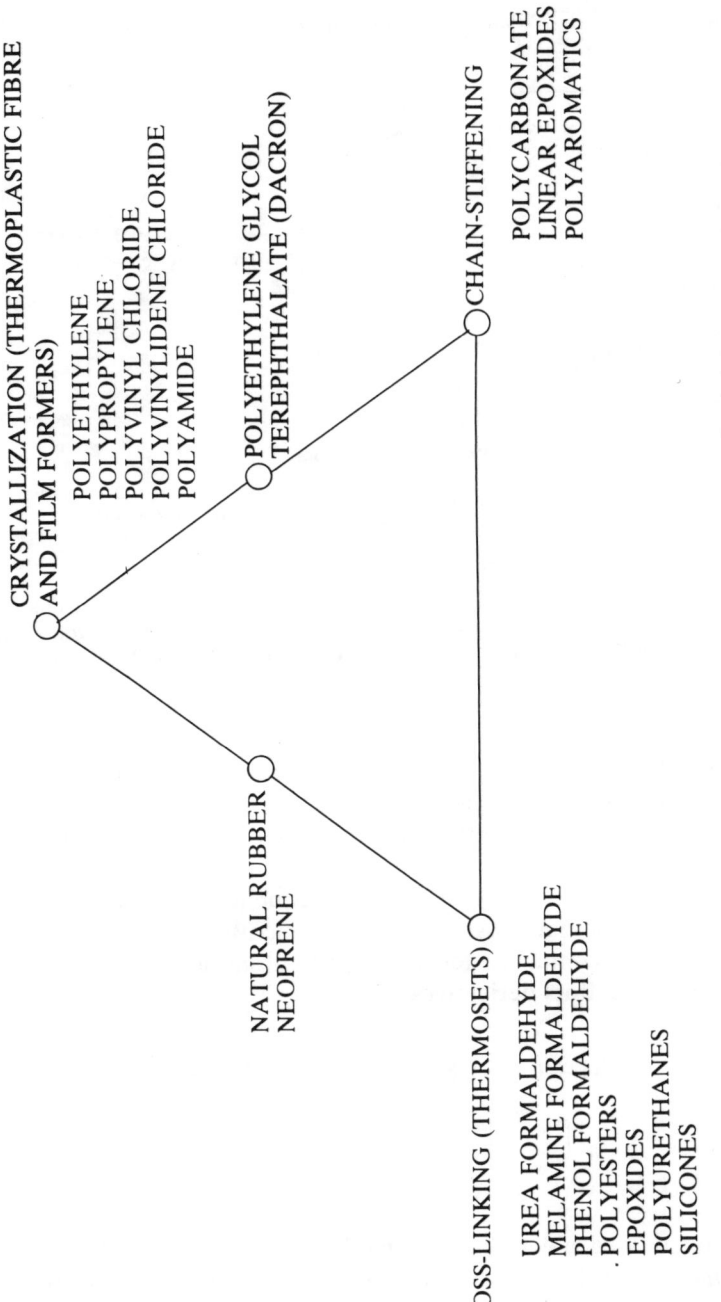

CRYSTALLIZATION (THERMOPLASTIC FIBRE AND FILM FORMERS)

POLYETHYLENE
POLYPROPYLENE
POLYVINYL CHLORIDE
POLYVINYLIDENE CHLORIDE
POLYAMIDE

POLYETHYLENE GLYCOL TEREPHTHALATE (DACRON)

CHAIN-STIFFENING

POLYCARBONATE
LINEAR EPOXIDES
POLYAROMATICS

NATURAL RUBBER
NEOPRENE

CROSS-LINKING (THERMOSETS)

UREA FORMALDEHYDE
MELAMINE FORMALDEHYDE
PHENOL FORMALDEHYDE
POLYESTERS
EPOXIDES
POLYURETHANES
SILICONES

Fig. 8.4. Principles of polymer structure which influence thermal stability [8.7]. (Reprinted with permission from *Thermal Stability of Polymers* by R. T. Conley (Ed.), 1970. Copyright Marcel Dekker.)

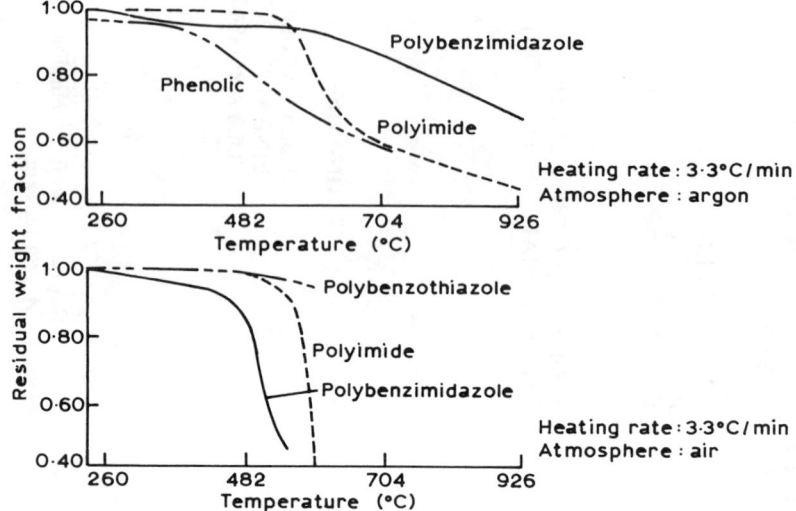

Fig. 8.5. Thermogravimetric analysis of several thermally stable resins in argon and in air [8.7]. (Reprinted with permission from *Thermal Stability of Polymers* by R. T. Conley (Ed.), 1970. Copyright Marcel Dekker.)

reaction is difficult. This difficulty has been increased in the last decade with the advent of heat resistant polymers which are required to combat the severe environments often encountered in high-speed flight and related areas [8.9].

Depolymerization or depropagation is sequential elimination (unzipping) with monomer loss. It most frequently occurs when the polymer backbone contains a tertiary carbon atom, as is the case with polystyrene (PS) and its derivatives

$$\cdots -CH_2-CH-CH_2-CH-CH_2-CH- \cdots$$

In the unmodified polymer of formaldehyde known as polyformaldehyde there is a hydroxyl end group from which molecules of monomer are easily lost.

Table 8.1.
Summary of polymer degradation data [8.9]

Polymer	Monomer yield (%)	Energy of activation (kcal/mol)	Temperature for half-life of about 30 min (°C)
Polyoxymethylene	100	—	—
Poly(α-methyl styrene)	100	—	—
Polytetrafluoroethylene)	96	81	510
Poly(methyl methacrylate)	95	52	330
Polymethacrylonitrile	85	—	—
Poly(α-methyl styrene)	45	56	360
Polystyrene	41	55	360
Polychlorotrifluoroethylene	26	57	380
Polyisobutene	20	49	350
Polybutadiene	20	62	410
Poly(ethylene oxide)	4	46	350
Poly(propylene oxide)	3	20	300
Poly(methyl acrylate)	1	34	330
Polyacrylonitrile	0	58	390
Polypropylene	0	63	400
Polyethylene	1	—	—
Poly(butyl methacrylate)	0	High yield of isobutene	—
Poly(vinyl chloride)	0	32	260
Poly(vinyl acetate)	0	17	270

Reprinted with permission from Achhammer, G. G., Tyron, M., & Kline, G. M., *Kunststoffe–Plastics*, **49** (1959) 600.

$$\cdots -CH_2O-CH_2O-\boxed{CH_2-O}H \longrightarrow \cdots -CH_2O-CH_2OH + CH_2O$$
$$\text{formaldehyde}$$

It is thus obvious that blocking of the terminal hydroxyl group will stabilize the polymer, but there are some difficulties in achieving sufficiently high extents of reaction.

Besides main chain scission and depolymerization some other reactions [8.10–8.15] may occur, as we may see in Fig. 8.6.

Thermal Degradation of PVC

Poly(vinyl chloride) has many applications in construction such as: flooring, roofing, cladding, window and door frames, window shades,

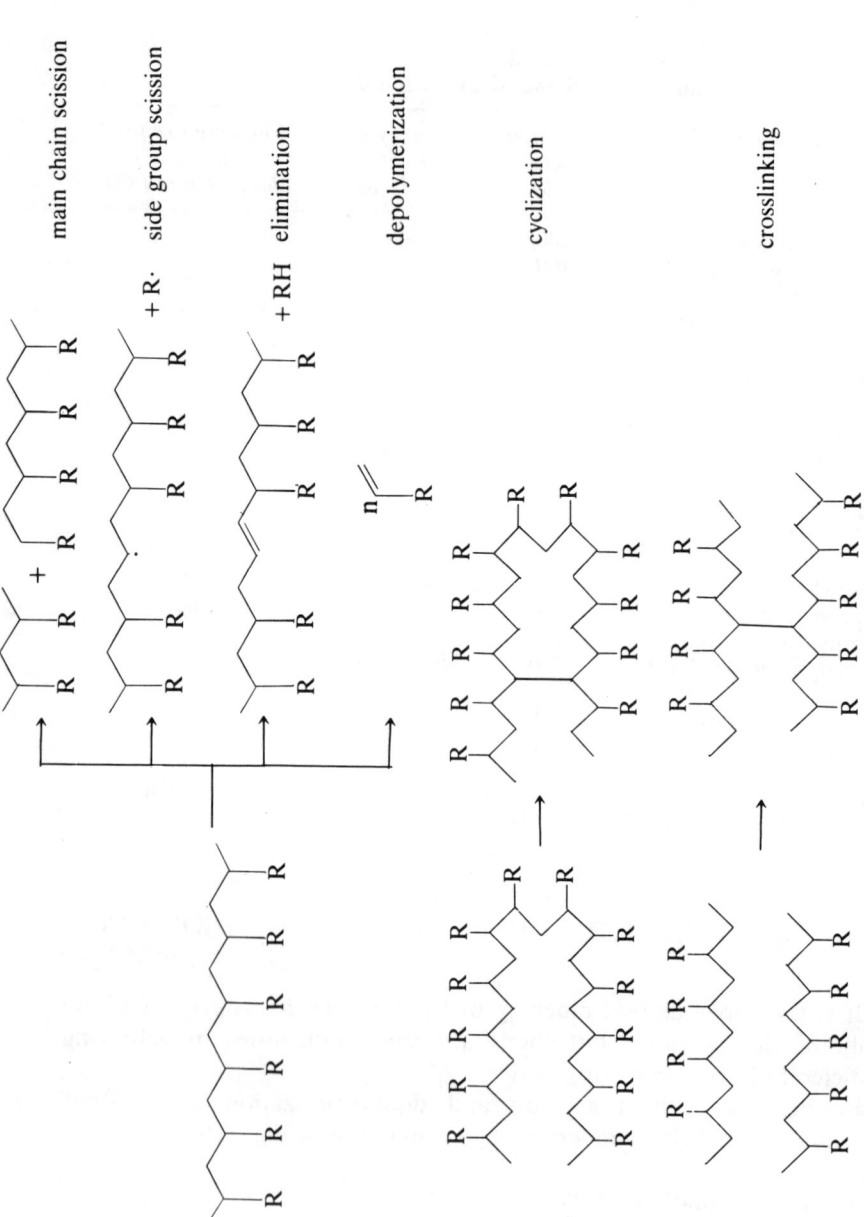

Fig. 8.6. Schematic depiction of reactions occurring during thermal degradation of polymers [8.1]. (Reprinted with permission from *Polymer Degradation* by W. Schnabel, 1981. Copyright C. Hanser Verlag.)

main chain scission

side group scission

elimination

depolymerization

cyclization

crosslinking

blinds, awnings, wallcoverings, carpet backing, pipes and fittings, rain gutters [8.16].

Its big advantage is the high variability of properties which permit its application in rigid or soft plasticized items; there is no other polymer which shows such a broad variety of changes in modification by plasticization as PVC. However, all polymers of vinyl chloride (VC) are quite unstable against heat and light, which leads to extensive changes in the polymer microstructure. This degradation is dehydrochlorination (elimination of hydrochloric acid) and is characterized by discoloration of the polymer and in turn by unfavourable effects on the mechanical, optical and electrical properties.

The fundamental fact of PVC thermal degradation is the elimination of hydrochloric acid at relatively low temperatures (140°C) or under the influence of light. In the first stage, this process leads to the formation of one double bond [8.17–8.20]:

$$\cdots-CH_2-\underset{\underset{Cl}{|}}{CH}-CH_2-\underset{\underset{Cl}{|}}{CH}-\cdots \longrightarrow$$

$$\cdots-CH_2-\underset{\underset{Cl}{|}}{CH}-CH=CH-\cdots + HCl$$

The first step is followed by a zipper-like splitting off of more HCl molecules; the result is a system of conjugated double bonds, a polyene [8.21].

$$\cdots-CH=CH-CH=CH-CH=CH-\ldots$$

Because of the fact that the double bonds are chromophoric groups, with increasing length of these polyene sequences the light absorption of the polymer is shifted towards longer wavelengths from UV to the region of visible light. The elimination of HCl can be observed visually by the colour change of the polymer. This change occurs from yellow through orange, red, brown, until black.

The dehydrochlorination and the colour change during PVC degradation are used for most experimental investigations in the field of PVC and its copolymers.

The most important stage in the elimination of HCl from PVC is the initial step which requires a relatively high activation energy.

In the literature various micro-structural irregularities are discussed as weak sites of the dehydrochlorination, such as:

—chain ends from initiator residues or unsaturated end groups;
—presence of branches (tertiary carbon);
—random unsaturation;
—oxidation structures;
—head to head units [8.22].

Mayer [8.23], having studied various PVC models, concluded that mainly branches with tertiary chlorine atoms or random unsaturations

$$
\begin{array}{c}
CH_2\!-\!CH_3 \\
| \\
\cdots -CH_3-C-CH_2-CH_2- \cdots \\
| \\
Cl
\end{array}
$$

should be discussed as initial points for the thermal elimination of HCl from PVC backbones. Also that the double bonds inside the chains reduce the thermal stability more markedly than when they are located at the end of the macromolecular chain.

Besides the structural irregularities already mentioned various oxygen-containing structures in PVC are discussed by Geddes [8.24]. They may be formed during the polymerization process by traces of oxygen or by oxidation of the polymer.

Up to now there is no experimental proof for the existence of head to head units.

Stivala & Reich [8.25] showed that structural order obtained by mechanical orientation in some macromolecular compounds and ordered (tactic) structures resulting from stereospecific polymerizations give rise to enhanced oxidative, thermal and chemical stability. Head to head structures generally inhibit initial thermal breakdown more readily than head to tail arrangements (e.g. PVC). The same authors mentioned the following effects of branching on the stability of polymers:

(1) decreasing half-decomposition temperature $T_{d/2}$ (temperature where 50% weight loss occurs upon heating polymers);
(2) increasing oxidation rate.

So far, the discussion has mainly centred around heat-induced degradation of PVC. It should be understood, however, that heat is but one form of energy capable of initiating a degradation chain of reactions and that the same or a similar sequence of degradation steps can be initiated by other energy sources, e.g. photo-initiation by UV radiation, gamma radiation, or other high energy radiation. It is generally accepted that all the modes of attack are inter-related and inter-dependent. That is to say, PVC already has had some amount of thermal history (during processing), however slight, and is much more susceptible to damage resulting from outdoor ageing and weathering than if it had not been previously heat treated. The reverse is also true. PVC items that have first been exposed to UV energy in the 300–400 nm wavelength range (terrestrial sunlight at sea level) will discolour earlier under the influence of heat than the same objects that have not been first exposed outdoors.

As already mentioned, the elimination of HCl (dehydrochlorination) is accompanied by a colour change, which is the least desired effect for construction and other practical applications. With increased scission of HCl the colour becomes more intense.

The experimental basis of the investigations of the polyene sequences in degraded PVC is the spectroscopy in the visible and UV range. Such polymers show a number of not very well resolved absorption maxima in their spectra. The wavelength of these maxima is nearly independent of the type of polymer and the kind and mechanism of degradation. The spectra can be analysed assuming that the observed absorption maxima are related to the main absorption bands of the polyene sequences with different lengths. The overall spectrum is therefore formed by the overlapping of the spectra of the different polyenes. It is known that between the number n of double bonds in the sequence and the wavelength λ of the absorption maximum at the largest band of a polyene the so-called square root law is valid:

$$\lambda = k\sqrt{n} + k' \tag{8.1}$$

It is also known that the extinction coefficient ε_n of the main band of a polyene is directly proportional to the number of double bonds (n):

$$\varepsilon_n = \text{const} \times n \tag{8.2}$$

On the basis of those assumptions and the Lambert–Beer law it is possible to calculate the frequency of the polyene sequences of

Table 8.2.
Thermal and physical properties of polyvinyl chloride [8.27]

	Rigid PVC	Flexible PVC	Chlorinated PVC
Heat deflection temperature @ 1820 kPa (°C)	65	—	105
Maximum resistance to continuous heat (°C)	60	35	90
Coefficient of linear expansion (°C × 10^{-5})	6·0	12·5	7·0
Compressive strength (kPa)	68 950	6 895	103 400
Flexural strength (kPa)	89 600	—	103 400
Impact strength (Izod: cm·N/cm of notch)	27	—	—
Tensile strength (kPa)	44 800	10 340	55 150
Elongation (%)	50	200	5
Hardness: Rockwell	—	—	R120
Specific gravity	1·4	1·3	1·22

Reprinted with permission from *Plastics versus Corrosives* by R. B. Seymour, 1982. Copyright John Wiley.

different length for the conversion x:

$$H'_n = \frac{\log J_0/J}{dnxC_p} \tag{8.3}$$

where C_p = concentration of the polymer in the testing solution, J_0 and J are the light intensities, and d = the cell path.

The value of H'_n gives a relative measure for the frequency of a polyene sequence with n double bonds in a degraded PVC sample.

These and similar calculations can be used to determine in a semi-quantitative way the frequency distribution of polymer sequences in degraded PVC or vinyl chloride polymers.

In normal PVC the polyene sequences consist of a few to about fifteen double bonds; several of the authors reported similar rather short polyene sequences in the initial stage of PVC degradation [8.22]. Table 8.2 presents the thermal and physical properties of PVC.

Thermal Degradation of Epoxy Polymers

Cross-linked (cured or hardened) epoxy polymers are formed by the reaction of a difunctional or polyfunctional cross-linking agent (hardening) with the functional or polyfunctional epoxide. There are several different types of commercially available epoxides all of which may be cured with a variety of hardeners such as aliphatic and aromatic diamines, acid anhydrides and dicarboxylic acids. It is therefore possible to formulate a very large number of epoxy structures with thermal stabilities dependent on the particular epoxy–hardener combinations.

Bishop & Smith [8.28] quoted studies for heat ageing characteristics of 66 different epoxy–hardener–filler combinations by baking disc-shaped samples for 1000 h in a convection oven at 180°C. The samples were weighed and measured before and after baking to determine weight loss and size change. The nature of the epoxides was not disclosed, and the only useful information which can be deduced is that acid anhydrides tend to give cured resins of greater thermal stability than those hardened with diamines.

Ehlers [8.29] suggested that epoxides such as I and II below should form more stable polymers than common epoxies based on di-phenylolpropane or partially condensed phenol-formaldehyde systems.

I   II

Kovarskaya & Zhigunova [8.30] have considered the thermal and thermo-oxidative degradation of cured epoxide and epoxy/phenol polymers: they showed that the latter were the more heat stable under both oxidizing and inert atmospheres. Neiman *et al.* [8.31] have made a fairly extensive study of the thermal degradation of two epoxylated diphenylolpropanes having MW of 500 and 2000 by pyrolysis in vacuo.

Volatile degradation products of the epoxides and of the polymers formed from them by cross-linking with polyamine and maleic anhydride were identified by a combination of gas chromatography and chemical methods. Methane (CH_4), propylene ($CH_2\text{=}CH\text{—}CH_3$), carbon monoxide (CO), formaldehyde (CH_2O), acetaldehyde ($CH_3\text{—}CHO$), and water were the principal products identified; also, large quantities of carbon dioxide (CO_2) were evolved by the polymer hardened with maleic anhydride. The authors postulated the following mechanisms for the scission of epoxide groups:

$$\cdots -R-O-CH_2-CH-CH_2 \qquad > \cdots -R-O-CH_2\cdot + \cdot CH-CH_2$$

(the end epoxy group)   I   II

$$\cdots -R-O-CH_2\cdot \longrightarrow \cdots -R\cdot + CH_2O \qquad \text{(formaldehyde)}$$
I

$$\cdot CH-CH_2 \longrightarrow \cdot CH_2-C-H \longrightarrow CH_3-C\cdot$$

II   III

$$CH_3-CH + \text{\small\textasciitilde}R\cdot$$
polymer
(acetaldehyde)

$$CH_3-C\cdot + \text{\small\textasciitilde}RH$$
polymer
III

(methane)
$$CH_4 + CO + \text{\small\textasciitilde}R\cdot$$
polymer

Although this reaction scheme satisfactorily explains the formation of most of the volatile products found by the mentioned authors, it gives little indication of the mechanisms of the thermal breakdown of completely cured epoxy polymers. Some electron paramagnetic resonance data support the theory that degradation of epoxy polymers occurs through free radical mechanisms.

Bishop & Smith [8.28] concluded that impurities and additives may influence that thermal stability; for this reason it is recommended that in degradation studies the epoxies and hardeners be carefully characterized to establish the basic degradation reactions. Some thermal and physical properties of this group of polymers are presented in Table 8.3.

It seems quite clear that additional detailed studies are necessary to further define the plausible degradation routes for the thermal degradation of epoxy polymers.

Table 8.3.
Thermal and physical properties of epoxy polymers [8.27]

	Epoxy plastic (EP)	Glass-filled EP	Glass-spheres-filled EP
Heat deflection temperature @ 1820 kPa (°C)	140	150	115
Maximum resistance to continuous heat (°C)	120	135	110
Coefficient of linear expansion (°C × 10^{-5})	2·5	2·0	2·5
Compressive strength (kPa)	120 000	206 850	82 740
Flexural strength (kPa)	124 100	103 400	41 370
Impact strength (Izod: cm·N/cm of notch)	53·4	53·4	10·6
Tensile strength (kPa)	51 710	87 740	41 370
Elongation (%)	5	4	1
Hardness: Rockwell	M90	M105	—
Specific gravity	1·2	1·8	0·8

Reprinted with permission from *Plastics versus Corrosives* by R. B. Seymour, 1982. Copyright John Wiley.

The thermal stability of epoxy polymers is chiefly affected by [8.7, 8.32]:

—the structure of the polymer;
—the type of curing agent;
—the chlorine content;
—the curing schedule.

Thermal Degradation of Poly(Urea-Formaldehyde)

Urea-formaldehyde polymers find practical utilization mainly in the form of network polymers. The synthesis is normally carried out in two separate operations. The first one involves the formation of a linear oligomer (low molecular weight fusible, soluble polymer) and the second operation involves curing reactions which lead to the cross-linked polymer.

Various types of aminopolymers based on urea and formaldehyde are produced commercially but they may be classified broadly into unmodified and modified polymers.

Unmodified urea-formaldehyde polymers are not suitable for use in surface coating formulations because they are insoluble in common solvents, do not interact readily with other polymers and are comparatively unstable. These limitations are substantially overcome when the polymers are modified by alcohols (e.g. *n*-butanol).

Urea-formaldehyde polymers have relatively poor heat resistance and discolour and degrade easier than other polymers.

The cross-linked microstructure of this polymer may be represented:

$$\cdots -U-F-U-F-U-F-\overset{\overset{\displaystyle F}{|}}{U}-F-U- \cdots$$

$$\cdots -U-\overset{\overset{\displaystyle F}{|}}{F}-U-F-\underset{\underset{\displaystyle F}{|}}{U}-F-\underset{\underset{\displaystyle F}{|}}{U}-F-\underset{\underset{\displaystyle F}{|}}{U}- \cdots$$

The major uses of this polymer in the construction industry are: thermal insulating foam, particle-board binder and coatings.

The use of urea-formaldehyde foams increased markedly in North America in the mid-1970s, as many homeowners insulated the walls of their residences in order to reduce the cost of heating and cooling. Urea-formaldehyde foams have the lowest thermal conductivities of insulation commonly used to retrofit the walls of residences. However, shrinkage of foams after installation results in splits, cracks and voids where air may circulate, reducing the thermal efficiency of the insulated wall.

Rossiter *et al.* [8.33], in an early National Bureau of Standards (NBS) review of the properties and performance of urea-formaldehyde foam insulations, evidenced that these foams may be susceptible to hydrolytic degradation. Similar findings that foams would crumble apart under laboratory exposure to combined elevated temperature and high humidity conditions have been reported by the Building Technology Laboratory of Finland.

The hydrolytic degradation of urea-formaldehyde foams under conditions of elevated temperature and humidity is accompanied by an emission of formaldehyde, and an increase in either the temperature or the humidity of the exposure results in an increase in the level of released formaldehyde. The emission of this aldehyde may occur by two pathways:

(a) an initial short term burst caused by release of free formaldehyde or hydrolysis of chain-end N-methylol groups of urea-formaldehyde polymer; and

(b) a slow, longer term evolution caused by degradation of the urea-formaldehyde polymer backbone.

Because of concerns of adverse effects on the health of building occupants exposed to the aldehyde emitted from foams, a ban on its use as residential insulation has been enacted in USA and in Canada.

Because of the susceptibility of urea-formaldehyde foam insulations to hydrolytic degradation, their durability under certain environmental conditions has been considered questionable. However, some foams may be more stable than others under severe temperature and humidity conditions. In a more recent paper, Rossiter *et al.* [8.33] describe the results of a study that examines the effect on the cellular microstructure of urea-formaldehyde foam insulations from exposure to combined elevated temperature and humidity conditions. The changes observed in volume and mass at 60°C are presented in Figs. 8.7 and 8.8.

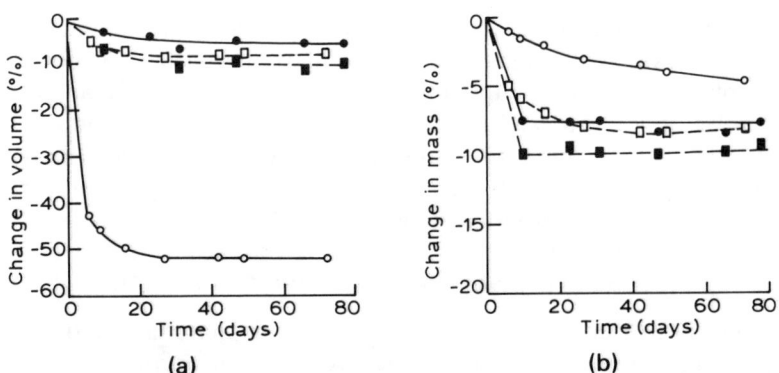

Fig. 8.7. Changes in mass and volume of samples (a) and (b) versus time under moist and dry conditions at 60°C (140°F): O—, sample (a) (60°C/75% RH); ●—, sample (a) (60°C/dry); □—, sample (b) (60°C/75% RH); ■, sample (b) (60°C/dry) [8.33]. (Reprinted with permission from *Thermal Insulation, Materials and Systems for Energy Conservation in the 80s* by F. A. Govan, D. M. Greason & J. D. McAllister (Eds). Copyright ASTM.)

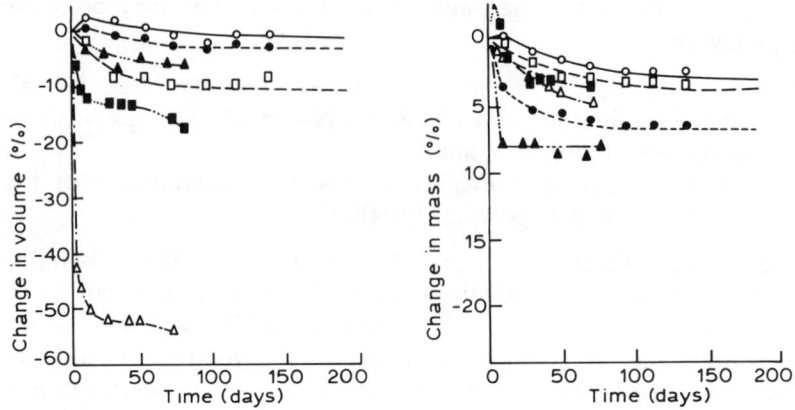

Fig. 8.8. Changes in mass and volume of sample (a) for various conditions of exposure: O—, 30°C/75% RH; ● – – –, 40°C/53% RH; □——, 40°C/75% RH; ■··——···, 50°C/75% RH; △·—·—, 60°C/75% RH; ▲··—··—··, 60°C/dry [8.33]. (Reprinted with permission from *Thermal Insulation, Materials and Systems for Energy Conservation in the 80s* by F. A. Govan, D. M. Greason & J. D. McAllister (Eds). Copyright ASTM.)

PHOTODEGRADATION

Most commercial organic polymers undergo chemical modification upon irradiation with ultra-violet (UV) light, because they possess chromophoric groups (groups able to absorb light) as regular components or as impurities. Since the spectrum of sunlight contains a portion of UV light, photoreactions are usually induced when organic polymers are subjected to outdoor exposures. Generally, photoreactions in commercial high polymers are harmful: they cause changes in colour and embrittlement.

It is appropriate to emphasize here two important aspects:

(a) the specific interaction of light with organic materials, and
(b) the randomness of photochemical reactions in polymers.

First aspect: light absorption in a molecule consists of a *specific interaction* of a certain chromophoric group with a photon of given energy. The remainder of the molecule remains unaffected during the absorption act. Knowledge about absorptivities of chromophores is the polymer chemist's tool for attaining photochemical selectivity. In other words, if a polymer chain is supposed to be ruptured at a certain position on irradiation, this goal can be achieved by synthetically introducing an appropriate chromophore at that place in the polymer backbone.

Second aspect: light is absorbed statistically by the chromophore groups it contains. As far as practical applications of macromolecular compounds are concerned, the sun is the most important light source. The spectrum of sunlight penetrating the earth's surface ranges from 290 to 3000 nm.

The spectral distribution depends on atmospheric conditions and the latitude. Somewhat less than 10% of the sunlight at the surface of the earth is UV light, about 50% is visible and 40% is infra-red (IR) light. For laboratory and industrial irradiations various types of lamps are available. Frequently, mercury lamps are used: low pressure lamps with two intense lines at 184·9 and 253·8 nm and medium pressure lamps with a great number of lines, the most intense one corresponding to 366 nm. For preparative purposes high pressure Hg lamps are the most suitable because of their high emission intensities. The most frequently used lines are those at 254, 265, 313, 366, 436 and 546 nm. Carbon arc and xenon arc lamps are utilized quite often in devices for accelerated weathering tests, etc.

More recently, powerful lasers have become commercially available. They emit coherent and monochromatic light. High power lasers are operated in a pulsed mode, in many cases, with adjustable pulse repetition rates.

The absorption of light is a prerequisite for the occurrence of photochemical processes. Saturated compounds possessing bonds such as [8.34]:

$$\begin{array}{l} \ce{>C-C<} \\ \ce{>C-H} \\ \ce{-O-H} \\ \ce{>C-Cl} \text{ etc.,} \end{array}$$

absorb light at $\lambda < 200$ nm. Carbonyl ($\ce{>C=O}$) groups and conjugated double bonds ($\ce{>C=C-C=C<}$) absorb above $\lambda \leqslant 200$ nm and have absorption maxima between 200 and 300 nm. Figure 8.9 shows absorption spectra of several polymers. It should be pointed out that only a small number of the important polymers are capable of absorbing solar radiation. However, quite frequently commercial

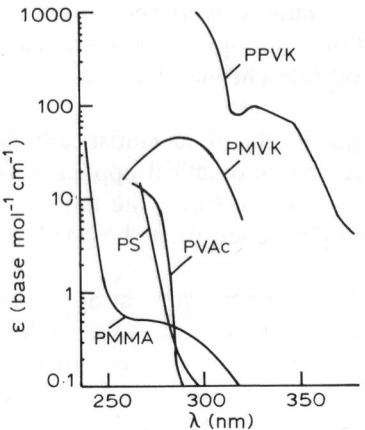

Fig. 8.9. Absorption spectra of several polymers. PPVK, poly(phenyl-vinyl ketone); PMVK, poly(methyl-vinyl ketone); PVAc, PS, PMMA [8.1]. (The spectra of PS and PVAc were taken from Golemba, F. J. & Guillet, J. E., *J. Paint Technol.*, **41** (1969) 315.)

polymers contain impurities capable of absorbing sunlight. This explains, in most cases, the instability of polymers which, according to their microstructure (chemical constitution), should be resistant to solar radiation. The chance of an absorbed photon to induce a chemical change in the molecule depends principally on the photophysical processes following the absorption act.

The physical processes involved in photodegradation include:

(a) absorption of light by the material;
(b) electrical excitation of the molecules;
(c) deactivation by radiative or radiationless energy transitions or by energy transfer to some acceptor.

When the lifetime of the excited state is sufficiently long, the species can participate in various chemical transformations.

Phenomenologically, the absorption of light can be described by the Lambert–Beer law, as mentioned previously.

As shown in Table 8.4, the maximum sensitivity of several polymers (as determined by the bond dissociation energies) is in the range of 290–400 nm.

Although the atmosphere of the earth filters out the UV portion of

Table 8.4.

Wavelength of UV radiation (energy of a photon) at which various polymers have maximum sensitivity [8.4]

Polymer	Wavelength (nm)	Energy (kcal/mol)
Styrene–acrylonitrile copolymer	290, 325	99, 88
Polycarbonate	295, 345	97, 83
Polyethylene	300	96
Polystyrene	318	90
Polyvinyl chloride	320	89
Polyester	325	88
Vinyl chloride–vinyl acetate copolymer	327, 364	87, 79
Polypropylene	370	77

Reprinted with permission from *Polymer Degradation* by T. Kelen, 1983. Copyright Van Nostrand Reinhold Co.

the solar radiation, the actinic range (290–400 nm) of solar UV radiation is about 6% of the total radiation of the sun which reaches the surface of the earth (Fig. 8.10).

The absorption of light results in an electronic transition between two energy levels in the absorbing molecule. This absorbed energy is exactly equal to the energy of a light quantum:

$$\Delta E = h\nu \tag{8.4}$$

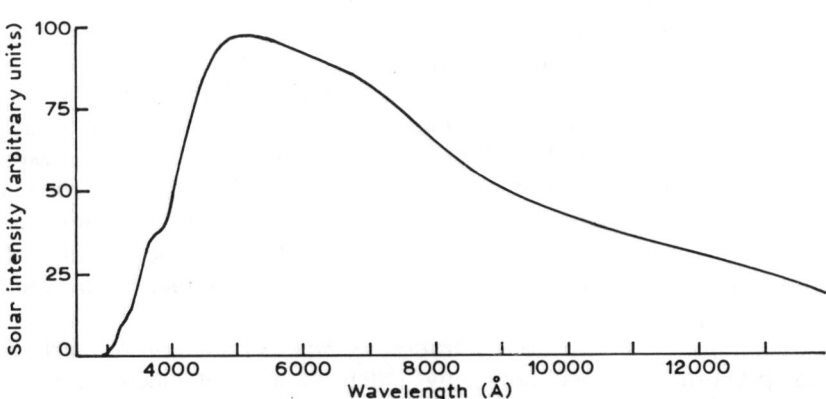

Fig. 8.10. UV and visible spectrum of sunlight at noon in midsummer in Washington, D.C. [8.4]. (Reprinted with permission of Elsevier Publishing Co.)

where h is Planck's constant and v is the frequency of the absorbed light:

$$v = c^*/\lambda = c^* v^* \tag{8.5}$$

where c^* is the velocity, λ is the wavelength, and v^* is the wave number of the absorbed light (a possible set of values and dimensions: $h = 6 \cdot 62 \times 10^{-27}$ erg s; v s^{-1}; $c^* = 3 \times 10^{10}$ cm/s; λ cm; v^* cm^{-1}).

The energy absorption produces an excited state of the molecule; two types of such states can be distinguished:

(a) The singlet state (S), in which the spins of electrons are (remain) paired,
(b) The triplet state (T), in which they are unpaired.

The ground state (S_o) is almost always a singlet state. The excitation of a molecule from the ground state to the first excited (S_1) singlet state is represented as follows:

$$S_o \rightarrow S_1$$

The higher excited singlet states are represented as:

$$S_2, S_3, S_4, \ldots$$

The excited molecule can lose its excess energy by vibrational relaxation and emission (fluorescence) and the emission of light by the $T_1 \rightarrow S_o$ transition is called phosphorescence [8.4].

The chemical process of photodegradation includes (1) chain scission or cross-linking, and (2) oxidation (photo-oxidation), which introduces carbonyl, carboxyl, hydroxyl, or peroxide groups into the polymer and which may also result in polymer chain scission. The chain rupture or radical formation in the various photochemical processes is often followed by embrittlement due to cross-linking. However, secondary reactions, especially in the presence of oxygen, cause further degradation of the polymer. Mechanical properties, such as tensile strength, elongation and impact strength, may deteriorate drastically. Surface crazing can also be a sign of UV-induced degradation [8.35].

Some polymers show discoloration as well as reduction of mechanical properties, others show only a deterioration of mechanical properties. This degradation can be lessened by the incorporation of photostabilizers (UV absorbers) into the polymer.

Photodegradation of PVC

The weathering of PVC is greatly affected by its thermal history (synthesis, processing). It has been observed that those PVC compounds which have been mixed for long periods or at high temperatures show poor outdoor performance.

Moritz [8.36] has postulated that the surface finish has an important bearing on the weathering performance of PVC. Blemishes, pores and cracks on the surface are potential sites for degradation. Also light is scattered on a matt surface and therefore will be strongly absorbed whereas a smooth glassy surface will reflect a greater part of the radiation. Moreover, dust will cling more easily to a rough surface and is not so easily washed off by rain as it would be from a smooth pore-free finish. The effects of the nature of the surface on weathering performance can obviously be equally important in other polymers.

Even with thermal stabilizers, PVC does not perform well outdoors unless steps have been taken to protect it from the damaging effects of sunlight.

Loss of impact strength is generally accompanied by a darkening of colour although bleaching is sometimes observed. An interrelationship between loss of strength and development of colour is suggested by results obtained by Lutz [8.37]. Experimental studies were carried out on a PVC modified with an acrylic polymer (which is an impact modifier) and titanium dioxide. A comparison of performance at three quite different sites showed that exposure in Arizona causes more severe degradation than is the case for exposure in Pennsylvania or Florida, and this indicates the importance of solar radiation in determining the extent of weathering (Fig. 8.11). The samples behave relatively well because of the presence of rutile (TiO_2), a pigment with a high degree of opacity which effectively stabilizes the degradation of PVC.

The use of plastic window frames, made primarily of hard, modified, extruded PVC, has spread in the past 2–5 years from West Germany, its place of origin, to the surrounding countries. Wooden windows are plagued with fungal attack and hence high maintenance costs, and this has certainly helped increase the market share for PVC window frames.

Figure 8.12 graphically depicts the rapid market share growth of PVC versus other typical window frame materials in West Germany. Today, approximately 50% of all windows installed in Germany and in

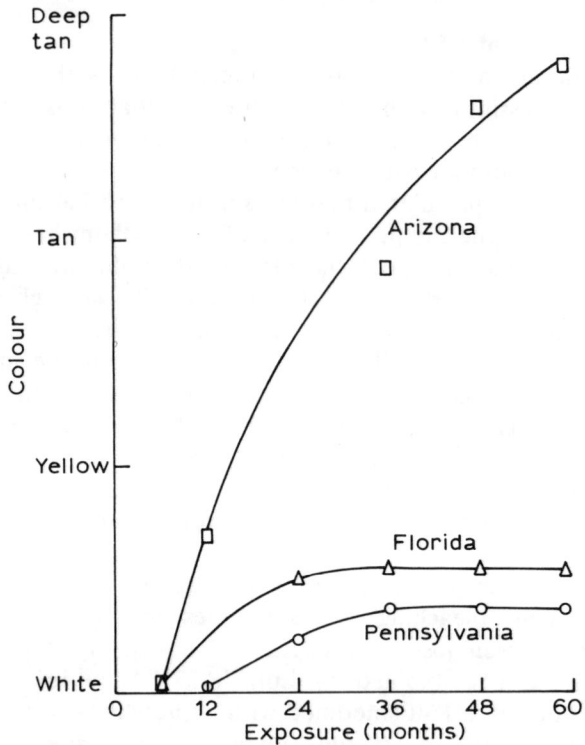

Fig. 8.11. Discoloration of white PVC with time at various exposure sites [8.16]. Reprinted from *Polym. Plast. Technol. Engng,* **11**(1) (1978) 55–80, by courtesy of Marcel Dekker Inc.

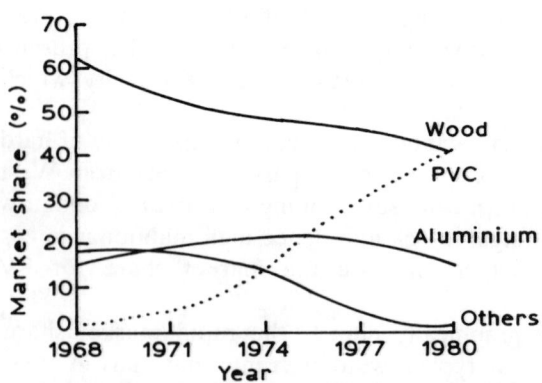

Fig. 8.12. Window profile materials in West Germany [8.38]. (Reprinted with permission from *J. Vinyl Technol.,* **5**(3) (1983) 135–42. Copyright Society of Plastics Engineers.)

some other European countries are made of polymers, one half of which are installed in older renovated buildings.

Different pigments are used to study their influence upon the weatherability of the PVC window frames. The investigations quoted by Svane [8.39] from the Technology Institute of Denmark, show that changes in appearance of white or light greyish PVC after approximately 20 years of natural exposure in an industrial environment were very moderate and probably would continue to be for several years in the future.

There are some very basic differences in construction between European windows and those typically produced in North America. The double-hung window so familiar to us is virtually unknown in Europe where the 'tilt-turn' design is prevalent. The European windows appear massive in construction when compared to their American counterparts.

The components of a typical rigid PVC window frame compound are PVC, impact modifier, stabilizers, lubricants, UV absorbers, antioxidant and fillers. The most widely used PVC is the suspension type having a K value in the 65–70 range. Mass and emulsion polymerized polymers can also be used, but it is important that the polymers be free of residual surfactants or other additives able to absorb moisture. A higher MW (K value 68) is preferred for its better mechanical properties.

Pfeiffer [8.38] showed that good weathering results have been obtained with EVA (ethylene–vinyl acetate copolymer). Chlorinated PE is widely used for both white and brown profiles, while acrylic modifiers incorporated at the 8–10% level are becoming increasingly popular in Europe. These acrylic impact modifiers appear to provide weatherability equal to EVA types along with the positive aspects of lower shrinkage, higher output and higher Vicat softening point. When the levels of the pigment (TiO_2) do not exceed 3% the addition of a suitable UV absorber will have a positive effect on the weathering of rigid PVC compounds.

The addition of 0·2–0·3% of Tinuvin P has been shown to produce good results. A standard formulation typical of those used today in Europe to produce dark brown shades, using a Ba–Cd stabilized system, and a comparative tin carboxylate stabilized formulation are presented in Tables 8.5 and 8.6.

Table 8.7 lists the European countries producing PVC windows along with production volumes and current market shares. Most of

Table 8.5.
Typical brown compound—Ba/Cd stabilized [8.38]

S–PVC K-value 65–68, 6% EVA as graft polymer	100	phr[a]
Ba/Cd stabilizer (polyol free)	2·5	phr
Decyl-diphenyl phosphite	0·5	phr
Epoxidized soya bean oil	1·0	phr
Wax-ester	0·4	phr
12-Hydroxy-stearic acid	0·4	phr
Irganox® 1076 (hindered phenol antioxidant)	0·2	phr
Tinuvin® P (benzotriazole UV absorber)	0·25	phr
Cromophtal® Brown 5R (Pigment Brown 23)	0·7–0·9	phr
Pigment Blue 15	0·03	phr

[a] phr = per hundred.
Reprinted with permission from *J. Vinyl Technol.*, **5**(3) (1983) 135. Copyright Society of Plastics Engineers.

these countries were much later than Germany with the introduction of vinyl windows and therefore the market shares are still relatively small. Significant growth possibilities exist however in France, the United Kingdom, Benelux and Scandinavian countries. Adoption of coloured PVC windows in the more severe climatic zones, such as Spain, Italy and the Mediterranean, has been hampered by the problem of heat absorption and subsequent distortion of the frames.

Pfeiffer mentioned also the resulting effect on temperature build-up using different shading pigments to produce a brown shade (Table 8.8).

Table 8.6.
Brown compound—Sn carboxylate stabilized [8.38]

S–PVC/EVA	100	phr[a]
Dialkyl tin carboxylate (liquid)	2·5	phr
Solid paraffin wax	0·2–0·4	phr
Solid, partly oxidized PE-wax	0·2–0·4	phr
Irganox® 1076 (hindered phenol antioxidant)	0·1	phr
Tinuvin® P (benzotriazole UV absorber)	0·25	phr
Irgamon® F 131 (acrylic processing aid)	1·0	phr
Cromophtal® Brown 5R (Pigment Brown 23)	0·7–0·9	phr
Pigment Blue 15	0·03	phr

[a] phr = per hundred.
Reprinted with permission from *J. Vinyl Technol.*, **5**(3) (1983) 135. Copyright Society of Plastics Engineers.

Table 8.7.
PVC window profiles approximate market size—Europe, 1980 [8.38]

Country	PVC windows annual production $(10^6 m)$	Market share (%)	PVC tons (tonnes)
W. Germany	6·0	40	140·0 (inc. export)
Austria	0·3	30	6·5
Denmark	0·4	10	6–7·0
Benelux	0·25	10	5·0
Italy	<0·25	3	1·0 (+4 000 export)
France	<0·25	3	3·0 (inc. export)
UK	?	4	3–4·0
Total	7·5	4	170–180·0

Reprinted with permission from *J. Vinyl Technol.*, **5**(3) (1983) 135. Copyright Society of Plastics Engineers.

The European technology for producing dark brown PVC window frames has undergone considerable improvement over the past 10 years. However, much work remains to be done here because of the substantial climatic differences between parts of the North American and the major European window markets.

Table 8.8.
Maximum surface temperature of coloured PVC window profiles[a] [8.38]

Test specimen profile		Max. surface temperature between 11 and 13 h
Brown window profile	Not shaded	54°C
Brown window profile	Shaded with phthalocyanine blue	57°C
Brown window profile	Shaded with carbon black	58°C
White profile		43°C
Yellow profile		46°C
Dark grey profile		58°C

[a] Ambient air temperature 30°C (mid-day), Pfeffingen, Switzerland.
Reprinted with permission from *J. Vinyl Technol.*, **5**(3) (1983) 135. Copyright Society of Plastics Engineers.

J. T. Lutz, Jr, [8.16] shows that the TiO$_2$ level is frequently raised to increase the potential for shielding the PVC from UV radiation. The data in Fig. 8.13 indicate that the optimum level of TiO$_2$ is about 10%. This optimum may vary for specific formulations depending on stabilizer type and amount, impact modifier type and level, and the presence or absence of a UV absorber. Since TiO$_2$ has the role of a filler, increasing the concentration beyond 10% begins to have a detrimental effect on impact strength. The beneficial effect of in-

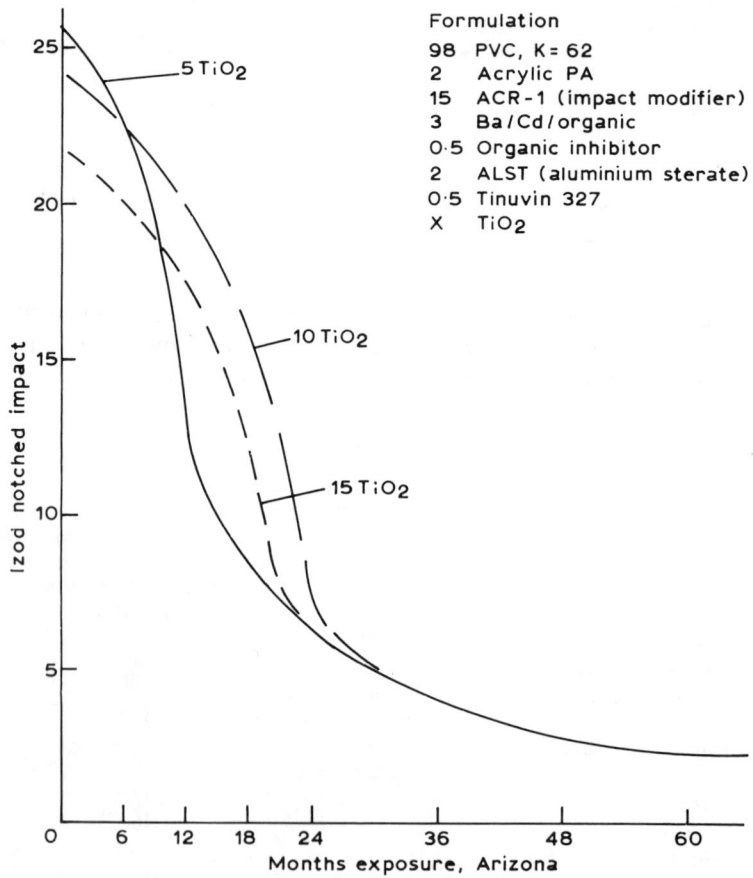

Fig. 8.13. Weatherability of white PVC. Effect of TiO$_2$ level [8.16]. Reprinted from *Polym. Plast. Technol. Engng*, **11**(1) (1978) 55–80, by courtesy of Marcel Dekker Inc.

creased TiO$_2$ concentration vanishes when exposures are extended beyond 2 years. Retention of the white colour is somewhat better over the entire exposure period when increased amounts of TiO$_2$ are used. This does not appear to be a function of protection of the PVC, but rather due to the presence of more pigment.

Pigments, per se, must be carefully chosen for their heat and light stability and their effects on the mechanical properties of PVC items. Currently most exterior PVC articles are limited to white or pastel colours. Dark colours convert sunshine into higher temperatures that cause PVC to distort.

The influence of the modifier (for impact) in the PVC mixture is important although not readily understood.

All modifiers of acrylic type excel in both colour and impact retention. Chlorinated PE, depending on level, will retain impact strength, but will rapidly discolour due to early and excessive chalking (Table 8.9). The all-acrylic modifiers resist chalking due to the inherent stability of the acrylic polymer in the surface of the PVC mixture.

Some studies [8.18, 8.40] on polymer sequences produced during the degradation of PVC were performed.

Davidson & Meek [8.41] have used mass spectrometry to examine the photo-induced dehydrochlorination of PVC films both in the absence and presence of TiO$_2$. Hydrochloric acid evolution was only observable when light of $\lambda < 300$ nm was used. The authors also observed that the presence of TiO$_2$ affords some protection. When light of $\lambda > 300$ nm is used photodegradation of added stabilizers is observed.

By the use of IR spectroscopy it is shown that PVC photodegrades in the presence of air to give carbonyl compounds, hydro peroxides and polyenes if light having $\lambda < 300$ nm is used. Added TiO$_2$ can retard these processes.

Polyethylene Photodegradation

PE is a polymer of great commercial importance because of its processibility, useful properties and relatively low cost. However, unstabilized PE readily photo-oxidizes outdoors and for this reason a great deal of attention has been given to weathering studies. The importance of the shortest wavelengths of solar radiation has been established and the various modes of initiation, the nature of transient species and their roles are well understood.

Polymeric Building Materials

Table 8.9.
Effect of impact modifier on colour retention[a] [8.16]

Pigment	Modifier	Bristol, Pennsylvania, exterior exposure, years to colour change			
		Slight	Definite	Severe	Chalking
Green	ACR-2	1/2	3	4	3–4
Phthalo green/ furnace black/ Cd yellow/ TiO$_2$G	CPE	<1/2	1/2	1	1
Blue	ACR-2	1/2	3	4	3–4
Phthalo blue/ furnace black/ TiO$_2$	CPE	<1/2	1/2	1	1/2–1
Black	ACR-2	2	4	>5	4–5
Furnace black/ TiO$_2$	CPE	1/2	3	4	4–5
Olive grey	ACR-2	2–3	2–3	3–4	3–4
TiO$_2$/Cd orange/ furnace black	CPE	<1/2	<1/2	1	1/2–1

[a] Formulation: 85 PVC, $k = 69$; 15 modifier; 4 Ba/Cd/organic; 0·5 organic inhibitor; 2 GMS; colourant as indicated.
Reprinted from *Polym. Plast. Technol. Engng*, **11**(1) (1978) 55–80, by courtesy of Marcel Dekker.

Furneaux *et al.* [8.42] have exposed horizontally at the Joint Tropical Trials and Research Establishment, Innisfail, and at Cloncurry, Queensland, Australia, low density polyethylene (LDPE) plaques backed with aluminium, with the metal surface facing the ground.

LDPE exposure to sunlight in the presence of air leads to the photo-oxidation of the polymer established by the modification of the microstructure and of the mechanical properties. The main transformation of the macromolecular chain consists in formation of carbonyl and vinyl groups.

The authors recorded IR spectra of microformed 150 nm sections. The tangent-baseline technique was employed to determine the absorbances of the peaks at 1715 cm^{-1} (carbonyl) and 910 cm^{-1} (vinyl) using the peak at 1375 cm^{-1} as an internal standard. The ratio

$1715\,\text{cm}^{-1}/1375\,\text{cm}^{-1}$ was presented as the carbonyl index and the ratio $910\,\text{cm}^{-1}/1375\,\text{cm}^{-1}$ as the vinyl index.

The carbonyl concentration through LDPE plaques weathered for various intervals is shown in Fig. 8.14. Considering the exposed surface, the carbonyl concentration falls from what appears to be a high surface amount to a level which shows little change with penetration, until, at a depth of about 0·8 mm, a rapid drop in carbonyl concentration is observed; a similar profile is observed as the rear surface is penetrated.

The concentration of unsaturated (vinyl) groups in weathered plaques is shown in Fig. 8.15. The pattern is similar to that obtained for carbonyl except that in the middle of the specimen, the vinyl content at first increases, then decreases, as the exposure time is lengthened, whereas the carbonyl content increases with exposure period.

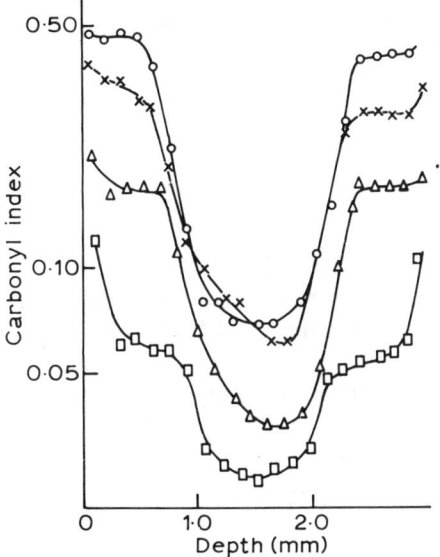

Fig. 8.14. Carbonyl concentration through weathered LDPE plaques: □, +67 days (200 kWh) and x, 188 days (670 kWh) at Innisfail. △, 83 days (410 kWh) and ○, 173 days (1054 kWh) at Cloncurry [8.42]. (Reprinted with permission from *Polym. Degrad. Stabilization,* **3** (1981) 431. Copyright Applied Science Publishers.)

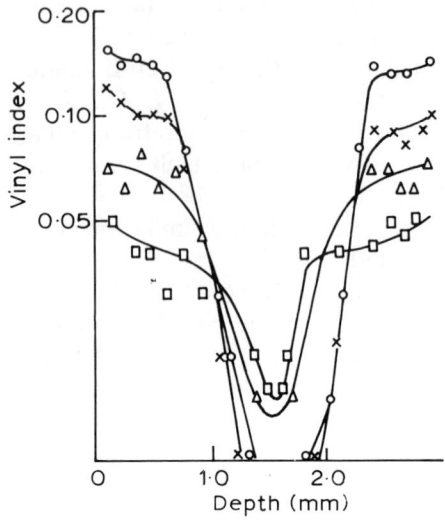

Fig. 8.15. Vinyl concentration through weathered LDPE plaques: □, 67 days (200 kWh) and x, 188 days (670 kWh) at Innisfail; △, 83 days (410 kWh) and ○, 173 days (1054 kWh) at Cloncurry [8.42]. (Reprinted with permission from *Polym. Degrad. Stabilization,* **3** (1981) 431. Copyright Applied Science Publishers.)

Following the results obtained, Furneaux *et al.* considered that the photo-oxidation profile observed is primarily determined by the ability of oxygen to diffuse into the LDPE during the effective daylight period.

The oxidation mechanism of polyethylene can be described as follows (RH = PE):

Initiation $\quad RH \xrightarrow{hv} R^{\cdot} + H^{\cdot}$

Propagation $\quad R^{\cdot} + O_2 \xrightarrow{k_2} ROO^{\cdot}$

$\qquad\qquad\qquad ROO^{\cdot} + RH \xrightarrow{k_3} ROOH + R^{\cdot}$

Termination $R^{\cdot} + R^{\cdot} \xrightarrow{k_4}$

$\left.\begin{array}{l} \\ ROO^{\cdot} + R^{\cdot} \xrightarrow{k_5} \\ \\ ROO^{\cdot} + RO_2^{\cdot} \xrightarrow{k_5} \end{array}\right\}$ inert products

Branching $ROOH \xrightarrow[k_7]{h\nu} RO^{\cdot} + {}^{\cdot}OH$

If the photo-oxidation of PE proceeds according to this sequence of reactions then the rate of oxygen consumption is given by the expression:

$$\frac{-d[O_2]}{dt} = \frac{k_2 k_3 [RH][O_2][k_1]^{1/2}}{k_2 k_5^{1/2}[O_2] + k_3 k_4^{1/2}[RH]} \qquad (8.6)$$

Cunliffe & Davis [8.43] have developed equations to describe the influence of oxygen diffusion on the rate of photo-oxidation of polyethylene plaques. The observed and calculated rates show a relatively weak distance dependence to a depth of the order of 1 nm, beyond which they are diffusion controlled.

Many applications of polyethylene as garden equipment, horticultural films, crates, stadium seating, etc., require a degree of outdoor stability. Unstabilized polyolefins show little resistance to sunlight. For example, LDPE exposed in Central Europe in the spring showed an immediate loss of strength (Fig. 8.16) whereas samples exposed in the fall time lasted appreciably longer and indicated the dependence of durability on quality of sunlight.

Photodegradation of Polycarbonate

Polycarbonate sheeting is being used more and more to replace glass as glazing material in factories, schools and commercial buildings, primarily because of its greater resistance to breakage (impact strength is 250 times that of safety glass of the same thickness) and high transparency (slightly less than window glass).

PC, based on bisphenol A, is a tough transparent thermoplastic material which is widely used where these properties are important, e.g. as street lamp diffusers, visors, vandal-resistant glazing, riot shields, housing for lamps, aircraft windows, etc.

The mechanical properties of a macromolecular compound are to a major extent determined by the MW characteristics. Measurements of

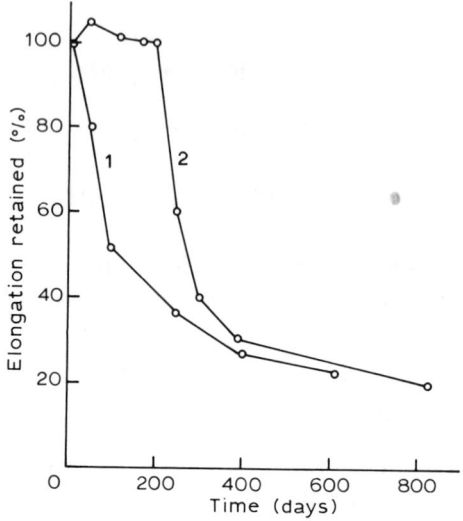

Fig. 8.16. Seasonal exposure effects on the weathering of LDPE: (1), sample exposed in spring; (2) sample exposed in autumn [8.44]. (Reprinted with permission from *Weathering of Polymers* by A. Davis & D. Sims, 1983. Copyright Applied Science Publishers.)

the exposed materials are expressed as MW_t/MW_O, the ratio of MW after exposure time t to MW of the control specimen, and are given for two tropical sites in Fig. 8.17.

It can be seen that whereas the unstabilized and thermally stabilized PC suffered a significant drop in MW after only 6 months' exposure, it is only after 24 months that appreciable changes occurred in the UV stabilized grade. It is also clear from these results that the incorporation of a thermal stabilizer does not improve the weathering performance. No significant difference could be detected between the severity of the tropical wet and tropical dry sites in their degradative effect on polycarbonates as far as MW changes are concerned. This is in contrast to the impact strength results which indicate that hot/dry is a more severe environment.

It has been found by Yamasaki & Blaga [8.45] at the Division of Building Research of the National Research Council of Canada that when UV stabilized PC sheeting was subjected to outdoor weathering on a test rack, microscopic cracks developed on the surface after about

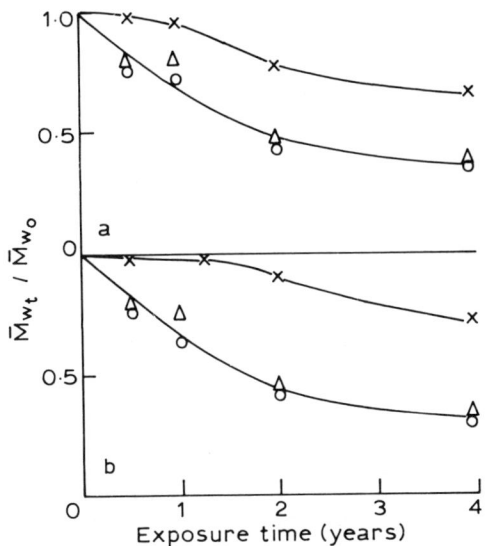

Fig. 8.17. Fall in MW of polycarbonate surface on exposure: O, unstabilized. △, heat stabilized. x, UV stabilized. (a) Hot/dry site (b) hot/wet site [8.14]. (Reprinted with permission from *Weathering of Polymers* by A. Davis & D. Sims, 1983. Copyright Applied Science Publishers.)

23 months. These cracks decreased the degree of transparency of PC. The formation of such surface microcracks was investigated with the use of a scanning electron microscope.

The literature shows that PC undergoes photochemical degradation when subjected to solar radiation. Laboratory exposure of PC at 82°C has shown that photochemical processes like the microcracking observed at DBR are confined to the irradiated surface. Under artificial weathering, PC sheeting underwent an abrupt decrease in its tensile strength and elongation to failure.

Between other tests, in order to characterize the effect of outdoor and artificial weathering on the tensile properties of PC, changes in elongation at yield and net elongation at break (net ultimate elongation), and tensile stress at yield and at break (ultimate tensile stress), were determined for each exposure. Some results are illustrated in Fig. 8.18.

Yamasaki & Blaga concluded that the reduction in the tensile stress and elongation to failure is due to the reduction in MW of the surface

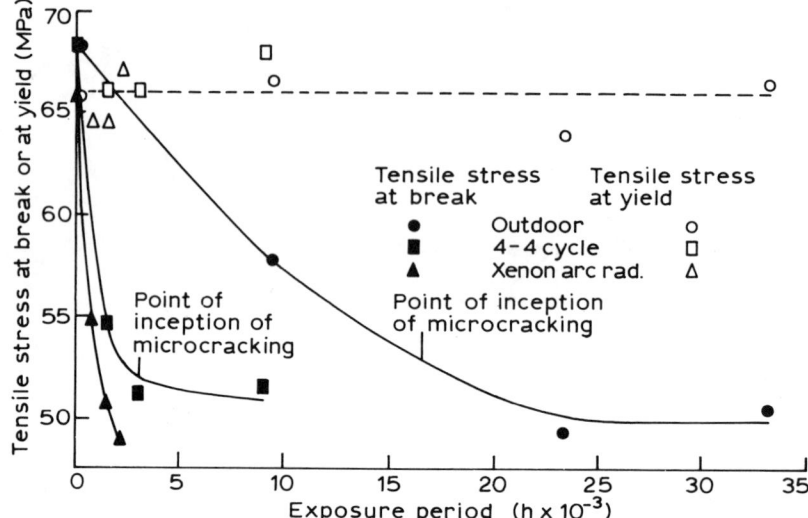

Fig. 8.18. Changes in tensile stress at break and at yield of PC on exposure to outdoor and artificial weathering conditions [8.45].

layer. Also, that the ranges of temperature and moisture conditions that prevailed during the exposures did not significantly influence the change in MW.

Blaga [8.46] mentioned that the degradation of polymers in the outdoor environment is believed to involve a very complex reaction initiated by light and propagated by oxidative reactions; thermal processes induce subsequent steps whose rates may be affected by the presence of moisture or the participation of pollutants.

The absorbed UV light energy causes the dissociation of bonds (mostly C—C and C—H) in the molecules of a polymer by a homolytic process to produce free radicals as the primary photochemical products. This event, with or without the participation of oxygen, can lead subsequently to one or more of the following chemical changes:

(a) chain scission with the formation of small molecules;
(b) cross-linking;
(c) elimination of low molecular weight compounds (water, hydrochloric acid, carbon monoxide, carbon dioxide, etc.);
(e) depolymerization; or
(f) photohydrolysis.

Light of different wavelengths of the solar spectrum can provide different modifications simultaneously in the same polymer.

As already mentioned, oxidative photodegradation of polymers involves processes of initiation, propagation and termination. The most frequently occurring photo-induced chemical degradations of polymers are chain scission and cross-linking, with concomitant formation of oxygen-containing functional groups such as carbonyl, carboxyl, hydroxyl and peroxides, which can be assessed easily by IR spectroscopy.

It is believed that carbonyl photocleavage processes may be the major cause of photodegradation in carbonyl-containing polymers as well as in most macromolecular hydrocarbons that always contain at least traces of carbonyl functions as a result of thermal oxidation during processing. If a ketone carbonyl group is located in the main chain or on adjacent carbons of a side chain, the degradation may occur by two well known scission mechanisms, Norrish I and Norrish II:

Norrish I:

$$\cdots -CH_2-CH_2-\underset{\underset{O}{\|}}{C}-CH_2-CH_2-\cdots \xrightarrow{h\nu}$$

$$\cdots -CH_2-CH_2\cdot + \cdot\underset{\underset{O}{\|}}{C}-CH_2-CH_2-\cdots$$

Norrish II:

$$\cdots -CH_2-CH_2-CH_2-\underset{\underset{O}{\|}}{C}-CH_2-\cdots \xrightarrow{h\nu}$$

$$\cdots -CH=CH_2\cdot + CH_3-\underset{\underset{O}{\|}}{C}-CH_2-\cdots$$

Both of these processes produce a decrease in the molecular weight, resulting in deterioration of the properties.

POLYMER RADIODEGRADATION

Until recently, most of the investigators in the field of radiative degradation have directed their attention to studies on the effect of high energy radiation, in the presence or absence of oxygen on polymer chemical and physical properties.

A basic parameter which determines the radiodegradation of polymers is the type of radiation involved in this process. Radiations composed of particles with higher energies than those encountered in binding electron orbitals are referred to as high energy radiations, e.g. γ-rays, high energy electron beams, neutrons, etc.

Radioactive cobalt (Co^{60}), which has a half-life of 5·3 years, is widely used as a source of γ-radiation. This has a penetration of at least several cm and an intensity range of 0·1–1 Mrad/h (1 Mrad = 10^6 rad = 10^8 erg/g = 10 γ/g = 6·25 × 10^{19} eV/g). Electron accelerators (resonant transformers, linear accelerators, electrostatic accelerators) are excellent sources of high energy electron beams. The dose rate here is much higher than that of Co^{60} rays, and is typically 1 Mrad/s. However, the high energy electron beam penetration is much lower than that of X-rays. The less important type of radiation is X-rays. In order to evaluate the efficiency of radiation in causing chain scission or linking, it is customary to express the radiation yield in terms of a G value. This represents the average number of events or changes of a given type for absorption of 1000 eV of energy [8.47].

Exposure of macromolecular compounds to high energy radiation results in a number of physical and chemical changes. These modified properties are not due to any change in nuclear structure, but arise from new electronic configurations and hence new chemical reactions. In some respects such changes can also be induced by exposure to ultra-violet light, but the ability of high energy radiation to penetrate organic materials, and interact almost at random with orbital electrons, confers on high energy radiation a much more extensive scope.

The chemical modifications induced in polymers are not inherently different from those seen when low molecular weight compounds are irradiated, but, in the case of high polymers, a relatively small proportion of chemical change can, when applied at the appropriate positions, result in very important changes in physical properties, e.g. the cleavage of a linear macromolecule at only one point, corresponding to a chemical modification of perhaps one bond in 10^5 bonds, can greatly affect such an obvious property as viscosity.

Chemical modifications which occur in high polymers may be separated into the following two groups: (a) modifications proportional to the radiation dose; (b) modifications which depend considerably on dose rate as well as dose.

The first occur when each separate radiation event such as ionization or excitation leads to a single localized modification such as a molecular scission. The latter often involve a chain reaction.

Examples of the former are *cross-linking* and *degradation* of saturated linear polymers; examples of the latter are radiation-induced polymerization, graft polymerization, curing of polyesters, epoxies, etc. In comparison with these processes which may be initiated by common chemical means, high energy radiation has a number of distinct advantages such as the following:

(1) it can penetrate and induce reactions in the solid state, over a wide range of temperatures and environments;
(2) by a variety of well established techniques many of the intermediate steps can be followed, and the results are readily reproducible;
(3) most radiation treatments of polymers leave no radioactive residue, so that after exposure, they can be handled with full safety.

The effectiveness of radiation in modifying high polymers is influenced by the presence of:

(a) type of radiation used;
(b) the macrostructure of the polymer (e.g. degree of crystallinity, etc.);
(c) presence of oxygen, catalysts traces, additives, solvent, etc.

During irradiation, polymers may either degrade or cross-link depending on their microstructure. Both processes take place simultaneously, and the classification in Table 8.10 merely indicates the process that appears to predominate.

There has been some disagreement, possibly due to the presence at times of some additives. Also, a change in irradiation conditions may at times shift a polymer from one group to the other.

In the case of cross-linking, the MW of the polymer increases with increasing dose until a three-dimensional network is formed:

Linear polymer Irradiation Cross-linked polymer

Table 8.10.
Predominant processes in irradiated polymers

Scission	*Cross-linking*
Polyisobutene	Polyethylene
Poly(methyl methacrylate)	Polystyrene
Poly(vinylidene chloride)	Polypropylene
Poly(ethylene terephthalate)	Polyacrylates
Poly(vinyl chloride)	Polyamides
Poly(tetrafluoroethylene)	Polyesters
Cellulose and derivatives	Natural rubber
	Synthetic rubbers (except
	polyisobutylene)
	Poly(vinyl alcohol)
	Polysiloxanes

Depending on the frequency of the cross-links the properties will be similar to bi- or three-dimensional macromolecular compounds.

When scission predominates in an irradiated polymer, the MW decreases as dose increases. In some cases a solid polymer may be transformed into a low molecular fluid.

Teflon, which is poly(tetrafluoroethylene), degrades rapidly; when a bulk solid specimen is irradiated in the presence of oxygen, it loses strength and becomes powderlike.

Polymers which predominantly degrade are often characterized by the presence of two large side groups:

$$\cdots \left[-CH_2-\underset{\underset{R_2}{|}}{\overset{\overset{R_1}{|}}{C}}- \right]_n \qquad \text{where } R_1 \text{ and } R_2 \text{ are side groups}$$

as is the case of poly(methyl methacrylate) (Plexiglas) and poly-isobutylene:

$$
\left[\begin{array}{c} CH_3 \\ | \\ -CH_2-C- \\ | \\ COOCH_3 \end{array} \right]_n \quad ; \quad \left[\begin{array}{c} CH_3 \\ | \\ -CH_2-C- \\ | \\ CH_3 \end{array} \right]_n
$$

PMMA   Polyisobutylene

PMMA resists large doses at room temperature although it will ultimately crack and come apart. When it is heated during irradiation, evolution of gas swells it to a foaming mass.

Several theories have been advanced to explain why scission predominates in some high polymers and cross-linking in others. It seems that groups R_1 and R_2 produce a steric strain which weakens the bonds of the main chain.

Also, most of the polymers which degrade have low heats of polymerization, which correlates with a tendency to form monomer on pyrolysis and undergo scission during irradiation. Table 8.11 illustrates a fair correlation between heat of polymerization and tendency to depolymerize to monomer [8.2].

It may also be surmised that the presence of side groups results in steric hindrance. If a chain is broken, the two formed ends cannot rejoin, as can occur more easily if there are only H atoms as in PE.

Table 8.11.
Cross-linking versus scission from the point of view of heat of polymerization [8.2]

Polymer	Predominant effect of radiation	Heat of polymerization (kcal/mol)	Monomer yield on pyrolysis (wt%)
Polyethylene	Cross-linking	22	0·025
Polypropylene	Cross-linking	16·5	2·0
Poly(methyl acrylate)	Cross-linking	19	2·0
Polystyrene	Cross-linking	17	40·0
Polyisobutylene	Scission	10	20
Poly(methyl methacrylate)	Scission	13	100

Reprinted with permission from *Radiation Chemistry of Monomers, Polymers and Plastics* by J. E. Wilson, 1974. Copyright Marcel Dekker.

The effect of main chain scission can be established by measurements of viscosity or decreases in mechanical properties.

In PMMA (Plexiglas), fairly small doses appear to produce little effect on a block specimen. On standing for long periods however, a series of internal cracks appear. Furthermore, some side chains are detached and remain as individual molecules or radicals. On subsequent heating the polymer becomes softer, so that these molecules can migrate together to form bubbles, and the entire specimen becomes foamed.

The behaviour of PTFE (Teflon) (poly(tetrafluoroethylene)) is unusual. Teflon is a crystalline polymer of high thermal stability yet it degrades extremely readily when irradiated—as we said before, in the presence of oxygen. It may be surmised that breaks in the main chain are in a state of strain. Since the number of chains holding crystallites together is limited, the resultant damage can be great. It has also been found that this degradation results in an increase in the degree of crystallinity.

The cross-linking mechanism may possibly involve the production of macroradicals at neighbouring sites on adjacent chains, accompanied by the loss of molecular hydrogen as follows [8.2]:

$$\cdots -CH_2-CH_2-CH_2-CH_2- \cdots \sim\!\!\sim\!\!\longrightarrow$$
$$\text{(PE)}$$

$$\cdots -CH_2-CH_2-\overset{\cdot}{C}H-CH_2- \cdots + H\cdot$$
$$\text{Macroradical I}$$

$$H\cdot + \cdots -CH_2-CH_2-CH_2-CH_2- \cdots \sim\!\!\sim\!\!\longrightarrow$$
$$\text{(PE)}$$

$$\cdots -CH_2-\overset{\cdot}{C}H-CH_2-CH_2- \cdots + H_2$$
$$\text{Macroradical II}$$

$$\cdots -CH_2-CH_2-\overset{\cdot}{C}H-CH_2- \cdots +$$
$$\text{(I)}$$

$$\cdots -CH_2-\overset{\cdot}{C}H-CH_2-CH_2- \cdots \longrightarrow$$
$$\text{(II)}$$

$$\cdots -CH_2-CH_2-\underset{|}{CH}-CH_2- \cdots$$
$$\cdots -CH_2-CH-CH_2-CH_2- \cdots$$

The effects of such cross-linking are:

—a reduction of the number of separate macromolecules;
—an increase in the average size and degree of branching;
—higher viscosity and degree of entanglement.

As the dose is increased, there is a greater probability of such links being formed between macromolecules which have already been joined elsewhere.

It is possible to form extended networks, theoretically of infinite MW. Such networks are no longer soluble, though they will swell in appropriate solvents. They are referred to as gels; the still soluble fraction, partly branched, is termed the 'sol'. As the irradiation proceeds, the 'gel' fraction rises rapidly, and, if there is no simultaneous scission, tends to unity. The point at which the irradiated polymer first shows any gel fraction can be calculated.

Above the gel point, we have a sol and a gel fraction, the latter constituting the network structure with very distinctive properties. The density of cross-links in this network can be defined either in terms of the number per unit weight, or as MW between cross-links.

As the dose increases, Young's modulus (E) tends to higher values than predicted by the theory. For the same density of cross-linking, similar elastic moduli are obtained, although T_g and the tensile strength do differ.

We may encounter cross-linking between the macromolecules of the same polymer (A—A cross-linking):

$$\cdots -A-A-A-A-A-A-A-A-A- \cdots$$
$$\cdots -A-A-A-A-A-A-A-A-A- \cdots$$

or in the case of polyblends (A—B cross-linking):

$$\cdots -A-A-A-A-A-A-A-A- \cdots$$
$$\cdots -B-B-B-B-B-B-B-B- \cdots$$

The *presence of oxygen* changes the effect of radiation on polymers; with thin polymeric fibres, oxygen diffuses in and causes an effect throughout the film. For thicker films, the oxidation may take place only at the surface of the film. When PE is irradiated in oxygen, cross-linking still takes place, as shown by gel fraction determination. However, the nature of the mechanism has changed considerably, because even prolonged irradiation does not render the polymer infusible. Possibly the links consist of peroxide bridges between the macromolecules, which decompose when the polymer is heated. In addition, water and other groups are formed including carbonyl, carboxyl and hydroxyl. Some of the reactions involved may be:

$$PE \rightsquigarrow\longrightarrow \cdots -CH_2-CH_2-\overset{\cdot}{C}H-CH_2- \cdots \rightsquigarrow + O_2 \longrightarrow$$

$$\overset{\displaystyle O-O\cdot}{\underset{|}{}}$$
$$\cdots -CH_2-CH_2-CH-CH_2- \cdots \rightsquigarrow\longrightarrow$$

$$\cdots -CH_2-CH_2-\underset{\underset{O}{\parallel}}{C}-CH_2- \cdots + HO\cdot$$

More frequently, however, the oxygen reacts with the polymer radical (macroradical), giving a peroxide or hydroperoxide radical which subsequently can be decomposed by warming.

It is of interest that in radiobiology, the oxygen effect is known to enhance the sensitivity of many biological systems by several times.

With proton or alpha (α) radiation where the radicals are formed in closer proximity, the oxygen effect is greatly reduced.

CHEMICAL DEGRADATION

Chemical reactions start spontaneously when certain chemicals are brought into contact with macromolecular compounds. Commonly, the rate of chemical reactions is strongly dependent on temperature, which implies that thermal and chemical processes overlap.

Reactions of polymers with molecular oxygen are of general importance because oxygen is ubiquitous. Therefore, oxidative processes cause problems not only in outdoor exposures of plastics (weathering) but also in processing.

Frequently, oxidative degradation proceeds according to free radical chain reactions initiated by UV light, radiation, mechanical stress, etc.

Because air pollution has become a worldwide problem, the behaviour of plastics used outdoors towards pollutant gases (SO_2 and NO_2) has gained importance and has become the subject of numerous studies.

The term 'chemical stability' is commonly used in a broad sense covering both chemical and physical interaction of low molecular weight reactants with polymers. For certain chemical reactions, e.g. hydrolysis, the crystalline regions have been found to be impervious to chemical agents.

Generally *solvolysis* reactions concern the breaking of C—X bonds, X designating hetero (noncarbon) atoms, i.e. O, N, P, S, Si or halogen. Of primary interest are solvolysis reactions of polymers

Table 8.12.
Hydrolysis of linear polymers (typical examples) [8.1]

Main-chain linkage under attack	Products of hydrolysis	Examples
$-\overset{\mid}{C}-\overset{\mid}{\underset{\underset{O}{\|}}{C}}-O-\overset{\mid}{C}-\overset{\mid}{C}-$ Carboxylic acid ester	$-\underset{\underset{O}{\|}}{C}-OH + HO-\overset{\mid}{C}-$	Polyester
$-O-\overset{O(-)}{\underset{\underset{O}{\|}}{P}}-O-\overset{\mid}{\underset{\mid}{C}}-$ Phosphoric acid ester	$-O-\overset{O(-)}{\underset{\underset{O}{\|}}{P}}-OH + HO-\overset{\mid}{\underset{\mid}{C}}-$	Nucleic acids (DNA)
$-\overset{\mid}{\underset{\mid}{C}}-O-\overset{\mid}{\underset{\mid}{C}}-$ Ether, glycoside	$-\overset{\mid}{\underset{\mid}{C}}-OH + HO-\overset{\mid}{\underset{\mid}{C}}-$	Polyether, poly-saccharides (cellulose, amylose, etc.)
$-\overset{\mid}{\underset{\mid}{C}}-\overset{\mid}{\underset{\underset{O}{\|}}{C}}-\overset{\mid}{\underset{H}{N}}-\overset{\mid}{\underset{\mid}{C}}-$ Amide (peptide)	$-\overset{\mid}{\underset{\mid}{C}}-\overset{\mid}{\underset{\underset{O}{\|}}{C}}-OH + H_2N-\overset{\mid}{\underset{\mid}{C}}-$	Polyamides, proteins, polypeptides
$-\overset{\mid}{\underset{\mid}{C}}-O-\overset{\mid}{\underset{\underset{O}{\|}}{C}}-\overset{\mid}{\underset{H}{N}}-\overset{\mid}{\underset{\mid}{C}}-$ Urethane	$-\overset{\mid}{\underset{\mid}{C}}-OH + CO_2 + H_2N-\overset{\mid}{\underset{\mid}{C}}-$	Polyurethanes
$-\overset{\mid}{\underset{\mid}{Si}}-O-\overset{\mid}{\underset{\mid}{Si}}-$ Siloxanes	$-\overset{\mid}{\underset{\mid}{Si}}-OH + HO-\overset{\mid}{\underset{\mid}{Si}}-$	Polydialkyl-siloxanes

containing heteroatoms in the main chain, because in these cases, solvolysis implies main chain rupture as indicated by the next reaction:

$$.. \text{—C—X—C—} ... + YZ \rightarrow ... \text{—C—X—Z} + \text{Y—C—} ..$$

Common solvolysis agents (YZ) are water (hydrolysis), alcohols (alcoholysis), acids (acidolysis), ammonia (ammonolysis), etc.

Regarding polymer degradation, hydrolysis (YZ = HOH) has received prominence. At pH ≤ 7, hydrolysis is initiated by a protonation process which is followed by the addition of water (H_2O) and cleavage of the ester linkage (Table 8.12) [8.1].

Table 8.13 shows the stability of selected polymers against solvolytic agents at ambient temperature [8.1].

Table 8.13.
Stability of selected polymers against solvolytic agents [8.1]

Polymer	Acidic media	Alkaline media[a]
Polyethylene	+	+
Polypropylene	+	+
Poly-1-butene	+	+
Polyisobutene	+	+
Polystyrene	+	+
Polytetrafluoroethylene	+	+
Polytrifluorochloroethylene	+	+
Polyvinyl fluoride	+	+
Polyvinyl chloride (unplasticized)	+	+
Polyvinyl chloride (plasticized)	−	−
Polymethyl methacrylate	−	−
Polyacrylonitrile	−	−
Polyoxymethylene	−	−
Polysulphones	−	−
Polyamides	−	−
Polycarbonates	−	−
Polyethylene terephthalate	−	−
Polyurethanes	−	−
Phenol–formaldehyde resins	+	−
Polydimethylsiloxanes	−	−
Natural rubber	+	+
Butyl rubber	+	+
Unsaturated polyesters (UPE)	+	−

[a] + satisfactory; − unsatisfactory.
Reprinted with permission from *Polymer Degradation* by W. Schnabel, 1981. Copyright C. Hanser Verlag.

Table 8.14.
Ozonization of water soluble polymers [8.1]

Polymer	Ozonization time (h)	Ozone consumed (mg/g polymer)	MW
Poly(ethylene oxide)	0	0	8000
	2	836	250
Polyvinyl alcohol	0	0	28 000
	4	368	460
Polyvinyl pyrrolidone	0	0	27 000
	4	1 273	560
Sodium polyacrylate	0	0	410 000
	4	860	250
Polyacrylamide	0	0	280 000
	4	910	340

Reprinted with permission from *Polymer Degradation* by W. Schnabel, 1981. Copyright C. Hanser Verlag.

Ozonolysis. The rate of ozone reaction with saturated materials is rather low, and therefore of minor importance for straightforward degradation processes (Table 8.14). However, the reaction of ozone (O_3) with unsaturated compounds might serve to initiate auto-oxidation processes.

BIODEGRADATION

It is a well known fact that living organisms not only synthesize biopolymers such as proteins, nucleic acids, polysaccharides (including cellulose) but are also capable of degrading them.

The general mechanism of degradation of polymers into the small molecules employed by nature is a chemical one. Living organisms are capable of producing *enzymes* which can attack natural polymers. The attack is usually specific with respect to both the enzyme/biopolymer couple and the site of attack.

The majority of synthetic polymers are rather inert towards biological attack although in principle being biodegradable.

Micro-organisms play an eminent role in the decomposition of organic material of all kinds including biopolymers. There is a large body of micro-organisms, such as *fungi, bacteria* and *actinomycetes,* which are distributed ubiquitously around the earth. Under appropriate conditions the growth of micro-organisms will occur simultaneously with the decomposition of any biological species after the latter has perished.

The growth of micro-organisms depends on the pH and other factors such as:

—temperature,
—availability of mineral nutrients,
—oxygen concentration,
—humidity (the presence of water being a prerequisite).

Commonly, the microbial degradability of synthetic polymers is studied by growth tests on solid agar media. Test fungi and/or test bacteria are introduced together with the polymer in the form of films,

Table 8.15.
Microbial degradation of commercial plastics [8.1]

Product	Growth rate[a] on agar plates
Polyisobutene, poly-4-methyl-1-pentene, polymethylmethacrylate, poly(vinyl butyral), polyformaldehyde, poly(vinyl ethyl ether), cellulose acetate, bisphenol A polycarbonate, ABS terpolymer	0
Poly(vinyl acetate), styrene–butadiene block copolymer poly(vinylidene chloride), poly(ethylene terephthalate), polystyrene, polypropylene	1 (less than 10% covered)
Polyethylene	2 (10–30% covered)
Plasticized polyvinyl chloride	3 (30–60% covered)
Polyurethane	4 (60–100% covered)

[a] Standard tests with *Aspergillus niger, Aspergillus flavus, Chaetomium globosum* and *Penicillium funiculosum.*
Reprinted with permission from *Polymer Degradation* by W. Schnabel, 1981. Copyright C. Hanser Verlag.

granules, plaques or powder. The agar media contain all nutrients necessary for microbial growth except a carbon source. The tests are run over a definite time (usually 3 weeks). Growth rates are classified according to the fraction of the gel surface covered with colonies, i.e. 0 = no growth; 1 = 10% covered; 2 = 10–30% covered; 3 = 30–60% covered; 4 = 60–100% covered.

Table 8.16.
Typical macroradicals generated by mechanically induced main-chain scission at 77 K [8.1]

Polymer macroradicals	
Polyethylene	$-\overset{\text{H}}{\underset{\text{H}}{\text{C}}}-\overset{\text{H}}{\underset{\text{H}}{\text{C}}}\cdot$
Polypropylene	$-\overset{\text{CH}_3}{\underset{\text{H}}{\text{C}}}-\overset{\text{H}}{\underset{\text{H}}{\text{C}}}\cdot \qquad \cdot\overset{\text{CH}_3}{\underset{\text{H}}{\text{C}}}-\overset{\text{H}}{\underset{\text{H}}{\text{C}}}-$
Polyvinyl alcohol	$-\text{CH}_2-\overset{\text{H}}{\underset{\text{OH}}{\text{C}}}\cdot \qquad \cdot\overset{\text{H}}{\underset{\text{H}}{\text{C}}}-\overset{\text{H}}{\underset{\text{OH}}{\text{C}}}-$
Polytetrafluoroethylene	$-\overset{\text{F}}{\underset{\text{F}}{\text{C}}}-\overset{\text{F}}{\underset{\text{F}}{\text{C}}}\cdot$
Polymethyl methacrylate	$-\overset{\text{H}}{\underset{\text{H}}{\text{C}}}-\overset{\text{CH}_3}{\underset{\text{R}}{\text{C}}}\cdot \qquad \cdot\overset{\text{H}}{\underset{\text{H}}{\text{C}}}-\overset{\text{CH}_3}{\underset{\text{R}}{\text{C}}}-$ $-\overset{\text{CH}_3}{\underset{\text{R}}{\text{C}}}-\overset{\text{H}}{\underset{\cdot}{\text{C}}}-\overset{\text{CH}_3}{\underset{\text{R}}{\text{C}}}-\overset{\text{H}}{\underset{\text{H}}{\text{C}}}-\text{H} \qquad \text{R:}\ \underset{\text{O}-\text{CH}_3}{\overset{}{\text{C}}}{=}\text{O}$

As many micro-organisms are capable of producing hydrolyases (enzymes catalysing hydrolysis), it is assumed that polymers containing hydrolysable groups in the main chains would be especially prone to microbial attack. In fact, only aliphatic polyesters, polyethers, poly-methanes, and polyamides exhibit a general sensitivity towards commonly occurring micro-organisms. Table 8.15 provides some information on the microbial degradation of certain plastics.

MECHANICAL DEGRADATION

Mechanical degradation of polymers, in its broader sense, comprises a large field, covering fracture phenomena as well as chemical changes

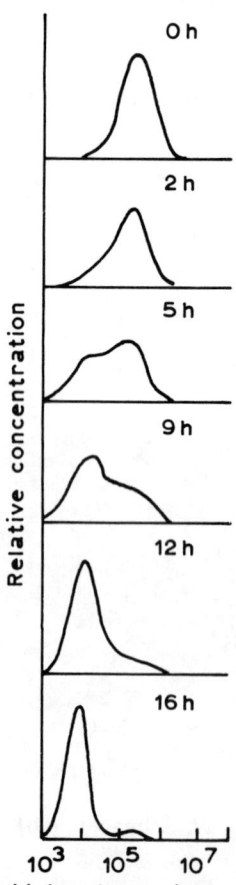

Fig. 8.19. Main-chain scission of PS upon milling. Gel permeation chromatograms indicating the change of MW durring milling time (time of treatment in hours as indicated on the graphs) [8.1]. (Reprinted with permission from *Polymer Degradation* by W. Schnabel, 1981. Copyright C. Hanser Verlag.)

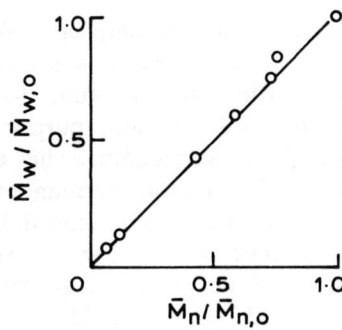

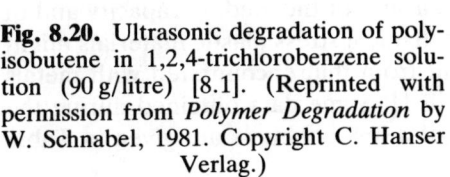

Fig. 8.20. Ultrasonic degradation of polyisobutene in 1,2,4-trichlorobenzene solution (90 g/litre) [8.1]. (Reprinted with permission from *Polymer Degradation* by W. Schnabel, 1981. Copyright C. Hanser Verlag.)

Fig. 8.21. Schematic illustration of bond rupture in the amorphous regions of semi-crystalline polymers [8.1]. (After Peterlin.)

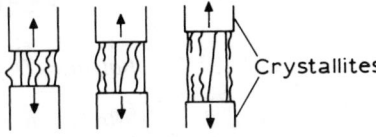

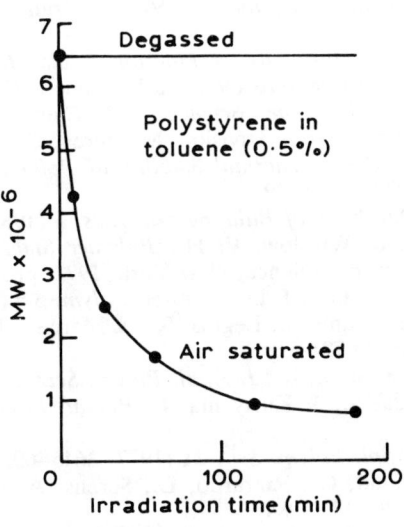

Fig. 8.22. Ultrasonic degradation of PS in toluene solution at ambient temperature. Average MW versus time of irradiation. Polymer concentration 1%; frequency 0·4 MHz; intensity 50 W [8.1]. (After Weissler.)

induced by mechanical stress. Because of their macromolecular structure, plastics possess in many respects rather unique physical properties that account for their utilization as raw engineering materials for various purposes. Nevertheless, there exist load limits, i.e. the plastics engineer has to be aware of the loading capacity and of the fact that under permanent mechanical stress plastic materials might exhibit a behaviour quite different from that encountered with metals and other mineral raw materials [8.1]. Some data obtained during the study of mechanical degradation of a few polymers are given in Table 8.16 and Figs 8.19–8.22.

REFERENCES

[8.1] Schnabel, W., *Polymer Degradation*. C. Hanser Verlag, München, Wien, 1981.
[8.2] Wilson, J. E., *Radiation Chemistry of Monomers, Polymers and Plastics*. Marcel Dekker, New York, 1974, p. 370.
[8.3] Ranby, B., & Rabek, F., *The Effects of Hostile Environments on Coatings and Plastics*, ed. D. P. Garner & G. A. Stahl. ACS Symposium Series 229, Washington, 1983, pp. 291–308.
[8.4] Kelen, T., *Polymer Degradation*. Van Nostrand Reinhold Co., New York, 1983.
[8.5] Hoffman, J. D., *Durability of Macromolecular Materials*, ed. R. K. Eby. ACS Symposium Series 95, Washington, 1979, p. 2.
[8.6] Askeland, D. R., *The Science and Engineering of Materials*. Brooks/Cole Engineering Division, Monterey, CA., 1984, p. 11.
[8.7] Conley, R. T. (Ed.), *Thermal Stability of Polymers*. Marcel Dekker, New York, 1970, pp. 1–49.
[8.8] Eurin, Ph., *Durability of Building Materials*, **1** (1982) 161–8.
[8.9] Loan, L. D., & Winslow, F. H., *Polymer Stabilization*, ed. W. L. Hawkins. Wiley-Interscience, New York, 1972, pp. 117–53.
[8.10] Gupta, M. C., & Nath, J. D., *J. Appl. Polym. Sci.*, **25** (1980) 1017–28.
[8.11] Daoust, D., Bormann, S., Legras, R., & Mercier, J. P., *Polym. Engng Sci.*, **21**(11) (1981) 721–6.
[8.12] Tsuchiya, Y., & Sumi, K., *J. Appl. Polym. Sci.*, **21** (1977) 975–80.
[8.13] Ito, M., Okada, S., & Kuriyama, I., *Radiat. Phys. Chem.*, **16** (1980) 481–5.
[8.14] Yano, S., *J. Appl. Polym. Sci.*, **21** (1977) 2645–60.
[8.15] Cacioli, P., Moad, G., Rizzardo, E., Serelis, A. K., & Solomon, D. H., *Polym. Bull.*, **11** (1984) 325–8.
[8.16] Lutz, J. T., Jr, *Polym. Plast. Technol. Engng*, **11**(1) (1978) 55–80.
[8.17] Fujikura, Y., Suzuki, T., & Matsumoto, M., *Polym. J. (Jap.)*, **40**(3) (1984) T95–T100.

[8.18] Owen, E. D., Pasha, I., & Moayyedi, F., *J. Appl. Polym. Sci.*, **25** (1980) 2331–8.

[8.19] Littimer, R. P., & Kroenke, W. J., *J. Appl. Polym. Sci.*, **25** (1980) 101–10.

[8.20] Naqvi, M. K., *JMS Rev. Macromol. Chem. Phys.*, **C25**(1) (1985) 119–55.

[8.21] Gerrard, D. L., & Maddams, W. F., *Makromol. Chem.*, **185** (1984) 1843–54.

[8.22] Braun, D., *Degradation and Stabilization of Polymers*, ed. G. Geuskens. John Wiley, New York, 1975, pp. 23–42.

[8.23] Mayer, Z., *J. Macromol. Sci.*, **C10** (1974) 263.

[8.24] Geddes, W. C., *Rubb. Chem. Technol.*, **40** (1967) 177.

[8.25] Stivala, S. S., & Reich, L., *Polym. Engng Sci.*, **20**(10) (1980) 654.

[8.26] Nass, L., *Encyclopedia of PVC, Vol. 1.* Marcel Dekker, New York, 1977, p. 275.

[8.27] Seymour, R. B., *Plastics versus Corrosives*. John Wiley, New York, 1982, pp. 135, 145.

[8.28] Bishop, D. P., & Smith, D. A., *Ind. Engng Chem.*, **59**(8) (1967) 32.

[8.29] Ehlers, G. F. L., *Polymer*, **1** (1960) 304.

[8.30] Kovarskaya, B. M., & Zhigunova, I. E., *Vysokomol. Soedin*, **1** (1959) 1531.

[8.31] Neiman, M. B., *et al.*, *J. Polym. Sci.*, **56** (1962) 383.

[8.32] Pitt, J. J., Pearce, P. J., Rosewarne, T. W., Davidson, R. G., Ennis, B. C., & Morris, C. E. M., *J. Macromol. Sci. Chem.*, **A17**(2) (1982) 227–42.

[8.33] Rossiter, W. J., Jr, Ballard, D. B., & Sleater, G. A., *Thermal Insulation, Materials and Systems for Energy Conservation in the 80's*, ed. F. A. Govan, D. M. Greason & J. D. McAllister. ASTM Publication, STP 789, 1983, pp. 665–87.

[8.34] Schnabel, W., & Kiwi, J., *Aspects of Degradation and Stabilization of Polymers*, ed. H. H. G. Jellinek. Elsevier, Amsterdam, 1978, p. 195.

[8.35] Cassidy, P. E., & Aminabhavi, T. M., *J. Macromol. Sci.-Rev. Macromol. Chem.*, **C21**(1) (1981) 89–133.

[8.36] Moritz, N., *Kunststoffe*, **65** (1975) 290.

[8.37] Menzel, G., *Plastverarbeiter*, **9** (1983) 213.

[8.38] Pfeiffer, T. R., *J. Vinyl Technol.*, **5** (1983), 135–42.

[8.39] Svane, P., *Proceedings of the Second International Conference on the Durability of Building Materials and Components*, Sept. 14–16, (1981), Gaithersburg, Maryland, p. 410.

[8.40] Delfosse, L., *J. Macromol. Sci. Chem.*, **A11**(8) (1977) 1491.

[8.41] Davidson, R. S., & Meek, R. R., *Polymer Photochemistry*, **2** (1982) 1–12.

[8.42] Furneaux, G. C., Ledbery, K. J., & Davis, A., *Polym. Degrad. and Stab.*, **3** (1981) 431–42.

[8.43] Cunliffe, A. V., & Davis, A., *Polym. Degrad. and Stab.*, **4** (1982) 11–37.

[8.44] Davis, A., & Sims, B., *Weathering of Polymers*, Applied Science Publ., London & New York, 1983, p. 162, 201.

[8.45] Yamasaki, R. S., & Blaga, A., *Materials and Structures* (*RILEM*), 10(58), (1977) pp. 197–204.
[8.46] Blaga, A., *Durability of Building Materials and Structures*, ed. P. J. Sereda & G. G. Litvan, ASTM, STP691, (1980), pp. 827–37.
[8.47] Shalaby, S. W., *J. Polym. Sci.*, 14 (1971) 419–58.

Index

Absorption (of light), 521–2
Acetal polymers. *See* Polyacetals
Acetate fibres, properties of, 69
Acid resistance, concrete–polymer
 composites, 232, 238, 251, 261
Acrylic fibres, properties of, 69
Acrylics
 adhesives, 372–4
 blend with PVC, 422
 chemical structure of, 372
 glass-fibre-reinforced, 423
 latex, 246–7
 processing methods for, 60
 roofing applications, 487
 sealants, 417–23
  advantages/disadvantages of,
   422
  properties of, 421
  typical formulation for, 418
Acrylonitrile–butadiene (NBR)
 copolymer, 242–3
Acrylonitrile–butadiene–styrene
 (ABS) copolymer, 89
 flammability of, 51, 347
 foams, 310, 313
  glass-reinforced, 313–14
  properties of, 312–13
 processing methods for, 60
 synthesis of, 16
 thermal properties of, 43
 toxicity of combustion products,
  328, 329
Addition polymerization, 10–17

Additives, 52–7
 requirements for, 52–3
 types of, 53
Adhesion
 forces acting in, 359
 meaning of term, 356
 theories of, 362, 364
 work of, 360–1
Adhesion promoters, 366–8
Adhesive bonded joints, composites,
 186–7
Adhesive bonding technology,
 scientific disciplines involved,
 358–9
Adhesives, 358–91
 acrylics, 372–4
 advantages of, 358, 363
 aminoplastics, 378–9
 cellulosics, 376–7
 classification of, 368, 369
 coupling agents used, 366–7
 early use of, 356
 effects of moisture on, 390–1
 heat resistance of, 382–4
 joints used, 385–7
  typical strength data, 387–9
 meaning of term, 358
 methods of use, 363
 phenolics, 377–8
 polyamides, 374
 polyaromatics, 382–4
 polycarbonates, 374–5
 polyesters, 382

Adhesives—*contd.*
　polyurethane, 375–6
　poly(vinyl acetal)s, 371–2
　poly(vinyl acetate), 368, 370
　poly(vinyl alcohol), 370–1
　poly(vinyl chloride), 372
　properties of, 363, 368–84
　requirements of, 361
　solar energy equipment, 473
Ageing (of polymers), 45–6
Alumina fibres, properties of, 130
Aluminium fibres, properties of,
　　135–7
Aminoplastics, 378–9
Ammonium bicarbonate, 289
Anaerobic adhesives, 373
Anionic polymerization, 15
Antimony trioxide, 348, 465
Anioxidants, 55, 56
Applications, degradation affected
　　by, 501
Aramid fibres
　fibre structure of, 157, 160
　molecular structure of, 156
　properties of, 154, 157–9
Asbestos cement
　impregnation of, 217–18
　properties of, 270, 277
　world production of, 262
Asbestos fibres, 150–2
　properties of, 150, 152, 153, 263,
　　266–7
　substitutes for, 263
　use in composites, 151
Asphalts, 478
Autophobic liquids, 360

Beryllium fibres, properties of, 135,
　　136
Bidimensional polymers, 3, 4–5
Biodegradation, 549–52
　meaning of term, 498
Bisphenol A, 374–5
Bitumens, modification with
　　polymers, 479–84
　EVA, 484
　SBR, 484
　SBS, 483–4

Block copolymers, 459
Blowing agents, 287–90
　chemical agents, 288–90
　physical agents, 288, 289
Boron filaments, properties of, 113,
　　130, 135–7
Boron nitride fibres, properties of,
　　130
Bouncing putty, 399
Branched polymers, 2–3, 4
Brittle polymers, 38
Building codes, fire requirements,
　　351–2
Butyl acrylate, concrete impregnated
　　with, 227, 229
Butyl rubber
　molecular structure of, 427–8
　permeability of, 48
　roofing applications of, 487
　sealants, 427–9
　　advantages/disadvantages of,
　　　429
　　typical formulation for, 427

Calendering, 63, 65–6
　equipment used, 67
Capillary porosity, 213
Carbon fibres
　advantage of, 143
　concrete reinforced with, 269, 270,
　　271, 272, 275
　fibrillar structure of, 143, 146, 160
　hybrid-system, use of, 143, 146–9
　manufacture of, 138–43
　polymers reinforced with, 128, 131,
　　146–9
　properties of, 130, 135, 136, 146,
　　157, 266–7
　stress–strain curve for, 146
Catalysts, ionic chain polymerization,
　　13–14
Cationic polymerization, 14–15
Ceiling temperature, meaning of
　　term, 500
Cellular polymers, 283–352
　blowing agents used, 287–90
　classification of, 284–7
　density of, 287

Cellular polymers—*contd.*
  flammability of, 323–8, 339,
    343–52
  manufacture of, 290–6
  polymers used, 283
  special foams, 296–304
  stabilization of, 292
  thermoplastic foams, 304–13
  thermosetting foams, 313–43
  US markets for, 284
  see also Polymer foams
Cellulose acetate
  foams, 313
  permeability of, 48
Cellulose fibres
  carbon fibres produced from,
    138–9, 140–2
  concrete reinforced with, 270
  properties of, 266–7
Cellulose triacetate, Mark–Houwink
    parameters for, 10
Cellulosics
  adhesives, 376–7
  flammability of, 347
  molecular structure of, 25, 376
  uses of, 40
Cement
  pore structure of, 209, 210
  properties of, 268
Chain polymerization, 12–13
  types of, 13
Chain scission
  determination of amount, 544
  radiation-induced, 542
Chemical bonds, adhesive action of,
    362–3
Chemical degradation, 546–9
  meaning of term, 498
Chlorinated polyethylene,
    weatherability of, 527
Chlorosulphonated polyethylene
  roofing applications for, 487
  sealants, 416–17, 418
    advantages/disadvantages of,
      418
    formulation for, 417
Chopped strand mat (CSM)
    composites, 178, 184
Classification (of polymers), 19–21

Closed-mould techniques, 168–70
Coatings, solar energy equipment,
    470–1
Cobalt, radioactive sources, 540
Combustion products, 350
  toxicity of, 328, 329
Composites, 74–200
  adhesively bonded joints in, 186–7
  advantages of, 76
  applications of, 76–7, 190–1, 192–4
  constructional use of, 187–94
  coupling agents used, 119–24
  definition of, 74, 75
  factors affecting mechanical
    properties of, 173–85
  fibre length effects, 178, 179
  fibre loading effects, 173–6
  fibre orientation effects, 176–7, 184
  fibre shape effects, 177–8
  fracture modes in, 183–4
  load-carrying joints in, 185–7
  main features of, 75
  mechanically fastened joints in,
    185–6
  orientation effects, 176–7
  polymers in, 77–113
  reinforcing agents used, 113–19
  sheets used, 192–3
  specific stiffness of, 182
  specific strength of, 182
  strength as function of fibre
    loading, 175
  technology used, 160–72
    methods listed, 162–3
  types of, 76
  Young's modulus as function of
    fibre loading, 174
Compression moulding, 66
  equipment used, 68
Concrete
  dynamic properties of, 249
  effect of polymers on, 209, 215, 221
  impregnation with monomer, 213,
    218–19
  modulus of, 221
  porosity of, 13, 212
  properties of, 188–9
  structure of, 209
  tensile strength of, 381

Concrete–polymer composites, 207–78
  see also Fibre-reinforced concrete; Polymer–cement concrete (PCC); Polymer concrete (PC); Polymer-impregnated concrete (PIC)
Condensation polymerization, 17–19
Consumer Product Safety Commission (CPSC—USA), on UF foams, 335
Contact angle, 360–1
Continuous filament glass fibres
  direct melt process for, 132–3
  marble process for, 131–2
Copolymerization, 15–16
Copolymers, crystallinity in, 25–6
Cork
  thermal conductivity of, 307
  toxicity of combustion products, 329
Cotton, properties of, 69
Coupling agents, 119–24, 366–7
  examples of, 122–3
  function of, 120–1
Creep behaviour
  polymer-impregnated concrete, 222, 226
  polyurethane foam, 318, 319
Critical surface tension of wetting, adhesive strength affected by, 365–6
Cross-linking, radiation-induced, 541, 542, 544–5
Crystalline melting temperature, 31, 33
  values listed, 35
Crystallinity, 26
  copolymers, 25–6
  factors affecting, 28–9
  natural fibres, 24
Crystallization, meaning of term, 34
Cyanoacrylate-based polymers, 373, 417–18

Dacron fibres, properties of, 159
Degradation
  biodegradation, 549–52

Degradation—*contd.*
  chemical degradation, 546–9
  effects of, 53
  factors affecting, 500–1, 512
  meaning of term, 496–7
  mechanical degradation, 552–4
  photodegradation, 520–39
  radiodegradation, 539–46
  requirements for stable polymer, 505
  thermal degradation, 503–20
Degree of polymerization (DP), meaning of term, 5–6
Denier, meaning of term, 133
Depolymerization, 504
Diallyl phthalate polymer, 78
Diamond lattice, 3, 4
Differential scanning calorimetry (DSC)
  polyurethane, 405–7
  stearic acid–PVA blends, 469
  stearic acid–PVC blends, 467–8
Dioctyl phthalate (DOP)
  flammability affected by, 349
  polymer-impregnated concrete affected by, 223, 227
Discoloration, 500
  poly(vinyl chloride), 511, 513, 525, 526
Dough moulding compound (DMC), 169–70
Dust particles, effect of, 499

E-glass fibres
  composition of, 137
  concrete reinforced by, 265
  properties of, 113, 130, 135–7, 146, 157, 266–7
  use in composites, 147
Elastometers, 33
Electron accelerators, 231, 540
Element-organic polymers, 20
Emery 400
  composition of, 464–5
  properties of, 464
Emulsion polymerization, 238
  advantages of, 238–9
  molecular mechanism of, 240

Emulsion polymerization—*contd.*
  particulate system in, 240
  typical recipes for, 239, 243, 244
Encapsulation materials, 457–69
Epoxy polymers
  adhesives, 379–82, 384
    blends used, 381
    moisture effects on, 390–1
    tensile strength of, 381, 384, 388
  characteristics of, 77, 78, 111
  concrete mixed with, 234–8
  curing agents used, 380
  flammability of, 347
  flooring, applications, 491–2
    advantages of, 492
  foams, 339–40, 341
    curing agents used, 339
    properties of, 340, 341
    typical formulation for, 339
  molecular structure of, 110
  physical properties of, 517
  polymer concretes made from, 259,
    261
  processing methods for, 60
  properties of, 112, 146, 380, 388
  syntactic foams, 300, 301, 340
  thermal degradation of, 514–17
    decomposition products from,
    515–16
    factors affecting, 516–17
  thermal properties of, 43, 517
  use in composites, 111, 113
Epoxy–phenolic polymers
  adhesives, 381, 384
  heat stability of, 515
Ethylene–propylene–diene
    terpolymer (EPDM), roofing
    uses, 486, 487
Ethylene–vinyl acetate copolymer
    (EVA)
  bitumen mixtures, 484
  encapsulant use of, 458, 460, 461
  weathering of, 455, 457, 527
Europe
  PVC window profiles, market in,
    529
  window construction compared
    with North American, 527

Extrusion processes, 59, 61–3
  components produced by, 61
  equipment used, 61, 62

Fabrication technologies, factors
    affecting, 58–9
Fatty acids, 430–1
  content in oils, 432
Fibre-reinforced composites (FRC),
    124–31
  factors affecting mechanical
    properties, 173–85
  types of, 126
Fibre-reinforced concrete, 262–78
  applications of, 272–3
  asbestos-fibre-reinforced, 270
  carbon-fibre-reinforced, 269–72,
    275
  cellulose-fibre-reinforced, 270
  dynamic properties of, 249
  Kevlar-reinforced, 277, 278
  mass applications for, 272
  nylon-reinforced, 275, 276, 278
  polypropylene-reinforced, 263–7,
    270, 274–5, 277
  precast applications for, 273
  properties of, 270, 271, 274–5, 276
  types of fibres used, 262
Fibre-reinforced foam laminates,
    301, 304
  facing sheets used, 301
  manufacture of, 303
Fibre-reinforced plastics (FRP),
    fibres used, 125, 127–31
Fibres
  dry spinning of, 70, 71
  fabrication methods for, 68, 70–1
  materials used, 66
  melt spinning of, 70, 71
  orientation of, 176–7
  properties listed, 69
  properties of, 266–7
  requirements for, 173
  structural features of, 67–8
  wet spinning of, 70, 71
Fibrous glass, thermal conductivity
    of, 307

Fibrous materials, properties of, 130
Filament winding technique, 167–8
  advantages of, 168
Fillers
  chemical composition of, 118
  classification of, 57, 114–15
  factors affecting selection, 116
  properties improved by, 57,
    116–17, 119
Fire behaviour, factors affecting, 350
Fire-retardant additives, 51, 345–9,
  465
Flaming drips, 350
Flammability
  cellular polymers, 323–8, 339,
    343–52
  method of determination, 323, 324
  polymers, 50, 346, 347
  polypropylene-reinforced concrete,
    267
  reduction of, 345
  wood products, 325
Flexible foams, definition of, 286
Flexigard composite, 447
Flooring applications, 490–3
Fluorinated ethylene–propylene
    copolymer (FEP)
  mechanical properties of, 446
  optical properties of, 443
  thermal properties of, 446
Fluoropolymers
  applications of, 90
  properties of, 42, 43, 90, 91, 92,
    443, 446
Foamed glass, thermal conductivity
    of, 307
Foaming (blowing) agents, 287–90
Foams
  ABS, 310, 312–13
  cellulose acetate, 313
  chemical structure of, 342
  epoxy, 339–40, 341
  phenol–formaldehyde, 329–31
  polyolefins, 310, 311–12
  polystyrene, 304–7
  polyurethane, 314–29
  poly(vinyl chloride), 308–10

Foams—*contd.*
  silicones, 340, 342–3
  thermal properties of, 43
  urea–formaldehyde, 331–9
  see also Cellular polymers; Polymer
    foams
Formaldehyde
  emission from UF foams, 335, 336,
    337, 339, 518–19
  health hazards of, 335–6, 338, 519
  odour of, 336, 337, 338
France, market building, 194, 196
Fringed-micelle theory, 26–7
Furan polymers, concrete mixed
    with, 238

Gas permeability, 47
  values listed, 48
Gases, properties of, 293
Gaskets, 393, 396, 397
Gel coat, role of, 165
Gel porosity, 213
Germany (West), window materials
    used, 525, 526
Glass
  mechanical properties of, 447
  optical properties of, 441, 443
  thermal properties of, 447
Glass fibres
  fibre structure of, 160
  foams reinforced with, 301–4
  hybrid-system, use of, 147–9
  manufacture of, 131–8
  properties of, 113, 130, 134–7, 157,
    266
  use in composites, 125, 127, 147
  see also E-glass . . . ; S-glass fibres
Glass reinforced foam, properties of,
  14, 313
Glass reinforced plastics (GRP)
  applications of, 190–1, 193
  mechanical properties of, 447
  optical properties of, 443
  properties of, 188–9, 517
  thermal properties of, 447
  weathering resistance of, 450,
    451

Glass transition temperature, 31,
 33–6
 adhesives affected by, 390
 values listed, 35
Glassy polymers, stress–strain
 behaviour of, 41
Glassy state, 32–3, 34
Glazing applications, 373, 419, 420
Glazing materials
 evolved products, effect of, 452–3
 mechanical properties of, 446–7
 optical properties of, 443
 thermal properties of, 446–7
Goland–Reissner bending moment
 factor, 387
Graft copolymers, 16, 459
Graphite fibres
 properties of, 113, 130
 see also Carbon fibres
Graphite lattice, 3, 4
GRS rubber
 emulsion polymerization of, 239
 synthesis of, 16
 see also Styrene–butadiene
 copolymer

Halogenated butyl sealants, 429–30
Hand lay-up process, 161, 163–6
 advantages/disadvantages of,
 165–6
 pressure bag method, 164
 requirements of, 161–2
Heat-resistant adhesives, 382–4
High-alumina cement, properties of,
 268
High-energy radiation, effect on
 polymers, 46
High-resilient foams, 286
Hot-melt adhesives/sealants, 428
Hot-melt butyl sealants, moisture
 transmission by, 429
Hot-press moulding technique, 168–9
House
 applications *See* Flooring . . . ;
 Glazing . . . ; Roofing . . . ;
 Window . . .
 futuristic design, 194, 195

Hybrid composites, 143, 146–60
 meaning of term, 149
Hydrolysis, 547
Hypalon
 roofing applications, 489
 sealants, advantages/disadvantages
 of, 418
 see also Chlorosulphonated
 polyethylene

Impact damage, 499
Impact resistance, fibre-reinforced
 concrete, 265–6, 267–8
Impact strength, factors affecting, 449
Infrared spectroscopy
 polyethylene weathering studies,
 532–3
 stearic acid–PVC blends, 465, 466
Initiators, concrete–polymer
 composites, 215
Injection moulding, 63
 advantage of, 63
 components produced by, 63
 composites made by, 172
 equipment used, 64
Interfacial free energy, 365

Kapton type H film, properties of,
 104
Kevlar fibres
 concrete reinforced by, properties
 of, 277, 278
 molecular structure of, 156
 properties of, 130, 134, 157–9,
 266–7
 spectral sensitivity of, 54

Ladder polymers, formation of, 472,
 473
Latex, definition of, 239
Latex rubber foams
 compression properties of, 312
 flammability of, 327
 manufacture of, 290

Latices
  acrylic, 246–7
  acrylonitrile–butadiene copolymer,
    242–3
  applications of, 243, 244, 246
  characteristic properties of, 241
  concrete mixed with, 238–41
  factors affecting properties, 241
  physical properties of, 242, 244,
    245, 246, 247
  polychloroprene, 243–4
  poly(vinyl acetate), 244–6, 247
  poly(vinyl chloride), 246
  styrene–butadiene copolymer, 242,
    247
Limiting oxygen index (LOI), 51, 345
  effect of fire-retardants on, 51, 347,
    348
  values quoted, 50, 346
Linear polymers, 2, 3–4
Liquid state, 33

Macrocomposite, definition of, 75
Mark–Houwink equation, 7
  parameters for, 10
Market building (France), composites
    used in, 194, 196
Mastics, 392, 393
Matrix, functions of, 173
Mechanical corrosion, 498–9
Mechanical degradation, 552–4
  meaning of term, 498
  polyisobutene, 553
  polystyrene, 552, 553
Mechanical joints, composites, 185–6
Mechanical properties (of polymers),
    36–41
  effects of temperature on, 41
Melamine–formaldehyde polymers,
    78, 79, 379
Mercury lamps, 521
Metal fibres, 152, 154
  properties compared with bulk,
    155–6
Methyl methacrylate
  concrete impregnated with,
    218–19, 224, 225, 226, 229

Methyl methacrylate—*contd.*
  concrete mixed with, 255–7
Methyl rubber, permeability of, 48
Micelles, 240
Microbial degradation, 550–2
Microcomposite, definition of, 75
Microspheres, composites prepared
    using, 300–1, 302, 340, 517
Mild steel, properties of, 188–9
Mineral polymers, 20–1
Mirrors, polymers used in, 453–7,
    471
Molecular weight (MW), 5–8
  thermal degradation effects on,
    504–5
  variation with polymerization type,
    14
Monomers
  addition polymerization, 12
  factors affecting stability of, 210–11
  polymer concrete systems, 210,
    215–21, 234
  volatilization during impregnation-
    polymerization, 214, 220
Morphological changes (in polymers),
    31–6
Mylar, 443

Natural rubber
  Mark–Houwink parameters for, 10
  permeability of, 48
Neoprene
  latex, 247
  roofing applications for, 487
Network polymers, 545
Norrish photochemical mechanisms,
    539
North America
  ban on UF foams, 339, 519
  urea–formaldehyde foams used,
    335, 336, 519
  window construction compared
    with European, 527
Nucleic acids, hydrolysis of, 547
Number average molecular weight
  meaning of term, 6
  values quoted, 7

Nylon, 90, 92–4
  adhesives, 374
  fibres
    concrete reinforced by, 275, 276, 278
    properties of, 69, 136, 137, 157, 266–7
  flammability of, 347
  glass transition temperature, 35
  melting temperature, 35
  molecular weights of, 7
  permeability of, 48
  toxicity of combustion products, 328
  use in composites, 80

Oak wood
  flammability of, 325
  toxicity of combustion products, 328
Occupational Safety & Health Administration (OSHA), limits on formaldehyde, 337
Ocean energy systems, applications for, 439
Oil-based caulking compounds, 430–3
  advantages/disadvantages of, 433
  oils used, 432
  typical formulation for, 433
Oils, composition of, 432
Optical characteristics, 441–5
Organic polymers, 1, 19–20
Oxygen, radiodegradation affected by, 545–6
Ozonolysis, 549

Paraffin chloride, 465
Particle characteristics, 114
Permeability, polymers, 46–9
Phase change materials, 463
  heat transfer mechanisms for, 462
  interactions with encapsulants, 465, 467–9
Phenol–formaldehyde (PF)
  applications of, 40, 77
  foams, 329–31, 351

Phenol–formaldehyde (PF)—*contd.*
  production of, 105, 108–9
  thermal stability of, 108
Phenolic polymers
  adhesives, 377–8
    durability of, 390
  characteristics of, 77, 78
  flammability of, 347
  polymer concretes made from, 259, 261
  processing methods for, 60
  thermal stability of, 508
Photodegradation, 520–39
  chemical process of, 524
  effects of, 520
  meaning of term, 497
  physical processes involved in, 522
  polycarbonate, 535–9
  polyethylene, 531–5
  poly(vinyl chloride), 525–31
Photovoltaic encapsulation systems
  construction elements of, 459
  designs of, 461
  polymers used, 458, 460–2
Physical adsorption, 359
Physical structure (of polymers), 21–31
Pigments, effects of, 54, 55
Piping, solar energy equipment, 469–70
Plasticizers, 55–7
Plastics technology, 52–71
Plexiglass, 20, 443
  see also Poly(methyl methacrylate) (PMMA)
Plywood
  flammability of, 325
  properties of, 188–9
Polluted atmospheres, polymers affected by, 500–3
Polyacetals, 94–5
  flammability of, 50
  thermal properties of, 43
  use in composites, 81
Polyacrylamide, 374
  Mark–Houwink parameters for, 10
  ozonolysis of, 549
Polyacrylate, flammability of, 50

Polyacrylics. *See* Acrylics
Polyacrylonitrile (PAN)
  carbon fibres produced from, 139,
    143, 144–5
  flammability of, 347
  glass transition temperature of, 35
  Mark–Houwink parameters for, 10
  melting temperature of, 35
  pyrolysis of, 472–3
  thermal degradation of, 509
Polyamides
  additives used, 93–4
  adhesives, 374
  applications of, 94
  fibres, properties of, 130
  hydrolysis of, 547
  molecular structure of, 90
  physical structure of, 21
  processing methods for, 60
  properties of, 92–4
  thermal properties of, 42, 43
  uses of, 40, 80
  see also Aramid . . . ; Kevlar . . . ;
    Nylon . . . ; X-500G . . .
Polyaromatics, adhesives, 382–4
Polybenzimidazole
  adhesives, 383, 384
  thermal stability of, 508
Polybenzothiazole, thermal stability
  of, 508
Polybutadiene
  bonding to glass, 367
  flammability of, 346
  permeability of, 48
  pyrolysis of, 472–3
  spectral sensitivity of, 54
  thermal degradation of, 509
Polybutene, piping applications, 470
Poly(butyl methacrylate), thermal
  degradation of, 509
Poly(butylene terephthalate),
  flammability of, 50
Polycarbonate (PC)
  adhesives, 374–5
  applications of, 98, 535
  film, stress–strain curves for, 39
  flammability of, 50
  mechanical properties of, 447

Polycarbonate (PC)—*contd.*
  molecular structure of, 98
  optical properties of, 443
  photodegradation of, 535–9
    chemical changes involved,
      538–9
    mechanical properties affected
      by, 537–8
  properties of, 98, 99
  spectral sensitivity of, 54
  thermal properties of, 43, 447
  toxicity of combustion products,
    329
  use in composites, 81
  UV sensitivity of, 525
  weathering of, 536–8
Poly(carborane siloxane),
  flammability of, 50
Polychlorofluoroethylene,
  flammability of, 346
Polychloroprene
  latex, 243–4
  molecular structure of, 424
  sealants, 424–5
    advantages/disadvantages of,
      426
    typical formulation for, 423
Polychlorotrifluoroethylene (PCTFE)
  permeability of, 48
  properties of, 91
  thermal degradation of, 509
Polycondensation, 17–19
Poly(cyano methylacrylate), 373,
  417–18
Polydialkylsiloxanes, hydrolysis of,
  547
Polydimethylsiloxane (PDMS)
  Mark–Houwink parameters for, 10
  molecular structure of, 20
Polydispersity (of polymers), 8–9
Polyesters
  adhesives, 382
  characteristics of, 77, 78
  concrete mixed with, 253–61
  fibres
    properties of, 69, 157, 159
    see also Dacron . . .
  flammability of, 347

Polyesters—*contd.*
  glass composites, 77
  hydrolysis of, 547
  molecular structure of, 106
  processing methods for, 60
  production of, 104–5
  properties of, 107, 108
  spectral sensitivity of, 54
  thermal properties of, 43
  thermal stability of, 108
  uses of, 40
  UV sensitivity of, 525
Poly(ether ether ketone),
    flammability of, 50
Poly(ether sulphone), flammability
    of, 50
Polyethers, 94
  hydrolysis of, 547
Poly(ethylene oxide) (PEO)
  ozonolysis of, 549
  thermal degradation of, 509
Polyethylene (PE)
  applications of, 83
  carbonyl concentration in, 533
  crystallinity of, 29
  flammability of, 50, 346, 347
  foams
    flammability of, 327
    properties of, 311–12
  glass transition temperature, 35
  infrared spectra of, 532–3
  Mark–Houwink parameters for, 10
  mechanical degradation of, 551
  melting temperature of, 35
  molecular structure of, 79
  molecular weights of, 7
  oxidation mechanism of, 534–5
  permeability of, 48
  photodegradation of, 531–5
  physical structure of, 21, 23–4
  piping applications for, 470
  processing methods for, 60
  properties of, 79, 82
  radiodegradation of, 543
  spectral sensitivity of, 54
  stress–strain behaviour of, 39
  synthesis of, 15
  thermal properties of, 43

Polyethylene (PE)—*contd.*
  uses of, 40, 80
  UV sensitivity of, 525
  vinyl concentration in, 534
  weathering of, 55, 532–5, 536
Poly(ethylene terephthalate) (PET)
  glass transition temperature of, 35
  mechanical properties of, 447
  melting temperature of, 35
  molecular weights of, 7
  optical properties of, 443
  permeability of, 48
Polyformaldehyde, 94–5
  flammability of, 346
  molecular structure of, 508–9
Polygermanes, 20
Polyimides
  adhesives, 383, 384
  flammability of, 50
  molecular structure of, 506
  production of, 101–3
  properties of, 103–4
  thermal stability of, 508
  use in composites, 81, 104
Polyisobutene, 425–6
  Mark–Houwink parameters for,
    10
  molecular structure of, 426
  radiodegradation of, 543
  sealants, 429
  thermal degradation of, 509
  ultrasonic degradation of, 553
  uses of, 40
  see also Butyl rubbers
Polyisobutylene. *See* Polyisobutene
Polyisocyanurate foams
  fire performance of, 351
  flammability of, 323–4, 325, 327
  properties of, 315, 318
  thermogravimetric analysis of, 324
  toxicity of combustion products,
    329
Polyisoprene
  flammability of, 346
  uses of, 40
Polymer, meaning of term, 2
Polymer–cement concrete (PCC),
    208, 233–50

Polymer–cement concrete (PCC)—
  contd.
  compressive strength, of, 235,
    236–7
  dynamic properties of, 249–50
  epoxy-modified, 234–8
  latex-modified, 238–41
  monomers used, 234
  preparation of, 247
  properties of, 236–7, 248–50, 251
  requirements for polymers in,
    247–8
Polymer concrete (PC), 208, 250–62
  aggregates used, 258
  applications of, 261, 262
  bonding mechanism for, 253–5
  compressive strength of, 251, 252,
    253
  economics of, 250
  flexural strength of, 251, 255
  hydrothermal stability of, 253,
    255
  mass production of, 259, 260
  monomers used, 250
  properties of, 251–62
  silane coupling agent used, 252
  stress–strain diagrams for, 257
Polymer foams, 283–352
  blowing agents used, 287–90
  bubble growth in, 295
    factors affecting, 295–6, 297
  classification of, 284–7
  density of, 287
  flammability of, 339, 343–52
  manufacture of, 290–6
  polymers used, 283
  special foams, 296–304
  stabilization of, 292
  thermoplastic foams, 304–13
  thermosetting foams, 313–43
  US markets for, 284
  see also Cellular polymers
Polymer-impregnated concrete (PIC),
  208–33
  acid resistance of, 232, 251
  applications of, 231, 233
  bending strength of, 223, 224, 227
  classification of, 231

Polymer-impregnated concrete
  (PIC)—*contd.*
  compressive strength of, 223,
    224–5, 227, 228, 251
  creep properties of, 222, 226
  economics of, 215
  extraction of PMMA from, 219,
    220
  factors affecting properties of, 212,
    218
  full-impregnation process for, 219
  K coefficient for, 230–1
  kinetics of impregnation for, 217
  manufacture of, 209–12
  modulus of, 221–2, 223, 225, 251
  monomer volatilization in, 214
  monomers used, 210, 215–21
  partial-depth impregnation of, 218,
    221
  polymer loading in, 216, 226, 228
  polymerization of monomer in,
    214–15, 231
  properties of, 221–31, 251
  radiopolymerization used, 214, 231
  shrinkage in, 216
  tensile strength of, 225, 251
Polymeric roofing membranes
  advantages of, 485–6
  disadvantages of, 486
Polymers
  ageing of, 45–6
  applications of, in
    flooring, 490–3
    glazing, 373, 419, 420
    roofing, 478–89
    solar energy systems, 438–75
  classification of, 19–21
  combustion of, 344–5
    see also Flammability
  composites, in, 77–113
  conversion to finished products,
    57–71
  crystalline regions in, 22
  degradation of, 496–553
  factors affecting growth of use, 2
  flammability of, 49–51
  folded-chain lamella theory for, 28
  fringed-micelle theory for, 26–7

Polymers—*contd.*
  geometrical shape classification of, 2–3
  mechanical properties of, 36–41
  molecular weight of, 5–8
  morphological changes in, 31–6
  permeability of, 46–9
  physical structure of, 21–31
  polydispersity of, 8–9
  processing of, 51–2, 59–71
  synthesis of, 9–19
  thermal properties of, 41–4
  toxicity of, 49
  weathering properties of, 44–5, 55
Polymethacrylonitrile, thermal degradation of, 509
Poly(methyl acrylate)
  radiodegradation of, 543
  thermal degradation of, 509
Poly(methyl methacrylate) (PMMA)
  adhesives, 373
  extraction from polymer-impregnated concrete, 219, 220
  flammability of, 50, 346, 347
  glass transition temperature of, 35
  Mark–Houwink parameters for, 10
  mechanical degradation of, 551
  mechanical properties of, 446
  melting temperature of, 35
  molecular structure of, 20
  optical properties of, 443
  radiodegradation of, 543, 544
  thermal degradation of, 506, 509
  thermal properties of, 43, 446
  uses of, 40
  UV absorption of, 522
  weathering resistance of, 44, 450, 454–6
Poly (α-methyl styrene), thermal degradation of, 506, 509
Poly(methyl vinyl ketone) (PMVK), UV absorption of, 522
Poly(4-methylpent-1-ene), flammability of, 346
Polyolefins
  flammability of, 346, 347
  foams, 310

Polyolefins—*contd.*
  foams—*contd.*
    properties of, 311–12
Polyoxymethylene, thermal degradation of, 509
Poly(phenyl vinyl ketone) (PPVK), UV absorption of, 522
Poly-*p*-phenylene, 506
Polyphenylene
  production of, 98, 100
  properties of, 101
  thermal stability of, 101, 102
  use in composites, 81, 101
Poly(phenylene oxide) (PPO), spectral sensitivity of, 54
Poly(phenylene sulphide)
  flammability of, 50
  thermal properties of, 42
Polypropylene (PP)
  applications of, 85
  atactic form, 30, 31
  carbon fibre reinforced, stress–strain curves for, 176
  concrete reinforced by
    applications of, 264
    flammability of, 267
    impact strength of, 265–6
    properties of, 270, 274–5, 277
  effect of coupling agent on filled polymer, 121, 124
  fibres, properties of, 69, 263, 266–7
  fibrillated films
    concrete reinforced by, 263, 264, 265
    disadvantages of, 265
    manufacture of, 264
  flammability of, 50, 346, 347
  flooring applications, 491
  foams, 310
  isotactic form, 29–30
  mechanical degradation of, 551
  molecular structure of, 83
  molecular weights of, 7
  permeability of, 48
  processing methods for, 60
  properties of, 84
  radiodegradation of, 543
  syndiotactic form, 30

Polypropylene (PP)—*contd.*
  thermal degradation of, 509
  thermal properties of, 42, 43
  toxicity of combustion products, 328
  uses of, 40, 80
  UV sensitivity of, 525
Poly(propylene oxide), thermal
    degradation of, 509
Polysaccharides, hydrolysis of, 547
Polysilanes, 20
Polysiloxane, uses of, 40
Polystyrene (PS)
  applications of, 88
  concrete impregnated with, 223,
    228
  flammability of, 50, 346, 347
  foams, 304–7, 351
    applications of, 306–7
    flammability of, 325, 327
    manufacture of, 305–6
    thermal conductivity of, 307
  glass transition temperature of, 35
  Mark–Houwink parameters for, 10
  mechanical degradation of, 552
  melting temperature of, 35
  molecular structure of, 88, 508
  permeability of, 48
  processing methods for, 60
  properties of, 88
  radiodegradation of, 543
  spectral sensitivity of, 54
  synthesis of, 15
  thermal degradation of, 509
  thermal properties of, 43
  toxicity of combustion products, 328
  ultrasonic degradation of, 553
  uses of, 40, 80
  UV absorption of, 522
  UV sensitivity of, 525
Polysulphide sealants, 408–16, 471
  applications of, 416
  curing agents used, 411–13
  fillers used, 410
  formulation for, 409
  fuel resistance of, 413
  mechanical properties of, 411, 412
  moisture transmission by, 429
  primers used, 414

Polysulphide sealants—*contd.*
  stress relaxation in, 415
Polysulphone, 95–8
  applications of, 96, 98
  molecular structure of, 96
  properties of, 96, 97
  spectral sensitivity of, 54
  use in composites, 81
Polytetrafluoroethylene (PTFE)
  applications of, 90
  flammability of, 50, 346
  mechanical degradation of, 551
  molecular structure of, 89
  properties of, 90, 91, 92
  radiodegradation of, 542
  thermal degradation of, 509
  thermal properties of, 42, 43
  use in composites, 81
  uses of, 40
Polyurethane (PU)
  adhesives, 375–6
  flammability of, 347
  flooring applications, 493
  foams, 314–29, 351
    advantages of, 315
    anisotropic properties of, 317,
      318
    applications of, 315
    blowing agents used, 292, 316
    chemistry of reaction, 315–16
    continuous production of, 292,
      293
    creep behaviour of, 318, 319
    diffusion of gases in, 321
    flammability of, 323–8, 330
    manufacture of, 314–15
    one-shot system, 293–5
    properties of, 316
    thermal conductivity of, 307,
      319–22
    thermogravimetric analysis of,
      324
    typical formulation for, 317
  hydrolysis of, 547
  molecular structure of, 20
  one-component sealants, 402, 404
  one-part adhesives, 375
  roofing applications, 487

Polyurethane (PU)—*contd.*
  sealants, 402–8, 471
    advantages/disadvantages of,
      406
    blends with other polymers,
      404–8
    formulation for, 403
    moisture transmission by, 429
    properties of, 403
    thermal transitions in, 407
    toxicity of combustion products,
      328, 329
    two-component sealants, 402–3,
      405
  two-part adhesives, 375, 376
  uses of, 40
Poly(vinyl acetal)s
  adhesives, 371–2
  molecular structure of, 371
Poly(vinyl acetate) (PVAc)
  adhesives, 368, 370
  copolymers, 245
  latex, 244–6, 247
  Mark–Houwink parameters for, 10
  spectral sensitivity of, 54
  thermal degradation of, 509
  thermal properties of, 43
  UV absorption of, 522
Poly(vinyl alcohol) (PVA)
  adhesives, 370–1
  chemical structure of, 370
  mechanical degradation of, 551
  molecular structure of, 20
  ozonolysis of, 549
  thermal properties of, 43
Poly(vinyl chloride) (PVC)
  adhesives, 372
  applications of, 85, 88, 509, 511
  asbestos-reinforced, 151
  dehydrochlorination of, 511
    factors affecting, 512
    mass-spectrometric studies, 531
    spectroscopic studies of, 513–14
  discoloration of, 511, 525, 526
  flammability of, 50, 346, 349
  flooring applications, 490–1
  foams, 308–10
    flammability of, 327

Poly(vinyl chloride) (PVC)—*contd.*
  foams—*contd.*
    manufacture of, 308
    properties of, 309
  impact modifier used, 531, 532
  latex, 246
  Mark–Houwink parameters for, 10
  molecular structure of, 85
  molecular weights of, 7
  permeability of, 48
  photodegradation of, 525–31
  physical properties of, 514
  pigments used, 528–31
  piping applications of, 470
  plasticizers in, 56, 57
  polyene sequences in degraded
    polymer, 513–14
  processing methods for, 60
  properties of, 85, 86–7
  roofing applications, 487
  spectral sensitivity of, 54
  synthesis of, 15
  thermal degradation of, 509–14
  thermal properties of, 43, 514
  toxicity of, 49
  toxicity of combustion products,
    328
  use in composites, 80
  uses of, 40
  UV light, effects of, 513
  UV sensitivity of, 525
  weathering of, 44, 86–7
Poly(vinyl fluoride) (PVF)
  mechanical properties of, 446
  optical properties of, 443
  properties of, 91
  thermal properties of, 43, 446
Poly(vinylidene chloride) (PVDC)
  permeability of, 48
  sealants, 471
  thermal properties of, 43
Poly(vinylidene fluoride) (PVDF),
  properties of, 91
Polyvinylpyrrolidone, ozonolysis of,
  549
Pore size distribution
  determination of, 212
  effect of polymer on, 229

Porosity, determination of, 212–13
Portland cement
  interaction with polymers, 254, 255
  properties of, 268
Pre-form moulding, 169
Processing
  degradation affected by, 501
  polymers, 51–2, 59–71
Pultruded composites, properties of, 183
Pultrusion moulding technique, 170–2
  injection moulding used, 171–2
Punking, meaning of term, 331
Pyroxene chain, 21

Radiation-induced polymerization
  advantages of, 541
  factors affecting, 541
Radical polymerization, 11–12
Radiodegradation, 539–46
  chemical modifications induced by, 540–1
  effect of oxygen on, 545–6
  meaning of term, 497–8
Radiopolymerization, 214, 231
Random chain scission, 504
Rayon fibres, properties of, 69
Reinforcement, mechanisms of, 172–6
Reinforcing agents, 113–19
Resorcinol–formaldehyde resin adhesives, 377–8
Rigid foams
  definition of, 286
  polyurethane, 315, 317–18
Roofing, 478–89
  components of, 478, 479
  market data, 489
  polymers used, 479, 484, 486–9
Room-temperature vulcanizing (RTV) sealants, 400
Rubber hydrochloride, permeability of, 48
Rubber state, 33
Rubberized asphalt, roofing applications, 487

S-glass fibres
  composition of, 137
  properties of, 135–7
Sandwich panels, 194, 197–200
  applications of, 199
  core of, 199
  requirements for, 198
  skin of, 198–9
Sapphire fibres, properties of, 136, 137
Saran, synthesis of, 15–16
Sealants, 391–433
  acrylics, 417–23
  applications of, 394
  backup materials used, 398
  butyl rubber, 427–9
  characteristics of, 392
  chlorosulphonated polyethylene, 416–17, 418
  classification of, 392–3
  curing agents used, 398
  failure of, 396–7
  fillers used, 397
  forms used, 392, 393
  joints used for, 394–6
  meaning of term, 357, 391, 392
  oil-based caulking compounds, 430–3
  pigments used, 398
  plasticizers used, 397–8
  polychloroprenes, 424–5
  polyisobutene, 429
  polysulphides, 408–16
  polyurethane, 402–8
  primers used, 398
  properties of, 398–433
  release agents used, 398
  silicones, 398–402
  solar energy equipment, 471–2
  types of, 358, 391
Seamless floors, 493
  on-site production of, 493
Sheet moulding compound (SMC), 169, 170
Silica fume, polymer–cement concrete affected by, 237
Silicate polymers, 21

Silicon carbide fibres, properties of, 130
Silicon nitride whiskers, properties of, 130
Silicones
  characteristics of, 78, 79
  foams, 340, 342–3
  roofing applications, 487
  sealants, 398–402, 471, 472
    advantages/disadvantages of, 401
    composition of, 399–400
    curing of, 400
    moisture transmission by, 429
    properties of, 402
Silk, properties of, 69
Singlet state, 524
Smoke-suppressants, 348–9
Sodium polyacrylate, ozonolysis of, 549
Solar energy equipment
  adhesives used, 473
  coatings used, 470–1
  components of, 442
  example of low-cost installation, 458
  piping used, 469–70
  sealants used, 471–2
Solar energy systems, applications for, 439
Solar heat collector covers, 438, 440–57
  components of, 440–1
  durability of, 449–57
  factors affecting, 441
  honeycomb used, 441, 445
  impact resistance of, 446–7, 449
  mechanical properties of, 445–9
  optical characteristics required, 441–5
  thermal properties of, 446–7, 448–9
Solar mirrors, polymers used in, 453–7, 471
Solar ponds, 473–4
  costs of, 474
  liners for, requirements for, 474
Solvolysis, 547–8

Spandex, properties of, 69
Spinnable polymers, 66–8
  properties of, 69
Spray-up moulding technique, 166–7
  advantages of, 166–7
  disadvantages of, 167
Stabilizers, 53–5
  role of, 211
Stearic acid
  interaction with polymers, 465, 467–9
  IR spectra of, 467
  thermal-storage, use of, 464–5
Steel fibres/wire, properties of, 113, 130, 266–7
Step polymerization, compared with chain polymerization, 13
Stepwise polymerization, 17–19
Stress–strain performance (of polymers), 37–9
Structural adhesive bonding, 385–91
Structural adhesives, definition of, 385
Structural foam process, 293–5
Structural foams, 298
Styrene, concrete mixed with, 255–7
Styrene–acrylonitrile copolymer (SAN), 88
  concrete impregnated with, 225–6
  toxicity of combustion products, 328
  UV sensitivity of, 525
Styrene–butadiene rubber (SBR)
  latex, 242, 247
  Mark–Houwink parameters for, 10
  sealants, 471
Styrene–butadiene–styrene block copolymers (SBS), 481–2
  bitumens affected by, 483–4
Sunlight
  PVC affected by, 513
  spectrum of, 521, 523
Sun-Lite, 443, 444, 451
Suntek, 443, 444, 446
Surface crazing, 524
Surfaces
  effect on adhesion, 364–8
  energy-classification of, 360

Syntactic foams
  composition of, 299–300
  properties of, 301
  structure of, 298–9
Synthesis (of polymers), 9–19

Tapes, sealant, 393, 396
Tedlar, 443, 444, 450, 451
Teflon
  weathering resistance of, 450, 451,
    452
  see also Polytetrafluoroethylene
    (PTFE)
Temperature, strength properties
    affected by, 503
Tensile strength, polymers, 37
Textile fibres, 66
  properties listed, 69
Thermal conductivity, foams, 307,
    319–22, 334
Thermal degradation, 503–20
  epoxies, 516–17
  meaning of term, 497
  poly(vinyl chloride), 509–14
  reactions occurring during, 504,
    510
  structural factors affecting, 507
  urea–formaldehyde, 517–20
Thermal properties (of polymers),
    41–4
Thermal storage modules, 462–9
  encapsulants used, 464, 465, 466
  heat transfer mechanisms for,
    462–3
  phase change material used,
    463–5
Thermogravimetric analysis (TGA),
    506, 508
Thermomechanical curves, 34–5
Thermoplastic polymers, 3–4
  adhesives, 368–77
  composites, 79, 80–1
  foams, 304–13
Thermoset polymers, 5
  adhesives, 377–84
  characteristics of, 78
  foams, 13–43

Thick sections, polymer-impregnated
    concrete, 213
Three-dimensional polymers, 3, 5
Titanium dioxide, PVC formulations,
    530–1
Titanium fibres, properties of, 135,
    136
Toluene diisocyanate, 293, 294, 317
Tough polymers, 38
Toxicity, polymers, 49
Transmittance characteristics, 444
Transparent polymeric pottants, costs
    of, 460
Triblock thermoplastic elastomers,
    481
Trimethylpropane trimethylacrylate
    (TMPTMA), 210
  properties of polymer-impregnated
    concrete affected by, 225
Triplet state, 524
Trovipor process, 308
Tufflak, 443
Tungsten filaments, properties of, 113
Two-phase model, 26–7

Ultrasonic degration, 553
Ultraviolet (UV) light, 54
  see also Photodegradation; Sunlight
Urea–formaldehyde (UF)
  adhesives, 378–9
  applications of, 518
  chemistry of reactions, 331–2
  composites, 79
  foams, 331–9
    additives used, 333
    banning in North America, 339,
      519
    cancer risk from, 337, 339
    degradation of, 334–5, 518–20
    formaldehyde emitted from, 335,
      336, 337, 339, 518–19
    fungus growth on, 335, 336–7
    hydrolytic degration of, 518–19
    manufacture of, 332, 333
    on-site production of, 333
    properties of, 333–4
    thermal conductivity of, 334

Urea–formaldehyde (UF)—*contd.*
   thermal degradation of, 517–20
   uses of, 40
Urea–melamine polymers, processing
   methods for, 60

Vinyl . . . . *See* Poly(vinyl . . .
Vinyl chloride–vinyl acetate
   copolymer, UV sensitivity of,
   525
Vinyl tiles, formulations for, 491
Viscoelastic materials, 32, 33
Viscosity average molecular weight,
   meaning of term, 7–8

Waste disposal problems, 503
Water permeability, 47
Weathering properties
   ethylene–vinyl acetate copolymer,
      455, 457, 527
   glass-reinforced plastics, 450, 451
   polycarbonate, 536–8
   polyethylene, 55, 532–3, 536
   polymers, 44–5, 55
   poly(methyl methacrylate), 44,
      450, 454–6
   polytetrafluoroethylene, 450, 451,
      452

Weathering properties—*contd.*
   poly(vinyl chloride), 86–7
   solar heat collector covers, 449–57
Weight average molecular weight,
   meaning of term, 6–7
Wind energy systems, applications
   for, 439
Window frames
   materials used, 525, 526
   PVC used, 525–9
Wood, toxicity of combustion
   products, 328, 329
Wool, properties of, 69

X-500G fibres
   molecular structure of, 156
   properties of, 157
X-ray diffraction scans, spinnable
   polymers, 25, 26
X-ray fibre diagram, 23
X-ray powder diffraction pattern, 22

Young's equation, 360
Young's modulus, polymer-
   impregnated concrete, 221–2,
   223, 225, 251

Zipper reactions, 511